Dictionary of Literary Biography

1 *The American Renaissance in New England,* edited by Joel Myerson (1978)

2 *American Novelists Since World War II,* edited by Jeffrey Helterman and Richard Layman (1978)

3 *Antebellum Writers in New York and the South,* edited by Joel Myerson (1979)

4 *American Writers in Paris, 1920-1939,* edited by Karen Lane Rood (1980)

5 *American Poets Since World War II,* 2 parts, edited by Donald J. Greiner (1980)

6 *American Novelists Since World War II, Second Series,* edited by James E. Kibler Jr. (1980)

7 *Twentieth-Century American Dramatists,* 2 parts, edited by John MacNicholas (1981)

8 *Twentieth-Century American Science-Fiction Writers,* 2 parts, edited by David Cowart and Thomas L. Wymer (1981)

9 *American Novelists, 1910-1945,* 3 parts, edited by James J. Martine (1981)

10 *Modern British Dramatists, 1900-1945,* 2 parts, edited by Stanley Weintraub (1982)

11 *American Humorists, 1800-1950,* 2 parts, edited by Stanley Trachtenberg (1982)

12 *American Realists and Naturalists,* edited by Donald Pizer and Earl N. Harbert (1982)

13 *British Dramatists Since World War II,* 2 parts, edited by Stanley Weintraub (1982)

14 *British Novelists Since 1960,* 2 parts, edited by Jay L. Halio (1983)

15 *British Novelists, 1930-1959,* 2 parts, edited by Bernard Oldsey (1983)

16 *The Beats: Literary Bohemians in Postwar America,* 2 parts, edited by Ann Charters (1983)

17 *Twentieth-Century American Historians,* edited by Clyde N. Wilson (1983)

18 *Victorian Novelists After 1885,* edited by Ira B. Nadel and William E. Fredeman (1983)

19 *British Poets, 1880-1914,* edited by Donald E. Stanford (1983)

20 *British Poets, 1914-1945,* edited by Donald E. Stanford (1983)

21 *Victorian Novelists Before 1885,* edited by Ira B. Nadel and William E. Fredeman (1983)

22 *American Writers for Children, 1900-1960,* edited by John Cech (1983)

23 *American Newspaper Journalists, 1873-1900,* edited by Perry J. Ashley (1983)

24 *American Colonial Writers, 1606-1734,* edited by Emory Elliott (1984)

25 *American Newspaper Journalists, 1901-1925,* edited by Perry J. Ashley (1984)

26 *American Screenwriters,* edited by Robert E. Morsberger, Stephen O. Lesser, and Randall Clark (1984)

27 *Poets of Great Britain and Ireland, 1945-1960,* edited by Vincent B. Sherry Jr. (1984)

28 *Twentieth-Century American-Jewish Fiction Writers,* edited by Daniel Walden (1984)

29 *American Newspaper Journalists, 1926-1950,* edited by Perry J. Ashley (1984)

30 *American Historians, 1607-1865,* edited by Clyde N. Wilson (1984)

31 *American Colonial Writers, 1735-1781,* edited by Emory Elliott (1984)

32 *Victorian Poets Before 1850,* edited by William E. Fredeman and Ira B. Nadel (1984)

33 *Afro-American Fiction Writers After 1955,* edited by Thadious M. Davis and Trudier Harris (1984)

34 *British Novelists, 1890-1929: Traditionalists,* edited by Thomas F. Staley (1985)

35 *Victorian Poets After 1850,* edited by William E. Fredeman and Ira B. Nadel (1985)

36 *British Novelists, 1890-1929: Modernists,* edited by Thomas F. Staley (1985)

37 *American Writers of the Early Republic,* edited by Emory Elliott (1985)

38 *Afro-American Writers After 1955: Dramatists and Prose Writers,* edited by Thadious M. Davis and Trudier Harris (1985)

39 *British Novelists, 1660-1800,* 2 parts, edited by Martin C. Battestin (1985)

40 *Poets of Great Britain and Ireland Since 1960,* 2 parts, edited by Vincent B. Sherry Jr. (1985)

41 *Afro-American Poets Since 1955,* edited by Trudier Harris and Thadious M. Davis (1985)

42 *American Writers for Children Before 1900,* edited by Glenn E. Estes (1985)

43 *American Newspaper Journalists, 1690-1872,* edited by Perry J. Ashley (1986)

44 *American Screenwriters, Second Series,* edited by Randall Clark, Robert E. Morsberger, and Stephen O. Lesser (1986)

45 *American Poets, 1880-1945, First Series,* edited by Peter Quartermain (1986)

46 *American Literary Publishing Houses, 1900-1980: Trade and Paperback,* edited by Peter Dzwonkoski (1986)

47 *American Historians, 1866-1912,* edited by Clyde N. Wilson (1986)

48 *American Poets, 1880-1945, Second Series,* edited by Peter Quartermain (1986)

49 *American Literary Publishing Houses, 1638-1899,* 2 parts, edited by Peter Dzwonkoski (1986)

50 *Afro-American Writers Before the Harlem Renaissance,* edited by Trudier Harris (1986)

51 *Afro-American Writers from the Harlem Renaissance to 1940,* edited by Trudier Harris (1987)

52 *American Writers for Children Since 1960: Fiction,* edited by Glenn E. Estes (1986)

53 *Canadian Writers Since 1960, First Series,* edited by W. H. New (1986)

54 *American Poets, 1880-1945, Third Series,* 2 parts, edited by Peter Quartermain (1987)

55 *Victorian Prose Writers Before 1867,* edited by William B. Thesing (1987)

56 *German Fiction Writers, 1914-1945,* edited by James Hardin (1987)

57 *Victorian Prose Writers After 1867,* edited by William B. Thesing (1987)

58 *Jacobean and Caroline Dramatists,* edited by Fredson Bowers (1987)

59 *American Literary Critics and Scholars, 1800-1850,* edited by John W. Rathbun and Monica M. Grecu (1987)

60 *Canadian Writers Since 1960, Second Series,* edited by W. H. New (1987)

61 *American Writers for Children Since 1960: Poets, Illustrators, and Nonfiction Authors,* edited by Glenn E. Estes (1987)

62 *Elizabethan Dramatists,* edited by Fredson Bowers (1987)

63 *Modern American Critics, 1920-1955,* edited by Gregory S. Jay (1988)

64 *American Literary Critics and Scholars, 1850-1880,* edited by John W. Rathbun and Monica M. Grecu (1988)

65 *French Novelists, 1900-1930,* edited by Catharine Savage Brosman (1988)

66 *German Fiction Writers, 1885-1913,* 2 parts, edited by James Hardin (1988)

67 *Modern American Critics Since 1955,* edited by Gregory S. Jay (1988)

68 *Canadian Writers, 1920-1959, First Series,* edited by W. H. New (1988)

69 *Contemporary German Fiction Writers, First Series,* edited by Wolfgang D. Elfe and James Hardin (1988)

70 *British Mystery Writers, 1860-1919,* edited by Bernard Benstock and Thomas F. Staley (1988)

71 *American Literary Critics and Scholars, 1880-1900,* edited by John W. Rathbun and Monica M. Grecu (1988)

72 *French Novelists, 1930-1960,* edited by Catharine Savage Brosman (1988)

73 *American Magazine Journalists, 1741-1850,* edited by Sam G. Riley (1988)

74 *American Short-Story Writers Before 1880,* edited by Bobby Ellen Kimbel, with the assistance of William E. Grant (1988)

75 *Contemporary German Fiction Writers, Second Series,* edited by Wolfgang D. Elfe and James Hardin (1988)

76 *Afro-American Writers, 1940-1955,* edited by Trudier Harris (1988)

77 *British Mystery Writers, 1920-1939,* edited by Bernard Benstock and Thomas F. Staley (1988)

78 *American Short-Story Writers, 1880-1910,* edited by Bobby Ellen Kimbel, with the assistance of William E. Grant (1988)

79 *American Magazine Journalists, 1850-1900,* edited by Sam G. Riley (1988)

80 *Restoration and Eighteenth-Century Dramatists, First Series,* edited by Paula R. Backscheider (1989)

81 *Austrian Fiction Writers, 1875-1913,* edited by James Hardin and Donald G. Daviau (1989)

82 *Chicano Writers, First Series,* edited by Francisco A. Lomelí and Carl R. Shirley (1989)

83 *French Novelists Since 1960,* edited by Catharine Savage Brosman (1989)

84 *Restoration and Eighteenth-Century Dramatists, Second Series,* edited by Paula R. Backscheider (1989)

85 *Austrian Fiction Writers After 1914,* edited by James Hardin and Donald G. Daviau (1989)

86 *American Short-Story Writers, 1910-1945, First Series,* edited by Bobby Ellen Kimbel (1989)

87 *British Mystery and Thriller Writers Since 1940, First Series,* edited by Bernard Benstock and Thomas F. Staley (1989)

88 *Canadian Writers, 1920-1959, Second Series,* edited by W. H. New (1989)

89 *Restoration and Eighteenth-Century Dramatists, Third Series,* edited by Paula R. Backscheider (1989)

90 *German Writers in the Age of Goethe, 1789-1832,* edited by James Hardin and Christoph E. Schweitzer (1989)

91 *American Magazine Journalists, 1900-1960, First Series,* edited by Sam G. Riley (1990)

92 *Canadian Writers, 1890-1920,* edited by W. H. New (1990)

93 *British Romantic Poets, 1789-1832, First Series,* edited by John R. Greenfield (1990)

94 *German Writers in the Age of Goethe: Sturm und Drang to Classicism,* edited by James Hardin and Christoph E. Schweitzer (1990)

95 *Eighteenth-Century British Poets, First Series,* edited by John Sitter (1990)

96 *British Romantic Poets, 1789-1832, Second Series,* edited by John R. Greenfield (1990)

97 *German Writers from the Enlightenment to Sturm und Drang, 1720-1764,* edited by James Hardin and Christoph E. Schweitzer (1990)

98 *Modern British Essayists, First Series,* edited by Robert Beum (1990)

99 *Canadian Writers Before 1890,* edited by W. H. New (1990)

100 *Modern British Essayists, Second Series,* edited by Robert Beum (1990)

101 *British Prose Writers, 1660-1800, First Series,* edited by Donald T. Siebert (1991)

102 *American Short-Story Writers, 1910-1945, Second Series,* edited by Bobby Ellen Kimbel (1991)

103 *American Literary Biographers, First Series,* edited by Steven Serafin (1991)

104 *British Prose Writers, 1660-1800, Second Series,* edited by Donald T. Siebert (1991)

105 *American Poets Since World War II, Second Series,* edited by R. S. Gwynn (1991)

106 *British Literary Publishing Houses, 1820-1880,* edited by Patricia J. Anderson and Jonathan Rose (1991)

107 *British Romantic Prose Writers, 1789-1832, First Series,* edited by John R. Greenfield (1991)

108 *Twentieth-Century Spanish Poets, First Series,* edited by Michael L. Perna (1991)

109 *Eighteenth-Century British Poets, Second Series,* edited by John Sitter (1991)

110 *British Romantic Prose Writers, 1789-1832, Second Series,* edited by John R. Greenfield (1991)

111 *American Literary Biographers, Second Series,* edited by Steven Serafin (1991)

112 *British Literary Publishing Houses, 1881-1965,* edited by Jonathan Rose and Patricia J. Anderson (1991)

113 *Modern Latin-American Fiction Writers, First Series,* edited by William Luis (1992)

114 *Twentieth-Century Italian Poets, First Series,* edited by Giovanna Wedel De Stasio, Glauco Cambon, and Antonio Illiano (1992)

115 *Medieval Philosophers,* edited by Jeremiah Hackett (1992)

116 *British Romantic Novelists, 1789-1832,* edited by Bradford K. Mudge (1992)

117 *Twentieth-Century Caribbean and Black African Writers, First Series,* edited by Bernth Lindfors and Reinhard Sander (1992)

118 *Twentieth-Century German Dramatists, 1889-1918,* edited by Wolfgang D. Elfe and James Hardin (1992)

119 *Nineteenth-Century French Fiction Writers: Romanticism and Realism, 1800-1860,* edited by Catharine Savage Brosman (1992)

120 *American Poets Since World War II, Third Series,* edited by R. S. Gwynn (1992)

121 *Seventeenth-Century British Nondramatic Poets, First Series,* edited by M. Thomas Hester (1992)

122 *Chicano Writers, Second Series,* edited by Francisco A. Lomelí and Carl R. Shirley (1992)

123 *Nineteenth-Century French Fiction Writers: Naturalism and Beyond, 1860-1900,* edited by Catharine Savage Brosman (1992)

124 *Twentieth-Century German Dramatists, 1919-1992,* edited by Wolfgang D. Elfe and James Hardin (1992)

125 *Twentieth-Century Caribbean and Black African Writers, Second Series,* edited by Bernth Lindfors and Reinhard Sander (1993)

126 *Seventeenth-Century British Nondramatic Poets, Second Series,* edited by M. Thomas Hester (1993)

127 *American Newspaper Publishers, 1950-1990,* edited by Perry J. Ashley (1993)

128 *Twentieth-Century Italian Poets, Second Series,* edited by Giovanna Wedel De Stasio, Glauco Cambon, and Antonio Illiano (1993)

129 *Nineteenth-Century German Writers, 1841-1900,* edited by James Hardin and Siegfried Mews (1993)

130 *American Short-Story Writers Since World War II,* edited by Patrick Meanor (1993)

131 *Seventeenth-Century British Nondramatic Poets, Third Series,* edited by M. Thomas Hester (1993)

132 *Sixteenth-Century British Nondramatic Writers, First Series,* edited by David A. Richardson (1993)

133 *Nineteenth-Century German Writers to 1840,* edited by James Hardin and Siegfried Mews (1993)

134 *Twentieth-Century Spanish Poets, Second Series,* edited by Jerry Phillips Winfield (1994)

135 *British Short-Fiction Writers, 1880-1914: The Realist Tradition,* edited by William B. Thesing (1994)

136 *Sixteenth-Century British Nondramatic Writers, Second Series,* edited by David A. Richardson (1994)

137 *American Magazine Journalists, 1900-1960, Second Series,* edited by Sam G. Riley (1994)

138 *German Writers and Works of the High Middle Ages: 1170-1280,* edited by James Hardin and Will Hasty (1994)

139 *British Short-Fiction Writers, 1945-1980,* edited by Dean Baldwin (1994)

140 *American Book-Collectors and Bibliographers, First Series,* edited by Joseph Rosenblum (1994)

141 *British Children's Writers, 1880-1914,* edited by Laura M. Zaidman (1994)

142 *Eighteenth-Century British Literary Biographers,* edited by Steven Serafin (1994)

143 *American Novelists Since World War II, Third Series,* edited by James R. Giles and Wanda H. Giles (1994)

144 *Nineteenth-Century British Literary Biographers,* edited by Steven Serafin (1994)

145 *Modern Latin-American Fiction Writers, Second Series,* edited by William Luis and Ann González (1994)

146 *Old and Middle English Literature,* edited by Jeffrey Helterman and Jerome Mitchell (1994)

147 *South Slavic Writers Before World War II,* edited by Vasa D. Mihailovich (1994)

148 *German Writers and Works of the Early Middle Ages: 800-1170,* edited by Will Hasty and James Hardin (1994)

149 *Late Nineteenth- and Early Twentieth-Century British Literary Biographers,* edited by Steven Serafin (1995)

150 *Early Modern Russian Writers, Late Seventeenth and Eighteenth Centuries,* edited by Marcus C. Levitt (1995)

151 *British Prose Writers of the Early Seventeenth Century,* edited by Clayton D. Lein (1995)

152 *American Novelists Since World War II, Fourth Series,* edited by James and Wanda Giles (1995)

153 *Late-Victorian and Edwardian British Novelists, First Series,* edited by George M. Johnson (1995)

154 *The British Literary Book Trade, 1700-1820,* edited by James K. Bracken and Joel Silver (1995)

155 *Twentieth-Century British Literary Biographers,* edited by Steven Serafin (1995)

156 *British Short-Fiction Writers, 1880-1914: The Romantic Tradition,* edited by William F. Naufftus (1995)

157 *Twentieth-Century Caribbean and Black African Writers, Third Series,* edited by Bernth Lindfors and Reinhard Sander (1995)

158 *British Reform Writers, 1789-1832,* edited by Gary Kelly and Edd Applegate (1995)

159 *British Short-Fiction Writers, 1800-1880,* edited by John R. Greenfield (1996)

160 *British Children's Writers, 1914-1960,* edited by Donald R. Hettinga and Gary D. Schmidt (1996)

161 *British Children's Writers Since 1960, First Series,* edited by Caroline Hunt (1996)

162 *British Short-Fiction Writers, 1915-1945,* edited by John H. Rogers (1996)

163 *British Children's Writers, 1800-1880,* edited by Meena Khorana (1996)

164 *German Baroque Writers, 1580-1660,* edited by James Hardin (1996)

165 *American Poets Since World War II, Fourth Series,* edited by Joseph Conte (1996)

166 *British Travel Writers, 1837-1875,* edited by Barbara Brothers and Julia Gergits (1996)

167 *Sixteenth-Century British Nondramatic Writers, Third Series,* edited by David A. Richardson (1996)

168 *German Baroque Writers, 1661-1730,* edited by James Hardin (1996)

169 *American Poets Since World War II, Fifth Series,* edited by Joseph Conte (1996)

170 *The British Literary Book Trade, 1475-1700,* edited by James K. Bracken and Joel Silver (1996)

171 *Twentieth-Century American Sportswriters,* edited by Richard Orodenker (1996)

172 *Sixteenth-Century British Nondramatic Writers, Fourth Series,* edited by David A. Richardson (1996)

173 *American Novelists Since World War II, Fifth Series,* edited by James R. Giles and Wanda H. Giles (1996)

174 *British Travel Writers, 1876-1909,* edited by Barbara Brothers and Julia Gergits (1997)

175 *Native American Writers of the United States,* edited by Kenneth M. Roemer (1997)

176 *Ancient Greek Authors,* edited by Ward W. Briggs (1997)

177 *Italian Novelists Since World War II, 1945-1965* edited by Augustus Pallotta (1997)

178 *British Fantasy and Science-Fiction Writers Before World War I,* edited by Darren Harris-Fain (1997)

Documentary Series

1 *Sherwood Anderson, Willa Cather, John Dos Passos, Theodore Dreiser, F. Scott Fitzgerald, Ernest Hemingway, Sinclair Lewis,* edited by Margaret A. Van Antwerp (1982)

2 *James Gould Cozzens, James T. Farrell, William Faulkner, John O'Hara, John Steinbeck, Thomas Wolfe, Richard Wright,* edited by Margaret A. Van Antwerp (1982)

3 *Saul Bellow, Jack Kerouac, Norman Mailer, Vladimir Nabokov, John Updike, Kurt Vonnegut,* edited by Mary Bruccoli (1983)

4 *Tennessee Williams,* edited by Margaret A. Van Antwerp and Sally Johns (1984)

5 *American Transcendentalists,* edited by Joel Myerson (1988)

6 *Hardboiled Mystery Writers: Raymond Chandler, Dashiell Hammett, Ross Macdonald,* edited by Matthew J. Bruccoli and Richard Layman (1989)

7 *Modern American Poets: James Dickey, Robert Frost, Marianne Moore,* edited by Karen L. Rood (1989)

8 *The Black Aesthetic Movement,* edited by Jeffrey Louis Decker (1991)

9 *American Writers of the Vietnam War: W. D. Ehrhart, Larry Heinemann, Tim O'Brien, Walter McDonald, John M. Del Vecchio,* edited by Ronald Baughman (1991)

10 *The Bloomsbury Group,* edited by Edward L. Bishop (1992)

11 *American Proletarian Culture: The Twenties and The Thirties,* edited by Jon Christian Suggs (1993)

12 *Southern Women Writers: Flannery O'Connor, Katherine Anne Porter, Eudora Welty,* edited by Mary Ann Wimsatt and Karen L. Rood (1994)

13 *The House of Scribner, 1846–1904,* edited by John Delaney (1996)

14 *Four Women Writers for Children, 1868–1918,* edited by Caroline C. Hunt (1996)

Yearbooks

1980 edited by Karen L. Rood, Jean W. Ross, and Richard Ziegfeld (1981)

1981 edited by Karen L. Rood, Jean W. Ross, and Richard Ziegfeld (1982)

1982 edited by Richard Ziegfeld; associate editors: Jean W. Ross and Lynne C. Zeigler (1983)

1983 edited by Mary Bruccoli and Jean W. Ross; associate editor: Richard Ziegfeld (1984)

1984 edited by Jean W. Ross (1985)

1985 edited by Jean W. Ross (1986)

1986 edited by J. M. Brook (1987)

1987 edited by J. M. Brook (1988)

1988 edited by J. M. Brook (1989)

1989 edited by J. M. Brook (1990)

1990 edited by James W. Hipp (1991)

1991 edited by James W. Hipp (1992)

1992 edited by James W. Hipp (1993)

1993 edited by James W. Hipp, contributing editor George Garrett (1994)

1994 edited by James W. Hipp, contributing editor George Garrett (1995)

1995 edited by James W. Hipp, contributing editor George Garrett (1996)

1996 edited by L. Kay Webster and Samuel W. Bruce, contributing editor George Garrett (1997)

Concise Series

Concise Dictionary of American Literary Biography, 6 volumes (1988-1989): *The New Consciousness, 1941-1968; Colonization to the American Renaissance, 1640-1865; Realism, Naturalism, and Local Color, 1865-1917; The Twenties, 1917-1929; The Age of Maturity, 1929-1941; Broadening Views, 1968-1988.*

Concise Dictionary of British Literary Biography, 8 volumes (1991-1992): *Writers of the Middle Ages and Renaissance Before 1660; Writers of the Restoration and Eighteenth Century, 1660-1789; Writers of the Romantic Period, 1789-1832; Victorian Writers, 1832-1890; Late Victorian and Edwardian Writers, 1890-1914; Modern Writers, 1914-1945; Writers After World War II, 1945-1960; Contemporary Writers, 1960 to Present.*

Dictionary of Literary Biography® • Volume One Hundred Seventy-Eight

British Fantasy and Science-Fiction Writers Before World War I

Dictionary of Literary Biography® • Volume One Hundred Seventy-Eight

British Fantasy and Science-Fiction Writers Before World War I

Edited by
Darren Harris-Fain
Shawnee State University

A Bruccoli Clark Layman Book
Gale Research
Detroit, Washington, D.C., London

Printed in the United States of America

The paper used in this publication meets the minimum requirements of American National Standard for Information Sciences–Permanence Paper for Printed Library Materials, ANSI Z39.48-1984. ™

Library of Congress Cataloging-in-Publication Data

British fantasy and science-fiction writers before World War I / edited by Darren Harris-Fain.
p. cm.–(Dictionary of literary biography; v. 178)
"A Bruccoli Clark Layman book."
Includes bibliographical references and index.
ISBN 0-8103-9941-5 (alk. paper)
1. English fiction–19th century–Bio-bibliography–Dictionaries. 2. Fantastic fiction, English–Bio-bibliography–Dictionaries. 3. Science fiction, English–Bio-bibliography–Dictionaries. 4. Authors, English–19th century–Biography–Dictionaries. 5. English fiction–20th century–Bio-bibliography–Dictionaries. 6. Authors, English–20th century–Biography–Dictionaries. 7. Fantastic fiction, English–Dictionaries. 8. Science fiction, English–Dictionaries. I. Harris-Fain, Darren. II. Series.
PR868.F3B75 1997
823' .087609'03–dc21 97-8790
[B] CIP

10 9 8 7 6 5 4 3 2 1

Contents

Plan of the Series

. . . Almost the most prodigious asset of a country, and perhaps its most precious possession, is its native literary product – when that product is fine and noble and enduring.

Mark Twain*

The advisory board, the editors, and the publisher of the *Dictionary of Literary Biography* are joined in endorsing Mark Twain's declaration. The literature of a nation provides an inexhaustible resource of permanent worth. We intend to make literature and its creators better understood and more accessible to students and the reading public, while satisfying the standards of teachers and scholars.

To meet these requirements, *literary biography* has been construed in terms of the author's achievement. The most important thing about a writer is his writing. Accordingly, the entries in *DLB* are career biographies, tracing the development of the author's canon and the evolution of his reputation.

The purpose of *DLB* is not only to provide reliable information in a convenient format but also to place the figures in the larger perspective of literary history and to offer appraisals of their accomplishments by qualified scholars.

The publication plan for *DLB* resulted from two years of preparation. The project was proposed to Bruccoli Clark by Frederick C. Ruffner, president of the Gale Research Company, in November 1975. After specimen entries were prepared and typeset, an advisory board was formed to refine the entry format and develop the series rationale. In meetings held during 1976, the publisher, series editors, and advisory board approved the scheme for a comprehensive biographical dictionary of persons who contributed to North American literature. Editorial work on the first volume began in January 1977, and it was published in 1978. In order to make *DLB* more than a reference tool and to compile volumes that individually have claim to status as literary history, it was decided to organize volumes by topic, period, or genre. Each of these freestanding volumes provides a biographical-bibliographical guide and overview for a particular area of literature. We are convinced that this organization – as opposed to a single alphabet method – constitutes a valuable innovation in the presentation of reference material. The volume plan necessarily requires many decisions for the placement and treatment of authors who might properly be included in two or three volumes. In some instances a major figure will be included in separate volumes, but with different entries emphasizing the aspect of his career appropriate to each volume. Ernest Hemingway, for example, is represented in *American Writers in Paris, 1920–1939* by an entry focusing on his expatriate apprenticeship; he is also in *American Novelists, 1910–1945* with an entry surveying his entire career. Each volume includes a cumulative index of the subject authors and articles. Comprehensive indexes to the entire series are planned.

The series has been further augmented by the *DLB Yearbooks* (since 1981) which update published entries and add new entries to keep the *DLB* current with contemporary activity. There have also been *DLB Documentary Series* volumes which provide biographical and critical source materials for figures whose work is judged to have particular interest for students. One of these companion volumes is entirely devoted to Tennessee Williams.

We define literature as the *intellectual commerce of a nation:* not merely as belles lettres but as that ample and complex process by which ideas are generated, shaped, and transmitted. *DLB* entries are not limited to "creative writers" but extend to other figures who in their time and in their way influenced the mind of a people. Thus the series encompasses historians, journalists, publishers, book collectors, and screenwriters. By this means readers of *DLB* may be aided to perceive literature not as cult scripture in the keeping of intellectual high priests but firmly positioned at the center of a nation's life.

DLB includes the major writers appropriate to each volume and those standing in the ranks behind them. Scholarly and critical counsel has been sought in deciding which minor figures to include and how full their entries should be. Wherever possible, useful references are made to figures who do not warrant separate entries.

Each *DLB* volume has an expert volume editor responsible for planning the volume, selecting the figures for inclusion, and assigning the entries. Volume editors are also responsible for preparing, where appropriate, appendices surveying the major periodicals and literary and intellectual movements for their volumes, as well as lists of further readings. Work on the series as a whole is coordinated at the Bruccoli Clark Layman editorial center in Columbia, South Carolina, where the editorial staff is responsible for accuracy and utility of the published volumes.

One feature that distinguishes *DLB* is the illustration policy – its concern with the iconography of literature. Just as an author is influenced by his surroundings, so is the reader's understanding of the author enhanced by a knowledge of his environment. Therefore *DLB* volumes include not only drawings, paintings, and photographs of authors, often depicting them at various stages in their careers, but also illustrations of their families and places where they lived. Title pages are regularly reproduced in facsimile along with dust jackets for modern authors. The dust jackets are a special feature of *DLB* because they often document better than anything else the way in which an author's work was perceived in its own time. Specimens of the writers' manuscripts and letters are included when feasible.

Samuel Johnson rightly decreed that "The chief glory of every people arises from its authors." The purpose of the *Dictionary of Literary Biography* is to compile literary history in the surest way available to us – by accurate and comprehensive treatment of the lives and work of those who contributed to it.

The *DLB* Advisory Board

Introduction

In Charles Robert Maturin's Gothic masterpiece *Melmoth the Wanderer* (1820), a Spaniard who is attempting to escape the Inquisition via tunnels beneath Madrid stumbles upon a mysterious underground chamber. Spying unrecognizable instruments, books and manuscripts, and the bones of a mammoth and elk along with human skeletons, he supposes the room to serve "some infernal ceremony"; he learns instead that the old man who enters the room is not a "grand wizard" but rather a physician and scientist.

The supernatural and the scientific often go hand in hand in European literature from the Renaissance to the Victorian era. John Donne proclaimed in the seventeenth century that with the new learning everything changed, but it did not change overnight. In the sixteenth and seventeenth centuries Galileo faced the indignation of the church for contradicting contemporary beliefs about the nature of the universe; in the second half of the nineteenth century Charles Darwin was widely attacked for suggesting that humanity emerged in a way other than that described in Genesis. While science, technology, and industrialization made increasingly impressive inroads from the Enlightenment to World War I, old verities were challenged.

With the Enlightenment and its aftermath came a slow but gradual emphasis on social improvement. Most significantly for literature, this involved efforts to increase literacy, which before the nineteenth century was for the most part limited to those of certain classes or stations. As a result, the growing readership of Europe sought material to read–and enterprising publishers were more than happy to supply it. In the eighteenth century this led to Grub Street and the rise of newspapers; in the nineteenth it led to penny dreadfuls and dime novels; and in the late nineteenth century and the early twentieth century this led to pulp magazines and other popular periodicals. The common denominator was a popular type of fiction that catered to tastes for exciting action and interesting characters rather than poetic language or philosophical themes.

Such literature tended to be escapist. With what would come to be known as fantasy and science fiction, however, readers could escape in reading about supernatural events or far-future settings while at the same time reading stories that addressed present concerns in veiled language or symbolism. It is noteworthy that fantasy and science fiction gradually began to be viewed as distinct modes of writing apart from the rest of literature at a time of social change in which the distinction between science and the supernatural was crucial.

Of course, poems, plays, stories, and novels drawing on what would become identified as fantastic or science-fiction tropes had always existed in European literature, and British literature is no exception. For instance, one of the earliest works of British literature, *Beowulf* (eighth century A.D.), features monsters and dragons as foes of the eponymous warrior-hero, while *Sir Gawain and the Green Knight* (fourteenth century) is about a knight's fateful encounter with a mysterious supernatural counterpart. Many of the stories in Geoffrey Chaucer's *The Canterbury Tales* (circa 1387–1400) also include supernatural or fantastic elements, side by side with more-realistic stories of medieval life.

With the Renaissance came a renewed interest not only in classical literature but in new ideas about art, literature, and the natural world. Part of this worldview, in contrast with the medieval conception of life as static, involved a belief in progress, not only spiritual but also artistic, intellectual, and material. Such a belief is manifested in the seminal work of a tradition with strong connections to fantasy and science fiction, utopian literature. In Sir Thomas More's *Utopia* (1516), imperfect England is juxtaposed to the seemingly perfect island of Utopia (a Latin coinage meaning "Nowhere"). The citizens of Utopia live according to natural law and a communistic social contract. Like much fantastic literature, despite its "unrealistic" character *Utopia* thus serves as a critique of the author's times.

Similarly, Edmund Spenser in *The Faerie Queene* (1590, 1596) presents not only an elaborately structured and beautifully written literary work but also a multilayered Protestant allegory of Elizabethan England. Inspired by the Arthurian legends that had existed in England for centuries, the literal story of the long poetic sequence concerns the various quests of the knights serving the fairy queen Gloriana; thus, the poem is inherently a work of fantasy, with its fairy knights and their amazing adventures. At the same time *The Faerie Queene* is also an instruction manual for noble behavior, a political allegory, and a religious allegory.

Elizabethan dramatists also relied on the fantastic when it suited their purposes–and since their purpose was generally to entertain, the use of incredible scenes or exotic locales and characters seemed natural. For instance, in *The Tragical History of Doctor Faustus* (first performed 1592, published 1604) Christopher Marlowe draws on the European legend of Faust, the man who in his thirst for knowledge sells his soul, not only to explore serious themes such as the price of knowledge or the sanctity of the soul but also to delight audiences with scenes of infernal machinations and supernatural events. Of course, the Faust legend has proven fertile ground for many writers, most notably Johann Wolfgang von Goethe, and in fantasy and science fiction Faust may be seen as the archetype for the necromancer or the mad scientist who pushes too far in his quest for knowledge or power and pays a terrible price. Victor Frankenstein of Mary Wollstonecraft Shelley's Gothic masterpiece is only one of many such ambiguous figures.

William Shakespeare also employed the fantastic frequently in his plays. *A Midsummer Night's Dream* (circa 1595–1596), for instance, includes fairies such as Oberon, Titania, and Puck among the cast of characters, and at one point the head of the unfortunate amateur actor Bottom is transformed into that of a donkey. *Hamlet* (circa 1601) and *Macbeth* (circa 1606) both feature ghosts as prominent parts of their respective stories (and, in *Macbeth,* witches as well), while *The Tempest* (circa 1611) includes a magician, Prospero, who lives on an exotic island with his daughter and who controls the fantastic creatures Ariel and Caliban. Also, the tempest of the title is created by Prospero. Aldous Huxley ironically appropriated the phrase "brave new world" for his well-known 1932 dystopia from this play, and *The Tempest* also served as the basis, with some modifications, for the classic science-fiction film *Forbidden Planet* (1956).

The Tempest is one of many literary works of the Renaissance to use the motif of the fantastic journey, which practically became a subgenre unto itself. Logically enough, the most common destination, from Cyrano de Bergerac to Francis Godwin to Daniel Defoe, was the Moon, as seen in works such as Godwin's *The Man in the Moone* (1638) and Defoe's anonymous *The Consolidator; or, Memoirs of Sundry Transactions from the World in the Moon* (1705). Not only are such works obviously playful literary excursions inspired by the growth of interest in Earth and what lies beyond it, but also by actual voyages of exploration from Marco Polo and Christopher Columbus to the eighteenth century.

For example, Francis Bacon in his utopian work *The New Atlantis* (1627) combines the tale of travel and discovery with a call to scientific endeavor. While more realistic, Defoe's *Robinson Crusoe* (1719) nonetheless features travel, an exotic setting, and a protagonist who survives isolation on an island by methodically getting to know his environment and rationally working toward his goals, much like a modern scientist.

While science rose in interest and importance, however, religion (which is also an aid to Crusoe's survival) continued to exert a powerful influence both in society in general and in literature. In specifically religious works, in fact, it is not surprising to find elements of the fantastic, since religion is fundamentally supernatural and otherworldly in its orientation. Thus, in John Milton's epic *Paradise Lost* (1667) the poet stages scenes with the angels in heaven and Satan and his demons in hell along with the biblical story of the Fall. Milton here expands on his biblical source to present a grand religious vision in which this world is only part of the cosmic scheme.

Another work that employs the fantastic to deal with religious subjects is John Bunyan's allegorical *The Pilgrim's Progress* (1678, 1684), which is about the efforts of a man named Christian to reach the Celestial City after passing through imaginatively depicted trials and temptations such as the Slough of Despond and Vanity Fair. While not as complex or rich as Spenser's *The Faerie Queene,* Bunyan's work nonetheless is worth reading, apart from its message, as a work of fantasy.

Such has certainly been the fate of Jonathan Swift's *Gulliver's Travels* (1726). Like Alexander Pope's *The Dunciad,* this is a biting satire of eighteenth-century society; and like Pope's poem, much of the satire now requires historical explanation. However, much of its satiric edge is still evident–after all, stupidity, hypocrisy, greed, and Yahooism have not disappeared since Swift's time. Still, for many readers the interest of *Gulliver's Travels* lies not in its satire but in its imaginative creations of tiny Lilliput, enormous Brobdingnag, the flying island of Laputa, and the intelligent horselike Houyhnhnms and the bestial humanlike Yahoos. In fact, as a work of fantasy *Gulliver's Travels* has often been classified as a book for children. While children can certainly enjoy the adventures of Gulliver as a giant in Lilliput and as a doll in Brobdingnag, marvel at the mysterious Laputa, and delight in the concept of talking animals (as the success of television shows such as *Mister Ed* and movies such as *Babe* continues to illustrate), only adults are likely to appreciate–if that is the word for it–Swift's scathing portrayal of human stupidity and venality.

Swift's masterpiece stands in sharp contrast to the Enlightenment ideals of progress and even perfectibility. So too does a literary subgenre that emerged in the second half of the eighteenth century: the Gothic novel. While the Enlightenment thinkers praised reason, Gothic writers pointed to humanity's strong streak of irrationality; while the Enlightenment promoted science, the Gothic reveled in the supernatural; while the Enlightenment fostered newness, the Gothic drew on antiquity; and while the Enlightenment stood for light, the Gothic world was one of darkness.

The first of these works, which named the genre in its subtitle, was Horace Walpole's *The Castle of Otranto: A Gothic Story* (1765). It also established many of the motifs of the form, including haunted castles, other supernatural phenomena (real or contrived), ancestral curses, fiendish villains, and threatened maidens. The formula, with its many variations, became extremely popular, and from the 1760s to the 1820s hundreds of Gothic novels confronted eager readers with their tales of horror and depravity. Among the best are Clara Reeve's *The Old English Baron* (1777), similar to *The Castle of Otranto* in setting and plot; William Beckford's *Vathek* (1786), which includes a satanic quest for pleasure in an orientalized setting; William Godwin's *Things as They Are; or, The Adventures of Caleb Williams* (1794), which unites Gothic paraphernalia with a call to social reform; and the works of Charlotte Dacre (writing under the pseudonym Rosa Matilda), including *Zofloya; or, The Moor: A Romance of the Fifteenth Century* (1806), which includes infernal pacts, seductions, betrayals, and the Inquisition as well.

Three of the major works of the Gothic period were produced by two of its best-known authors. In *The Mysteries of Udolpho* (1794) and *The Italian* (1797) Ann Radcliffe created two of the quintessential works of the form with their depictions of intense psychological states in environments or physical and moral peril, while both *The Italian* and Matthew Gregory Lewis's *Ambrosio; or, The Monk* (1796) feature diabolical men of the cloth (anti-Catholicism was a prominent feature of Gothic fiction) who wreak destruction in their drives for power.

While not Gothic per se, many works of the Romantic period reflect the influence of this popular subgenre. For instance, in Samuel Taylor Coleridge's poem *The Rime of the Ancient Mariner* (1798) the Mariner brings death to his crewmates and a curse on himself by killing an albatross, a bird lucky to seamen. Also, Percy Bysshe Shelley's fictional efforts, such as *Zastrozzi: A Romance* (1810) and *St. Irvyne; or, The Rosicrucian: A Romance* (1811) have strong Gothic elements.

The fashion for Gothic fiction naturally led to parody, which is the motivating force behind novels such as Jane Austen's *Northanger Abbey* (1818) and Thomas Love Peacock's *Nightmare Abbey* (1818). While the Gothic was beginning to diminish in strength by the 1810s, however, a few noteworthy efforts in this vein remained: Mary Wollstonecraft Shelley's *Frankenstein; or, The Modern Prometheus* (1818, revised 1831), John Polidori's *The Vampyre* (1819), Charles Robert Maturin's *Melmoth the Wanderer,* and James Hogg's *The Private Memoirs and Confessions of a Justified Sinner* (1824).

Frankenstein is important on several counts. First, it uses a Gothic atmosphere of horror with its story of a man who attempts to create human life from the parts of disembodied corpses. The concept is original to the Gothic, but it echoes older stories such as that of the golem in Jewish legend and the legend of Faust. Moreover, the story combines horror with contemporary science: Frankenstein creates and animates his monster not through supernatural means but through vivisection and electricity. For this reason Brian W. Aldiss and other critics have identified *Frankenstein* as the first genuine work of science fiction.

Like *Frankenstein,* Polidori's *The Vampyre* was the result of a friendly competition among the Shelleys, Polidori, and George Gordon, Lord Byron to create a Gothic work of horror. Polidori did not create the figure of the vampire, which had existed for many years in European folklore, but he gave it its first substantial literary basis. Later incarnations of the vampire in works such as Joseph Sheridan Le Fanu's "Carmilla" (1872) and Bram Stoker's *Dracula* (1897) owe a considerable debt to *The Vampyre.*

Another intriguing combination of Gothic atmosphere with existing legends appears in Maturin's idiosyncratic *Melmoth the Wanderer.* Here the Faust motif merges with the legend of the Wandering Jew in this story of an Irishman who, having sold his soul to the devil in exchange for immortality, must wander the earth in search of someone in desperate enough circumstances to take Melmoth's immortality, and with it his damnation, upon himself or herself. Damnation, both spiritual and temporal, is also the theme of Hogg's *The Private Memoirs and Confessions of a Justified Sinner,* a psychologically disturbing work.

Mary Wollstonecraft Shelley gave science fiction another enduring trope with *The Last Man* (1826), which concerns the only human survivor of a worldwide cholera epidemic. While she gives the idea a specifically Romantic twist, with the last man on earth a symbol of the isolation of the Romantic soul apart from society, the figure of one person or a small group surviving a major world catastrophe would find many later expressions in science fiction, particularly in works by British authors such as John

Wyndham and Samuel Youd (writing as John Christopher).

The figure of the last man in a radically changed world is also an appropriate symbol for slightly less radical shifts in the nineteenth century as British literature passed from the Romantic period to the Victorian age and as Britain was transformed by the Industrial Revolution, the increasing ascendancy of science (including the publication of Charles Darwin's *The Origin of Species* in 1859), and the consolidation of the British Empire. In fantasy and science fiction, these changes are manifested in various ways. Growing industrialism; shrinking horizons because of inventions such as the railroad, the telegraph, and the telephone; and a boom in popular publishing encouraged in part by increases in literacy caused by education reforms all contributed to the development of these literary forms. Fantasy continued to be written by a wide variety of writers, but often fantastic elements were employed either to make a point about contemporary society or simply to entertain readers. Thus, Victorian fantasy tends to be of two broad types: either as a vehicle of social, political, or religious critique or in light tales intended for a popular audience. Skillful writers, of course, could do both at the same time.

The same applies to science fiction of the period. Though the term *science fiction* did not gain wide currency until after an American editor named Hugo Gernsback applied it to the genre in 1929, a great deal of fiction that would now be called science fiction appeared in the nineteenth century. Much of it, like much general fiction of the time, was ephemeral in nature–the twentieth-century American science-fiction writer Theodore Sturgeon once admitted that 90 percent of science fiction was crap, but added that 90 percent of *everything* is crap–but much Victorian science fiction is of interest for later readers, and at the end of the century the early works of one of science fiction's great figures, H. G. Wells, would appear.

One example of light fantasy that nonetheless features elements of social criticism can be found in the works of an early Victorian writer named James Dalton. His anonymous *The Gentleman in Black* (1831), for instance, employs the common motif of the deal with the devil. Rather than a dark tone, as in Maturin's *Melmoth the Wanderer* or much of Marlowe's *The Tragical History of Doctor Faustus,* however, Dalton instead uses the conceit for comic purposes: the devil promises an Englishman and a Frenchman limitless wealth if only they can double the number of their sins each year–which proves more difficult than they had at first imagined, even though they start from one second of sin in their first year of the deal. A lawyer, however, finds some loopholes in the contract and manages to extricate them. Dalton's anonymous *The Invisible Gentleman* (1833) is also comic in tone. Here a man wishes for the power of invisibility, gets it, and finds it more trouble than he had thought. Such a use of fantasy to entertain while also teaching moral lessons is evident in Charles Dickens's *A Christmas Carol* (1843). This famous story of how the miserly Ebenezer Scrooge is turned from his selfish, mean-spirited ways by visits from the ghost of his old partner and the spirits of Christmas past, present, and future, though seldom read now, has nonetheless become a staple of the Christmas season in its many media adaptations.

The variety of works in John Sterling's posthumous collection *Essays and Tales* (1848) is representative of the many different types of fantasy to be found in Victorian literature. In the short novel *The Onyx Ring* (1838–1839), for instance, a man is given the opportunity to feel others' experiences as they feel them. "Zamor" concerns Alexander the Great; "Cydon" is about a spirit that guides a young man to the site of the torture of Prometheus; "Land and Sea" (1838) is about a strange sea creature; and "A Chronicle of England" is a fairy tale with allegorical elements, much like Spenser's *The Faerie Queene.*

Nor do such stories exhaust the imaginative possibilities explored by Victorian writers of fantasy. In *A Man Made of Money* (1848–1849) Douglas Jerrold takes another approach to the concept of the person whose wish is granted with comically disastrous consequences. In this novel a man who is constantly asked for funds from his wife and her daughters wishes he were made of money. What he intended idiomatically becomes literally true, and the physical and moral decline that accompanies his sudden wealth is both humorous and satiric.

While such fantasies are often light in tone, George MacDonald created a more strange and otherworldly atmosphere in his fantasies. Works such as *Phantastes* (1858), *At the Back of the North Wind* (1871), the stories collected in the ten-volume *Works of Fancy and Imagination* (1871), and *Lilith* (1895) all convey an impression of fantastic worlds related to but apart from the world of the reader and would prove widely influential on later fantasists such as C. S. Lewis and J. R. R. Tolkien. Also, like many works of British fantasy, much of MacDonald's work contains a strong theological strain. While not strictly allegorical like Bunyan's *Pilgrim's Progress,* MacDonald's fantasies are nevertheless informed by a deeply felt Christian sentiment and often involve elemental conflict between good and evil.

Christianity was also an important part of the worldview of Charles Kingsley. His best-known fantasy, *The Water-Babies* (1863), concerns a young boy who, because he did not have an opportunity to learn virtue before his untimely death, is given a second chance as a "water-baby." The novel is notable not only for its premise, which strikes one as an intriguing mixture of fantasy with the poetry of William Blake, but also for its equally interesting combination of evangelical Christianity and evolutionary theory, which Kingsley, a clergyman, enthusiastically supported.

In its story of a type of existence beyond death *The Water-Babies* also serves as an example of a popular Victorian type of fantasy, known by some critics as the "posthumous fantasy." At a time when many scientists were essentially confuting the belief in humanity's divine origins (and with that its divine destination) and when many biblical scholars challenged the divine inspiration of the Bible, it is natural that fiction would deal with the question of what happens when people die–and fantasy fiction seemed appropriately poised for such issues.

Finally, *The Water-Babies* is a good example of an interest in Victorian literature with children. Imaginative literature for children first emerged as a major endeavor in the second half of the nineteenth century. Much of it was boringly didactic, and most of it was concerned with moral development even if it included interesting stories and characters. This is the context in which a shy mathematics professor and amateur photographer named Charles Lutwidge Dodgson wrote, under the pseudonym Lewis Carroll, his masterpieces of fantasy literature. In particular, *Alice's Adventures in Wonderland* (1865) and *Through the Looking-Glass, and What Alice Found There* (1872) are among the best-known and most-influential works in the history of British fantasy. They have appealed to readers, young and old, for many reasons: their imaginative situations and characters, their anarchic wit, and the almost unflappable reactions of a level-headed girl named Alice to the bizarre people and events that she encounters. Also, the chaos of the Alice books served as a healthy antidote to their pedantic counterparts in Victorian children's literature, and stories and books for children now have more in common with Lewis Carroll than they do with the McGuffey Readers.

Carroll's books also reflected the social and political chaos of their times, which were a period of rapid change in Europe. Naturally, fiction in the late nineteenth century mirrored such tensions. Two related manifestations of concerns about political conflict are what scholars have called the future-war story and the imaginary-invasion story. Examples include Alfred Bate Richards's *The Invasion of England: A Possible Tale of Future Times* (1870); Sir George Chesney's anonymous "The Battle of Dorking," published in the popular *Blackwood's Magazine* in May 1871; and William Le Queux's *The Great War in England in 1897* (1894). Since such works include future settings and involve imaginative possibilities, they are inherently fantastic or science fictional in nature. Moreover, though they enjoyed only a brief period of popularity that effectively ended with the all-too-real experience of World War I, they exerted a significant influence on the development of science fiction. The key factor in this development was Wells's combination of the concepts of a future war and an imaginary invasion with the possibility of life on other planets in his novel *The War of the Worlds* (1898). Science fiction from Wells to movies such as *Independence Day* (1996) has been filled with stories of future wars and alien invasions ever since.

Another popular type of story in fantasy and science fiction of the late nineteenth and early twentieth centuries was the lost-race novel. Historically this subject is interesting, since by this time all parts of the globe had been explored, but not all of the world was thoroughly studied yet. The lost-race novel posits that in a remote, isolated part of the world there may exist a group of people, usually ancient and utterly untouched by modern civilization–which thus provided authors with an effective foil for social commentary along with a great deal of license for imaginative tales. Popular lost-race novels included Edward Bulwer-Lytton's *The Coming Race* (1871), Samuel Butler's utopian *Erewhon; or, Over the Range* (1872), H. Rider Haggard's supernatural *She* (1886), and Sir Arthur Conan Doyle's *The Lost World* (1911). Differences among these works illustrate some of the fundamental concerns of their times. For instance, Bulwer-Lytton's *The Coming Race* features a race of people living under England who enjoy an idyllic egalitarian society (in fact, women are not only men's social equals but their physical superiors). Butler's *Erewhon* is even more political in intent; and while Bulwer-Lytton does not think machines are incompatible with happiness, Butler attacks the machine age as well as evolution and deism. Butler, like William Morris later, advocates a return to an agrarian way of life–as does Richard Jefferies in *After London; or, Wild England* (1885), which combines a catastrophe narrative with a pastoral theme. Similarly, in *A Crystal Age* (1887) W. H. Hudson envisions, along with a society in which marriage does not exist, a postindustrial civilization closer to a nostalgic view of preindustrial England.

Other writers of this period also expressed reservations about modern society through fantasy and science fiction. As has often been the case, horror fiction served many writers as an especially good vehicle for dealing with the anxieties of contemporary life through the lens of dark fantasy. For example, many of the stories in Joseph Sheridan Le Fanu's collection *In a Glass Darkly* (1872) transpose modern fears into supernatural ones, while Charlotte Riddell's *The Uninhabited House* (1875), on the surface a conventional ghost story, delves deeply into issues of justice, as much horror fiction does.

Likewise, contemporary issues inform Percy Greg's science-fiction novel *Across the Zodiac: The Story of a Wrecked Record* (1880), which contrasts a fictional Martian society with that of Earth so that Greg may more or less directly express his views on a variety of subjects. For example, on Mars a utopian monarchy governs after a failed effort at communism, whose weaknesses are said to have led to its downfall. On Mars a scientific worldview is prevalent, which is presented as positive. Also, the Martians practice polygamy, and this and woman suffrage are among the other ideas discussed in the novel.

Another novel that uses a fictional society to address contemporary social problems or hypocrisies is Edwin A. Abbott's *Flatland: A Romance of Many Dimensions* (1884). The premise is that there are worlds of many different dimensions; the narrator, for instance, is "A. Square," who lives in the two-dimensional world of Flatland. Abbott describes the society of Flatland in great detail along with visions of one-dimensional and other-dimensional worlds. All of this contributes not only to an imaginative tour de force but also to some rather sharp social criticism and satire.

Less satiric but no less entertaining are the works of F. Anstey, one of the last of the masters of light Victorian fantasy. They include *Vice Versâ; or, A Lesson to Fathers* (1882), in which the personalities of a self-important businessman and his undisciplined son exchange bodies; *Tourmalin's Time Cheques* (1891), one of the first time-travel narratives; and *The Brass Bottle* (1900), a modern story about a man and his genie.

Authors such as Anstey obviously wrote about such things without believing that they could actually happen. Despite the rise of science in the nineteenth century, however, many people maintained a belief in the supernatural, whether in their religious faith or in other ways such as spiritualism, theosophy, and the occult. Occasionally such believers wrote fiction based on their views. Critics sometimes label these works "credulous romances" because their authors believe, to some extent at least, in the possibility of the marvels they describe. One such author was Marie Corelli, whose creativity with her own life story paralleled her literary efforts. *A Romance of Two Worlds* (1886) and *The Sorrows of Satan* (1895) are both credulous romances, as are some of the later fantasies of Sir Arthur Conan Doyle.

H. Rider Haggard did not expect anyone to take the fantastic events of *She* (1886) seriously, but it is an entertaining novel regardless. Here Haggard, also known for his hero Allan Quatermain, presents a lost race ruled by a powerful ancient figure named Ayesha, or She-Who-Must-Be-Obeyed. As if this were not enough, *She* also features reincarnation–another popular motif in Victorian fantasy, which was related to the "timeslip romance," in which a character travels in time and falls in love with someone in the past.

More disturbing is Robert Louis Stevenson's short novel *The Strange Case of Dr. Jekyll and Mr. Hyde* (1886), which is now widely revered as one of the classics of horror fiction. The well-known story about the transformation of a suave man of science into a brutish monster (though his monstrosity, unlike many media adaptations of the book, is more moral than physical in nature) serves to highlight contemporary discussions about humanity's animal nature, the civilizing role of society, and the then-new science of psychology.

Many writers of the late nineteenth century who did not devote their careers to fantasy or science fiction nonetheless wrote in these modes occasionally. For example, Richard Garnett wrote enough fantasies, usually historical in scope, to make a collection, *The Twilight of the Gods* (1888). Oscar Wilde employed fantasy frequently in his works for children as well as in his novel *The Picture of Dorian Gray* (1891), about a man who supernaturally maintains his youth in order to pursue a life of pleasure. Rudyard Kipling, like many writers of the period both prolific and diverse, often drew upon fantastic or science-fiction topics. His collections, such as *The Phantom 'Rickshaw and Other Tales* (1888), include several works of fantasy; *The Jungle Books* (1894, 1895) and *Just So Stories* (1902) rely upon the premise of talking, intelligent animals; and his related stories "With the Night Mail" (*McClure's,* July 1905) and "As Easy as ABC" (*London Magazine,* April 1912) are both science-fiction depictions of the future of mail delivery.

Garnett was not the only author to write fantasy stories with historical settings. Such works were popular, and one of the most successful writers in this mode was Edwin Arnold. While limited strictly as historical fiction, imaginative fantasies with a variety of historical settings, novels such as *The Wonderful Adventures of Phra the Phoenician* (1890) and *Lepidus the Centurian* (1900) are still worth reading as entertainment. Arnold also published a popular science-fantasy novel called *Lieut. Gullivar Jones: His Vacation*

(1905), which anticipates the Martian novels of Edgar Rice Burroughs.

Another prolific and diverse author of the period was Sir Arthur Conan Doyle. Best known as the creator of Sherlock Holmes, Doyle also wrote several works of science fiction and fantasy, including *The Captain of the Polestar and Other Stories* (1890), *The Lost World* (1911), and *The Poison Belt* (1913). Doyle's considerably different contemporary William Morris also wrote in these forms, both science fiction as in his utopian *News from Nowhere* (1890) and fantasy as in his medieval-influenced romances such as *The Story of the Glittering Plain* (1891) and *The Well at the World's End* (1896).

By the end of the nineteenth century many of the patterns of fantasy and science fiction were firmly established, and many subsequent efforts are thus variations or adaptations of existing subjects and formulas. Still, individual writers managed to create personal visions within these forms, such as Vernon Lee, George Du Maurier, and E. Nesbit. Also, another mode began to take shape in the 1890s–the scientific romance, a combination of the romance (with its unrealistic events and often exotic settings) with real or conceivably imaginable science. Many critics, such as Brian M. Stableford in *Scientific Romance in Britain 1890–1950* (1985), distinguish between scientific romance and science fiction; others see scientific romance as a particular type of science fiction.

George Griffith was one of the founders of the scientific romance. His *The Angel of the Revolution* (1894), with its airships, submarines, and other nineteenth-century wonders, is one of many future-war stories offered to British readers in the second half of the nineteenth century. Like predecessors such as "The Battle of Dorking" and the future-war narratives of William Le Queux, it features plenty of scenes of impressive weapons and enormous destruction. Unlike such reactionary predecessors, however, Griffith instead posits a future of substantial change, both scientific and political–here a socialist revolution. The sequel, *Olga Romanoff; or, The Syren of the Skies* (1894), continues in the same vein. Griffith's talents in the scientific romance were soon overshadowed by the more impressive talents of H. G. Wells, but Griffith is also remembered for historical fantasies such as *Valdar the Oft-Born* (1895) and the light science fantasy *A Honeymoon in Space* (1901).

By far the most popular and successful writer of fantasy and science fiction in the last decade of the nineteenth century was H. G. Wells. Possessing both a scientific education and a strong literary background, Wells in the early years of his career created masterpiece after masterpiece: *The Time Machine* (1895), *The Island of Doctor Moreau* (1896), *The Invisible Man* (1897), *The War of the Worlds* (1898), and *The First Men in the Moon* (1901) are all important, influential scientific romances. While Wells was not the first author to write about time travel, mad scientists, invisibility, or future wars, he either explained such situations pseudoscientifically, making them significant for the future development of science fiction, or he wrote about them so imaginatively and convincingly that the situations then became identified with these works. Though they have not acquired the importance of these novels, Wells also wrote other scientific romances, including *Tales of Space and Time* (1899), *When the Sleeper Wakes* (1899), *The Food of the Gods* (1904), *In the Days of the Comet* (1906), *The War in the Air* (1908), and *The Country of the Blind* (1911) as well as fantasies such as *The Wonderful Visit* (1895) and *The Sea Lady* (1902).

Another popular fantasist around the turn of the century and beyond was Arthur Llewelyn Jones Machen, whose frequently dark tales influenced many later writers of fantasy and horror. Novels of horror were especially popular; among the best known are Robert W. Chambers's *The King in Yellow* (1895); Bram Stoker's *Dracula,* which with Shelley's *Frankenstein* and Stevenson's *Dr. Jekyll and Mr. Hyde* has become one of the classics of horror fiction; the works of Robert S. Hichens; and even *The Turn of the Screw* (1898) by American expatriate Henry James. Other popular writers of fantasy and horror during this time include Margaret Oliphant and W. W. Jacobs. Some popular writers, such as M. R. James and George Oliver Onions, became identified with a particular type of fantasy and horror: the ghost story, which enjoyed a period of considerable interest at this time.

Between 1900 and 1918 Britain experienced a wide range of literary efforts, from romantic and traditional to modernistic and experimental. Similarly, in fantasy and science fiction in these years one can find works as diverse as M. P. Shiel's idiosyncratic *The Purple Cloud* (1901) and Joseph Conrad and Ford Madox Hueffer's *The Inheritors* (1901) in the same year. Other notable authors and works include J. M. Barrie, whose play *Peter Pan* (1904) established its popularity early on with its story of a boy who refuses to grow up and who lives in Never-Never-Land; G. K. Chesterton, whose novels *The Napoleon of Notting Hill* (1904) and *The Man Who Was Thursday* (1908) both unite supernatural elements with his social and religious concerns; well-known fantasists such as Lord Dunsany, Algernon Henry Blackwood, Marjorie Bowen, and Walter de la Mare; Kenneth Grahame with his fantasies for young readers, the best known of which is *The Wind in the Willows* (1908); and William Hope Hodgson, whose masterpieces of weird fiction include *The House on the Borderland* (1908) and *The Night Land* (1912).

The years 1900 to 1918 also saw early expressions of two significant types of science fiction that would find fuller developments later. In the first, "The Machine Stops" (1909) by E. M. Forster, is the dystopia, the nightmare version of the utopia. Forster's story is not a true dystopia, since it does not portray a planned totalitarian society but rather a seemingly paradisaical technological society that grinds to a halt when the machinery breaks. However, it is significant in the history of dystopian literature both for its critique of the technological utopianism of Wells in the early twentieth century and for its influence on later dystopias such as Aldous Huxley's *Brave New World* (1932) and George Orwell's *Nineteen Eighty-Four* (1949).

Another important work, though seldom read now, is John Davys Beresford's *The Hampdenshire Wonder* (1911), one of the first science-fiction works to employ the concept of the superman. While the nineteenth-century German philosopher Friedrich Nietzsche claimed that the superman was someone freed from the conventions of society–a notion adopted in different works by Bernard Shaw and D. H. Lawrence, among others–science fiction quickly picked up the concept and transformed it into a vision of a person not only intellectually but also physically superior to other human beings. Such superhumans could be the result of evolution, mutation, or experimentation, and the concept would offer many courses of development, from comic-book superheroes to mutants with various powers.

World War I (1914–1918) did not bring an end to fantasy and science fiction, but for many writers it did bring a subtle change of direction. While older thinkers such as Wells continued to preach a gospel of progress, many were disillusioned by the war that Wells wrongly predicted would end all war. Also, just as the involvement of the United States in the Great War proved crucial, so the growing influence of American fantasy and science fiction, especially through the popular pulp magazines of the 1920s and 1930s, would have a substantial impact on their further development in Britain. Before 1918, by and large, fantasy and science fiction were the province of any writer who felt their use appropriate; though this does not disappear, after 1918 fantasy and science fiction became more and more separated from other types of literature, more often than not seen as popular fiction as opposed to genuine literary forms. They would not begin to escape the literary ghetto of popular culture until the 1960s, and the idea that all fantasy and science fiction is poorly written escapist entertainment still has not been completely overcome.

– *Darren Harris-Fain*

Acknowledgments

This book was produced by Bruccoli Clark Layman, Inc. Karen L. Rood is senior editor for the *Dictionary of Literary Biography* series. George P. Anderson and Samuel W. Bruce were the in-house editors.

Administrative support was provided by Ann M. Cheschi and Brenda A. Gillie.

Bookkeeper is Joyce Fowler.

Copyediting supervisor is Laurel M. Gladden Gillespie. The copyediting staff includes Phyllis A. Avant, Patricia Coate, Jeff Miller, William L. Thomas Jr., and Allison Trussell.

L. Kay Webster and Jane M. J. Williamson are editorial associates.

Layout and graphics staff includes Marie L. Parker and Janet E. Hill.

Office manager is Kathy Lawler Merlette.

Photography editors are Julie E. Frick and Margaret Meriwether. Photographic copy work was performed by Joseph M. Bruccoli.

Production manager is Samuel W. Bruce.

Software specialist is Marie L. Parker.

Systems manager is Chris Elmore.

Typesetting supervisor is Kathleen M. Flanagan. The typesetting staff includes Pamela D. Norton and Patricia Flanagan Salisbury. Freelance typesetters include Melody W. Clegg and Delores Plastow.

Walter W. Ross, Steven Gross, and Mark McEwan did library research. They were assisted by the following librarians at the Thomas Cooper Library of the University of South Carolina: Linda Holderfield and the interlibrary-loan staff; reference-department head Virginia Weathers; reference librarians Marilee Birchfield, Stefanie Buck, Stefanie DuBose, Rebecca Feind, Karen Joseph, Donna Lehman, Charlene Loope, Anthony McKissick, Jean Rhyne, Kwamine Simpson, and Virginia Weathers; circulation-department head Caroline Taylor; and acquisitions-searching supervisor David Haggard.

The publishers acknowledge the generous assistance of William R. Cagle, director of the Lilly Library, Indiana University, and his staff, who provided many of the illustrations in this volume. Their work represents the highest standards of librarianship and research.

Dictionary of Literary Biography® • Volume One Hundred Seventy-Eight

British Fantasy and Science-Fiction Writers Before World War I

Dictionary of Literary Biography

Edwin A. Abbott
(1838 - 1926)

Darren Harris-Fain
Shawnee State University

BOOKS: *A Shakespearean Grammar: An Attempt to Illustrate Some of the Differences between Elizabethan and Modern English* (London: Macmillan, 1869; revised and enlarged, 1870; New York: Haskell House, 1972);

A Concordance to the Works of Alexander Pope (London: Chapman & Hall, 1875; New York: Appleton, 1875);

How to Write Clearly: Rules and Exercises on English Composition (Boston: Roberts, 1875);

Philochristus: Memoirs of a Disciple of the Lord (London: Macmillan, 1878; Boston: Roberts, 1878);

How to Parse: An Attempt to Apply the Principles of Scholarship to English Grammar (Boston: Roberts, 1880);

Three Lectures on Subjects Connected with the Practice of Education: Delivered in the University of Cambridge in the Easter Term, 1882, by Abbott, H. W. Eve, and Arthur Sidgwick (Cambridge: Cambridge University Press, 1883);

The Common Tradition of the Synoptic Gospels: In the Text of the Revised Version, by Abbott and W. G. Rushbrooke (London: Macmillan, 1884);

Flatland: A Romance of Many Dimensions, as A. Square (London: Seeley, 1884; enlarged, 1884; Boston: Roberts, 1885);

Francis Bacon: An Account of His Life and Works (London: Macmillan, 1885);

Philomythus, an Antidote against Credulity: A Discussion of Cardinal Newman's Essay on Ecclesiastical Miracles (London & New York: Macmillan, 1891);

The Anglican Career of Cardinal Newman, 2 volumes (London & New York: Macmillan, 1892);

English Lessons for English People, by Abbott and J. R. Seeley (Boston: Roberts, 1893);

Edwin A. Abbott (Hulton Getty Picture Collection Limited)

Latin Prose through English Idiom: Rules and Exercises on Latin Prose Composition, second edition, by Abbott and E. R. Humphreys (Boston: Allyn, 1896);

The Spirit on the Waters: The Evolution of the Divine from the Human (London: Macmillan, 1897);

St. Thomas of Canterbury: His Death and Miracles (2 volumes, London: Black, 1898; 1 volume, New York: AMS Press, 1980);

Clue: A Guide through Greek to Hebrew Scripture (London: Black, 1900);

The Corrections of Mark Adopted by Matthew and Luke (London: Black, 1901);

From Letter to Spirit: An Attempt to Reach through Varying Voices the Abiding Word (London: Black, 1903);

Paradosis; or, "In the Night in Which He Was (?) Betrayed" (London: Black, 1904);

Johannine Vocabulary: A Comparison of the Words of the Fourth Gospel with Those of the Three (London: Black, 1905);

Silanus, the Christian (London: Black, 1906);

Notes on New Testament Criticism (London: Black, 1907);

Notes on the New Testament (London: Black, 1907);

"The Son of Man"; or, Contributions to the Study of the Thoughts of Jesus (Cambridge: Cambridge University Press, 1910);

Light on the Gospel from an Ancient Poet (Cambridge: Cambridge University Press, 1912);

The Fourfold Gospel, 5 volumes (Cambridge: Cambridge University Press, 1913–1917).

OTHER: Francis Bacon, *Bacon's Essays,* 2 volumes, edited by Abbott (London: Longmans, Green, 1876).

Edwin A. Abbott was a man of many qualities. As a cleric, a biblical and literary scholar, and an author, he established a minor but nonetheless significant place in Victorian literature. While many scholars remember Abbott for his well-known debate with John Henry Newman, readers of fantasy and science fiction are more familiar with the novel that combines several of his various interests: *Flatland: A Romance of Many Dimensions* (1884).

Abbott was a private man, and little is known of his life. He was born in London in 1838 to an Anglican family and received a degree in divinity at St. John's College at Cambridge University. He spent two years as assistant master at the King Edward School in Birmingham, then from 1865 to 1889 served as headmaster at the City of London School, where he championed the study of English. After his educational career Abbott devoted himself to scholarship, aided by his 1875 appointment as a lecturer at Cambridge. His earliest publications included books on William Shakespeare and Alexander Pope and on grammar and rhetoric. He also edited a collection of Francis Bacon's essays and wrote a historical Christian novel, *Philochristus: Memoirs of a Disciple of the Lord* (1878). Neither these books nor the many books on theology and literature he later published, however, bear any resemblance to his best-known work, *Flatland.*

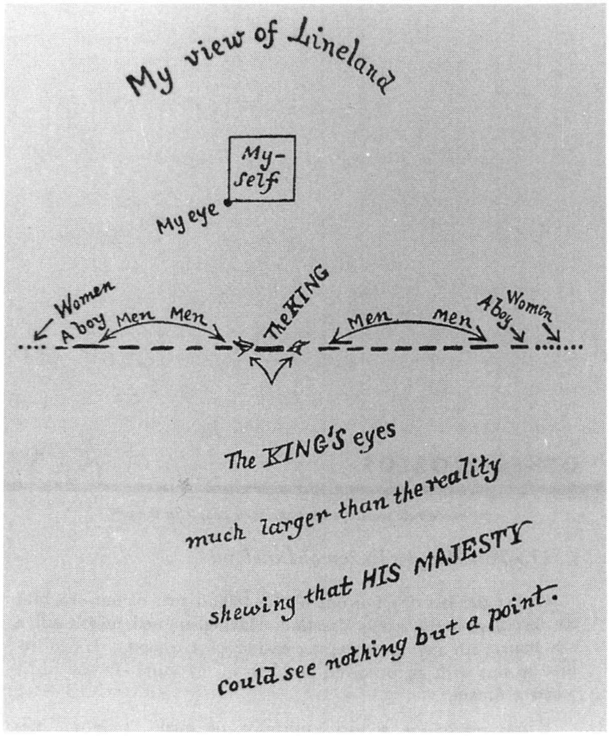

Abbott's illustration of Lineland from the perspective of the narrator, a square

In *Flatland* Abbott creates a fictional world of only two dimensions rather than three. Using his mathematical knowledge, he delineates a universe inhabited by two-dimensional beings. The narrator, a square, begins by introducing readers to his two-dimensional world, including its different types of life and its social order. A male's position in this social order is determined by angles and sides: for instance, the square narrator, a lawyer, is higher in the social hierarchy than triangles (who tend to be laborers, soldiers, and merchants) but lower than more-complex regular figures.

The highest figures in Flatland are those with so many sides and angles that they are close to being circles. Parents aspire to have children who possess more sides and angles than themselves; while the term *upward mobility* would have no literal meaning in a two-dimensional world, this is the intent. In addition, figures with irregular angles are considered outcasts and criminals. Thus, Abbott subtly satirizes Victorian class distinctions in the three-dimensional world.

Abbott also satirizes gender differences in *Flatland.* In this two-dimensional world males are geo-

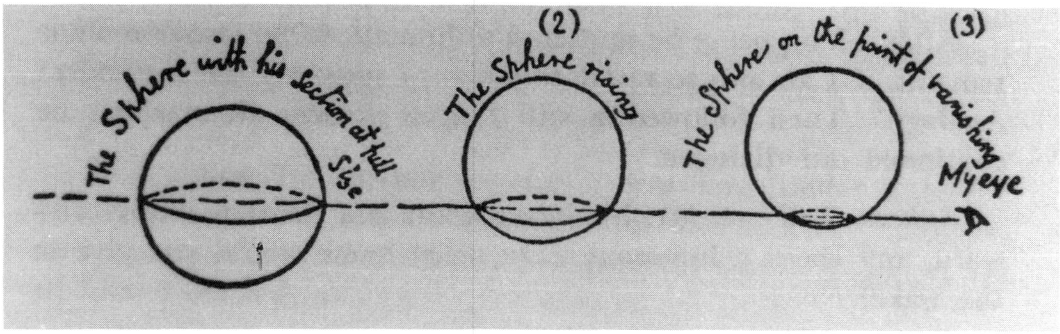

Abbott's illustration of the square's view of the sphere in Flatland

metric figures and females are closer to line segments. Since this society values multiple sides and regular angles, women obviously are second-class citizens in Flatland. However, they are nonetheless powerful figures, since their sharp forms can puncture and deflate the males. Given this fact and their emotionally volatile natures, females in Flatland are governed by sets of rules that limit their full participation in society. Clearly in *Flatland* Abbott critiques the patriarchal Victorian paternalism in relation to women.

Other social conventions are satirized in *Flatland* as well. For instance, since Flatland is a two-dimensional world, its inhabitants see each other as line segments or, if facing a female directly, a point. Thus, distinctions between people, though they exist, are difficult to perceive. Significantly, the ability to judge another's shape, and with that another's status, comes mainly with education.

The second part of the novel concerns the narrator's vision of a one-dimensional world called Lineland. More fantastic and less satiric than the first part of *Flatland,* this section describes what life might be like in such a world. This dream of Lineland is both highly imaginative and mathematically intriguing.

The tables are turned on the narrator in the last third of *Flatland,* in which the two-dimensional being encounters a being from Spaceland–in other words, a being from a three-dimensional world. The Spacelander, Lord Sphere, tries valiantly to explain the concept of three dimensions to someone who knows only two, but the square cannot understand and feels threatened by the Spacelander's seemingly supernatural view of events in Flatland granted by his three-dimensional perspective. As the Flatlander was attacked by the Linelanders, so the Flatlander here attacks the Spacelander–a sharp commentary on how people tend to respond to things beyond their understanding. The Spacelander then "lifts" the square to three-dimensional space, from which he can see Flatland from above. The narrator then accepts the notion of space and ascribes divinity to its inhabitants, which the Spacelander appropriately denies. Such an ascription is natural, however, since in the square's world the most powerful beings are those many-sided males closest to circles; a sphere must be much greater still, so the square thinks. Also, some readers have seen this encounter as a metaphor for humanity's limited understanding in its encounters with the divine–which is certainly plausible given Abbott's theological preoccupations.

Any notions of more than two dimensions are considered heretical in Flatland, so the narrator's experiences with the sphere lead him to trouble with the Flatland Council. Still, despite persecution and imprisonment, he continues to insist on the reality of a three-dimensional world, worlds of more than three dimensions, and a world revealed by the sphere called Pointland, inhabited by a single being satisfied with its own existence and unaware of the existence of others. This is the most religious part of *Flatland,* exhibiting Abbott's concerns about the nature of religious truth and understanding.

Contemporary reviewers were mixed in their reception of the novel. Some appreciated its imagination and satire, while others found the book tedious and didactic. Many readers were puzzled by what they took to be the metaphorical intent of the novel, unsure if Abbott was suggesting that God occupies a higher dimension and can become partially manifest on the human plane just as Lord Sphere can appear in Flatland as a cross section or whether the divine is more vaguely related to the natural world like a three-dimensional world would be related to one of two dimensions.

Though Abbott never clarified these issues in subsequent editions of *Flatland,* he defined his religious views in his many theological works, especially in his late entry into the Tractarian controversy. This debate centered around an Anglican priest and writer, John Henry Newman, who promoted traditionalism within the Church of England and, when instead the church continued to incorporate modern tendencies, converted to Roman Catholicism. In contrast, Abbott was comfortable with both his membership in the Church of England and a modern scientific worldview, and his *Philomythus, an Antidote against Credulity: A Discussion of Cardinal Newman's Essay on Ecclesiastical Miracles* (1891) is one of several responses to the more-traditional beliefs of Newman and others.

In his religious writings Abbott also distinguished between the miraculous, which he dismissed as untrue, and the supernatural, which he thought was different from the miraculous in being both above nature yet linked to nature. In this way one might see a connection between Abbott's "romance of many dimensions" and his theological beliefs.

Flatland has since inspired various influences and tributes. For instance, both Rudy Rucker's story "Message Found in a Copy of *Flatland*," collected in his *The 57th Franz Kafka* (1983), and the Hypernaut story in the *Tales of the Uncanny* issue of Alan Moore's 1993 comic-book miniseries *1963* draw on Abbott's fictional world as part of their own. Also, related books and sequels have been published, including *An Episode of Flatland* (1907) by the logician C. H. Hinton, *Sphereland* (1965) by Dionys Berger, and *The Planiverse* (1984) by A. K. Dewdney.

The combination of intriguing ideas and playfulness in *Flatland* has made it consistently popular with a select group of readers, and it has been reprinted often since its initial publication, usually under Abbott's name. Readers of fantasy have often appreciated its extreme otherworldliness, while science-fiction readers frequently have admired Abbott's logical extrapolations from the premise of the story.

In addition, the novel has been used by mathematics instructors to illustrate various concepts. While not as widely known as the fantasies of a professional mathematician who wrote as Lewis Carroll, *Flatland* still holds a unique place in the history of fantasy and science fiction.

Reference:

A. K. Dewdney, Introduction to Abbott's *Flatland: A Romance of Many Dimensions* (New York: Signet, 1984), pp. 7–23.

Grant Allen

(24 February 1848 – 28 October 1899)

Nicholas Ruddick
University of Regina

See also the Allen entries in *DLB 70: British Mystery Writers, 1860–1919* and *DLB 92: Canadian Writers, 1890–1920.*

Grant Allen (Hulton Getty Picture Collection Limited)

BOOKS: *Physiological Aesthetics* (London: Henry S. King, 1877; New York: Appleton, 1878);

The Colour-Sense: Its Origin and Development: An Essay in Comparative Psychology (London: Trübner, 1879; Boston: Houghton Osgood, 1879);

Anglo-Saxon Britain (London: Society for Promoting Christian Knowledge, 1881; New York: Young, 1881);

The Evolutionist at Large (London: Chatto & Windus, 1881; New York: Fitzgerald, 1881; revised edition, London: Chatto & Windus, 1894);

Vignettes from Nature (London: Chatto & Windus, 1881; New York: Fitzgerald, 1882);

The Colours of Flowers: As Illustrated in British Flora (London: Macmillan, 1882; London & New York: Macmillan, 1891);

Colin Clout's Calendar: The Record of a Summer, April–October (London: Chatto & Windus, 1883; New York: Funk & Wagnalls, circa 1883);

Flowers and Their Pedigrees (London: Longmans, Green, 1883; New York: Appleton, 1884);

Nature Studies, by Allen, Andrew Wilson, Thomas Foster, Edward Clodd, and R. A. Proctor (London: Wyman, 1883; New York: Funk & Wagnalls, 1883);

Biographies of Working Men (London: Society for Promoting Christian Knowledge, 1884; New York: Young, 1885);

Philistia, as Cecil Power, (3 volumes, London: Chatto & Windus, 1884; 1 volume, New York: Harper, 1884);

Strange Stories (London: Chatto & Windus, 1884);

Babylon, as Cecil Power (3 volumes, London: Chatto & Windus, 1885; 1 volume, New York: Appleton, 1885);

Charles Darwin (London: Longmans, Green, 1885; New York: Appleton, 1885);

Common Sense Science (Boston: Lothrop, circa 1886);

For Maimie's Sake: A Tale of Love and Dynamite (London: Chatto & Windus, 1886; New York: Appleton, 1886);

In All Shades: A Novel (3 volumes, London: Chatto & Windus, 1886; 1 volume, Chicago: Rand, McNally, 1888);

Kalee's Shrine, by Allen and May Cotes (Bristol: Arrowsmith, 1886; reprinted edition, London: Simpkin, Marshall, 1886; New York: New Amsterdam, 1897); republished as *The Indian Mystery* (New York: New Amsterdam, 1902);

The Sole Trustee (London: Society for Promoting Christian Knowledge, 1886);

The Beckoning Hand and Other Stories (London: Chatto & Windus, 1887);
A Terrible Inheritance (London: Society for Promoting Christian Knowledge, 1887; New York: Crowell, n.d.);
The Devil's Die: A Novel (3 volumes, London: Chatto & Windus, 1888; 1 volume, New York: Lovell, 1888);
Force and Energy: A Theory of Dynamics (London & New York: Longmans, Green, 1888);
A Half-Century of Science, by Allen and T. H. Huxley (New York: Fitzgerald, 1888);
This Mortal Coil: A Novel (3 volumes, London: Chatto & Windus, 1888; 1 volume, New York: Appleton, 1889);
The White Man's Foot (London: Hatchards, 1888);
Dr. Palliser's Patient (London: Samuel Mullen, 1889);
Falling in Love: With Other Essays on More Exact Branches of Science (London: Smith, Elder, 1889; New York: Appleton, 1890);
The Jaws of Death: A Novel (London: Simpkin Marshall, 1889; republished edition, London: Jarrold, 1896; New York: New Amsterdam, 1897);
A Living Apparition (London: Society for Promoting Christian Knowledge, 1889);
The Tents of Shem (3 volumes, London: Chatto & Windus, 1889; 1 volume, Chicago: Rand, McNally, 1889);
The Great Taboo (London: Chatto & Windus, 1890; New York: Burt, 1890);
Individualism and Socialism (Glasgow: Scottish Land Restoration League, circa 1890);
Wednesday the Tenth: A Tale of the South Pacific (Boston: Lothrop, circa 1890); republished as *The Cruise of the Albatross, or, When Was Wednesday the Tenth?: A Story of the South Pacific* (Boston: Lothrop, circa 1898);
Dumaresq's Daughter: A Novel (3 volumes, London: Chatto & Windus, 1891; 1 volume, New York: Harper, 1891);
Recalled to Life (Bristol: Arrowsmith, 1891; New York: Holt, 1891);
What's Bred in the Bone (London: Tit-Bits Offices, 1891; Chicago: Tucker, 1891);
The Attis of Caius Valerius Catullus: Translated into English Verse, with Dissertations on the Myth of Attis, on the Origins of Tree-Worship, and on the Galliambic Metre (London: David Nutt, 1892);
The Duchess of Powysland: A Novel (3 volumes, London: Chatto & Windus, 1892; 1 volume, New York: Munro, 1892);
The General's Will and Other Stories (London: Butterworth, 1892);
Science in Arcady (London: Lawrence & Bullen, 1892);
The Tidal Thames (London, Paris & Melbourne: Cassell, 1892);
An Army Doctor's Romance (London: Tuck, 1893);
Blood Royal: A Novel (London: Chatto & Windus, 1893; New York: Cassell, 1893);
Ivan Greet's Masterpiece and Other Stories (London: Chatto & Windus, 1893);
Michael's Crag (London: Leadenhall Press, 1893; Chicago: Rand, McNally, 1893);
The Scallywag (3 volumes, London: Chatto & Windus, 1893; 1 volume, New York: Cassell, circa 1893);
At Market Value: A Novel (2 volumes, London: Chatto & Windus, 1894; 1 volume, New York & Chicago: Neely, 1894);
The Lower Slopes: Reminiscences of Excursions Round the Base of Helicon, Undertaken for the Most Part in Early Manhood (London: Elkin Mathews & John Lane, 1894; Chicago: Stone & Kimball, 1894);
Post-Prandial Philosophy (London: Chatto & Windus, 1894);
Under Sealed Orders: A Novel (1 volume, New York: Collier, 1894; 3 volumes, London: Chatto & Windus, 1895);
The British Barbarians: A Hill-Top Novel (London: John Lane, 1895; New York: Putnam, 1895);
The Desire of the Eyes and Other Stories (London: Digby, Long, 1895; New York: Fenno, circa 1895);
In Memoriam: George Paul Macdonell (London: Lund, 1895);
The Story of the Plants (London: George Newnes, 1895); republished as *The Plants* (New York: Review of Reviews, 1909);
The Woman Who Did (London: John Lane, 1895; Boston: Roberts, 1895);
A Bride from the Desert (New York: Fenno, circa 1896);
Moorland Idylls (London: Chatto & Windus, 1896);
A Splendid Sin (London: White, 1896; New York: Buckles, 1899);
An African Millionaire: Episodes in the Life of the Illustrious Colonel Clay (London: Richards, 1897; New York: Arnold, 1897);
Cities of Belgium (London: Richards, 1897; New York: Wessels, 1900); republished as *Belgium: Its Cities,* 2 volumes (Boston: Page, 1903);
The Evolution of the Idea of God: An Inquiry into the Origins of Religion (London: Richards, 1897; New York: Holt, 1897);
Florence (London: Richards, 1897; New York: Wessels, 1900; revised and enlarged edition, London: Richards, 1906);

Paris (London: Richards, 1897; New York: Wessels, 1900; revised and enlarged edition, London: Richards, 1906);
Tom, Unlimited: A Story for Children, as Martin Leach Warborough (London: Richards, 1897);
The Type-Writer Girl, as Olive Pratt Rayner (London: C. Arthur Pearson, 1897); as Grant Allen (New York: Street & Smith, 1900);
Flashlights on Nature (New York: Doubleday & McClure, 1898; London: George Newnes, 1899);
The Incidental Bishop: A Novel (London: C. Arthur Pearson, 1898; New York: Appleton, 1898);
Linnet: A Romance (London: Richards, 1898; New York: Grosset & Dunlap, 1900);
Venice (London: Richards, 1898; New York: Wessels, 1900);
The European Tour: A Handbook for Americans and Colonists (London: Richards, 1899; New York: Dodd, Mead, 1899);
Miss Cayley's Adventures (London: Richards, 1899; New York: Putnam, 1899);
Rosalba, The Story of Her Development: With Other Episodes of the European Movement, More Especially as They Affected the Monti Berici Near Vicenza, as Olive Pratt Rayner (London: C. Arthur Pearson, 1899; New York: Putnam, 1899);
Twelve Tales: With a Headpiece, a Tailpiece, and an Intermezzo: Being Select Stories (London: Richards, 1899);
Hilda Wade: A Woman with Tenacity of Purpose, by Allen and Sir Arthur Conan Doyle (New York & London: Putnam, 1900);
The New Hedonism (New York: Tucker, 1900);
Plain Words on the Woman Question (Chicago: Harman, 1900);
County and Town in England, Together with Some Annals of Churnside (London: Richards, 1901; New York: Dutton, 1901);
In Nature's Workshop (London: George Newnes, 1901; New York: Mansfield, 1901);
Sir Theodore's Guest and Other Stories (Bristol: Arrowsmith, 1902);
Evolution in Italian Art, edited by J. W. Cruikshank (London: Richards, 1908; New York: Wessels, 1908);
The Hand of God and Other Posthumous Essays, edited by Clodd (London: Watts, 1909);
The Reluctant Hangman and Other Stories of Crime, edited by Tom Schantz and Enid Schantz (Boulder: Aspen Press, 1973).

OTHER: *The Miscellaneous and Posthumous Works of Henry Thomas Buckle,* 2 volumes, edited by Allen (London: Longmans, Green, 1885);
Gilbert White, The Natural History of Selbourne, edited by Allen (London & New York: John Lane, 1900);
"The Thames Valley Catastrophe," in *Science Fiction by Gaslight: A History and Anthology of Science Fiction in the Popular Magazines, 1891–1911,* edited by Sam Moskowitz (Cleveland & New York: World, 1968), pp. 53–68.

In the chapter of his *Experiment in Autobiography* (1934) called "Fairly Launched at Last," H. G. Wells looks back at the influence upon him at the beginning of his literary career. He recalls in particular "that once much reviled and now rather too much forgotten writer, Grant Allen" and notes that he had never "made a fair acknowledgment of a certain mental indebtedness" to Allen. Grant Allen, the older by eighteen years, had had, in Wells's own words, "an earlier infection of that same ferment of biology and socialism that was working in my blood" and had been as disturbed by "social injustice and sexual limitation" as Wells himself was to be. Yet fair acknowledgment still eludes Grant Allen, who, if he is not completely forgotten today, is exhumed usually to be reviled afresh. Motivated by misreadings of *The Woman Who Did* (1895), the feminist novel responsible for Allen's notoriety in the 1890s (and which Wells himself reviewed scathingly), such modern exhumations are usually undertaken by those who probably would have been not only charmed by the man himself, but also in agreement with his ideas. It is true that in the field of science fiction, which accounts for only a tiny fraction of his prodigious output, Allen's contribution has never been entirely forgotten. Yet his importance in the formation of the tradition of British scientific romance still remains to be fairly acknowledged.

Charles Grant Blairfindie Allen was born on 24 February 1848 at Alwington, near Kingston, Ontario, the second and only surviving son of Joseph Antisell Allen, a Protestant clergyman of Irish background, and Charlotte Catherine Ann Grant, daughter of the fifth baron of Longueuil, Quebec, whose family originated in Blairfindie, Scotland. The family moved to New Haven, Connecticut, when Allen was about thirteen, where he was taught by a tutor from Yale. He also spent time at the Collège Impériale in Dieppe, France, and at King Edward's School, Birmingham, England. In 1867 he was elected to a postmastership at Merton College, Oxford University, and graduated from there with a B.A. in classics in 1871. While still an undergraduate he married, but his first wife, an invalid, died soon thereafter. After spending some time teaching school, he remarried in 1873 and took up the

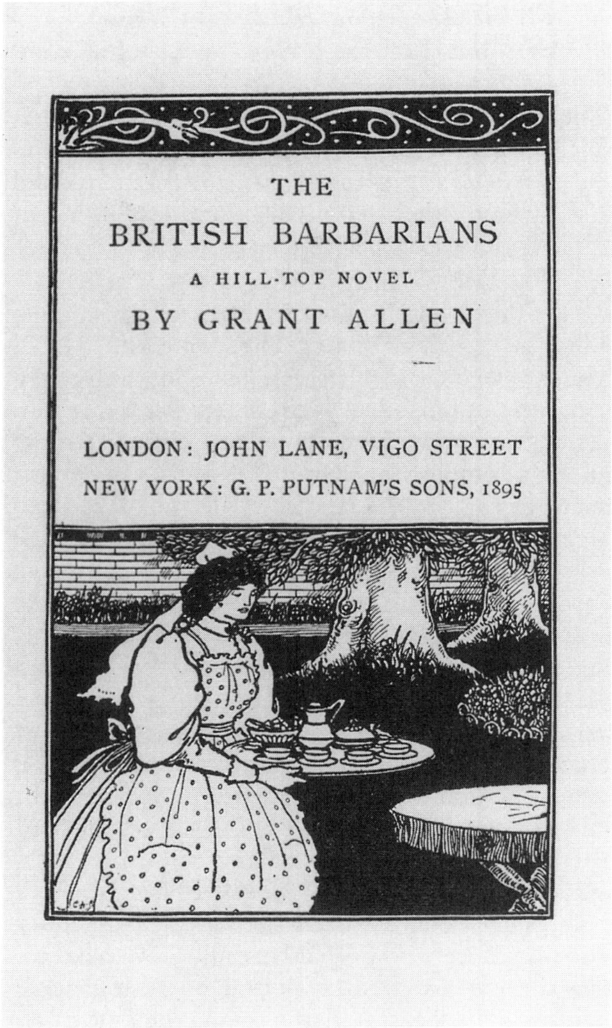

THE

BRITISH BARBARIANS

A HILL-TOP NOVEL

BY GRANT ALLEN

LONDON: JOHN LANE, VIGO STREET

NEW YORK: G. P. PUTNAM'S SONS, 1895

Title page for Allen's 1895 novel in which a twenty-fifth-century time traveler visits fin de siècle England

impressive-sounding chair of Mental and Moral Philosophy at the experimental Government College for black students in Spanish Town, Jamaica, founded in the wake of a rebellion. He was three years in the tropics before the college, always struggling, failed. His experiences in Jamaica made a profound expression on him, and he wrote with characteristic forthrightness, "I have no exaggerated sympathy with the blacks, yet I must say the way they are regarded by the whites is simply shameful." He then returned to England and began to try to make a living from his writing, first as a popularizer of evolutionary science–he was a disciple of Herbert Spencer–and later on a variety of other subjects. Having long suffered from bronchitis, and, latterly, an undiagnosed illness (probably malaria), he died on 28 October 1899. He was survived by his second wife, Ellen (née Jerrard), and their only son, Jerrard Grant, born in July 1878.

Though his professional writing career was not much more than twenty years long, Grant Allen was enormously productive. He published more than thirty novels, half a dozen volumes of short stories, a volume of poetry, and well over thirty nonfiction books–much work published in periodicals remains uncollected–on subjects as varied as Anglo-Saxon Britain, Charles Darwin and evolutionary theory, aesthetics, anthropology, botany, socialism, the Woman Question, and the British countryside, as well as a series of popular historical guidebooks to European cities. His science fiction consists of a small number of short stories written for the most part early in his career and the novel *The British Barbarians* (1895).

His colonial background gave Allen an outsider's perspective on late-Victorian England at the height of its imperial power, a perspective clarified by his own remarkable qualities of mind. Richard Le Gallienne recalled the "prodigious range and accuracy and instantaneous readiness of his memory," so that Allen's friends, lacking a suitable reference work on any given subject, would only have to "look it up in Grant." Little escaped his scrutiny: he had the trained eye of a naturalist or a detective, and his most characteristic action when he talked was to twirl a little pocket microscope between his finger and thumb. He was a democrat and a supporter of Irish Home Rule; a Fabian socialist who espoused the nationalization of land; a champion of women's rights who would "rather have sex relations unfettered than bound by conventions which involve woman's degradation"; and an atheist who was fascinated with tracing his intuition that religion grew from false psychology. Yet for all the radicalism of his views he was a man of peace and a gradualist, one of whose axioms was "I have never believed in fighting. I believe in permeation." Incapable of dissembling in an age of hypocrisy and double standards, he was in Le Gallienne's words "one of those true patriots who do their country the service of differing from it on every possible occasion." Andrew Lang called him "the most versatile, beyond comparison, of any man in our age."

His friend and biographer Edward Clodd noted that "philosophy and science, which Allen treated as one, were his first loves"; and indeed Allen himself noted that he was a fiction writer by default. His first story, "Our Scientific Observations on a Ghost," published in *Belgravia* in July 1878, used fiction as a convenient way of dramatizing his conviction that it would be impossible for us to recognize a ghost as such, even if we saw one. It appeared under the pseudonym "J. Arbuthnot Wilson" because Allen feared for his credibility as a science writer if his name became associated with fiction. But there was far more of a market for J. Arbuthnot Wilson's stories than for Grant Allen's scientific articles, so that fiction became for him a way of paying the bills so that he could devote his real energies to projects such as a theory of dynamics (*Force and Energy,* 1888) and an attempt to account anthropologically for the origins of religion (*The Evolution of the Idea of God,* 1897). But though Allen had no illusions about his importance as a fiction writer, noting in 1899 how his achievement consisted of picking up "crumbs that fall from the table of the Hardys, the Kiplings, the Merediths, and the Wellses," some of his fiction still retains interest, especially that into which his philosophical and scientific interests spill over. It is not surprising, therefore, that Allen has never been entirely forgotten in the field of science fiction.

In his introduction to *Strange Stories* (1884), Allen speaks almost apologetically of the "mere short sensational tales" that he has endeavored to give "some slight tinge of scientific or psychological import and meaning," noting also that almost all the stories "have their germ or prime motive in some scientific or quasi-scientific idea." Though Allen's own valuation of his tales is for the most part a fair one, *Strange Stories* is important as perhaps the first significant collection of science-fiction stories by a British author. (They might be more accurately called science-fantasy stories, or short scientific romances, though none of these categories was yet extant.) A frequent intention behind the stories, as with for example "The Mysterious Occurrence in Piccadilly," in which an Oxford professor sees what is apparently the wraith of his future son-in-law, seems to be the attempt to show how an apparently fantastic event has a rational or scientific explanation. Another common theme concerns what Allen, in an essay of that title in *Post-Prandial Philosophy* (1894), calls "The Romance of the Clash of Races," which he describes as "a momentous theme, for it encloses in itself half the problems of the future." It may be seen in "The Reverend John Creedy," in which the protagonist, a British-educated black clergyman, relapses into barbarism when he returns to Africa. Such scenarios, supported by Allen's wide reading in anthropology, seem racist to the modern reader; yet Allen's clearly expressed sympathies for the victims of imperialism ran very much against the orthodox ideology of the time.

Two stories in particular from *Strange Stories* have more lasting resonance in the field of science fiction. In "Pausodyne: A Great Chemical Discovery," first published in *Belgravia Christmas Annual* of 1881, the narrator encounters in the London of 1881 a mysterious stranger who turns out to be his great-great-uncle, Jonathan Spottiswoode, who mysteriously disappeared a century earlier. Spottiswoode had been conducting experiments with a powerful anesthetic, Pausodyne, which has the effect of suspending the animation of living things–including himself for a hundred years when he accidentally broke a bottle of the potion. The interest of the story resides in its examination of the psychological consequences of temporal displacement. Spottiswoode, an eighteenth-century man, is such a stranger in the strange land of late-Victorian London that he is considered mad and is ultimately driven to suicide, though not before curiously "infecting" his descendant, the narrator, with his alienation.

"The Child of the Phalanstery," first published in *Belgravia* in July 1884, is set in a quasi-Fourierian utopian commune of the future, in which all infants who are born defective in some way are subject to euthanasia. The story interrogates the idea voiced in the epigraph that in "a well-organized state of society . . . poor wee cripples . . . would be quietly put out of their misery while they were still babies" and so avoid a life of misery to themselves and others. Olive and Clarence are the parents of a baby that is born with twisted ankles, and is so doomed to die by chloroform. As the baby expires, Olive herself dies of a broken heart, and Clarence must suffer the grief of a double bereavement. The story makes at least two important and connected points. The first is that no conceivable future level of human progress is likely to free humanity of tragedy and misery. Who cannot not help but sympathize with Olive, who "would rather be a miserable naked savage in a wattled hut . . . than dwell among all the enlightenment, and knowledge, and art, and perfected social arrangements of phalansteric England without her child"? The second is that though this future society has replaced worship of a deity with a pure humanism, in the end the human need for consolation for the pain of existence remains, and the prayers of the phalanstery seem remarkably familiar: "The Cosmos is infinite, and man is but a parasite upon the face of the least among its satellite members."

The novel *Kalee's Shrine,* published in "shilling shocker" format in 1886, though only marginally science-fictional, is a good example of the way that Grant Allen's imagination was fired by the desire to rationalize apparently supernatural occurrences. (Though May Cotes is credited with joint authorship, it seems that she merely provided the seed of

THE THAMES VALLEY CATASTROPHE. 681

and began to gain upon them visibly. It was an awful sight. I cannot pretend to describe it. The poor creatures on the road, men and animals alike, rushed wildly, despairingly on; the fire took them from behind, and, one by one, before the actual sea reached them, I saw them shrivel and melt away in the fierce white heat of the advancing inundation. I could not look at it any longer. I certainly could not descend and court instant death. I felt that my one chance was to strike across the downs, by Stoke Poges and Uxbridge, and then try the line of northern heights to London.

Oh, how fiercely I pedalled! At Farnham Royal (where again nobody seemed to be aware what had happened) a rural policeman tried to stop me for frantic riding. I tripped him up, and rode on. Experience had taught me it was no use telling those who had not seen it of the disaster. A little beyond, at the entrance to a fine park, a gatekeeper attempted to shut a gate in my face, exclaiming that the road was private. I saw it was the only practicable way without descending to the valley, and I made up my mind this was no time for trifling. I am a man of peace, but I lifted my fist and planted it between his eyes. Then, before he could recover from his astonishment, I had mounted again and ridden on across the park, while he ran after me in vain, screaming to the men in the pleasure-grounds to stop me. But I would not be stopped; and I emerged on the road once more at Stoke Poges.

"THE POOR CREATURES RUSHED WILDLY DESPAIRINGLY ON."

Near Galley Hill, after a long and furious ride, I reached the descent to Uxbridge. Was it possible to descend? I glanced across, once more by pure instinct, for I had never visited the spot before, towards where I felt the Thames must run. A great white cloud hung over it. I saw what that cloud must mean: it was the steam of the river, where the lava sucked it up and made it seethe and boil suddenly. I had not noticed this white fleece of steam at Cookham, though I did not guess why till afterwards. In the narrow valley where the Thames ran between hills, the lava flowed over it all at once, bottling the steam beneath; and it is this imprisoned steam that gave rise in time to the subsequent series of appalling earthquakes, to supply forecasts of which is now the chief duty of the Seismologer Royal; whereas, in the open plain, the basalt advanced more gradually and in a thinner stream, and therefore turned the whole mass of water into white cloud as soon as it reached each bend of the river.

At the time, however, I had no leisure to think out all this. I only knew by such indirect signs that the flood was still advancing, and, therefore, that it would be impossible for me to proceed towards London by the direct route *viâ* Uxbridge and Hanwell. If I meant to reach town (as we called it familiarly), I must descend to the valley at once, pass through Uxbridge streets as fast as I could, make a dash across the plain, by what I afterwards knew to be Hillingdon (I saw it then as a nameless

Vol. xiv.—86

Page from "The Thames Valley Catastrophe," which first appeared in the December 1897 issue of The Strand Magazine

the plot, and that the novel was written by Allen alone.) In a prologue Olga Trevelyan, the baby daughter of a British commissioner in India, is stolen by her ayah and dedicated to the goddess Kalee (Kali), whose votaries, such as the notorious stranglers the Thugs, were supposed to be as bloodthirsty as the goddess herself. The rest of the novel takes place many years later in England, where the now-beautiful, twenty-one-year-old Olga, courted by the handsome oculist Alan Tennant, starts to behave in a strange, possibly maniacal way. (She has always had the uncanny habit of sleeping with her eyes open.) Under the influence of a celebrated mesmerist Mr. Keen, and of hashish given by a doctor to quiet her nerves, she apparently strangles Norah, a close friend. All the time Olga has been under the surveillance of a retired Anglo-Indian, Sir Donald Mackinnon, who, a prey to Celtic superstitiousness, has come to the conclusion that the lovely British girl is in reality a crazed and murderous devotee of

Kalee. But Alan manages to revive Norah and saves the day, accusing the irrationalists Mackinnon and Keen of feeding Olga poisonous suggestions, leading the susceptible but entirely innocent young woman to "the unconscious working out of your own wild hints and hideous imaginings." Thus, the lurid actions of Olga are not supernaturally induced, but are explicable by means of the new psychology of the unconscious, while a medical explanation is even provided for Olga's ocular peculiarities while asleep.

Wells in *Experiment in Autobiography* notes that he and Grant Allen "were in the tradition of Godwin and Shelley." Individual temperaments aside, this is a radical tradition central to the development of British science fiction. It begins with Mary Wollstonecraft Shelley–Godwin's daughter, Shelley's wife, and the creator of *Frankenstein*–and continues through Wells himself, George Orwell, Olaf Stapledon, William Golding, and Doris Lessing. One of the significant features of this tradition is a sort of subversive anthropology offered by the evolutionary perspective. In it the categories of savage and civilized are perpetually being interrogated and relativized–Golding's *Lord of the Flies* (1954) is perhaps the best-known exemplifying text. The tradition itself crystallized in the late nineteenth century, an age in which imperialists were using Darwinist ideas to justify their actions morally, even though the grounds for doing so were biologically indefensible. So Wells's Martians in *The War of the Worlds* (1898), whose hypertrophied brains and dependence upon machines reveal them to be embodiments of future human evolution, ruthlessly colonize the London suburbs and exterminate their less-advanced inhabitants–just as Englishmen had done in Tasmania. The Martians are thwarted not by any offense to morality but by their biology, which is ill-adapted to the local microbes. From an anthropological perspective, to be a member of an advanced civilization had come to mean only to be able to ravage the savage more savagely than he can ravage himself. Under such circumstances "the conquest of the earth," as Joseph Conrad's Marlow put it in *Heart of Darkness* (1902), "which mostly means the taking it away from those who have a different complexion or slightly flatter noses than ourselves, is not a pretty thing when you look into it too much."

Grant Allen had a particular interest in anthropology, and the impact on him of J. G. Frazer's *The Golden Bough* (1890) was immediate. Allen's novel *The Great Taboo,* published in 1890, the same year as *The Golden Bough* began to appear, clearly expresses the powerful impact of Frazer's comparative anthropology on Allen, especially in regard to taboos, primitive theories of the soul, and vegetation ceremonies. Though the plot of the novel is largely sentimental, involving the experiences of a young couple who are marooned among cannibals on Boupari, a South Sea island, Allen is clearly fascinated with the way that primitive taboos operate, and he frequently raises the question of the survival of these irrational constraints, often in disguise, in civilized society. So at the end of the novel, after the rescued heroine's aunt expresses horror at her being on an island alone with a young man not her husband, the narrator comments wryly: "Taboos, after all, are much the same in England as in Boupari."

In *Heart of Darkness* Conrad favorably contrasts primitive society, restrained by taboo, with modern civilization embodied by the completely unrestrained Kurtz. For Grant Allen, however, the irrationality of primitive taboo was exactly equivalent to various modern irrational beliefs, and so any society in which such beliefs flourished was not truly civilized at all. It was not until the success of the highly controversial *The Woman Who Did,* which ran to twenty editions the first year, that Allen was finally able to publish a fictional work that more fully explored the survival of taboo in modern civilization. This novel, *The British Barbarians,* is Allen's major contribution to British science fiction.

Published in 1895 the novel concerns Bertram Ingledew, a nomologist (a student of the laws of mind) from the twenty-fifth century, who travels back in time to the Surrey of the 1890s in order to gather "materials for a *History of Taboo,* from its earliest beginnings in the savage stage to the fully developed European complexity." When confronted with protests that "England . . . is a civilized country, and taboos are institutions that belong to the lowest and most degraded savages," Ingledew, taken aback, retorts that as far as his fellow nomologists are concerned, "England has always been regarded with the greatest interest as the home and center of the highest and most evolved taboo development." Most of the novel concerns itself with Ingledew's exposure of middle-class late-Victorian society as riddled with irrational taboos of every kind, and particularly in relation to laws of property and sexual customs. Ingledew notes, for example, that "nowhere . . . had he seen any system which entailed in the end so much misery on both sexes, though more particularly on the women, as that system of closely tabooed marriage, founded upon a broad basis of prostitution and infanticide, which has reached its most appalling height of development in hypocritical and puritan England." Ingledew's nobility, and his final murder from motives of sexual jealousy (a passion he, the product of a truly civilized, rational

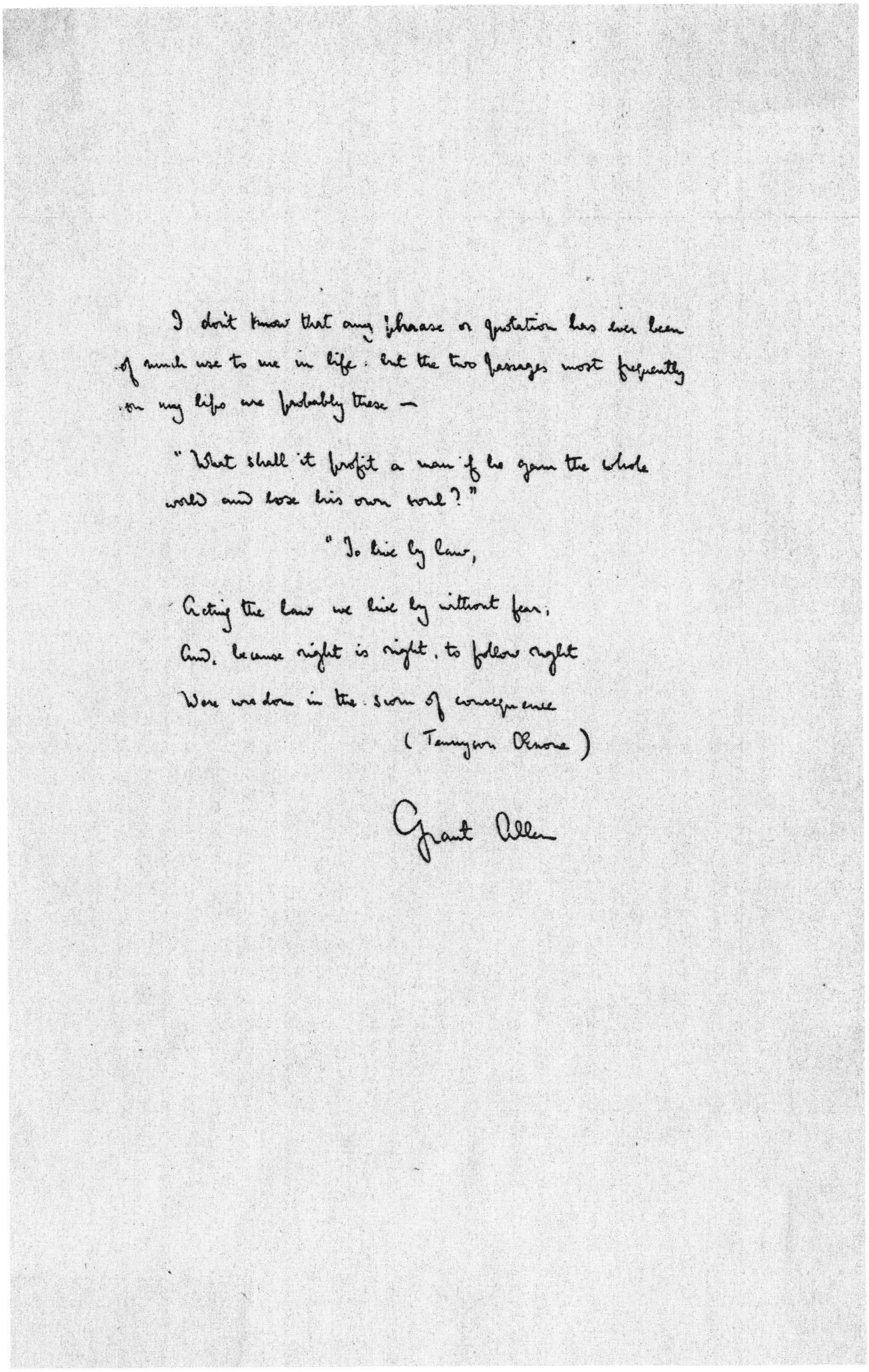

I don't know that any phrase or quotation has ever been of much use to me in life: but the two passages most frequently on my lips are probably these —

"What shall it profit a man if he gain the whole world and lose his own soul?"

"To live by law,
Acting the law we live by without fear;
And, because right is right, to follow right
Were wisdom in the scorn of consequence
(Tennyson Oenone)

Grant Allen

Manuscript reproduced as the epigraph for Edward Clodd's Grant Allen: A Memoir *(1900)*

society can understand only at the intellectual level), render the work a powerful satire on the smug hypocrisies of the age.

The British Barbarians, like *The Woman Who Did,* is an example of that rarity in English literature, the *Tendenzroman,* the novel written to achieve a clear purpose, to which literary charactåristics, such as characterization and plot, take a subordinate place. Allen himself subtitled *The British Barbarians* a "Hill-Top Novel," by which he meant that the novel was a sort of secular Sermon on the Mount, in which all other artistic considerations were subordinated to the expression of a truth emerging from an elevated perspective on the human condition. (Allen had been living in a house called "Hilltop" in the rural retreat of Hindhead, Surrey, since 1893.) Wells, writing on the novel in the *Saturday Review,* was even more harsh on *The British Barbarians* than he had been on *The Woman Who Did.* He described the novel as "redolent of bad taste and bad English, destitute alike of dramatic incident and character analysis" and called Allen "a philosopher who masquerades as a novelist, violating the conditions of art that his gospel might win notoriety." Yet in Ingledew we can recognize a forerunner of several of Wells's more notable creations: the Time Traveler with his evolutionary perspective on human society; the Invisible Man as the mysterious man from nowhere who brings chaos to an English village; and even the Martian astronomers whose vast and cool intellects cut Victorian imperialists down to size. Indeed, so close in spirit were Allen and Wells in 1895 that Ingvald Raknem, in *H. G. Wells and His Critics* (1962), charged Wells with having plagiarized *The British Barbarians* for his fantasy *The Wonderful Visit* (also published in the fall of 1895), in which an angel descends on an English village with disastrous consequences. David Y. Hughes later revealed that Raknem's charges were unfounded; but the remarkable similarities between the works that certainly exist only show just how very similar were the two men's interests at the time–and that any debt was primarily Wells's to Allen.

"The Thames Valley Catastrophe," Allen's most important science-fiction story, was first published in *The Strand Magazine* in December 1897. Like Wells's "The Star," which was published in the same month, it is a catastrophe fiction, though on a smaller scale than Wells's global disaster scenario. A fissure occurs without warning in the Thames Valley, and a huge sea of molten lava spreads rapidly eastward toward London. The narrator, at the time a lowly post-office clerk on a bicycle tour, retrospectively recounts how he furiously pedaled back to London just ahead of the lava in order to save his family, which he succeeds in doing in the nick of time. The whole valley of the Thames, including London and its low-lying suburbs, is inundated and its unsuspecting inhabitants incinerated. The lava later hardens to become a lake of basalt so deep that only part of the dome of St. Paul's Cathedral is visible, while in the rearranged geography north and west of the London area newly dammed rivers create lakes that drown Oxford and Newbury. Allen's motive must almost certainly have been to strike at British complacency in the year of Queen Victoria's Diamond Jubilee by visiting disaster on its heartland, as Sir George T. Chesney had done with "The Battle of Dorking" (1871) and as Wells himself was to do in *The War of the Worlds.* That Allen described a natural disaster rather than an invasion scenario is perhaps partially explained by his own oft-expressed hatred of London, the great metropolis of the modern world–"this human desert, this beflagged and macadamized man-made solitude," as he once put it–as the antithesis of the natural world that he loved.

"Wolverden Tower," first published in *The Illustrated London News* and later collected in *Twelve Tales* (1899), is apparently an exception in Allen's oeuvre in that it seems to be a story in which there is no rational explanation for the supernatural events that take place. Maisie Llewellyn, a Welsh rector's daughter and Pre-Raphaelite beauty, visits an old Kentish manor house whose Saxon church has recently been disfigured by a new tower. A strange, witchlike old woman, Bessie, "recognizes" Maisie as an appropriate incarnation of one of the three virgins whose willing blood sacrifices will ensure that the tower stands firm against thunder and lightning. The other two virgins appear as ghosts to Maisie and impel her to throw herself off the tower. But she is rescued at the brink by an Oxford undergraduate, and a strange winter thunderstorm destroys the tower, crushing Bessie beneath it. Beneath the surface of these fantastic events, however, the story is very much one with Allen's other works. Bessie, described as "a sort of modern ghoul, a degenerate vampire," embodies the persistence of superstition, which is finally destroyed at the same time as the tower, representing the hideous tastelessness of the modern world, comes crashing down. Allen is also careful to point out that the church itself occupies a site far more ancient than Christianity, and on the same spot once stood a temple of Woden, a shrine of Hercules, a grove of Nodens, a Stone Circle of the Host of Heaven, and so on back into the mists of antiquity. As an anthropologist he respects the numinosity of the site, even while he cannot abide the degenerate superstitiousness that exploits it.

Grant Allen's importance in the field of science fiction is not traceable to any particular work, but rather to his influence as one of the leading liberal intellectuals of the period who was respected by both scientists and humanists alike. As Edward Clodd neatly phrased it, "the man was greater than his books." Yet if one had to choose Allen's most important science-fiction works, then they would undoubtedly be *The British Barbarians* and "The Thames Valley Catastrophe." *The British Barbarians,* and its connection to Wellsian science fiction, still awaits proper recognition.

Brian Stableford in his *Scientific Romance in Britain 1890–1950* (1985) notes that Grant Allen's career "anticipated that of Wells in many ways, albeit feebly" and dismisses *The British Barbarians* as "frankly artless." But this aesthetic objection is unfair–just as Wells's was–for *The British Barbarians,* like *The Woman Who Did,* was intended as a didactic novel the interest of which is in the careful development of a thesis. Neither novel makes any attempt to achieve more than a minimal level of realism. The modern reader will probably find that, though *The British Barbarians* is far below the achievement of Wells's great scientific romances, it has actually survived better than *The Wonderful Visit,* with which it can most usefully be compared. It is more important to affirm, however, that Allen opened up the field for Wells's scientific romances in many ways. In a letter to Wells of 4 October 1895, quoted by David Y. Hughes, Allen reveals that *The British Barbarians* was written six years earlier but rejected by his usual publisher, Chatto & Windus. It was only the arrival on the publishing scene of John Lane and the success of *The Woman Who Did* (which also had to wait several years for publication) that Allen was able to get published works in which he could express his outrage at the social and sexual injustices of late-Victorian Britain. If one is sensitive to the cultural politics of the late nineteenth century, one can see that it was Allen's uncompromising honesty and hard-won personal prestige that blazed the trail, making it so much easier for Wells to publish his own "Hill-Top" novels, the great scientific romances, from the mid 1890s on.

Biographies:

Edward Clodd, *Grant Allen: A Memoir. With a Bibliography* (London: Richards, 1900);

Richard Le Gallienne, *Grant Allen* (New York: Tucker, 1900).

References:

Edward Clodd, "Grant Allen," in his *Memories* (New York: Putnam, 1916), pp. 21–36;

David Y. Hughes, "H. G. Wells and the Charge of Plagiarism," *Nineteenth Century Fiction,* 21 (1966–1967): 85–90;

Ingvald Raknem, *H. G. Wells and His Critics* (Trondheim, Norway: Universitetsforlaget/George Allen & Unwin, 1962), pp. 417–419;

Nicholas Ruddick, *Ultimate Island: On the Nature of British Science Fiction* (Westport, Conn. & London: Greenwood Press 1993), pp. 106–109;

Brian Stableford, *Scientific Romance in Britain 1890–1950* (New York: St. Martin's Press, 1985), pp. 118–119;

H. G. Wells, *Experiment in Autobiography: Discoveries and Conclusions of a Very Ordinary Brain (Since 1866)* (New York: Macmillan, 1934), pp. 461–466.

F. Anstey
(Thomas Anstey Guthrie)

(8 August 1856 - 10 March 1934)

Darren Harris-Fain
Shawnee State University

See also the Anstey entry in *DLB 141: British Children's Writers, 1880–1914.*

BOOKS: *Vice Versâ; or, A Lesson to Fathers* (London: Smith, Elder, 1882; New York: Appleton, 1882; revised edition, London: Smith, Elder, 1883; revised, 1884; revised and corrected edition, London: Newnes, 1901);

The Black Poodle and Other Tales (London: Longmans, Green, 1884; New York: Appleton, 1884);

The Giant's Robe (London: Smith, Elder, 1884; New York: Appleton, 1884);

The Tinted Venus: A Farcical Romance (Bristol: Arrowsmith / London: Simpkin, Marshall, 1885; New York: Lovell, 1885);

A Fallen Idol (London: Smith, Elder, 1886; Philadelphia: Lippincott, 1886);

Burglar Bill, and Other Pieces for the Young Reciter (London: Bradbury, Agnew, 1888); enlarged as *Mr. Punch's Young Reciter: Burglar Bill, and Other Pieces* (London: Bradbury, Agnew, 1892; enlarged, 1897);

The Pariah (3 volumes, London: Smith, Elder, 1889; 1 volume, Philadelphia: Lippincott, 1889);

Voces Populi (London & New York: Longmans, Green, 1890);

The Talking Horse, and Other Tales (New York: Lowell, 1891; London: Smith, Elder, 1892);

Tourmalin's Time Cheques: A Farcical Extravagance (Bristol: Arrowsmith / London: Simpkin, Marshall, Hamilton, Kent, 1891; New York: Appleton, 1891); republished as *The Time Bargain; or, Tourmalin's Cheque Book* (Bristol: Arrowsmith / London: Simpkin, Marshall, Hamilton, Kent, 1905);

Mr. Punch's Model Music-Hall Songs and Dramas (London: Bradbury, Agnew, 1892; New York: National Book Company, 1892);

The Travelling Companions: A Story in Scenes (London & New York: Longmans, Green, 1892);

F. Anstey (Thomas Anstey Guthrie) in 1930 (photograph by Whitlock)

Voces Populi, Second Series (London & New York: Longmans, Green, 1892);

Mr. Punch's Pocket Ibsen: A Collection of Some of the Master's Best-Known Dramas Condensed, Revised, and Slightly Re-arranged for the Benefit of the Earnest Student (London: Heinemann, 1893; New York & London: Macmillan, 1893); enlarged as *The Pocket Ibsen* (London: Heinemann, 1895);

The Man from Blankley's, and Other Sketches (London & New York: Longmans, Green, 1893);

Under the Rose: A Story in Scenes (London: Bradbury, Agnew, 1894);
Lyre and Lancet: A Story in Scenes (London: Smith, Elder, 1895; New York & London: Macmillan, 1895);
The Statement of Stella Maberly, Written by Herself, anonymous (London: Unwin, 1896; New York: Appleton, 1896); as F. Anstey (London: Unwin, 1896);
Baboo Jabberjee, B.A. (London: Dent, 1897); published as *Baboo Hurry Bungsho Jabberjee, B.A.* (New York: Appleton, 1897);
Puppets at Large: Scenes and Subjects from Mr. Punch's Show (London: Bradbury, Agnew, 1897);
Paleface and Redskin, and Other Stories for Boys and Girls (London: Richards, 1898; New York: Appleton, 1912);
Love among the Lions: A Matrimonial Experience (London: Dent, 1898; New York: Appleton, 1899);
The Brass Bottle (London: Smith, Elder, 1900; New York: Appleton, 1900);
A Bayard from Bengal (London: Methuen, 1902; New York: Appleton, 1902);
Only Toys! (London: Richards, 1903);
Salted Almonds (London: Smith, Elder, 1906);
Vice Versâ: A Farcical Fantastic Play (London: Smith, Elder, 1910; Boston: Baker, 1910);
The Brass Bottle: A Farcical Fantastic Play (London: Heinemann, 1911);
Percy and Others (London: Methuen, 1915);
In Brief Authority (London: Smith, Elder, 1915; New York: Doran, 1916);
The Last Load: Stories and Essays (London: Methuen, 1925);
The Man from Blankley's: A Comedy of the Early Nineties (London: Hodder & Stoughton, 1927);
Humour and Fantasy (London: John Murray, 1931; New York: Dutton, 1931);
A Long Retrospect (London & New York: Oxford University Press, 1936).

OTHER: "A Very Bad Quarter of an Hour," in *In a Good Cause,* edited by Margaret Susan Tyson-Amherst (London: Wells Gardner, Darton, 1885), pp. 137–146;
"Some Detached Thoughts on Omnibuses," in *Alma Mater's Mirror,* edited by Thomas Spencer Baynes and Lewis Campbell (Edinburgh: Edinburgh University Press, 1887), pp. 167–172;
"Three Wishes," by Anstey and Brander Matthews, in *With My Friends: Tales Told in Partnership,* by Matthews (Freeport, N.Y.: Books for Libraries Press, 1891; London & New York: Longmans, Green, 1891), pp. 233–284;
"Whom the Gods Hate Die Hard," in *The Fate of Fenella,* 3 volumes, by Anstey and others (London: Hutchinson, 1892), III: 195–230;
"Some Fellow Passengers of Mine," in *Souvenir of the Charing Cross Hospital Bazaar,* edited by Herbert Beerbohm Tree (London: Nassau, 1899), pp. 11–15;
"Two Anecdotes Concerning Pictures," in *The Press Album,* edited by Thomas Catling (London: John Murray, 1909), pp. 175–176;
"Winnie: An Everyday Story," in *In a Good Cause: Stories and Verse on Behalf of the Hospital for Sick Children* by Anstey and others (London: John Murray, 1909), pp. 1–78;
Molière, *The Would-Be Gentleman,* adapted by Anstey (London: Secker, 1926);
Molière, *The Imaginary Invalid,* adapted by Anstey (London: Hodder & Stoughton, 1929);
Molière, *Four Molière Comedies,* adapted by Anstey (London: Hodder & Stoughton, 1930);
Molière, *Three Molière Plays: Tartuf* [*sic*], *Scapin the Trickster, The School for Wives,* adapted by Anstey (London: Oxford University Press, 1933);
"The Game of Adverbs," in *Another Book of Miniature Plays,* edited by Theodore Johnson (Boston: Baker, 1934), pp. 97–108.

SELECTED PERIODICAL PUBLICATIONS–UNCOLLECTED: "The Decay of the British Ghost," *Longman's Magazine* (January 1884);
"My First Book," *Burning Bush,* 1 (4 June 1928): 25–27.

Thomas Anstey Guthrie, writing under his pseudonym, F. Anstey, was one of the most popular authors of humorous fantasies of the late Victorian period and the early twentieth century. His work is among the best of a particular type of British fantasy, in which the supernatural intrudes into the everyday with comically disastrous results. Though enjoyable as entertainment, such works also include subtle satire on contemporary foibles and unobtrusive moral instruction. In the case of Anstey, his works often poke fun at the pretensions of social class in ways similar to the works of Mark Twain, whom he admired.

Born in London on 8 August 1856, Thomas Anstey Guthrie was named after his father, a military tailor; his mother was Augusta Amherst Austen Guthrie. In his autobiography Anstey described his childhood as happy, his father as successful, and his mother as a "gentlewoman." His childhood, shared with two brothers and a sister, was filled with humor–both that of his parents and in his reading, which included magazines such as *Punch* as well as Edward Lear's *Book of Nonsense* (1846). His boarding-school experiences at Surbiton (1867–1872) were less pleasant and provided him with material for his satire of boarding-school life in his first novel, *Vice Versâ; or, A Lesson to Fathers* (1882). The verisimilitude of these scenes prompted C. S. Lewis to call it "the only truthful school story in existence" in his *Surprised by Joy: The Shape of My Early Life* (1955).

Anstey was then educated at King's College School in London (1872–1875) and at Trinity Hall, Cambridge University, where he began writing seriously. He unsuccessfully attempted to found a humor magazine called the *Home Journal* and contributed to other Cambridge periodicals. Also while at Cambridge he began his first novel, originally a series called Turned Tables in the *Cambridge Tatler*. He continued to publish installments of the series privately after the *Cambridge Tatler* ceased publication, but he set it aside for three years following the death of his mother in August 1877.

As a young man he was heading toward a career in law and was called to the bar in 1881, but instead, after his success with his first novel in 1882, he joined the staff of the British humor magazine *Punch,* which shaped his style and subjects to a considerable extent. The pseudonym F. Anstey emerged from a printer's error for the proposed penname "T. Anstey" in his first professionally published story, "Accompanied on the Flute" (*Mirth,* 1878). Anstey enjoyed a long association with the magazine and published in others as well. A few of his books are republications of material he wrote for the magazine. For example, *Voces Populi* (1890) is a collection of Anstey's columns for *Punch* dealing with various aspects of contemporary life. A second collection of columns was published in 1892.

Vice Versâ; or, A Lesson to Fathers was an enormous success and established his career. The story involves a magical exchange of personalities between a self-assured businessman and his son, a student; thus, the father finds himself in the son's body, and vice versa. Most of the comedy focuses on the father's ordeal of having to attend prep school, with its various boredoms, pettinesses, and humiliations, and on the son's understandably poor management of his father's business affairs. When proper identities are restored, each has learned a needed lesson.

Whether it served as a deliberate model or not, the best-selling *Vice Versâ* was the prototype for a story type that has appeared in other works since, from the children's novel *Freaky Friday* (1972), by Mary Rogers, to Craig Lucas's 1990 play *Prelude to a Kiss.* None of these, however, possesses Anstey's distinctive charm–part of which must surely include the fact that the chaos that results from this magical exchange is all the more enjoyable in the staid context of the Victorian era.

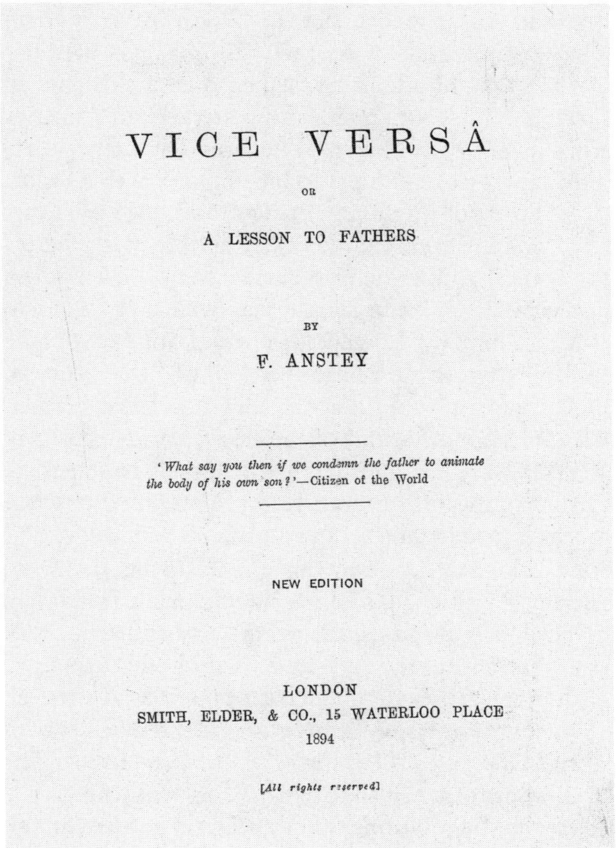
VICE VERSÂ

OR

A LESSON TO FATHERS

BY

F. ANSTEY

'What say you then if we condemn the father to animate the body of his own son?'—Citizen of the World

NEW EDITION

LONDON
SMITH, ELDER, & CO., 15 WATERLOO PLACE
1894

[All rights reserved]

Title page for a revised edition of Anstey's first novel, a humorous fantasy in which a father and son exchange bodies

Staid and *Victorian* accurately describe Anstey's novel *The Giant's Robe* (1884), one of his few attempts to write "serious" fiction. The poor success of this novel and *The Pariah* (1889) illustrates two facts: first, Anstey was more talented as a fantasist than as a realistic novelist, despite his aspirations; second, such aspirations to be respected as a "serious" writer apart from his success as a fantasist reveals a growing literary divide in the late nineteenth century between "popular" and "serious" literature. H. G. Wells's distinction between his own scientific romances and his realistic novels also highlights this expanding divide.

The stories in *The Black Poodle and Other Tales* (1884) and *The Talking Horse, and Other Tales* (1891) are also comic in intent, although Anstey was usually more successful in longer forms. In addition, two of the stories in *The Black Poodle*–"The Curse of the Catafalques" and "The Wraith of Barnjum"–parody the popular contemporary fascination in Victorian

England with ghost stories. In "The Wraith of Barnjum," for instance, a man who thinks he is haunted by the ghost of a man he pushed over a cliff tries to turn the ghost into a sideshow attraction. Another story in *The Black Poodle,* "The Siren," is a more standard fantasy of a siren who falls in love with a sailor. "The Story of the Sugar Prince" and "A Toy Tragedy" are children's fantasies. The title character of the former thinks he possesses a majestic kingdom but in reality is just a cookie that eventually is eaten, while "The Toy Tragedy" involves dolls that come to life. There is no obvious moral to "The Story of the Sugar Prince" (although critics have offered various interpretations of the story), but most of Anstey's works for children, like much nineteenth-century children's literature, were intended to teach moral lessons while telling an interesting story.

Like many Victorian ghost stories, some of Anstey's fantasies, and much Victorian fantasy in general, are based on the premise of contemporary characters interacting with ancient gods or historical figures, either in modern times or via time travel. Two such novels by Anstey are *The Tinted Venus: A Farcical Romance* (1885) and *A Fallen Idol* (1886). The first, appropriately subtitled "A Farcical Romance," concerns the relationship between a young man and the Roman goddess of love, who comes to life when the protagonist places an engagement ring on the finger of a statue of Aphrodite. (Anstey got the idea from Robert Burton's *Anatomy of Melancholy,* 1621). In *A Fallen Idol* a young artist has to deal with an evil supernatural being residing in a Jain idol. Anstey is best known for writing comic fantasies, and *A Fallen Idol* includes many humorous scenes, but Anstey could also write effective horror fiction, as this novel and several of his later short stories demonstrate.

In 1887 Anstey, who had been writing for *Punch* for years, was invited to join the magazine's well-known "Table," where writers and editors dined in the magazine's offices every Wednesday and discussed their work. A lifelong bachelor and a private man who resisted interviews, he nonetheless cherished his spot at the Table, which he retained even after he stopped writing as a regular columnist for *Punch.*

Tourmalin's Time Cheques: A Farcical Extravagance (1891) is one of the first works to deal with potential paradoxes created by time travel. Published four years before H. G. Wells's better-known novel of time travel, *The Time Machine*, Anstey's work, like other time-travel narratives such as Mark Twain's *A Connecticut Yankee in King Arthur's Court* (1889), lacks a "scientific" explanation of how travel through time is possible (here it is achieved through magical checks), and thus is closer to fantasy than science fiction. It is significant in the history of science fiction, however, because of its treament of the problems of time travel. Here the protagonist inadvertently winds up with two fiancées and several people mad at him for various things he has done in his confused time travels. Anstey's resolution of the confusion is disappointing, but the imaginative story redeems the weak ending.

Equally imaginative are many of the stories collected in *The Talking Horse.* The horse in the title story does indeed talk; it also disrupts its owner's love life and in general makes a nuisance of itself. Both "Tommy's Hero" and "The Good Little Girl" involve children who are granted wishes by fairies, only to find that one should sometimes be careful what one wishes for.

Some critics have commented on the dramatic nature of Anstey's fiction. In fact, both *Under the Rose* (1894) and *Lyre and Lancet* (1895) are subtitled "A Story in Scenes." Also, Anstey adapted two of his most popular novels as plays: *Vice Versâ: A Farcical Fantastic Play* in 1910 and *The Brass Bottle: A Farcical Fantastic Play* in 1911. Finally, toward the end of his life Anstey was active in writing adaptations of plays by Molière.

An unusual work for Anstey is his anonymous *The Statement of Stella Maberly, Written by Herself* (1896). It is not uncommon in modern fantasy to find works in which the disruption of conventional reality is as much a reflection of a character's psychology as it is a depiction of a fantastic world, but in Anstey's work the use of such a device is striking in contrast with his other efforts. Here many of the events that seem fantastic to the protagonist can be explained as delusions caused by her schizophrenia. Such a topic is not unusual, however, for the period, in which there was considerable interest in the new field of psychology. This final effort at a "serious" novel, however, proved as unsuccessful as *The Giant's Robe* and *The Pariah.*

The Brass Bottle (1900) was Anstey's most popular fantasy after *Vice Versâ.* Like *A Fallen Idol,* the novel brings to life a spirit from Eastern myth, in this case a "jinnee," or genie, clearly inspired by *The Thousand and One Nights.* The tone of this work is closer to the lightheartedness of *Vice Versâ* or *The Tinted Venus,* though, in its depiction of the chaos that ensues as a modern-day person comes into contact with forces beyond his control. Much of the comedy revolves around how the generosity of the grateful genie, who wants to make his master richer and more powerful than anyone else, comes into conflict with the protagonist's comfortable middle-class existence. There is also an element of horror in the story: the genie, upset at the protagonist's resistance to his plans, malevolently turns on him. *The Brass Bottle* was a likely influence on E. Nesbit's *Five Children and It* (1902), which refers to the book.

Anstey also enjoyed a modest reputation as a writer for children, and his efforts often contain fantastic elements. For instance, in *Only Toys!* (1903) two children who think they have outgrown their toys and Santa Claus are reduced in size by Santa himself so that they, as is typical in much of Anstey's work, may learn a necessary lesson. The lesson is a common one in much children's literature of the period: to enjoy one's childhood while it lasts, because the pressures and responsibilities of adult life come all too soon.

Unlike the stories in Anstey's earlier collections, some gathered in *Salted Almonds* (1906) are dark in tone rather than humorous. For instance, in a story reminiscent of those of Wells, "The Lights of Spencer Primmett's Eyes," a man becomes radically and horrifically changed by an unusual pair of glasses. The title of "The Gull," about a man who mistakenly believes that his dead lover has been reincarnated as a seagull, obviously has a double meaning. This shift in tone in some stories is also apparent in Anstey's later collection *The Last Load: Stories and Essays* (1925).

Still, many of the stories in *Salted Almonds* are classic Anstey. "Why I Have Given Up Writing Novels" is about a man whose characters literally come to life. In "The Adventure of the Snowing Globe" a coward gets a chance to redeem himself when he finds himself inside a snow globe and has to slay a dragon. Like many of Anstey's fantasies, some stories have unsympathetic characters receiving a deserved comeuppance, as in "Mrs. Brassington-Claypott's Children's Party" and in "At a Moment's Notice," in which a smug young man is temporarily transformed into an organ-grinder's monkey.

In Brief Authority (1915), like much of Anstey's fantasy, shows the interaction between the everyday and the fantastic, but here the pattern is reversed. While his fantasy usually features the appearance of the fantastic into the realistic world of contemporary England, the contemporary protagonist of *In Brief Authority* instead finds herself in the strange, wondrous world of the fairy tales of the Brothers Grimm. For a time she rules as queen of this fantastic realm, but when her relatives arrive in this märchenland and try to reshape it in the mold of their priggish Victorian values, chaos predictably ensues until the true queen is crowned.

In Brief Authority was Anstey's last major fantasy, and its limited success, which Brian M. Stableford suspects was caused by the poorly timed publication of a book rooted in Germanic folklore during World War I, may have prompted him to turn his energies elsewhere. In the remaining two decades of his life he served as a volunteer during World War I; wrote a few other books, mostly compilations of his contributions to *Punch;* worked on his adaptations of Molière; published his final collection, *The Last Load,* and an omnibus volume, *Humour and Fantasy* (1931), which like *In Brief Authority* was critically acclaimed but sold poorly; and wrote his autobiography, posthumously published as *A Long Retrospect* (1936). He died of pneumonia in London on 10 March 1934.

THE JINNEE AND HORACE RETURN TO VINCENT SQUARE.
(Page 286.)

Frontispiece for Anstey's The Brass Bottle *(1900), in which a genie disrupts his master's comfortable middle-class life*

Though his reputation was already in decline by the time of his death and he is now little known to modern readers, especially readers outside of Britain, F. Anstey was one of the best fantasists of his generation. Much of his popularity rested on his lighthearted satires of Victorian and Edwardian respectability, which may date his works somewhat, but he still deserves to be read not only for his historical significance but also for his humor, his occasional use of horror, and his highly imaginative stories.

Bibliography:

Martin John Turner, *A Bibliography of the Works of F. Anstey (Thomas Anstey Guthrie)* (London: Stockwell, 1931).

References:

Beth D. Fleischmann, "F. Anstey (Thomas Anstey Guthrie), Late Victorian Humorist," dissertation, University of South Carolina, 1981;

Roger Lancelyn Green, "A Neglected Novelist: 'F. Anstey,'" *English,* 11 (Summer 1957): 178–181;

Brian M. Stableford, "F. Anstey," in *Supernatural Fiction Writers: Fantasy and Horror*, volume 1, edited by E. F. Bleiler (New York: Scribners, 1985), pp. 287–292.

Papers:

The major collection of F. Anstey's papers is held at the British Library in London.

Edwin L. Arnold

(14 May 1857 - 1 March 1935)

Richard Bleiler
University of Connecticut

BOOKS: *A Summer Holiday in Scandinavia* (London: Sampson Low, Marston, Searle & Rivington, 1877);

On the Indian Hills: or, Coffee-Planting in Southern India (London: Sampson Low, Marston, Searle & Rivington, 1881; expanded edition, London: Sampson Low, Marston, Searle & Rivington, 1893);

Coffee: Its Cultivation and Profit (London: Wittingham, 1886);

Bird Life in England (London: Chatto & Windus, 1887);

England as She Seems: Being Selections from the Notes of an Arab Hadji (London: Warne, 1888);

The Wonderful Adventures of Phra the Phoenician (London: Chatto & Windus, 1891 [1890]; New York: Harper, 1890); republished as *Phra the Phoenician* (New York: Burt, 1890; London: Newnes, 1910);

The Constable of St. Nicholas (London: Chatto & Windus, 1894);

The Story of Ulla and Other Tales (London: Longmans, Green, 1895);

Lepidus the Centurion: A Roman of Today (London: Cassell, 1901; New York: Crowell, 1901);

Lieut. Gullivar Jones: His Vacation (London: Brown, Langham, 1905; New York: Arno, 1975); republished as *Gulliver of Mars* (New York: Ace, 1965);

The Soul of the Beast (London: Macmillan, 1960).

OTHER: *The Opium Question Solved,* edited by Arnold (London: Partridge, 1882);

Sir Edwin Arnold, *The Arnold Poetry Reader,* with a memoir and notes by Arnold (London: Kegan Paul, 1920).

Were it not for his three fantastic novels and one collection of short stories, Edwin L. Arnold would be remembered only as the older son of Sir Edwin Arnold, a poet and Orientologist noted for being among the first to introduce the English to Buddhism. These subjects do not figure in the works of Edwin L. Arnold, whose fiction is idiosyncratic and flawed by most literary standards. He was neither a successful creator of original or memorable characters nor a particularly successful storyteller, and his prose tends toward flatness. Nevertheless, his fiction is also oddly original and innovative; he had a fertile imagination, and his ideas are often ahead of their time. He remains, however, a pioneer, a historical curiosity rather than a landmark, for though his ideas were original, he did little with them. It remained to others to develop, expand, and exploit his suggestions.

Edward Lester Linden Arnold, called Lester by friends and family, was born on 14 May 1857 in Swanscombe; his middle names were derived from the family of his mother, Katherine. As a child he accompanied his father on his travels and lived for a time in India. Nevertheless, he did his writing in England, basing his first publications on his experiences abroad: *A Summer Holiday in Scandinavia* (1877), *On the Indian Hills: or, Coffee-Planting in Southern India* (1881), and *Coffee: Its Cultivation and Profit* (1886). At age twelve he became interested in natural history and became an enthusiastic amateur ornithologist; this led to the publication of *Bird Life in England* (1887). He married twice, first to Constance Boyce, with whom he had a daughter, then to Jessie Brighton in 1919.

It is not known why Arnold chose to write fantastic fiction, but during the end of the Victorian era the works of Mary Shelley, Jules Verne, Madame Blavatsky, H. Rider Haggard, Andrew Lang, Richard Jefferies, and Samuel Butler were widely known and read. There was also a general interest in history, spiritualism, reincarnation, and immortality, all of which appeared in *The Wonderful Adventures of Phra the Phoenician* (1890), his first novel, the narrative of which begins shortly before the Romans invaded Britain. Phra is a Phoenician trader whose curious fate is to return to life at random three-hundred- and four-hundred-year intervals. His adventures begin in the Achaean isles, where he purchases and frees the captive British slave girl Blodwen, whom

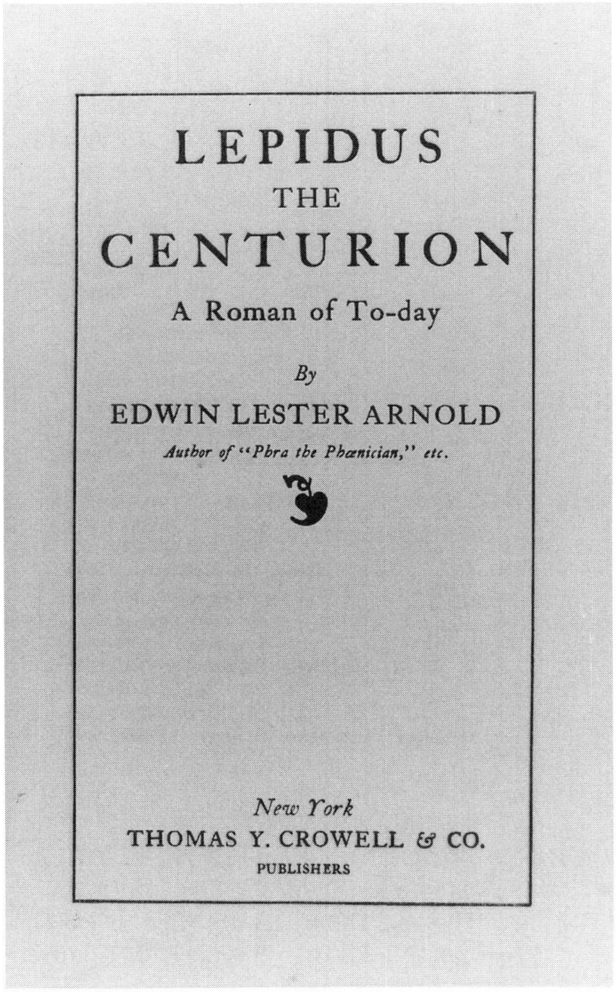

LEPIDUS
THE
CENTURION

A Roman of To-day

By

EDWIN LESTER ARNOLD

Author of "Phra the Phœnician," etc.

New York
THOMAS Y. CROWELL & CO.
PUBLISHERS

Frontispiece and title page for Edwin L. Arnold's second fantasy novel, in which a Roman centurion becomes involved in a nineteenth-century love triangle

he marries in Britain. When the Romans invade, the false Druid Dhuwallon betrays Phra and sees him sacrificed as a traitor. Phra awakens four hundred years later at the height of the Roman occupation and finds that Blodwen has tattooed a magical design onto his body that allows him to transcend ordinary death. His bravery and strength lead to a position of centurion in the guard of the corrupt Electra, but the British rebel, and while crossing a swollen river during the flight, Phra angers Electra when he will not release the slave girl Numidea. Electra thus cuts the rope, releasing Phra and Numidea into the water. Phra awakens in a monk's cell, having been left there by King Alfred, in time to side with King Harold's Saxons in their battle against the Normans. Though the Saxons are defeated, Phra lives happily for a time with Editha, Blodwen's reincarnation, until he refuses to answer questions for the Domesday Book and must flee. He awakens in 1376, becomes involved in the war against France, and meets the fair Isobel, yet another reincarnation of Blodwen. She disguises herself to accompany Phra onto the battlefield and perishes even as the English are victorious, and Phra dejectedly returns to England bearing Edward III's announcement of triumph. He sleeps in an old tomb, then awakens to deliver the message to the court of Elizabeth, from which he flees in embarrassment. He takes lodgings with a scholar and inventor of a steam engine, John Faulkener, whose beautiful daughter Elizabeth is yet another of Blodwen's reincarnations. Phra writes his history; Elizabeth and he announce their engagement; and Faulkener's jealous Spanish servant poisons them both. As he pens his dying words, Phra has a vision of Blodwen awaiting him.

Though the narrative is ultimately about the enduring power of love, *The Wonderful Adventures of Phra the Phoenician* achieves a life of its own only sporadically. The narrative is stodgy and enormously overlong, but it does have moments of interest. Arnold attempted to portray Phra's circumstances realistically within the fantastic framework, and there are occasionally acute observations:

> I would pass a sturdy yokel leading out his teams to plowing, and when I spoke to him it made my ears tingle to hear how antique Roman went hand in hand with ancient British, and good Norman was linked upon his lips with better Saxon! That polyglot youth, knowing no tongue but one, was most scholarly in his ignorance. To him 'twas English that he spoke; but to me, who had lived through the making of that noble speech, who knew each separate individual quantity that made that admirable whole, his jargon was most wonderful!

There are also details that convincingly illustrate the passage of time, as when Phra notices that the estuary in which he sailed during the pre-Roman times has by the Elizabethan days filled with silt. Furthermore, the book is important typologically: it is not the earliest novel to attempt to present extended life realistically from the viewpoint of the immortal. Eugène Sue's *Le Juif errant* (translated as *The Wandering Jew,* 1844–1845) dates from 1844 to 1845, and there are earlier examples found in such Gothic novels as Charles Robert Maturin's *Melmoth the Wanderer* (1820), but *The Wonderful Adventures of Phra the Phoenician* is nevertheless one of the first mainstream novels to take such an approach. It stands as ancestor behind works as diverse as George Griffith's *Valdar the Oft Born* (1895), Charles Godfrey Leland's *Flaxius: Leaves from the Life of an Immortal* (1902), Virginia Woolf's *Orlando* (1928), and Gerald Kersh's "Whatever Happened to Corporal Cuckoo?" (1953).

Lepidus the Centurion: A Roman of Today (1901), Arnold's second fantastic novel, also involves reincarnation and extended sleep and depicts an eighth-century Roman centurion, Lepidus, reawakened into the British society of the late nineteenth century. Lepidus's doings are recounted by Louis Allanby, a well-to-do socialite whose attitude toward Lepidus combines fascination, envy, and progressive frustration, for Lepidus is a larger figure than can be safely contained by Allanby's genteel society. Indeed, Lepidus's uninhibited nature and fearless physicality enable him not only to react quickly to presumed danger and to master supposedly unmanageable stallions but also to pursue the female guests on Allanby's estate, in particular Allanby's girlfriend, Priscilla, whom Lepidus believes to be the reincarnation of his Roman love, Prisca. After Lepidus sets fire to Allanby's house in the hopes of escaping with Priscilla, he and Allanby come to blows. They duel over the right to possess Priscilla then opt to let her decide which of them she favors, agreeing that the loser will commit suicide. Priscilla chooses the staid Allanby, and Lepidus drinks a fatal draught, dying in the tomb in which Allanby awakened him. Priscilla and Allanby are left with powerful memories and heightened personalities as a result of their exposure to him.

Though there is nothing original in Arnold's idea of an awakened visitor from the past–Mary Shelley wrote such a story in her unpublished "Valerius: The Re-animated Roman"–Arnold's thematic undercurrents provide a clear statement of the sexual tensions and fears of the late-Victorian Englishman, for the centurion is the dark, amoral, and utterly irresistible foreigner against whom no Englishman can physically compete and whom no Englishwoman can sexually withstand. More intriguing, Lepidus and Allanby share an unusual psychic bond: not only do they resemble each other, but at times they even share memories. Lepidus somehow learns English instantly from Allanby, and, as Lepidus waxes, Allanby wanes:

> There he was, as graciously irresistible as I was supine and helpless. My star seemed to have set in the splendour of that southern sun that had ridden suddenly into our skies. Never in my life had I felt so worthless, so washed out, so mentally sapped and soulless, as I did at this time. As for the Roman, he grew fat on my leanness, he waxed magnificent by my eclipse, and the strangest thing was that I recognised there at the breakfast-table, as I had done before so often in his talk, my own figures of speech, and in his knowledge (often blunderingly drawn upon) my very own experiences. In fact, there could not be any doubt to me, as I stared and listened, that Lepidus was using my mental raiment as certainly as he was inhabiting my best tweed suit, and though of the two perhaps the homespun fitted him the best, yet, as I have said, he did marvellously well in both.

Nevertheless, though Arnold's themes and ideas are occasionally powerful and intriguing, his treatment of them is novelistically weak and conservative, concentrating on superficial descriptions of Lepidus's disruptions of the social systems operating in Allanby's hitherto placid world.

The majority of the action in Arnold's final fantastic novel, *Lieut. Gullivar Jones: His Vacation* (1905), occurs on Mars, which Jones, an officer in the U.S. Navy, reaches by standing upon a magic carpet and wishing he were on Mars. The Martian slave, An, introduces Jones to a decadent and lazy race, reminiscent of the Eloi in H. G. Wells's *The Time Machine* (1895), and Janus soon falls in love with the lovely Princess Heru. When she is abducted to be the plaything of a barbarous tribe, he sets out to rescue her, and during his travels he encounters sentient plants, boats that are grown rather than built, and a frozen land of the dead whose description is among Arnold's most imaginative:

> All the opposite cliffs, rising sheer from the water, were in light, their cold blue and white surfaces rising far up into the black starfields overhead. Looking at them intently from this vantage-point I saw without at first understanding that along them horizontally, tier above tier, were rows of objects, like–like–why, good Heavens, they were like men and women in all sorts of strange postures and positions! Rubbing my eyes and looking again I perceived with a start and a strange creepy feeling down my back that they *were* men and women!–hundreds of them, thousands, all in rows as cormorants stand upon sea-side cliffs, myriads and myriads now I looked about, in every conceivable pose and attitude but never a sound, never a movement among the vast concourse.

After accomplishing two additional, supposedly impossible, tasks, Jones rescues his princess, though he fails to marry her: an invasion of the barbarians forces him back to Earth, where his story is assumed to be the result of too much alcohol.

As Jones's first name indicates, the novel is a descendant of Jonathan Swift's *Gulliver's Travels* (1726), and Arnold occasionally essays mild satire. When Jones attempts to describe the "Munroe [*sic*] Doctrine" to a Martian guide, he states, "Oh, it is simple enough, and put into plain language means you must not touch anything that is mine, but ought to let me share anything you have of your own." These satiric exchanges are few, however, and the story is far more interesting for combining the elements of fairy tales, fantastic Victorian boys' adventures, and a barely veiled adult eroticism that is constantly evident in Jones's perpetual leering at the charms of the semiclothed Martian women in general and Princess Heru in particular. *Lieut. Gullivar Jones* has often been claimed as an inspiration for Edgar Rice Burroughs's Martian novels, though the evidence is inconclusive at best. Finally, *Lieut. Gullivar Jones* is a work that is better in summary than it is to read. It is loosely written, containing an enormous number of ideas introduced and forgotten; there are hints of a fellow earthman blown to Mars via a volcano and episodes involving magical prophecies, reanimated corpses, ghosts, and overgrown ruins, all casually introduced and dropped without integration into the story.

Though Arnold lived for another thirty years and died on 1 March 1935, *Lieut. Gullivar Jones* was his last novel, and his fiction is today generally neglected, and remembered only by specialists and literary historians. This situation is not likely to change, for despite Arnold's occasional strokes of imagination, he was a weak writer and did not possess the ability necessary to develop a successful story. Even Arnold's short fiction, some of which is mildly fantastic–reincarnation figures in "Rutherford the Twice-Born," collected in *The Story of Ulla and Other Tales* (1895)–is as flaccid and fustian as the novels. Nevertheless, as an early explorer of themes that would dominate the fiction of the next generation of fantasists, Arnold remains a true pioneer.

References:

E. F. Bleiler, *The Guide to Supernatural Fiction* (Kent, Ohio: Kent State University Press, 1983);

Bleiler, *Science Fiction: The Early Years* (Kent, Ohio: Kent State University Press, 1990);

Richard Lupoff, "Edwin Lester Linden Arnold," in *Twentieth-Century Science Fiction Writers,* third edition, edited by Noelle Watson and Paul E. Schellinger (Chicago: St. James Press, 1991), pp. 16–17;

Robert Reginald, "The Wonderful Adventures of Phra the Phoenician," in *Survey of Modern Fantasy Literature,* edited by Frank Magill (Englewood Cliffs, N.J.: Salem Press 1983), pp. 2156–2158;

Frederick Shroyer, "Lieut. Gullivar Jones: His Vacation," in *Survey of Science Fiction Literature,* edited by Magill (Englewood Cliffs, N.J.: Salem Press, 1979);

Brooks Wright, *Interpreter of Buddhism to the West: Sir Edwin Arnold* (New York: Bookman, 1957).

John Davys Beresford

(7 March 1873 - 2 February 1947)

Richard Bleiler
University of Connecticut

See also the Beresford entry in *DLB 162: British Short-Fiction Writers, 1915–1945.*

John Davys Beresford (Hulton Getty Picture Collection Limited)

BOOKS: *The Early History of Jacob Stahl* (London: Sidgwick & Jackson, 1911; Boston: Little, Brown, 1911; New York: Doran, 1911);

The Hampdenshire Wonder (London: Sidgwick & Jackson, 1911); republished as *The Wonder* (New York: Doran, 1917);

A Candidate for Truth (London: Sidgwick & Jackson, 1912; Boston: Little, Brown, 1912; New York: Doran, 1912);

Goslings (London: Heinemann, 1913); republished as *A World of Women* (New York: Macaulay, 1913);

The House in Demetrius Road (London: Heinemann, 1914; New York: Doran, 1914);

H. G. Wells (London: Nisbet, 1915; New York: Holt, 1915);

The Invisible Event (London: Sidgwick & Jackson, 1915; New York: Doran, 1915);

The Mountains of the Moon (London & New York: Cassell, 1915);

Poems by Two Brothers, by Beresford and Richard Beresford (London: Macdonald, 1915);

These Lynnekers (London & New York: Cassell, 1916; New York: Doran, 1916);

House-mates (London & New York: Cassell, 1917; New York: Doran, 1917);

W. E. Ford: A Biography, by Beresford and Kenneth Richmond (London: Collins, 1917; New York: Doran, 1917);

God's Counterpoint (London: Collins, 1918; New York: Doran, 1918);

Nineteen Impressions (London: Sidgwick & Jackson, 1918);

The Jervaise Comedy (London: Collins, 1919; New York: Macmillan, 1919);

An Imperfect Mother (London: Collins, 1920; New York: Macmillan, 1920);

Revolution: A Novel (London: Collins, 1921); republished as *Revolution: A Story of the Near Future in England* (New York: Putnam, 1921);

Signs and Wonders (Waltham Saint Lawrence, U.K.: Golden Cockerel Press, 1921; New York: Putnam, 1921);

The Prisoners of Hartling (London: Collins, 1922; New York: Macmillan, 1922);

Taken from Life, by Beresford and Emil O. Hoppe (London: Collins, 1922);

The Imperturbable Duchess and Other Stories (London: Collins, 1923);

Love's Pilgrim (London: Collins, 1923; Indianapolis: Bobbs-Merrill, 1923);

Unity (London: Collins, 1924; Indianapolis: Bobbs-Merrill, 1924);

The Monkey Puzzle (London: Collins, 1925; Indianapolis: Bobbs-Merrill, 1925);
That Kind of Man (London: Collins, 1926); republished as *Almost Pagan* (Indianapolis: Bobbs-Merrill, 1926);
The Decoy (London: Collins, 1927);
The Tapestry (London: Collins, 1927; Indianapolis: Bobbs-Merrill, 1927);
The Instrument of Destiny: A Detective Story (London: Collins, 1928; Indianapolis: Bobbs-Merrill, 1928);
Writing Aloud (London: Collins, 1928);
All or Nothing (London: Collins, 1928; Indianapolis: Bobbs-Merrill, 1928);
The Meeting Place and Other Stories (London: Faber & Faber, 1929);
Real People (London: Collins, 1929);
Love's Illusions (London: Collins, 1930; New York: Dutton, 1930; New York: Viking, 1930);
Seven Bobsworth (London: Faber & Faber, 1930);
An Innocent Criminal (London: Collins, 1931; New York: Dutton, 1931);
The Old People (London: Collins, 1931; New York: Dutton, 1932);
The Next Generation (London: Benn, 1932);
The Middle Generation (London: Collins, 1932; New York: Dutton, 1933);
The Inheritor (London: Benn, 1933);
The Camberwell Miracle (London: Heinemann, 1933);
The Young People (London: Collins, 1933; New York: Dutton, 1934);
The Case for Faith-Healing (London: Allen & Unwin, 1934);
Peckover (London: Heinemann, 1934; New York: Dutton, 1935);
On a Huge Hill (London: Heinemann, 1935);
Blackthorn Winter and Other Stories (London: Hutchinson, 1936);
The Faithful Lovers (London: Hutchinson, 1936; New York: Furman, 1936);
Cleo (London: Hutchinson, 1937);
Unfinished Road (London: Hutchinson, 1938);
What I Believe (London: Heinemann, 1938);
Strange Rival (London: Hutchinson, 1939);
Snell's Folly (London: Hutchinson, 1939);
The Idea of God (London: Clarke, 1940);
Quiet Corner (London: Hutchinson, 1940);
"What Dreams May Come . . ." (London: Hutchinson, 1941);
A Common Enemy (London: Hutchinson, 1942);
The Benefactor (London: Hutchinson, 1943);
The Long View (London: Hutchinson, 1943);
Men in the Same Boat, by Beresford and Esmé Wynne-Tyson (London: Hutchinson, 1943);
If This Were True (London: Hutchinson, 1943);
The Riddle of the Tower, by Beresford and Wynne-Tyson (London: Hutchinson, 1944);
The Prisoner (London: Hutchinson, 1946);
The Gift, by Beresford and Wynne-Tyson (London: Hutchinson, 1947).

Although John Davys Beresford wrote more than sixty books, many of which received respectful reviews from contemporary reviewers, he is today virtually forgotten, with only his early fantastic fiction preserving his memory from complete extinction. This general neglect is not entirely undeserved, for Beresford was notably uneven as a writer. Similar characters and themes appeared in many of his novels, but worse than this repetition was his habit of permitting his enthusiasms and obsessions to dominate his fiction to the point where narration and characters were subverted, the narrative becoming a tract presenting a right-thinking individual contending with oppositional straw men before emerging victorious and vindicated. Despite these flaws, he wrote several intelligent, original, and insightful novels that remain vital and relevant to modern readers. Although he was not a particularly profound thinker, he read heavily in H. G. Wells and Bernard Shaw, letting their observations inspire him, and his best work thus often contains acute social commentary. In addition, Beresford's books never deteriorated into empty sensationalism and easy sentimentality but consistently attempted to grapple with serious themes, speculating seriously and philosophically about the nature of society and the future of humanity. Finally, although they have been to some extent overshadowed by his novels, Beresford's brief and elliptical short stories contain some of his finest writing.

John Davys Beresford was born 7 March 1873 at Castor, a small Northamptonshire village near Peterborough, where his father, J. J. Beresford, was a minor canon and precentor of the Peterborough cathedral. When he was three he contracted infantile paralysis, which left him permanently crippled and necessitated his using a cane for the rest of his life; he blamed his condition on the carelessness of a nurse, and, although it never availed him, in his later life he argued fervently in favor of the validity of faith healing. (He argued equally fervently for socialism, pacifism, Christianity, Eastern mysticism, women's rights, vegetarianism, occultism, and the need for fundamental educational and governmental reforms.) He was educated at Oundle and at King's School, Peterborough, and at age eighteen went to London to be apprenticed to an architect. He left architecture to work, with some success, as a bookseller, but by 1908 he had become a contribu-

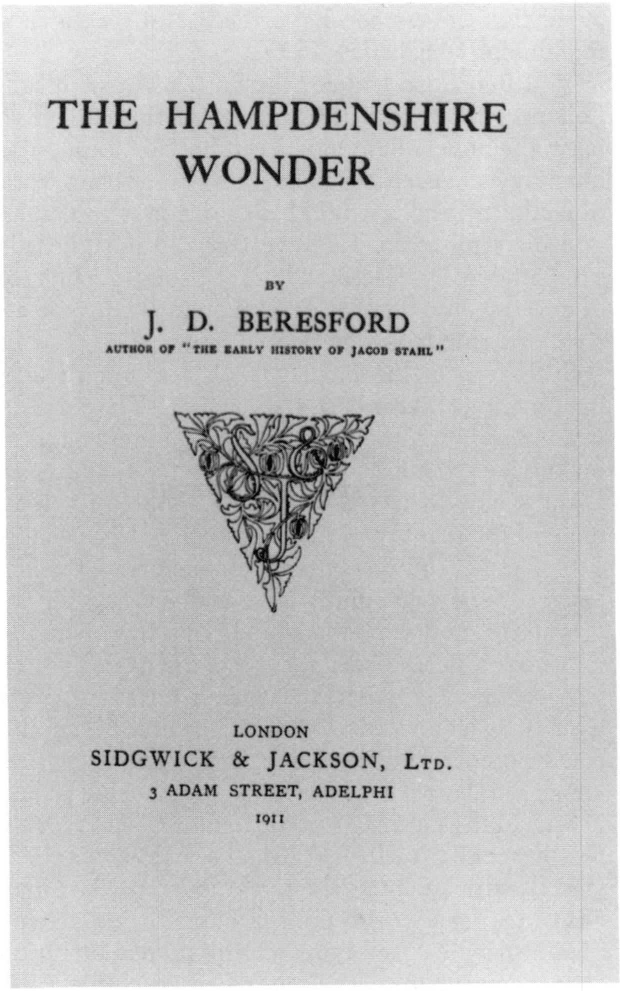

THE HAMPDENSHIRE WONDER

BY

J. D. BERESFORD

AUTHOR OF "THE EARLY HISTORY OF JACOB STAHL"

LONDON
SIDGWICK & JACKSON, LTD.
3 ADAM STREET, ADELPHI
1911

Title page for Beresford's second novel, in which a young boy's prodigious intellect alienates his community

tor to *Punch* and a reviewer for the *Westminster Gazette.* His first two novels were published in 1911. The first, *The Early History of Jacob Stahl,* was the first volume of a heavily autobiographical trilogy featuring the maturation of the crippled Jacob Stahl. It was well received by contemporary reviewers and brought him immediate fame. The second was *The Hampdenshire Wonder,* the fantastic novel for which Beresford is today remembered.

The Hampdenshire Wonder is an oddly elliptical, frequently ironic, and semiallegorical novel describing the brief life of Victor Stott, the "Hampdenshire Wonder," who is an intellectual superman. He possesses an enormous bald head, a gaze that disconcerts all, and an intellect beyond human comprehension. In addition, he is virtually emotionless, and when he speaks he is contemptuous of contemporary science and culture, largely unable to explain his thoughts to contemporary humanity because humanity simply does not have the words necessary for comprehension. Ginger Stott, the Wonder's intellectually ordinary father (a cricketer, perhaps in homage to Wells, whose father once played cricket professionally), could not endure life with the infant Wonder and fled, leaving the Wonder to be raised by his mother, who fears and worships him. They live in housing donated by the local squire, the scientist Henry Challis, whose enormous library the four-year-old Wonder rapidly consumes. The local priest, the fanatical Percival Crashaw, hates the Wonder for his atheistic views and attempts to force the boy to attend the local school. In a public examination the Wonder demolishes Crashaw and humiliates his educational committee. The narrator, a journalist and would-be writer who has sporadically encountered the Wonder from his earliest years, enters the story for the final time when the Wonder is about eight. He notices that the Wonder dislikes the hydrocephalic village idiot, the only person not disturbed by the Wonder's gaze, and then the Wonder is found drowned. The coroner rules the death accidental, but all clues point toward Crashaw as murderer, though the narrator would prefer to believe that the idiot committed the crime.

The Hampdenshire Wonder contains a superficially simple story line, but it is far from a simple book, a fact noted by Beresford's contemporaries. The reviewer for *The New York Times* stated that "one thing at least seems fairly certain–it is a book about which people will disagree." Adding depth to the narrative is a carefully established linkage between the Wonder, humanity, and the idiot: the Wonder is as far superior to humanity as humanity is to the idiot; the humans are repelled and annoyed by the idiot, even as the Wonder must endure ordinary humanity; and none of the parties can communicate meaningfully with the other. On an allegorical level the novel is a description of pure intellect destroyed by bigoted religiosity, a theme that appears sporadically in many of Beresford's later works. (As Brian Stableford has noted in his essay on Beresford, "one cannot imagine that Beresford's father read the book with any pleasure.") Furthermore, though Challis is thoroughly demolished in his one discussion with the ultrarational Wonder, he gets what is in essence the last word in a long speech reminiscent of Wells's fictional exhortations:

> Oh! pity the child . . . for whom there could be no mystery. Is not mystery the first and greatest joy of life? Beyond the gate there is unexplored mystery for us in our childhood. When that is explored, there are new and wonderful possibilities beyond the hills, then beyond the seas, beyond the known world, in everyday chances and movements of the unknown life in which we are circumstanced.

> Surely we should all perish through sheer inanity, or die desperately by suicide if no mystery remained in the world. Mystery takes a thousand beautiful shapes; it lurks even in the handiwork of man, in a stone god, or in some mighty, intricate machine, incomprehensibly deliberate and determined. The imagination endows the man-made thing with consciousness and powers, whether of reservation or aloofness; the similitude of meditation and profundity is wrought into stone. Is there not source for mystery to the uninstructed in the great machine registering the progress of its own achievement with each solemn, recurrent beat of its metal pulse?

Challis's closing speech not only provides a clear statement of one of the novel's major thematic concerns, it is successful in forcing a revaluation of the character of the deceased Wonder, "the child . . . for whom there could be no mystery," who has hitherto been presented as unnerving rather than sympathetic. The novel remains powerful and thought-provoking. In the field of fantastic fiction it is significant for being the first important novel about an intellectual superman, and it directly inspired E. V. Odle's *The Clockwork Man* (1923) and Olaf Stapledon's superior *Odd John* (1935).

Beresford's next fantastic novel was *Goslings* (1913), whose American title, *A World of Women,* provides a better indication of its story line, for it describes an England in which a plague from Asia has killed virtually all the men. The surviving women, who have hitherto been little more than pampered pets, are thus abruptly forced to learn to fend for themselves. The majority of *Goslings* recounts the travails of the Gosling family, who come to exemplify England and English society. Mr. Gosling, one of the few men to survive the plague, promptly abandons his middle-aged wife and two daughters to live a life of sensuality with several younger women. Despite his desertion, the Gosling women escape from London to the farming country, but Mrs. Gosling (her first name is never given) is severely limited; without a man to tell her what to do and give direction to her life, and unable to comprehend the social changes around her, she slowly withers and dies. The Gosling daughters, Millie and Blanche, ultimately reach a collective farm and are given the opportunity to create meaningful lives for themselves. Millie, however, becomes sexually frustrated: annoyed because the farm's mechanic, the ascetic Jasper Thrale, repels her advances, Millie briefly leaves the collective to get pregnant by a hedonistic butcher living on an adjoining farm. Upon her return she is not ostracized as an unwed mother but resumes her work, further proof that the women have successfully created a functioning socialistic society.

If Beresford had concluded *Goslings* with Millie's pregnancy and the success of the collective farm, the novel might have been initially unpopular but may have been hailed later as a pioneering work of feminism, and it would have done much to maintain his reputation. However, he did not give his sympathetically described feminist society a chance to develop, and he concluded the novel with the arrival of a ship from the United States. The crew of the ship is male, for the plague mutated upon reaching America, striking both sexes equally. The Englishwomen abandon their newly gained independence, make plans to marry the American men, and hope that the new society will be more equal in its treatment of the sexes than the dead society had been.

Despite this disappointing conclusion, the majority of *Goslings* is solidly imagined and capably extrapolated, and the body of the book is convincing as a novel. Beresford's social commentary, which emphasizes what patriarchal society does to oppress women as well as what women are capable of and can accomplish without men, is perceptive and remains as relevant as it was when the novel was published. In the field of fantastic fiction *Goslings* is a direct descendant of M. P. Shiel's *The Purple Cloud* (1901), which depicts the activities of the last man alive in a dead world, though Beresford eschewed Shiel's private cosmology; a more immediate influence was probably Sir Arthur Conan Doyle's *The Poison Belt* (1913), which explored many of the same themes. In demonstrating how women can survive and even flourish in a world without men *Goslings* is almost as ardently feminist as Joanna Russ's "When It Changed" (1972) and *The Female Man* (1975), and in its descriptions of the actions of the plague survivors, virtually alone in a world full of rotting corpses, *Goslings* is a worthy precursor of Stephen King's *The Stand* (1978).

The biting social commentary of *Goslings* is absent from Beresford's *Nineteen Impressions* (1918), a collection of short fiction thematically linked by narrative experiments in sight and impressions. "The Criminal" describes a man on trial for horrible crimes; descriptions of him disagree, and a picture shows nothing in the dock, yet something is hanged. "A Case of Prevision" concerns Jessop, whose hallucination involves seeing a precipice where others see rock; a landslide proves him right. "The Little Town," "The Empty Theatre," and "Lost in the Fog" are allegorical, using theaters and small towns to comment upon human life and society. "Powers of the Air" is narrated by a man who refuses to leave the safety of his house to venture into the night, for he knows the forces of the winds will destroy him; he attempts to warn a townsman, but the man ventures outside and returns more alive than before–confronting fears has untold rewards. The most thought-provoking story in *Nineteen Impressions,* and perhaps Beresford's single finest story, is "The Misanthrope," the account of a man who lives in isolation on an island with no mirrors. He has the ability to see the real natures of people in their faces, and it is never attractive. Though brief, "The Misanthrope" is a remarkably bleak condemnation of humanity.

Revolution (1921), Beresford's next romance, has as its direct inspiration the fears of unions and strikes that gripped post–World War I Britain. The book begins with the threat of a general strike: the charismatic and altruistic Isaac Perry has succeeded in uniting the labor unions, demanding that the wealthy surrender their properties to the laboring classes. Perry is seen but once, however, for the novel is narrated predominantly through the experiences of young Paul Leaming, who became shell-shocked during the war. This experience has made him an ardent pacifist; he is the diametric opposite of his belligerent father, able to recognize and propose viable solutions to social problems that other characters believe to be insoluble. After Perry is anonymously assassinated, the strike and accompanying revolution spread from London to the village of Fynemore, where the Leamings live. Paul realizes what must be done in order to avoid bloody class warfare and convinces Lord Fynemore to surrender his estates and properties to the farming people. At a town meeting the roughneck Jem Oliver proposes a largely totalitarian government, and when Mr. Leaming objects, Oliver casually shoots him. Paul nevertheless forgives Oliver this murder, forgiving him again when he kills a farmer who will not surrender his property to the state. The new classless system, however, is not immediately successful; a general shortage of supplies and a prohibition against hoarding cause the villagers to idealize their deposed government. Starting in Fynemore with Lord Fynemore and his son, the former government gradually and treacherously reclaims much of its lost power. Paul inadvertently discovers the Fynemores' plots and is briefly imprisoned to prevent him from warning Oliver of his danger. Nevertheless, the Fynemores' treachery begets only tragedy, for although Oliver is killed while resisting the government's forces, he first kills Fynemore's son. The restored government, however, is as doomed as the classless government, and *Revolution* concludes with Paul electing to go to London to preach his vision of peace and universal brotherhood.

The arguments of *Revolution* proved quite controversial. Beresford's English audience disliked the novel's explicitly pacifistic and socialistic messages, and for the American edition he wrote a special foreword acknowledging that "a reading of my message has stimulated a violent antagonism in many of my reviewers" and arguing that European society was ending, for traditional European social groups could no longer compromise. A modern audience is likely to agree with his social messages while finding *Revolution* deficient as a novel. The character of Paul, a man willing to defend his father's murderer over his father's bleeding body, is wholly unconvincing, and the other characters are no more than types representing political beliefs but having no existence apart from advancing Beresford's arguments. Despite these many flaws, *Revolution* is not an unintelligent book. Its fiction may be negligible, but its arguments are not easily dismissed.

The preachiness of *Revolution* is completely absent from *Signs and Wonders* (1921), Beresford's second collection of short fiction, which includes some of his finest work. The wonders include visions of elephants walking down the Thames embankment during a German bombardment in "Enlargement," the disastrous exchange of senses between a modern man and his ancestor in "The Cage," and a vision of an enormous eye in the sky observing humanity in "Signs and Wonders." A grim sense of humor enlivens several of the stories: "As the Crow Flies" describes a man remembered for the great novel he talked about but never wrote; and "Reference Wanted!" is about a man accused of plagiarism who is nevertheless unable to remember the great writers he studied. "The Miracle" is a love story describing a woman entering hell to reclaim her stricken husband's soul; unlike Orpheus, she succeeds. "The Appearance of Man: A Play out of Time and Space" is a short playlet in which the creation of the world and the death of the universe occur in the stage's background, while in the foreground two men and a woman engage in an utterly banal conversation. It is thematically similar to "A Negligible Experiment," in which parallels are established between a sculptor working with clay and a dream in which a nova destroys Earth. Two of the stories feature grim retributions. "The Perfect Smile" is the story of Douglas Owen, who has twice evaded the punishment he deserves because of his charming smile; the third time his smile betrays him, and he must pay the price for his crimes. Even grimmer and more horrifying is "Young Strickland's Career," in which Strickland is so determined to foretell his son's career that he buys a crystal ball. He soon gives it away, for the only vision he received was of hell and of a scarecrow digging turnips. Following World War I, Strickland goes to France to look for his missing son and learns of the essential validity of his vision.

For nearly eight years after the publication of *Signs and Wonders* Beresford avoided using significant fantastic elements in his longer fiction. During this time he wrote essays discussing religion, utopias, spiritualism, love, Indian art, reincarnation, and dogma. His novels became increasingly didactic and obsessional and deteriorated markedly in quality, their characters becoming little more than mouthpieces espousing his beliefs. In addition, and paralleling this general decline in narrative technique, he began to repeat himself. Characters similar to Paul Leaming of *Revolution*–ascetic, almost saintly men whose claims are initially ignored by callous unbelievers but who represent a new path to a higher truth or truths–appeared in Beresford's longer fiction with increasing frequency. When they bothered to review his work, the reviewers nearly always remarked upon the decline in his narrative skills, and his books gradually ceased to have American editions.

Most of Beresford's novelistic flaws are evidenced in *Real People* (1929), the novel with which he resumed writing fantastic fiction. A society story, the majority of it depicts the activities of the lovely Joanna Moore, who gets engaged to and ultimately marries the upper-class Bobby Messenger rather than permit the brilliant but domineering lower-class mathematician Michael Ross to possess her. The fantastic element concentrates on Joanna's father, Dr. Charles Moore, who gradually discovers that he is clairvoyant, able not only to diagnose complex medical problems with a mere glance but also to predict unanticipated death. Despite Joanna's mildly feminist behavior and some intriguing discussions between Bobby; his twin sister, Blanche; and Michael, *Real People* has virtually no intellectual content and is one of Beresford's weakest works.

The Camberwell Miracle (1933) is little more than an extended argument for the validity of faith healing and against a traditional British medical establishment that will not accept its documented cures. The story centers around Rosemary Henderson, a beautiful, intelligent, and sensitive woman, who contracted infantile paralysis when her parents went on vacation and left her with an inexperienced nurse. Her father, Sir Geoffrey, is a doctor; representative of the worst aspects of the traditional British medical establishment; he is unwilling to accept that a former classmate, Dr. Martin Davies, has been bringing about remarkable cures of the hitherto incurable. Davies is modest and lives quietly and religiously, desiring only to treat the ill and lame and refusing to make any claims to his powers. He nevertheless becomes notorious when the British press reveals his cures, and the British Medical Association, led by Sir Geoffrey, strikes his name from their rolls. At the saintly Davies's trial several witnesses testify in his behalf, including a cured Rosemary, and he is reinstated while the side of Sir Geoffrey is discredited.

"What Dreams May Come . . ." (1941) and *A Common Enemy* (1942) are more science fiction than fantasy, concentrating on the establishments of ideal societies and socialistic world brotherhoods. In both, Beresford used World War II to initiate the changes. In the first novel young David Shillingsford's lower-class English childhood is exceptional only for its recurrent and remarkably vivid dreams of an idyllic land and its people. After a building collapses on him during a German bombing raid he awakens in Oion, the land of his dreams. It is a bucolic, sparsely populated place existing either in the far future or on a parallel world (Beresford is deliberately unclear). Through extended conversations with the inhabitants, Shillingsford learns of its peace, acceptance, and love. In Oion all can do as they wish for as long as they wish. Telepathy exists; instruction is peaceful and self-motivated; conflict is nonexistent; and life spans are lengthy. Oionite bodies are hairless, light, and different in the arrangement of their internal organs. After learning much of the superior Oionite way of life, Shillingsford awakens in a hospital back on war-torn Earth, but as proof that his experiences were not a dream his body is that of an Oionite. He attempts to preach the Oionite gospel of universal love and responsibility that will enable its practitioners to enter Oion, but he has minimal success. Instead, he is arrested for subverting one of His Majesty's officers and put into a mental hospital. He dies on Earth and reawakens in Oion.

A Common Enemy details the establishment of a socialistic government on a war-torn Earth when something happens that makes the continuation of war impossible. There are massive geographical upheavals, for the Earth's axis and orbit have somehow shifted. Mr. Walter Campion, an official in the English treasury, realizes that the destruction of the old way of life can mark the beginning of a new way. He is instrumental in establishing a provisional government and in visiting the scattered English villages, each of which operates under a different system of government. He is successful in preventing the reestablishment of private ownership of factories, and the novel concludes with his meeting the like-minded representatives from the Continent.

In both *"What Dreams May Come . . ."* and *A Common Enemy* Beresford's intentions are remarkably

Beresford with Esmé Wynne-Tyson, who collaborated on The Riddle of the Tower *(1944), his last work of fantasy (courtesy of Jon Wynne-Tyson)*

at variance with his achievements. Although he clearly believed in the virtues of his utopias and presents them intelligently, a less enthralled modern reader is likely to find his seemingly idyllic lands to be thoroughgoing dystopias. Their cultures are cold and hostile places that repress all individualism. They have no room for artists or dreamers, and they have destroyed the majority of their books. Furthermore, both societies were established through the wholesale extermination of unbelievers and doubters, and in both ideal societies the penalty for acting individually is death through exile or execution.

Beresford's last fantastic work was *The Riddle of the Tower* (1944), a collaboration with Esmé Wynne-Tyson, an actress and friend of Noël Coward. In it Mr. Begbie, a small publisher, receives a manuscript called "Automatism" and learns from it that the majority of people respond automatically to life rather than thinking about events. After a German bomb explodes near him, Mr. Begbie finds himself possessing an almost immortal consciousness, witnessing the creation of the Earth and the growth and establishment of various cultures on it. He finds himself incarnated in the body of Aakisti, an inhabitant of an island on which peace and tranquillity are the way of life; this culture is brought down by treachery from outside, and the island sinks into the sea. Much time passes, and in the far future he finds himself witnessing a new order, an absolutely totalitarian society that exterminates all malcontents. This society is also destroyed from outside, and Begbie's consciousness witnesses the final stages of humanity, insectlike beings operating within a hive mind. He then reawakens to find himself back in London, having been unconscious for two days, and he discusses his experiences with the author of "Automatism," deciding that his vision is of a potential rather than an actual future.

As a work of ideas *The Riddle of the Tower* is indebted to Wells's *The Time Machine* (1895) and to Stapledon's *Last and First Men* (1930) and *Star Maker* (1937), and Beresford's novel is inferior to his literary inspirations. It is flat and undeveloped; more significantly, it is uninspired, containing little that he had not already offered at length in his earlier works. Though he concluded his writing career with two additional novels, *The Riddle of the Tower* is his last fantastic work–a sad conclusion to a career that began with powerfully original and thought-provoking novels.

Beresford died in Bath on 2 February 1947 at the age of seventy-three, survived by his wife and four children. His obituaries were unanimous in praising the Jacob Stahl trilogy and *The Hampdenshire Wonder,* but the remainder of his prose was generally dismissed. The obituary in the 4 February 1947 London *Times* provided what is probably the most accurate summation of his career, stating in part that never after the Jacob Stahl books did his novels "attain quite the same level of interest and power. In his concluding phase, indeed, the interest he had always displayed in abnormal psychology lost something of imaginative substance, and he gave expression to a quality of mystical thought that at moments seemed to come uncomfortably near to the facile."

Beresford is today largely unstudied, with only *The Hampdenshire Wonder* and, to a much lesser extent, *Goslings* and his short-story collections preventing the complete extinction of his memory. Apart from Brian Stableford, the only critic to have written anything substantial on Beresford is Frank Swinnerton, whose most pertinent criticism appears in *The Georgian Scene* (1934):

> All [Beresford's books] are written with precision and scrupulousness; all are reflective and without colour. If it is difficult to read them with excitement, it is impossible to read them without respect and (if one is oneself a novelist) without admiration for the skill, the practised and finished craft of their opening, development, and resolution.
>
> But to Beresford his own books are but so many pamphlets in which he has discussed the ideas cautiously and relishingly turned in his head since last he wrote. He has written them, perhaps, to try his hand at a different technical problem, to tell some episode which has occurred within hearsay or within his memory, to set forth an intellectual or scientific conception which happens to be of greater general interest than its originator supposes. . . . Conscience, courage, and patience are the qualities which one associates with Beresford; not brilliance.

It is improbable that Beresford's work will ever be rediscovered to be read with pleasure. He remains a perfect example of a writer whose best work was written early but who continued to write for decades thereafter, unable to recapture the excitement and energy that appeared in his earlier volumes. Few novelists are remembered for being enduringly brilliant, and there is no shame in a writer being remembered for possessing the virtues of "conscience, courage, and patience."

Bibliography:

Helmut E. Gerber, "J. D. Beresford: A Bibliography," *Bulletin of Bibliography and Magazine Notes,* 21 (January-April 1956): 201–204.

References:

A. St. John Adcock, *Gods of Modern Grub Street: Impressions of Contemporary Authors* (London: Sampson Low, Marston, 1923);

E. F. Bleiler, *The Guide to Supernatural Fiction* (Kent, Ohio: Kent State University Press, 1983);

Bleiler, *Science Fiction: The Early Years* (Kent, Ohio: Kent State University Press, 1990);

Brian Stableford, *"The Hampdenshire Wonder,"* in *Survey of Science Fiction Literature,* edited by Frank N. Magill (Englewood Cliffs, N.J.: Salem Press, 1979), pp. 945–949;

Stableford, "J. D. Beresford," in *Supernatural Fiction Writers: Fantasy and Horror,* volume 1, edited by Bleiler (New York: Scribners, 1985), pp. 457–461;

Stableford, *"The Riddle of the Tower,"* in *Survey of Science Fiction Literature,* volume 4, edited by Magill (Englewood Cliffs, N.J.: Salem Press, 1979), pp. 1780–1783;

Frank Swinnerton, *The Georgian Scene: A Literary Panorama* (New York: Farrar & Rinehart, 1934);

Harold Williams, *Modern English Writers: Being a Study of Imaginative Literature, 1890–1914* (New York: Knopf, 1919).

Algernon Henry Blackwood
(14 March 1869 – 10 December 1951)

Darren Harris-Fain
Shawnee State University

See also the Blackwood entries in *DLB 153: Late-Victorian and Edwardian British Novelists, First Series* and *DLB 156: British Short-Fiction Writers, 1880–1914: The Romantic Tradition.*

BOOKS: *The Empty House and Other Ghost Stories* (London: Eveleigh Nash, 1906; New York: Vaughan, 1915);

The Listener and Other Stories (London: Eveleigh Nash, 1907; New York: Vaughan & Gomme, 1914);

John Silence: Physician Extraordinary (London: Eveleigh Nash, 1908; Boston: J. W. Luce, 1909);

Jimbo: A Fantasy (London: Macmillan, 1909; New York: Macmillan, 1909);

The Education of Uncle Paul (London: Macmillan, 1909; New York: Paget, 1909);

The Lost Valley and Other Stories (London: Eveleigh Nash, 1910; New York: Vaughan & Gomme, 1914);

The Human Chord (London: Macmillan, 1910; New York: Macmillan, 1911);

The Centaur (London: Macmillan, 1911; New York: Macmillan, 1912);

Pan's Garden: A Volume of Nature Stories (London: Macmillan, 1912; New York: Macmillan, 1912);

A Prisoner in Fairyland (The Book That Uncle Paul Wrote) (London: Macmillan, 1913; New York: Macmillan, 1913);

Ten Minute Stories (London: John Murray, 1914; New York: Dutton, 1914);

Incredible Adventures (London: Macmillan, 1914; New York: Macmillan, 1914);

The Extra Day (London: Macmillan, 1915; New York: Macmillan, 1915);

Julius Le Vallon: An Episode (London: Cassell, 1916; New York: Dutton, 1916);

The Wave: An Egyptian Aftermath (London: Macmillan, 1916; New York: Dutton, 1916);

Day and Night Stories (London: Cassell, 1917; New York: Dutton, 1917);

Algernon Blackwood in 1949

The Promise of Air (London: Macmillan, 1918; New York: Dutton, 1918);

Karma: A Reincarnation Play (London: Macmillan, 1918; New York: Dutton, 1918);

The Garden of Survival (London: Macmillan, 1918; New York: Dutton, 1918);

The Wolves of God, and Other Fey Stories, by Blackwood and Wilfred Wilson (London: Cassell, 1921; New York: Dutton, 1921);

The Bright Messenger (London: Cassell, 1921; New York: Dutton, 1922);

Episodes Before Thirty (London: Cassell, 1923; New York: Dutton, 1924); republished as *Adventures Before Thirty* (London: Cape, 1934);
Tongues of Fire and Other Sketches (London: Jenkins, 1924); republished as *Tongues of Fire and Other Stories* (New York: Dutton, 1925);
Through the Crack (London: French, 1925);
Ancient Sorceries and Other Tales (London & Glasgow: Collins, 1927);
The Dance of Death and Other Tales (London: Jenkins, 1927; New York: Lincoln MacVeagh for the Dial Press, 1928); republished as *The Dance of Death and Other Stories* (London: Pan, 1963);
Sambo and Snitch (Oxford: Blackwell, 1927; New York: Appleton, 1927);
Mr. Cupboard (Oxford: Blackwell, 1928);
Dudley and Gilderoy: A Nonsense (London: Benn, 1929; New York: Dutton, 1929);
Strange Stories (London: Heinemann, 1929; New York: Arno, 1976);
Full Circle (London: Mathews & Marrot, 1929);
By Underground (Oxford: Blackwell, 1930);
Short Stories of To-day and Yesterday (London: Harrap, 1930);
The Parrot and the--Cat (Oxford: Blackwell, 1931);
The Willows and Other Queer Tales (London & Glasgow: Collins, 1932);
The Italian Conjuror (Oxford: Blackwell, 1932);
Maria (of England) in the Rain (Oxford: Blackwell, 1933);
Sergeant Poppett and Policeman James (Oxford: Blackwell, 1934);
The Fruit Stoners: Being the Adventures of Maria among the Fruit Stoners (London: Grayson & Grayson, 1934; New York: Dutton, 1935);
The Fruit Stoners (Oxford: Blackwell, 1935);
Shocks (London: Grayson & Grayson, 1935; New York: Dutton, 1936);
How the Circus Came to Tea (Oxford: Blackwell, 1936);
Not Long for This World (New York: Telegraph, 1936);
The Tales of Algernon Blackwood (London: Secker & Warburg, 1938; New York: Dutton, 1939);
The Adventures of Dudley and Gilderoy (New York: Dutton, 1941; London: Faber & Faber, 1941);
The Doll, and One Other (Sauk City, Wis.: Arkham House, 1946);
In the Realm of Terror (New York: Pantheon, 1957);
Selected Tales of Algernon Blackwood (London: John Baker for the Richards Press, 1964); republished as *Tales of Terror and the Unknown* (New York: Dutton, 1965); republished as *The Insanity of Jones and Other Tales* (Harmondsworth: Penguin, 1966);
The Magic Mirror: Lost Tales and Mysteries by Algernon Blackwood, edited by Michael Ashley (N.p.: Equation, 1989).

PLAY PRODUCTIONS: *The Starlight Express*, by Blackwood and Violet Pearn, London, Kingsway Theatre, 29 December 1915;
The Crossing, by Blackwood and Bertram Forsyth, London, Comedy Theatre, 29 September 1920;
Through the Crack, by Blackwood and Pearn, London, Everyman Theatre, 27 December 1920;
White Magic, by Blackwood and Forsyth, Toronto, Hart House Theatre, 21 November 1921;
The Halfway House, by Blackwood and Elaine Ainley, London, Victoria Palace, 5 December 1921;
Max Hensig, by Blackwood and Frederick Kinsey Peile, London, Gate Theatre Studio, 18 September 1929.

RADIO SCRIPTS: *The Blackmailers*, BBC London Regional, 11 July 1934;
The Empty House, BBC Empire Programme, 4 September 1934;
A Suspicious Gift, BBC National Programme, 29 September 1934;
By Water, BBC Empire Programme, 30 October 1934;
First Hate, BBC Regional Programme, 7 July 1935;
The Chinese Picture, BBC Regional Programme, 1 September 1935;
King's Evidence, BBC National Programme, 27 June 1936;
The Curse of the Wolf, BBC National Programme, 12 September 1937;
Two of a Kind, BBC Home Service, 17 December 1939;
Transition, BBC Home Service, 25 December 1939;
The Laughter of Courage, BBC Home Service, 25 December 1939;
A Boy and His Bag, BBC Home Service, 25 December 1939;
By Proxy, BBC Home Service, 29 February 1940;
The Voice, BBC Home Service, 17 April 1940;
The Destruction of Smith, BBC Home Service, 6 September 1940;
Violence, BBC Forces Programme, 23 October 1941;
Told in a Mountain Cabin, BBC Home Service, 13 April 1942;
The Magic Mirror, BBC Home Service, 13 December 1942;
Running Wolf, BBC General Forces Programme, 12 October 1944;
The Castlebridge Cat, BBC Home Service, 25 October 1944;

It's About Time, BBC General Forces Programme, 26 February 1945;
In a Glass Darkly, BBC General Overseas Service, 23 July 1945;
Lock Your Door, BBC Home Service, 6 May 1946;
The Woman's Ghost Story, BBC Third Programme, 23 September 1947;
The Secret Society, BBC Home Service, 22 October 1947;
A Dangerous Gift, BBC Light Programme, 3 January 1949;
The Occupant of the Room, BBC General Overseas Service, 21 December 1949.

SELECTED PERIODICAL PUBLICATIONS–UNCOLLECTED: "A Mysterious House," *Belgravia,* 69 (July 1889): 98–107;
"The Kit-Bag," *Pall Mall Magazine,* 42 (December 1908): 699–705;
"The Message of the Clock," *Nash's Magazine,* 2 (June 1910): 793–795;
"The Singular Death of Morton," *Tramp,* 2 (December 1910): 230–236;
"La Mauvaise Riche," *Westminster Gazette,* 30 November 1912, p. 3; and *Saturday Westminster Gazette,* 30 November 1912, p. 11;
"The Paper Man," *Saturday Westminster Gazette,* 9 October 1915, pp. 13–14;
"Onanonanon," *English Review,* 32 (March 1921): 248–254;
"Explorers' Ghost Stories," *Occult Review,* 42 (July 1925): 16–28;
"The Chocolate Cigarettes," *Number Six Joy Street* (September 1928): 105–125;
"Dreams and Fairies," *Bookman Christmas Supplement,* 77 (December 1929): 173–175;
"Cosmic Thrillers," *Time and Tide,* 11 (20 December 1930): 1606–1607;
"At a Mayfair Luncheon," *Wind–sor Magazine,* 495 (March 1936): 495–500;
"You Can't Tell Ghost Stories on the Radio," *Radio Times* (11 December 1936): 11;
"The Genesis of Ideas," *Writer,* 50 (February 1937): 35–37, 63;
"The Man-Eater," *Thrilling Mystery,* 6 (March 1937): 32–40;
"The Reformation of St. Jules," *Bystander,* 136 (29 December 1937): 500–502;
"The Magic Mirror," *Bystander,* 137 (16 March 1938): 431–432; and *Weird Tales,* 32 (September 1938): 282–288;
"Our Former Lives," *Prediction,* 12 (May 1947): 400–402;
"Roman Remains," *Weird Tales,* 40 (March 1948): 87–95;

Cover for Blackwood's first book, in which the title story was inspired by his investigation of haunted houses (courtesy of the Lilly Library, Indiana University)

"The Midnight Hour," *Queen* (24 November 1948): 37–64.

Algernon Henry Blackwood is one of the most respected writers in the fields of fantasy and horror fiction, and Everett F. Bleiler has called him "the foremost British supernaturalist of the twentieth century." Widely admired for his many ghost stories and works of weird fiction, Blackwood presented occult stories filled with suspenseful plots, haunting settings, and haunted characters. His work demonstrates that dark fantasy, though disturbing, can also be elegantly written, and he influenced many writers after him, including Robert Bloch and Ray Bradbury.

The fourth of five children, Blackwood was born on 14 March 1869 in Shooter's Hill, Kent, to Sir Stevenson Arthur Blackwood, who later served as financial secretary of the post office, and Harriet Sydney Dobbs. Both parents were from aristocratic backgrounds and of strong evangelical beliefs. His family moved two years later to Crayford in Kent, then in 1881 to Beckenham. He received an excellent although sporadic education, first at Wellington College in Berkshire, then at the School of the Moravian Brotherhood in Germany, then at Edinburgh University in 1888 and 1889. Blackwood rebelled against his privileged background, most notably by rejecting his family's fundamentalism in favor of an interest in Eastern religions, the theosophical ideas of Madame Helena Blavatsky, and the occult; a significant event in his intellectual development was his reading of Patanjali's *Yoga Aphorisms* in 1886 and later of the Bhagavad Gita. It is often difficult to tell whether Blackwood subscribed to the ideas presented in his fiction, but at any rate he clearly felt he was exploring possibilities.

Apart from his own voracious reading, Blackwood was not a diligent student and rejected the expected professional track, instead turning to various jobs and business enterprises in Canada and the United States as a young man and eventually a career as a writer. He never married. What did prove influential in his family background was the time his family spent on holidays in rural England and Scotland, which heightened the appreciation of nature that permeates Blackwood's work. Furthermore, during these outings his father would tell ghost stories, a form in which Blackwood would excel.

Blackwood's first published story, "A Mysterious House," appeared in the July 1889 issue of *Belgravia.* The piece reflects his interest in ghost stories along with his association with the Society for Psychical Research, with whom he investigated several houses said to be haunted. This early effort already shows some of Blackwood's characteristic approaches: while it includes many elements of the conventional ghost story, the real focus of the story is not on the frightening happenings but on the fright of the narrator. Such a depiction of psychological responses to the supernatural is common in Blackwood's work.

In April 1890 Blackwood moved with his father's encouragement to Canada, which the two men had visited in 1887; he stayed for two and a half years. After losing a job with the Canadian Pacific Railway that his father helped him get, he struggled for a few months at odd jobs and then found work with *The Methodist Magazine;* he also became a charter member of the Toronto branch of the Theosophical Society in 1891. Later in the year he entered into a short-lived partnership involving a dairy farm, and later he joined another partnership with a hotel. Both were initially profitable but later fell into decline, and in 1892 Blackwood moved to New York City. There he worked as a reporter for the *New York Sun* and, starting in 1895, *The New York Times.* His experience in the city, in contrast to the open spaces of Canada, reinforced his preference for nature over "civilization." After years of newspaper reporting and more business ventures, he returned to England in 1899 just before he turned thirty.

Blackwood, encouraged by William Butler Yeats, joined the Hermetic Order of the Golden Dawn in 1900. Though his own brand of nature-based mysticism differed from the ideas of the Golden Dawn, it is likely that his participation in this group encouraged his writing, as did his friend Angus Hamilton, whom Blackwood had met and befriended when both worked for the *New York Sun* and who showed Blackwood's stories to Maude M. C. ffoulkes at Eveleigh Nash. This led to the publication of Blackwood's first book, *The Empty House and Other Ghost Stories* (1906).

Despite their supernatural content, several stories in this collection are obviously based on Blackwood's personal experience. These include the title story, influenced by his investigation of supposedly haunted houses, as well as "A Haunted Island," based on his visit to an island in Canada. The title story also includes another theme that appears frequently in Blackwood's fiction, reincarnation: the narrator at one point sees a dead man with the narrator's face, making him believe either that he is seeing a vision of himself in a previous life or else hallucinating. Blackwood's subjects are interesting, but he had not yet really found his voice, and the stories are fairly conventional. Despite some positive reviews, including one from Hilaire Belloc, *The Empty House* is generally an apprentice effort.

More accomplished and confident in tone is Blackwood's second collection, *The Listener and Other Stories* (1907), which was also well received. It includes one of his best-known efforts, "The Willows." This story is also based on personal experience, here a canoeing trip Blackwood made down the Danube River in 1900 with his close friend Wilfred Wilson. The two characters land on an island in the Danube and soon experience strange feelings, in particular a sense of doom. Among the willows they find bizarre noises and sights and are threatened by shadowlike beings. After a terrifying pursuit, the men manage to escape from these beings, which are evidently from another realm or dimension. H. P. Lovecraft and others have acclaimed "The Willows" as one of the best weird tales in the field.

The title story is also noteworthy. Another story of a haunted house, it breaks away from established conventions of the ghost story through its narration in the form of a diary, which helps Blackwood in describing the character's psychological state convincingly. The volume also includes "The Insanity of Jones (A Study in Reincarnation)," which tells of how a man who learns that a coworker had once tortured and killed him in their former lives achieves his revenge four centuries later.

Blackwood's next book, *John Silence: Physician Extraórdinary* (1908), consists of five related stories about a type of character commonly described as a psychic detective–a combination of the detective figure of crime fiction with the investigator of occult phenomena found in much weird fiction, from Joseph Sheridan Le Fanu's Dr. Hesselius to William Hope Hodgson's Carnacki the Ghost Finder. For Blackwood's readers the combination worked, and the success of the book, which became a best-seller, enabled Blackwood to abandon his business ventures and dedicate himself to a full-time career as a writer.

John Silence occasionally offers rational explanations for seemingly supernatural events–as in "Ancient Sorceries," about the residents of a French village who appear to change into catlike creatures each night–but usually the suggestion is present that the phenomena he investigates truly are beyond the purview of rational scientific understanding. Examples include "A Psychical Invasion," in which Silence is called upon to exorcise a haunted house, and "Secret Worship," in which he must rescue a man who has gotten in over his head in dealing with satanic powers. The other stories in the volume include "The Nemesis of Fire," about unusual ritual sacrifices, and "The Camp of the Dog," a tale of lycanthropy.

Blackwood's ambiguous beliefs regarding the occult in the context of his somewhat religious views about nature receive further treatment in his next two books, *Jimbo: A Fantasy* and *The Education of Uncle Paul.* Both were published in 1909, the year Blackwood moved to Bôle, Switzerland, where he remained for six years.

Jimbo is about a boy whose spirit becomes trapped in an abandoned house following a head injury and his efforts to escape. In the process his spirit learns to fly, and the book contains captivating accounts of flight. Children are also important in *The Education of Uncle Paul,* about a man prone to mystical experiences who comes to Canada to care for his sister's children and instead joins with them to see the world through the magical eyes of a child. This results in some fantastic adventures with the children, who are more in touch with the numinous reality of life and thus can experience things most adults cannot. For instance, at one point in the novel Paul and the children sink into the earth and witness the movement of the winds in their underground origin:

JOHN SILENCE

Physician Extraordinary

BY

ALGERNON BLACKWOOD

AUTHOR OF

"THE LISTENER" "THE EMPTY HOUSE" ETC.

LONDON

EVELEIGH NASH

FAWSIDE HOUSE

1908

Title page for Blackwood's third story collection, which features a psychic detective who investigates supernatural phenomena (courtesy of the Lilly Library, Indiana University)

> The winds moved in their sleep, and awoke. In loops, folds, and spirals of indescribable grace they slowly began to unwrap themselves from the tree stems with a million little delicate undulations; like thin mist trembling, and then smoothing out the ruffled surface of their thousand serpentine eddies, they slid swiftly upwards from the moss and ferns, disentangled themselves without effort from roots and stones and bark, and then, reinforced by countless thousands from the lower branches, they rose up slowly in vast coloured sheets towards the region of the tree tops.

This passage also provides a representative example of the gracefulness of Blackwood's mature style.

The Lost Valley and Other Stories (1910) includes one of Blackwood's best-known works, "The Wendigo." Inspired by his experiences in the woods of Canada, the story is about a mysterious beast and the protagonist's growing awareness of its presence in the forest. The story is widely praised for how it masterfully builds suspense and how it creates a sense of

'The souls of the unhappy dead find . . . peace'
Page 46

'Edward . . . had suffered an incredible accession to his existing personality'
Page 208

Illustrations by W. Graham Robertson in the first edition of Blackwood's The Lost Valley and Other Stories *depicting scenes in "The Lost Valley" (left) and "The Terror of the Twins" (right)*

strangeness. Other stories in the volume include "The Lost Valley," in which the supernatural is simultaneously threatening and consolatory, and "The Terror of the Twins," a classic example of Blackwood's ability to convey the psychology of fear.

The Human Chord (1910) is based on an extremely fantastic premise: the efforts of one man to find four people who, through manipulating their voices together, can create a "human chord" capable of changing matter. In Blackwood's thinking the natural and the spiritual are interconnected, so conceivably certain sounds could have marvelous results. In addition to the strangeness of the concept, the novel gains its interest in the conflict between those who wish to control the universe and those who prefer to respect nature by leaving it alone.

While Blackwood wrote what is often called weird fiction and is sometimes labeled as a writer of horror fiction, in much of his work what he wanted to evoke was not horror but awe, and this purpose is evident in *The Centaur* (1911). Also based on personal experience–in this case a trip to the Caucasus Mountains in 1910–the novel describes the protagonist's mystical experience, part of which includes a vision of centaurs, while in the Caucasus Mountains with a mysterious pair of Russians who are more than they seem. This experience does not lead the protagonist to reject rationality, since instinct should be tempered by one's higher faculties, but it does cause him to reject modern civilization, which generally views experiences such as his with skepticism and scorn.

Pan's Garden: A Volume of Nature Stories (1912) is a representative indication of Blackwood's preoccupations, and it includes one of his best-known stories, "The Man Whom the Trees Loved." The story is about a man who enters into a literal communion with the trees on his estate; this horrifies his wife, but he not only accepts it but welcomes it. In the end, he is indistinguishable from the forest–a powerful symbol of Blackwood's ideas about humanity's spiritual connection with the natural world. These ideas are evident as well in "Special Delivery," in which a man who possesses a strong awareness of the power of nature is saved from a landslide that otherwise could have killed him. The powers of nature are also illustrated in stories such as "The Transfer," "The Temptations of the Clay," and "Sand"; in all of these nature is depicted almost as a sentient force. *Pan's Garden* also includes stories that are concerned more with supernatural experiences than with nature, such as "The Attic," "Clairvoyance," and "The Return."

Blackwood returned to the subject of childhood in his sequel to *The Education of Uncle Paul,* titled *A Prisoner in Fairyland (The Book That Uncle Paul Wrote)* (1913). Again the uncle joins with the children in experiencing magical events, such as retrieving stardust on the Starlight Express. Its success led to its adaptation as the play *The Starlight Express* by Blackwood and Violet Pearn in 1915. (Despite good reviews and music by Sir Edward Elgar, however, the play itself was not successful.)

In his collection *Ten Minute Stories* (1914) Blackwood returns to the figure of the centaur in "Imagination," where he writes, "His Centaur was to stand for instinct (the animal body close to Nature) combined with, yet not dominated by, the upright stature moving towards deity." Here, as in *The Centaur* and elsewhere in his work, Blackwood presents this combination of rationality and spiritual openness as ideal. Hence characters in this collection deal with such strange phenomena as astral life after death ("The Deferred Appointment") and extrasensory powers ("You *May* Telephone from Here").

Another story in the collection, "Entrance and Exit," involves the phenomenon of people disappearing into and returning from another dimension. The place where this happens is said to be haunted, but as a young clergyman discovers, the place is actually a portal to another realm. He disappears also and emerges months later, obviously having had a kind of ecstatic, indescribable experience.

Spirituality is also prominent in the stories collected in *Incredible Adventures* (1914), in particular the difference between genuine religiosity and the conventions of organized religion. For instance, "The Damned" is about a haunted house filled with souls whose damnation has resulted from the religious intolerance of the previous residents of the house. In "The Sacrifice" a man who has endured a series of devastating events is close to rejecting his spirituality when an unusual encounter with two people convinces him of its worth. These themes are also present in "The Regeneration of Lord Ernie," which concerns a man's recognition of humanity's innate spirituality and power and its connection with nature through his participation in a wild pagan ritual, and "A Descent into Egypt," which explores a similar spiritual growth. While little happens in such stories, they succeed in establishing a mood of fantastic strangeness and wonder along with an impression of psychological verisimilitude.

Blackwood again returned to childhood as a subject in *The Extra Day* (1915) and *The Promise of Air* (1918), but neither is among his best books. Like the Uncle Paul books, *The Extra Day* is about an open adult who participates with children in their imaginative adventures, here their experiences outside of time. *The Promise of Air,* on the other hand, includes much speculation about psychology and spirituality but is more realistic than fantastic.

Other novels Blackwood completed during World War I include *Julius Le Vallon: An Episode* (1916), *The Wave: An Egyptian Aftermath* (1916), and *The Garden of Survival* (1918). Both *Julius Le Vallon* and *The Wave* are stories involving reincarnation and love triangles–perhaps a reflection of his relationship at this time with Mabel "Maya" Stuart-King, who with her husband had accompanied Blackwood during his first visit to Egypt in 1912. Their relationship remained close if platonic even as she was married to others. Also, during the war Blackwood volunteered as a writer of war propaganda, worked in Europe as a war correspondent for the London *Times,* and even did espionage for British intelligence. (He was fluent in both French and German.)

Blackwood published another collection of short fiction, *Day and Night Stories* (1917). As with all of Blackwood's collections, most of the stories are supernatural in nature, but as is often the case with the weird tale, sometimes science-fiction elements are present. This is true of "A Victim of Higher Space," which is also notable for the return of John Silence. The story draws on the speculations about the fourth dimension of the American logician C. H. Hinton and concerns, like "Entrance and Exit" in *Ten Minute Stories,* disappearances into another dimension. While this is almost mystical in "Entrance and Exit," here the experience is terrifying, because not only does the character Silence is trying to help have no control of when this happens (the shift is caused by a type of "vibration" rather than passage through a dimensional portal), but also he has horrifying visions, similar to those depicted in Edwin A. Abbott's *Flatland: A Romance of Many Dimensions* (1884), of worlds of even more dimensions. After the character reappears in India, Silence has figured out how to prevent further disappearances.

Frontispiece for Blackwood's Pan's Garden: A Volume of Nature Stories *(illustration by W. Graham Robertson)*

Another piece in *Day and Night Stories,* "Cain's Atonement," reflects Blackwood's experiences in the war. Another story featuring reincarnation, it tells of how a soldier has a vision of a debt incurred in a previous life to his cousin, whom he saves in battle by sacrificing his life.

In 1917 Blackwood's older brother died, and the author attempted to deal with his grief in the novella *The Garden of Survival.* Here the story of grief concerns a man who loses his wife after one month of marriage; however, his grief is transfigured by the fact that he continues to experience her presence. The survival of the personality after death is a common theme in Blackwood's work, from his early ghost stories to three works–"A Desert Episode," "The Occupant of the Room," and "Transition"–in *Day and Night Stories,* to his later fiction.

After World War I Blackwood returned to England and began staying with a series of different friends. At the home of Henry Ainley, an actor, he met and befriended Lord Dunsany, his eminent contemporary fantasist. His friendship with Ainley was more influential, however. Blackwood had long been interested in the theater and with Violet Pearn had adapted some of his fiction–*A Prisoner in Fairyland* as *The Starlight Express* and *The Wave* as *Karma: A Reincarnation Play* (1918)–and his acquaintance with Ainley inspired him to write other plays. These include *The Crossing* (1920), about people who return from the dead; *Through the Crack* (1920), another collaboration with Pearn, which was based largely on *The Education of Uncle Paul;* and *White Magic* (1921), about a witch who helps a childless couple achieve their dream.

Another friendship, with Wilfred Wilson, led to the publication of another collection, *The Wolves of God, and Other Fey Stories* (1921), on which the two collaborated. A unifying theme in these stories is animal nature–either within wild animals or in people. For instance, the title story, "Running Wolf," and "The Valley of the Beasts" all concern animals or animal instincts, while in "The Empty Sleeve" a man's violent passions take feline form. Such passions are also dealt with in "First Hate" and "Vengeance Is Mine."

Another expression of Blackwood's belief that one should possess both reason and instinct in balance can be found in his novel *The Bright Messenger* (1921), a sequel to *Julius Le Vallon.* One character in the novel says, "If we lived closer to nature we might get better results. . . . Primitive man, I'm convinced, did get certain results, but he was a poor instrument. Modern man, in some ways, is a better, finer instrument to work through, only he is blind to the existence of any beings but himself. A bridge, however, might be built, I feel." Despite such optimism, however, the tone of this novel is darker than much of Blackwood's work, perhaps a reflection of the general sense of deflation following the war. Just as T. S. Eliot described such a shift in *The Waste Land* (1922) and other writers indicated it in various postwar works, so one character in *The Bright Messenger* says, "The war, above all the Peace, shook his optimism. If it did not wholly shatter his belief in human progress, it proved such progress to be so slow that his Utopia faded into remotest distance, and his dream of perfectibility became the faintest star in his hitherto bright sky of hope."

Blackwood published an autobiographical volume, *Episodes Before Thirty,* in 1923 and another collection, *Tongues of Fire and Other Sketches,* in 1924. One of the stories in this volume, "The Pikestaffe Case," is yet another approach to the concept of other dimensions. A combination scientist-necromancer experiments with reaching a fourth dimension, here envisioned as both a physical plane and a higher state of consciousness. A different type of consciousness is portrayed in "The Other Woman," in which a man believes that his mother resides within him as an "inner woman."

In addition to the autobiography and his fiction, in the 1920s Blackwood began to devote much of his time to writing articles and reviews. Later, in the 1930s and 1940s, he would also spend a great deal of time writing and reading for BBC radio broadcasts, and later yet he was a frequent guest on British television programs.

In the late 1920s Blackwood also devoted much of his energies to writing for children. He wrote *Sambo and Snitch* (1927), a children's novel, for Ainley's children, Patsy and Henry Jr., and Lady Cynthia Asquith convinced him to contribute to the anthologies she edited for children and adults. One of the best efforts from this period is the children's fantasy novel *Dudley and Gilderoy: A Nonsense* (1929), in which a cat and a parrot sneak out of their house and travel by train from the country to London. The book was a great success and led to a few sequels. Blackwood's last novel for children, *The Fruit Stoners: Being the Adventures of Maria among the Fruit Stoners* (1934), is about a young girl who witnesses the animation of the pits of different fruits.

Even as Blackwood's productivity began to diminish slightly in the late 1920s and the early 1930s, his reputation continued to expand, and many of his earlier works were republished in volumes such as *Ancient Sorceries and Other Tales* (1927), *The Dance of Death and Other Tales* (1927), *Strange Stories* (1929), and *The Willows and Other Queer Tales* (1932). His collection *Shocks* (1935) collects stories published between 1925 and 1935. One of these is "The Man Who Lived Backwards," about an unhappily married man who has a psychic experience in which he has the chance to influence his younger self and thus prevent his marriage. However, an inner voice tells him that to do so would be to create a devastating change in the workings of fate—a mystical variation on a common theme in science-fiction time-travel stories. Instead, some time after he returns from this experience his wife leaves him for another man, leaving him free to pursue his true love. "Full Circle" also shows a character encountering an earlier self.

Some of the tales in *Shocks* are realistic horror stories, but many have a supernatural element. "Elsewhere and Otherwise" and "The Adventures of Tornado Smith" both deal with experiences in other dimensions or out of time; "A Threefold Cord" is another reincarnation story; and "Dr. Feldman" and "Chemical" are both ghost stories.

An illustration for the British edition of Blackwood's Dudley and Gilderoy: A Nonsense, *a fantasy novel for children about the travels of a parrot and a cat*

Though Blackwood wrote less as he grew older, he remained active. In addition to his participation in radio and television broadcasts, he continued to travel. For instance, in 1936 he engaged in a walking tour of the Continent that took him through Austria, Italy, France, and Spain, where he witnessed some fighting in the Spanish civil war.

When World War II began in 1939 Blackwood again volunteered his services as a spy; he was turned down, which was hardly surprising since he was now seventy. He continued to write for radio, however, and his life was impacted by the war directly when, in October 1940 during the Battle of Britain his nephew's house in Hempstead, where he was staying, was hit. Blackwood was not present at the time, but many of his papers were destroyed.

During the war Blackwood also wrote a handful of stories. Two of these, "The Doll" and "The Trod," were published in 1946 by Arkham House as *The Doll, and One Other*. "The Doll" is a tale of revenge achieved through a doll magically invested with strange powers, and the plot of "The Trod" concerns a woman whose mother years ago disappeared into fairyland. The story focuses on whether her daughter will follow her. In 1949, at age eighty, Blackwood was honored as a Commander of the British Empire. The following year he suffered a mild stroke, and he died of cerebral thrombosis and arteriosclerosis on 10 December 1951.

Blackwood's work is of great historical significance in the fields of fantasy fiction, horror, and the weird tale. Despite the contemporary interest in these subgenres, however, many readers are unfamiliar with his work. Critical attention to Blackwood has been limited, although scholars such as Mike Ashley and S. T. Joshi have done much to promote his reputation. Given his status as one of the major writers of fantastic fiction in the first half of the twentieth century, it is likely that Blackwood will continue to be read and studied for years to come.

Bibliographies:

John Robert Colombo, *Blackwood's Books* (Toronto: Hounslow Press, 1981);

Mike Ashley, *Algernon Blackwood: A Bio-Bibliography* (New York, Westport, Conn. & London: Greenwood Press, 1987).

Biographies:

Maude M. C. ffoulkes, "Algernon Blackwood," in her *My Own Past* (London: Cassell, 1915), pp. 186–202;

Mike Ashley, *Algernon Blackwood: A Bio-Bibliography* (New York, Westport, Conn. & London: Greenwood Press, 1987).

References:

Mike Ashley, "The Road to *The Empty House*," *Fantasy Commentator,* 5 (Fall 1985): 166–175;

Hilaire Belloc, Review of *John Silence: Physician Extraordinary, Morning Post,* 17 September 1908, p. 2; reprinted in *Horror: 100 Best Books,* edited by Stephen Jones and Kim Newman (New York: Carroll & Graf, 1988), pp. 67–68;

Everett F. Bleiler, *The Guide to Supernatural Fiction* (Kent, Ohio: Kent State University Press, 1983), pp. 51–59;

Bleiler, Introduction to *Best Ghost Stories of Algernon Blackwood* (New York: Dover, 1973), pp. v–x;

Julia Briggs, "A Scientific Spirit: Mesmerism, Drugs and Psychic Doctors," in her *Night Visitors: The Rise and Fall of the English Ghost Story* (London: Faber & Faber, 1977), pp. 59–63;

Stuart Gilbert, "Algernon Blackwood: Novelist and Mystic," *Transition* (Paris), 23 (July 1935): 89–96;

Derek Hudson, "A Study of Algernon Blackwood," in *Essays and Studies,* new series, volume 14 (London: John Murray, 1961), pp. 102–114;

S. T. Joshi, "Algernon Blackwood: The Expansion of Consciousness," in his *The Weird Tale: Arthur Machen, Lord Dunsany, Algernon Blackwood, M. R. James, Ambrose Bierce, H. P. Lovecraft* (Austin: University of Texas Press, 1990), pp. 87–132;

Russell Letson Jr., "The Approaches to Mystery: The Fantasies of Arthur Machen and Algernon Blackwood," dissertation, Southern Illinois University, August 1975;

H. P. Lovecraft, *Supernatural Horror in Literature* (New York: Abramson, 1945);

Peter Penzoldt, "Algernon Blackwood," in his *The Supernatural in Fiction* (London: Nevill, 1952), pp. 228–253;

David Punter, "Algernon Blackwood," in *Supernatural Fiction Writers: Fantasy and Horror,* volume 1, edited by Bleiler (New York: Scribners, 1985), pp. 463–470;

Dorothy Scarborough, *The Supernatural in Modern English Fiction* (New York: Putnam, 1917);

Jack Sullivan, "The Visionary Ghost Story: Algernon Blackwood," in his *Elegant Nightmares: The English Ghost Story from Le Fanu to Blackwood* (Athens: Ohio University Press, 1978), pp. 112–129;

Edward Wagenknecht, "Algernon Blackwood," in his *Seven Masters of Supernatural Fiction* (New York: Greenwood Press, 1991), pp. 69–94.

Papers:

Collections of Blackwood's papers are located at the BBC Written Archives at Caversham, Reading, Berkshire; the BBC Play Library, Broadcasting House, London; the British Library, London; the Huntington Library, San Marino, California; Special Collections, Boston College Library; Berg Collection, New York Public Library; the August Derleth Collection, State Historical Society of Wisconsin, Madison; and the Thomas Fisher Rare Book Library, University of Toronto. Mike Ashley of Chatham, Kent, possesses a substantial collection of Blackwood's letters.

Lewis Carroll
(Charles Lutwidge Dodgson)

(27 January 1832 – 14 January 1898)

Darren Harris-Fain
Shawnee State University

See also the Carroll entries in *DLB 18: Victorian Novelists After 1885* and *DLB 163: British Children's Writers, 1800–1880.*

Lewis Carroll, holding a camera lens

BOOKS: *A Syllabus of Plane Algebraical Geometry, Part I,* as Charles Lutwidge Dodgson (Oxford & London: Parker, 1860);

The Formulae of Plane Trigonometry, as Dodgson (Oxford & London: Parker, 1861);

A Guide to the Mathematical Student, Part I, as Dodgson (Oxford: Parker, 1864);

The Dynamics of a Particle, anonymous (Oxford: Vincent, 1865);

Alice's Adventures in Wonderland (London: Macmillan, 1865; New York: Appleton, 1866);

An Elementary Treatise on Determinants, as Dodgson (London: Macmillan, 1867);

The Fifth Book of Euclid Treated Algebraically, as Dodgson (Oxford & London: Parker, 1868);

Phantasmagoria and Other Poems (London: Macmillan, 1869);

Through the Looking-Glass, and What Alice Found There (London: Macmillan, 1871; Boston: Lee & Sheppard / New York: Lee, Sheppard & Dillingham, 1872);

The New Belfry of Christ Church, Oxford, as D. C. L. (Oxford: Parker, 1872);

The Vision of the Three T's, as D. C. L. (Oxford: Parker, 1873);

The Blank Cheque: A Fable, as D. C. L. (Oxford: Parker, 1874);

Suggestions as to the Best Method of Taking Votes, as C. L. D. (Oxford: Hall & Stacy, 1874); revised as *A Method of Taking Votes on More Than Two Issues,* anonymous (Oxford, 1876);

The Hunting of the Snark: An Agony in Eight Fits (London: Macmillan, 1876; Boston: Osgood, 1876);

Euclid and His Modern Rivals, as Dodgson (London: Macmillan, 1879; New York: Dover, 1973);

Doublets: A Word-Puzzle (London: Macmillan, 1879);

Rhyme? And Reason? (London: Macmillan, 1883);

Supplement to "Euclid and His Modern Rivals," as Dodgson (London: Macmillan, 1885);

A Tangled Tale (London: Macmillan, 1885);

Thirty Years in a Curatorship, by One Who Has Tried, as Dodgson (Oxford: Baxter, 1886);

The Game of Logic (London & New York: Macmillan, 1886);

Alice's Adventures Under Ground: Being a Facsimile of the Original Ms. Book Afterwards Developed into "Alice's Adventures in Wonderland" (London & New York: Macmillan, 1886);

Curiosa Mathematica, Part I: A New Theory of Parallels, as Dodgson (London: Macmillan, 1888);

The Nursery "Alice" (London: Macmillan, 1889; New York: Dover, 1966);

Sylvie and Bruno (London & New York: Macmillan, 1889);

Eight or Nine Wise Words about Letter-Writing (Oxford: Emberlin, 1890);

Syzygies and Lanrick: A Word-Puzzle and a Game (London: Clay, 1893);

Curiosa Mathematica, Part II: Pillow-Problems Thought Out during Sleepless Nights, as Dodgson (London: Macmillan, 1893);

Sylvie and Bruno Concluded (London & New York: Macmillan, 1893);

Symbolic Logic, Part I: Elementary, as Dodgson (London & New York: Macmillan, 1896);

The Lewis Carroll Picture Book: A Selection from the Unpublished Writings and Drawings of Lewis Carroll, Together with Reprints from Scarce and Unacknowledged Work, edited by Stuart Dodgson Collingwood (London: Unwin, 1899); republished as *Diversions and Digressions* (New York: Dover, 1961);

Feeding the Mind (London: Chatto & Windus, 1907);

Further Nonsense Verse and Prose, edited by Langford Reed (London: Unwin, 1926; New York: Appleton, 1926);

The Collected Verse of Lewis Carroll (London: Macmillan, 1932; New York: Macmillan, 1933); republished as *The Humorous Verse of Lewis Carroll* (New York: Dover, 1960);

For the Train, edited by Hugh J. Schonfield (London: Archer, 1932);

The Rectory Umbrella and Mischmasch (London: Cassell, 1932; New York: Dover, 1971);

The Russian Journal and Other Selections from the Works of Lewis Carroll, edited by John Francis McDermott (New York: Dutton, 1935);

The Complete Works of Lewis Carroll (London: Nonesuch, 1939; New York: Modern Library, 1939);

Lewis Carroll, Photographer, edited by Helmut Gernsheim (London: Parrish, 1949; New York: Dover, 1969);

The Diaries of Lewis Carroll, 2 volumes, edited by Roger Lancelyn Green (London: Cassell, 1953; New York: Oxford University Press, 1954);

Useful and Instructive Poetry (London: Bles, 1954; New York: Macmillan, 1954);

Symbolic Logic and The Game of Logic (New York: Dover, 1958; London: Constable, 1959);

Pillow Problems and A Tangled Tale (New York: Dover, 1958; London: Constable, 1959);

Lewis Carroll Observed: A Collection of Unpublished Photographs, Drawings, Poetry, and New Essays, edited by Edward Guiliano (New York: Potter, 1976);

The Wasp in the Wig: A Suppressed Episode of Through the Looking-Glass and What Alice Found There, edited by Guiliano and Martin Gardner (New York: Lewis Carroll Society of North America, 1978);

The Complete Illustrated Works of Lewis Carroll, edited by Guiliano (New York: Avenel, 1982).

Editions: *The Annotated Snark,* edited by Martin Gardner (New York: Simon & Schuster, 1962);

The Annotated Alice: Alice's Adventures in Wonderland *and* Through the Looking-Glass, revised edition, edited by Gardner (New York: Potter, 1970; Harmondsworth, U.K.: Penguin, 1970);

Alice in Wonderland: Authoritative Texts of Alice's Adventures in Wonderland, Through the Looking-Glass, The Hunting of the Snark, *Backgrounds, Essays in Criticism,* edited by Donald J. Gray (New York: Norton, 1971);

Lewis Carroll's Symbolic Logic: Symbolic Logic by Lewis Carroll: Part I, Elementary, 1896, Fifth Edition; Part II: Advanced, Never Previously Published; Together with Letters from Lewis Carroll to Eminent Nineteenth-Century Logicians and to His "Logical Sister," and Eight Versions of the Barber-Shop Paradox, edited by William Warren Bartley III (New York: Potter, 1977);

Lewis Carroll's The Hunting of the Snark, edited by James Tanis and John Dooley (Los Altos, Cal.: Kaufmann, 1981).

Few writers of fantasy have managed to permeate their own cultures as did Charles Lutwidge Dodgson, a mathematician and amateur photographer who wrote children's books under the pseudonym Lewis Carroll. Along with memorable characters who have taken on lives beyond their fictional sources–such as King Arthur and the knights of the Round Table, Robin Hood and his Merry Men, Frankenstein's monster, Dracula, and Sherlock Holmes, among others–are such Carroll creations as the Mad Hatter, the Red Queen, and of course Alice. People who have never read either of Carroll's best-known books are nonetheless somewhat familiar with the adventures of the little girl who fell down a rabbit hole and passed through a looking glass and found "curiouser and curiouser" things.

Charles Lutwidge Dodgson was born on 27 January 1832 in Daresbury, Cheshire, the oldest son of the Reverend Charles Dodgson and Frances Jane Lutwidge Dodgson. He was the third of eleven children, and his ability to entertain children likely began with his younger siblings, for whom he invented games with nonsense elements that anticipate his fiction. In his early childhood Dodgson was educated at home. His father became rector at another parish in 1843, prompting the family's move to Croft in Yorkshire. The following year Dodgson began attending Richmond Grammar School, and in 1846 he started studies at Rugby. He was a good student but was unhappy in the public-school environment. In 1850, the year of his mother's death, he entered Christ Church, Oxford University, where he earned first-class honors in mathematics and second-class honors in classics before graduating in 1854. His logical and linguistic skills would become evident with the Alice books. In the middle of his college studies he had been granted a Studentship; it would provide a lifelong living, but to keep it Dodgson was required to take holy orders and remain celibate–neither of which he apparently found difficult.

In 1855 Dodgson was appointed a fellow at Christ Church and began lecturing on mathematics, the start of a long and fairly uneventful academic life. He completed his master's degree in 1857, and in 1861 he was ordained a deacon of the Church of England, making him officially the Reverend Charles Lutwidge Dodgson. His stammer, which he shared with many of his brothers and sisters, may have decided him against following his father into the next step; there were likely other personal reasons as well, such as his love of the theater. Still, occasionally Dodgson was called upon to officiate at religious ceremonies such as baptisms and funerals.

A lifelong bachelor, Dodgson devoted his time not spent in academic pursuits to reading, writing, and taking photographs, a hobby he took up starting in 1856. He admired many contemporary writers of the day, including Alfred, Lord Tennyson and William Makepeace Thackeray, and he often persuaded writers to be photographed. Dodgson also befriended young girls in the Oxford community, often making them the subjects of his photographs. He was at ease with children, joking with them and entertaining them with stories, songs, puzzles, and games. The same could not be said for Dodgson as a lecturer: his students routinely found him clear and knowledgeable but dry if not boring. Certainly the titles of his early mathematical publications give no indication of Dodgson's wit.

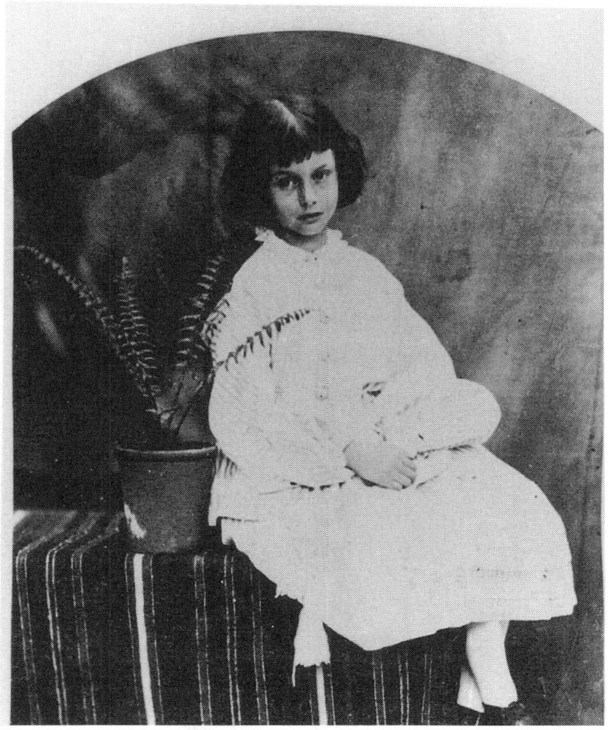

Alice Liddell, 1859

The origins of Carroll's Alice books are well known. Dodgson, whose interest in young girls was pronounced though proper, was particularly interested in a girl named Alice Liddell, who was nine years old when he began telling her a fantastic story about another girl named Alice. The event was a boat trip and picnic with Robinson Duckworth, a canon, and Alice, Lorina, and Edith Liddell, daughters of the dean of Christ Church who permitted Dodgson's ordination as deacon, on 4 July 1862. The children begged him to write the extemporaneous story down, and this served as the basis for his two pseudonymous books about his fictional Alice. As for the pseudonym itself, he first used it in 1856 in parodies published in various periodicals. He created it by transposing his first and middle names (Lutwidge Charles), translating them into Latin (Ludovicus Carolus), and then anglicizing them differently (Lewis Carroll). Dodgson completed the manuscript of his first novel, which he had titled "Alice's Adventures Under Ground," in 1864; at the urging of George MacDonald's children, it was published the following year as *Alice's Adventures in Wonderland*. Though many others have illustrated the Alice books, the best-known illustrations are those by John Tenniel that appeared in their original publications. Dodgson was very much a detail person and a perfectionist (he either threw away poorly printed copies of *Alice's Adventures in Wonderland* or sent them

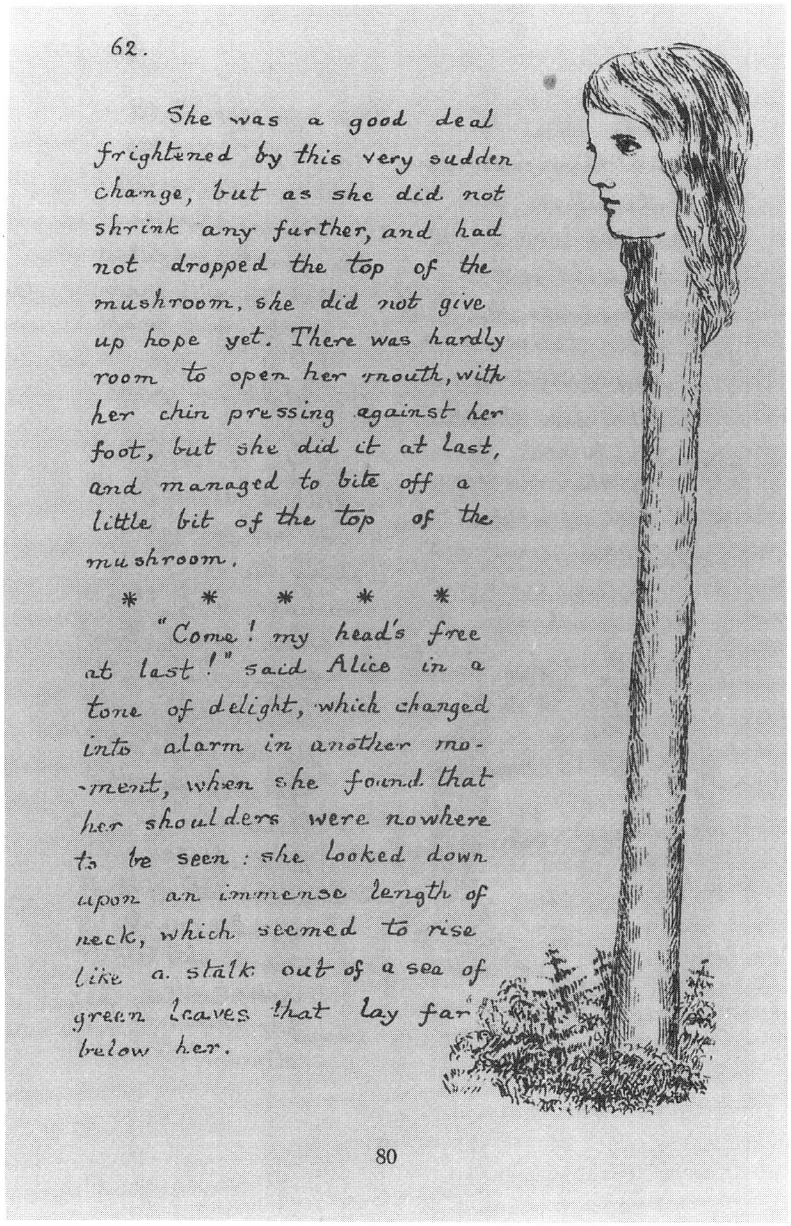

62.

She was a good deal frightened by this very sudden change, but as she did not shrink any further, and had not dropped the top of the mushroom, she did not give up hope yet. There was hardly room to open her mouth, with her chin pressing against her foot, but she did it at last, and managed to bite off a little bit of the top of the mushroom.

* * * * *

"Come! my head's free at last!" said Alice in a tone of delight, which changed into alarm in another mo-ment, when she found that her shoulders were nowhere to be seen: she looked down upon an immense length of neck, which seemed to rise like a stalk out of a sea of green leaves that lay far below her.

80

Page from "Alice's Adventures Under Ground," a manuscript book that Carroll presented to Alice Liddell, the basis for Alice's Adventures in Wonderland *(1865) (from* Alice's Adventures Under Ground: Being a Facsimile of the Original Ms. Book Afterwards Developed into "Alice's Adventures in Wonderland," *1886)*

to America), and Tenniel refused to work for him after *Through the Looking-Glass, and What Alice Found There* (1871).

The story of *Alice's Adventures in Wonderland* is also well known. A girl named Alice follows a White Rabbit underground, where she encounters a strange magical world in which, among other things, she meets the baffling Cheshire Cat and the oracular Caterpillar, attends a mad tea party, plays an antic game of croquet with the Queen of Hearts, and visits the trial of the Knave of Hearts for stealing the tarts. To summarize the plot, however, is to miss the rich playfulness of Carroll's narrative, with its bizarre characters and events, its puzzles and linguistic playfulness, and its combined sense of fun and fright.

Dodgson was a quiet, religious man, always somewhat embarrassed by his fame as Lewis Carroll (even to the point of returning any mail addressed to him as Lewis Carroll), but he had a rich sense of humor, as shown not only in his writings as Lewis Carroll but also in a story related to *Alice's Ad-*

ventures in Wonderland. After its publication he was invited to meet Queen Victoria, who exclaimed over the book and asked Dodgson to send her his next effort. This he dutifully did, and the queen received an inscribed copy of a book on mathematical determinants. Needless to say, she was not amused.

In 1867 Dodgson traveled to Russia with his friend Henry Liddon on his only trip beyond his familiar English surroundings. He recorded his responses to his travels in his journal, which have since been published in *The Russian Journal and Other Selections from the Works of Lewis Carroll* (1935), edited by John Francis McDermott. The following year his father died, which Dodgson took hard.

In 1869 Dodgson published another book as Lewis Carroll, *Phantasmagoria and Other Poems.* It was not as successful, though, as either *Alice's Adventures in Wonderland* or his next Lewis Carroll book, *Through the Looking-Glass, and What Alice Found There,* which, like *Alice's Adventures in Wonderland,* was both incredibly popular and critically acclaimed.

In this book Alice enters another world not by going down a rabbit hole but by passing through a mirror. The world she finds there, however, is not simply a reflection of the one she knows, for it is just as strange and magical as Wonderland. Here she meets characters such as Tweedledee and Tweedledum, Humpty Dumpty, and the White Knight along with the players in a chess match that provides the organizing principle of the story. Also, *Through the Looking-Glass,* like *Alice's Adventures in Wonderland,* contains poems–most notably Carroll's best-known poem, "Jabberwocky," and "The Walrus and the Carpenter."

Several qualities of *Alice's Adventures in Wonderland* and *Through the Looking-Glass* have contributed to their longevity. First is the books' success as fantasy: through Carroll's prodigious imagination his fantastic worlds come to life, vibrantly filled as they are with bizarre places, people, and happenings, all seen through the experience of a girl who is sensible enough to find these things silly and sometimes scary but who is young enough to take pretty much everything in stride as it comes. Supporting the fantasy is Carroll's playfulness with language. Second only to Edward Lear in the creation of nonsense words and verse, in his Alice books and elsewhere Carroll manages to intersperse delightfully nonsensical words into his stories. This succeeds in large part because, as Humpty Dumpty points out to Alice in his explication of "Jabberwocky," the words *do* make sense in a sort of distorted logic. Such frequently perverse games with logic in the books are yet another reason for why readers, young and old, have enjoyed Carroll's tales about Alice for more than a century.

Another Carroll work, *The Hunting of the Snark: An Agony in Eight Fits,* was published in 1876. Like the Alice books, it relies heavily on interesting characters and situations and on the use of nonsense. In this long comic poem ten figures in search of adventure hunt for a beast called a Snark; hunting the Snark, however, is difficult–plus it might end up being a Boojum. The seeming silliness of it all is part of the appeal, of course, but neither does the silliness obviate Carroll's wit.

In the 1880s Dodgson became increasingly more private. In 1880 he abandoned photography, and the following year, at age forty-nine, he resigned his lectureship in mathematics to devote himself to his studies and his writing. Although he supported his six unmarried sisters and helped to pay for relatives' education, he did not need much money himself (in fact, in 1880 he had requested that his salary be lowered since he was not teaching that much), and he apparently did not enjoy lecturing. However, he then accepted an appointment as curator of the Senior Common Room at Christ Church in 1882, a position he resigned ten years later.

Other works Dodgson published as Lewis Carroll include *Doublets: A Word-Puzzle* (1879); *Rhyme? And Reason?* (1883), a collection of humorous poems; *A Tangled Tale* (1885), which includes mathematical problems in the form of short stories; *The Game of Logic* (1886), which is more serious than its title indicates; *Eight or Nine Wise Words about Letter-Writing* (1890); and *Syzygies and Lanrick: A Word-Puzzle and a Game* (1893). In 1886 he also published a facsimile edition of his original manuscript for his first novel, the profits from which he donated to children's hospitals, and in 1889 he produced a version of his Alice books called *The Nursery "Alice"* for preschool readers.

Dodgson's other extended effort as Lewis Carroll appeared as two related books, *Sylvie and Bruno* (1889) and *Sylvie and Bruno Concluded* (1893). Readers and critics generally find these works less appealing than the Alice books and *The Hunting of the Snark,* probably because they are more deliberately moralistic than Carroll's earlier fantasies. While the early books had avoided the Victorian tendency to use children's literature primarily to teach values, the Sylvie and Bruno books are hurt by this tendency. In these books he uses fantasy–here the elements of fairy romance–not for their own sake but for allegorical purposes.

Dodgson died on 14 January 1898 of a bronchial infection while visiting some of his sisters in

Guildford, Surrey. That same year Alice Liddell Hargreaves, once the girl who inspired the Alice books and to whom Dodgson had given his illustrated manuscript "Alice's Adventures Under Ground" in 1864, had the manuscript auctioned for £15,400. At the time it was the highest price ever paid for a book in an auction in Britain; the same buyer acquired a Shakespeare First Folio for less.

Lewis Carroll's masterpieces, *Alice's Adventures in Wonderland* and *Through the Looking-Glass,* have not only endured in their own right. They have also been adapted many times on stage and screen; they have been alluded to in countless literary works, movies, television shows, and songs (even in the 1960s rock song "White Rabbit" by Jefferson Airplane); and they have inspired many other works of fantasy and science fiction, directly or indirectly. They have never gone out of print since their first publication, and it is likely that they, and to a lesser extent *The Hunting of the Snark,* will be read and enjoyed for generations to come.

Letters:

Stuart Dodgson Collingwood, *The Life and Letters of Lewis Carroll* (London: T. Fisher & Unwin, 1898);

A Selection from the Letters of Lewis Carroll to His Child-Friends, edited by Evelyn M. Hatch (London: Macmillan, 1933);

The Letters of Lewis Carroll, 2 volumes, edited by Morton N. Cohen and Roger Lancelyn Green (New York: Oxford University Press, 1979); abridged as *The Selected Letters of Lewis Carroll,* second edition (New York: Pantheon, 1988);

Lewis Carroll and the House of Macmillan, edited by Cohen and Anita Gandolfo (New York: Cambridge University Press, 1987).

Bibliographies:

Sidney Herbert Williams and Falconer Madan, *The Lewis Carroll Handbook,* revised by Roger Lancelyn Green and Denis Crutch (Folkestone, U.K. & Hamden, Conn.: Archon, 1979);

Edward Guiliano, *Lewis Carroll: An Annotated International Bibliography 1960–77* (Charlottesville: University Press of Virginia, 1980);

Rachel Fordyce, *Lewis Carroll: A Reference Guide* (Boston: G. K. Hall, 1988).

Biographies:

Isa Bowman, *The Story of Lewis Carroll* (London: Dent, 1899);

Florence Becker Lennon, *Victoria through the Looking-Glass: The Life of Lewis Carroll* (New York: Simon & Schuster, 1945);

Derek Hudson, *Lewis Carroll* (London: Constable, 1954);

Roger Lancelyn Green, *Lewis Carroll* (London: Bodley Head, 1960; New York: Walck, 1962);

Lennon, *The Life of Lewis Carroll* (New York: Collier, 1962);

Joanna Richardson, *The Young Lewis Carroll* (London: Parrish, 1964);

James Wood, *The Snark* Was *a Boojum: A Life of Lewis Carroll* (New York: Pantheon, 1966);

John Pudney, *Lewis Carroll and His World* (London: Thames & Hudson, 1976; New York: Scribners, 1976);

Hudson, *Lewis Carroll: An Illustrated Biography* (New York: Potter, 1977);

Anne Clark, *Lewis Carroll: A Biography* (New York: Schocken, 1979);

Morton N. Cohen, *Lewis Carroll: A Biography* (New York: Knopf, 1995).

References:

W. H. Auden, "Lewis Carroll," in his *Forewords and Afterwords by W. H. Auden,* edited by Edward Mendelson (New York: Random House, 1973), pp. 283–293;

Kathleen Blake, *Play, Games, and Sport: The Literary Works of Lewis Carroll* (Ithaca, N.Y.: Cornell University Press, 1974);

Harold Bloom, ed., *Lewis Carroll: Modern Critical Views* (New York & Philadelphia: Chelsea House, 1987);

Hélène Cixous, "Introduction to Lewis Carroll's *Through the Looking Glass* and *The Hunting of the Snark,* " translated by Marie Maclean, *New Literary History,* 13, no. 2 (1982): 231–251;

Beverly Lyon Clark, *Lewis Carroll* (Mercer Island, Wash.: Starmont, 1990);

Walter de la Mare, *Lewis Carroll* (London: Faber & Faber, 1932);

John Fisher, ed., *The Magic of Lewis Carroll* (New York: Simon & Schuster, 1973);

Edward Guiliano, ed., *Lewis Carroll: A Celebration–Essays on the Occasion of the 150th Anniversary of the Birth of Charles Lutwidge Dodgson* (New York: Potter, 1982);

Guiliano, ed., *Lewis Carroll Observed: A Collection of Unpublished Photographs, Drawings, Poetry, and New Essays* (New York: Potter, 1976);

Guiliano and James R. Kincaid, eds., *Soaring with the Dodo: Essays on Lewis Carroll's Life and Art* (Charlottesville: Lewis Carroll Society of North America/University Press of Virginia, 1982);

Richard Kelly, *Lewis Carroll,* revised edition (Boston: Twayne, 1990);

John F. Lehmann, *Lewis Carroll and the Spirit of Nonsense* (Nottingham: University of Nottingham Press, 1974);

Edmund Little, *The Fantasts: Studies in J. R. R. Tolkien, Lewis Carroll, Mervyn Peake, Nicolai Gogol, and Kenneth Grahame* (London: Avebury, 1984):

Robert Phillips, ed., *Aspects of Alice: Lewis Carroll's Dreamchild as Seen through the Critics' Looking-Glasses, 1865–1971* (New York: Vanguard, 1971);

Robert M. Polhemus, "Carroll's *Through the Looking-Glass* (1871): The Comedy of Regression," in his *Comic Faith: The Great Tradition from Austen to Joyce* (Chicago: University of Chicago Press, 1980), pp. 245–293;

Stephen Prickett, "Consensus and Nonsense: Lear and Carroll," in his *Victorian Fantasy* (Bloomington: Indiana University Press, 1979), pp. 114–149;

Donald Rackin, ed., Alice's Adventures in Wonderland: *A Critical Handbook* (Belmont, Cal.: Wadsworth, 1961);

Alexander L. Taylor, *The White Knight: A Study of C. L. Dodgson (Lewis Carroll)* (Edinburgh: Oliver & Boyd, 1952).

Papers:

Lewis Carroll's papers can be found at the British Museum in London; the Castle Arch Museum in Guildford, England; the Christ Church College Library, Oxford University; the Houghton Library at Harvard University; the Photographic Division, National Portrait Gallery, London; the Berg Collection at the New York Public Library; the Pierpont Morgan Library in New York; the Morris L. Parrish Collection, Princeton University Library; the Rosenbach Foundation in Philadelphia; and the Harry Ransom Humanities Research Center, University of Texas at Austin.

G. K. Chesterton

(29 May 1874 - 14 June 1936)

Kyle W. Friedow
Kent State University

See also the Chesterton entries in *DLB 10: Modern British Dramatists, 1900–1945; DLB 19: British Poets, 1880–1914; DLB 34: British Novelists, 1890–1929: Traditionalists; DLB 70: British Mystery Writers, 1860–1919; DLB 98: Modern British Essayists, First Series;* and *DLB 149: Late Nineteenth- and Early Twentieth-Century British Literary Biographers.*

G. K. Chesterton, 1935 (photograph by Howard Coster)

BOOKS: *Greybeards at Play: Literature and Art for Old Gentlemen. Rhymes and Sketches* (London: Johnson, 1900);

The Wild Knight and Other Poems (London: Richards, 1900; revised edition, London: Dent / New York: Dutton, 1914);

The Defendant (London: Johnson, 1901; New York: Dodd, Mead, 1902; enlarged edition, London: Dent, 1903);

Twelve Types (London: Humphreys, 1902); enlarged as *Varied Types* (New York: Dodd, Mead, 1903); abridged as *Five Types* (London: Humphreys, 1910; New York: Holt, 1911); also abridged as *Simplicity and Tolstoy* (London: Humphreys, 1912);

Robert Browning (London & New York: Macmillan, 1903);

G. F. Watts (London: Duckworth / New York: Dutton, 1904);

The Napoleon of Notting Hill (London & New York: John Lane/Bodley Head, 1904);

The Club of Queer Trades (London & New York: Harper, 1905);

Heretics (London & New York: John Lane/Bodley Head, 1905);

Charles Dickens (London: Methuen, 1906; New York: Dodd, Mead, 1906);

The Man Who Was Thursday (Bristol: Arrowsmith / London: Simpkin, Marshall, Hamilton, Kent, 1908; New York: Dodd, Mead, 1908);

All Things Considered (London: Methuen, 1908; New York: John Lane, 1908);

Orthodoxy (London: John Lane/Bodley Head / New York: John Lane, 1908);

George Bernard Shaw (London: John Lane/Bodley Head / New York: John Lane, 1909; enlarged, 1935);

Tremendous Trifles (London: Methuen, 1909; New York: Dodd, Mead, 1909);

The Ball and the Cross (New York: John Lane, 1909; London: Gardner, Darton, 1910);

What's Wrong with the World (London, New York, Toronto & Melbourne: Cassell, 1910);

Alarms and Discursions (London: Methuen, 1910; enlarged edition, New York: Dodd, Mead, 1911);

William Blake (London: Duckworth / New York: Dutton, 1910);

Appreciations and Criticisms of the Works of Charles Dickens (London: Dent / New York: Dutton, 1911);

The Innocence of Father Brown (London, New York, Toronto & Melbourne: Cassell, 1911; New York: John Lane, 1911);

The Ballad of the White Horse (London: Methuen, 1911; New York: John Lane, 1911); reissued with critical supplements (Kirkwood, Miss.: Catholic Authors Press, 1950);

Manalive (London, Edinburgh, Dublin, Leeds, New York, Leipzig & Paris: Nelson, 1912; New York: John Lane, 1912);

A Miscellany of Men (London: Methuen, 1912; enlarged edition, New York: Dodd, Mead, 1912);

The Victorian Age in Literature (London: Williams & Norgate, 1913; New York: Holt, 1913);

Magic: A Fantastic Comedy (London: Secker, 1913; New York & London: Putnam, 1913);

The Flying Inn (London: Methuen, 1914; New York: John Lane, 1914); the novel's songs were enlarged and included in *Wine, Water, and Song* (London: Methuen, 1915);

The Wisdom of Father Brown (London, New York, Toronto & Melbourne: Cassell, 1914; New York: John Lane, 1915);

The Barbarism of Berlin (London, New York, Toronto & Melbourne: Cassell, 1914); republished in *The Appetite of Tyranny, Including Letters to an Old Garibaldian* (New York: Dodd, Mead, 1915);

Letters to an Old Garibaldian (London: Methuen, 1915); republished in *The Appetite of Tyranny, Including Letters to an Old Garibaldian*;

Poems (London: Burns & Oates, 1915; New York: John Lane, 1916);

Wine, Water, and Song (London: Methuen, 1915);

The Crimes of England (London: Cecil Palmer & Hayward, 1915; New York: John Lane, 1916);

A Short History of England (London: Chatto & Windus, 1917; New York: John Lane, 1917);

Utopia of Usurers (New York: Boni & Liveright, 1917);

Irish Impressions (London, Glasgow, Melbourne & Aukland: Collins, 1919; New York: John Lane, 1920);

The Superstition of Divorce (London: Chatto & Windus, 1920; New York: John Lane, 1920);

The Uses of Diversity (London: Methuen, 1920; New York: Dodd, Mead, 1921);

The New Jerusalem (London: Hodder & Stoughton, 1920; New York: Doran, 1921);

Eugenics and Other Evils (London, New York, Toronto & Melbourne: Cassell, 1922);

What I Saw in America (London: Hodder & Stoughton, 1922; New York: Dodd, Mead, 1922);

The Ballad of St. Barbara and Other Verses (London: Cecil Palmer, 1922; London & New York: Putnam, 1923);

The Man Who Knew Too Much and Other Stories (London, New York, Toronto & Melbourne: Cassell, 1922; abridged edition, London & New York: Harper, 1922);

Fancies Versus Fads (London: Methuen, 1923; New York: Dodd, Mead, 1923);

St. Francis of Assisi (London: Hodder & Stoughton, 1923; New York: Doran, 1924);

The End of the Roman Road (London: Classic, 1924);

Tales of the Long Bow (London, New York, Toronto & Melbourne: Cassell, 1925);

The Everlasting Man (London: Hodder & Stoughton, 1925; New York: Dodd, Mead, 1925);

William Cobbett (London: Hodder & Stoughton, 1925; New York: Dodd, Mead, 1926);

The Catholic Church and Conversion (New York: Macmillan, 1926; London: Burns, Oates & Washburne, 1927; reissued, 1951; enlarged, 1960);

The Incredulity of Father Brown (London, New York, Toronto & Melbourne: Cassell, 1926);

The Outline of Sanity (London: Methuen, 1926; New York: Dodd, Mead, 1927);

The Queen of Seven Swords (London: Sheed & Ward, 1926; London & New York: Sheed & Ward, 1933);

The Return of Don Quixote (London: Chatto & Windus, 1927; New York: Dodd, Mead, 1927);

The Collected Poems of G. K. Chesterton (London: Cecil Palmer, 1927; New York: Dodd, Mead, 1932; revised editions, London: Methuen, 1933; New York: Dodd, Mead, 1966);

The Secret of Father Brown (London, Toronto, Melbourne & Sydney: Cassell, 1927; New York & London: Harper, 1927);

The Judgement of Dr. Johnson: A Comedy in Three Acts (London: Sheed & Ward, 1927; New York & London: Putnam, 1928);

Robert Louis Stevenson (London: Hodder & Stoughton, 1927; New York: Dodd, Mead, 1928);

Generally Speaking: A Book of Essays (London: Methuen, 1928; New York: Dodd, Mead, 1929);

The Poet and the Lunatics: Episodes in the Life of Gabriel Gale (London, Toronto, Melbourne & Sydney: Cassell, 1929; New York: Dodd, Mead, 1929);

The Thing (London: Sheed & Ward, 1929); republished as *The Thing: Why I Am a Catholic* (New York: Dodd, Mead, 1930);

G. K. C. as M.C., edited by J. P. de Fonseka (London: Methuen, 1929);

Four Faultless Felons (London, Toronto, Melbourne & Sydney: Cassell, 1930; New York: Dodd, Mead, 1930);

The Resurrection of Rome (London: Hodder & Stoughton, 1930; New York: Dodd, Mead, 1930);
Come to Think of It . . . (London: Methuen, 1930; New York: Dodd, Mead, 1931);
All Is Grist: A Book of Essays (London: Methuen, 1931; New York: Dodd, Mead, 1932);
Chaucer (London: Faber & Faber, 1932; New York: Farrar & Rinehart, 1932);
Sidelights on New London and Newer York and Other Essays (London: Sheed & Ward, 1932; New York: Dodd, Mead, 1932);
Christendom in Dublin (London: Sheed & Ward, 1932; New York & London: Dodd, Mead, 1932);
All I Survey: A Book of Essays (London: Methuen, 1933; New York: Dodd, Mead, 1933);
St. Thomas Aquinas (London: Hodder & Stoughton, 1933; New York: Sheed & Ward, 1933);
Avowals and Denials: A Book of Essays (London: Methuen, 1934; New York: Dodd, Mead, 1935);
The Scandal of Father Brown (London, Toronto, Melbourne & Sydney: Cassell, 1935; New York: Dodd, Mead, 1935);
The Well and the Shallows (London: Sheed & Ward, 1935; New York: Sheed & Ward, 1935);
As I Was Saying: A Book of Essays (London: Methuen, 1936; New York: Dodd, Mead, 1936);
Autobiography (London: Hutchinson, 1936; New York: Sheed & Ward, 1936);
The Paradoxes of Mr. Pond (London, Toronto, Melbourne & Sydney: Cassell, 1937; New York: Dodd, Mead, 1937);
The Coloured Lands (London: Sheed & Ward, 1938; New York: Sheed & Ward, 1938);
The End of the Armistice (London: Sheed & Ward, 1940; New York: Sheed & Ward, 1940);
The Common Man (London: Sheed & Ward, 1950; New York: Sheed & Ward, 1950);
The Surprise (London & New York: Sheed & Ward, 1952);
A Handful of Authors, edited by Dorothy Collins (London & New York: Sheed & Ward, 1953);
The Glass Walking-Stick and Other Essays from the Illustrated London News, 1905–1936, edited by Collins (London: Methuen, 1955);
Lunacy and Letters, edited by Collins (London & New York: Sheed & Ward, 1958);
The Man Who Was Orthodox: A Selection from the Uncollected Writings of G. K. Chesterton, edited by A. L. Maycock (London: Dobson, 1963);
The Spice of Life, edited by Collins (Beaconsfield: Darwen Finlayson, 1964; Philadelphia: Dufour, 1966);
Chesterton on Shakespeare, edited by Collins (Henley on Thames, Oxfordshire & Chester Springs, Pa.: Dufour, 1971);
The Apostle and the Wild Ducks, and Other Essays, edited by Collins (London: Elek, 1975);
The Spirit of Christmas: Stories, Poems, Essays, edited by Marie Smith (New York: Dodd, Mead, 1985);
G. K.'s Weekly: A Sampler, edited by Lyle W. Dorsett (Chicago: Loyola University Press, 1986);
Daylight and Nightmare: Uncollected Stories and Fables, edited by Smith (London: Xanadu, 1986; New York: Dodd, Mead, 1986).

Collection: *Collected Works of G. K. Chesterton,* 45 volumes (San Francisco: Ignatius / 1986–).

Gilbert Keith Chesterton was a prolific writer who wrote in such a variety of genres that he resists simple classification. Previous *Dictionary of Literary Biography* volumes have examined Chesterton as a poet, a dramatist, a traditional novelist, a mystery writer, an essayist, and a biographer. Because of these complexities Chesterton scholars and biographers offer such broad and inclusive descriptions as journalist, mystic artist, or critic to convey the essence of Chesterton. He is best known as a true man of letters whose greatness lies not merely in his mastery of a variety of forms but in his ability to apply these forms to a common theme and purpose–social change. He was also influential as a critic and writer of fantasy.

Fantasy may not be an obvious choice as an instrument of social change; in fact, fantasy is sometimes viewed with mistrust. In Chesterton's day fantasy was not a well-defined genre. It was instead considered part of the Romantic tradition that included gothic stories, nonsense verse, and fairy tales. As a critic Chesterton identified and defended fantastic forms, and he did so by emphasizing fantasy's positive aim and effect. One positive aim of fairy-tale fantasy is its role as an educational tool; a second is its ability to provoke or reawaken wonder in its reader, which Chesterton believed would prod the individual, and thus society, out of its paralysis.

Chesterton's ability to produce literature that provokes or recovers wonder in characters and readers is given special notice by J. R. R. Tolkien. In his essay "On Fairy-Stories" from *Tree and Leaf* (1965) Tolkien names a type of low fantasy after Chesterton. He describes "Chestertonian Fantasy" as symbolized by the strange word *mooreeffoc,* "used by Chesterton to denote the queerness of things that have become trite, when they are seen suddenly from a new angle." Chesterton became acquainted with *mooreeffoc–coffee room* spelled backward–in his studies of Dickens. In his *Charles Dickens* (1906) Chesterton recounts how this common word appeared strange to Dickens as he read it through the glass door from the other side. What Tolkien calls a Chestertonian fantasy, Chesterton calls Dickens's "eerie realism": the real becomes the fantastic. In *Charles Dickens* he explains "the principle that the most fantastic thing of all is often the precise fact." An example of this principle is articulated in *Orthodoxy* (1908): "It is one thing to describe an interview with a gorgon or a griffin, a creature who does not exist. It is another thing to discover that the rhinoceros does exist and then take pleasure in the fact that he looks as if he didn't." Chesterton, with his well-nurtured imagination, believed that this fanciful appreciation of life was healthy. People need balance in life, and wonder helps provide a balancing tension. In *Orthodoxy* Chesterton describes humanity as having a double need: "to be at once astonished at the world and yet at home in it," or more concisely, "the need for that mixture of the familiar and the unfamiliar which Christendom has rightly named romance."

Born in London on 29 May 1874, Chesterton relates that he was a child of the misunderstood Victorian age. Despite its repressive religious image, the Victorian age produced homes that were not religious. Chesterton writes in his *Autobiography* (1936) that the Victorian "was the first generation that ever asked its children to worship the hearth without the altar." Chesterton's family was not steeped in religion, nor were family members totally irreligious; they retained some Christian traditions, such as singing before meals, and Chesterton recalled being taken to hear sermons by the Reverend Stopford Brooke. Likewise, his childhood and family stood in opposition to another popular image of the Victorian home: the mad, tyrannical parent. Although such a parent might help sell autobiographies, Chesterton related that his family was "disappointingly respectable and even reasonable."

The influence that his family had on him is apparent in his *Autobiography* and his many biographies. His mother, Marie Louise Grosjean Chesterton, is described as an energetic, dominant, yet caring mother who enjoyed hosting and serving great meals. Unconcerned about her appearance, one trait she passed on to Gilbert, his mother is also credited with passing on her extraordinary wit to him and his brother, Cecil. The brothers sharpened this wit through continuous debate on a variety of topics (occasionally needing to be reminded about taking meals). Gilbert loved having a brother and is often quoted as remarking of Cecil's birth, "Now I shall always have an audience." Edward Chesterton, with his variety of interests and rich imagination, also had a great effect on Gilbert both personally and artistically. As the head of an established house agency and surveying firm Edward Chesterton was able to spend much time with his family. The family's happy home

Chesterton with his wife Frances, 1912

life included toffee making, painting, and listening to fairy tales and other stories. Of all these amusements, a toy theater Edward built became a formative influence on, as well as a philosophical metaphor for, his son's life and art.

Chesterton's earliest memories are of this toy theater; he remembers a young adventurer-hero poised on a bridge spanning a deep gorge about to cross it in order to save a princess imprisoned in a tower. He also vaguely recalled the presence of an evil king. Chesterton commented on the significance of that memory in his *Autobiography:* "All my life I have loved edges; and the boundary-line that brings one thing sharply against another. All my life I have loved forms and limits. . . . I have also a pretty taste in abysses and bottomless pits and everything else that emphasizes a fine shade of distinction between one thing and another." Furthermore, he believed that these proper distinctions are best made by children, who have not yet been "made ignorant by experience." Such distinctions include those between good and evil, creator and created, and reality and unreality. The last two distinctions were realized when Chesterton found he could reach down and, by touching the theater's characters, know that they were made of cardboard. Later, when Chesterton was threatened by solipsism, these earliest memories of a truth forgotten later in life helped return him to a healthy balance.

Chesterton's childhood was not lonely, but it was protected, which may have accounted for his slow development in school. Living at home, he was able to attend a local prep school, Colet House, and its parent school, St. John's. At school he was lonely and felt out of place. He grew physically at a rapid pace but was slow in developing mentally. At age ten he was as tall as or taller than most men, yet he had only learned to read at age nine. To the chagrin of his teachers, he showed no interest in his assignments and seemed to prefer to dream the days away. Thus Chesterton was left back at least twice. His peers perceived him as dull, and they made him the object of practical jokes, but these jokes never turned malicious, partly because of Chesterton's good-natured personality and partly because of his great size. Finally he caught on with a group of the younger boys led by E. C. Bentley. The turning point was a friendly playground fight between Chesterton and Bentley. When the battle ended, because of fatigue, the two combatants discovered each other's love for literature, and with this common bond Chesterton was accepted by the group and quickly became one of its leaders. With interests in literature, the boys formed the Junior Debating Club. Complete with its own magazine, the J.D.C. met regularly to discuss literary matters over elaborate teas and lunches.

Chesterton flourished in secondary school, and he eventually won the school's Milton Prize for a poem about Saint Francis Xavier. This blissful time quickly ended, however, and he was forced to choose between attending university or securing employment. Chesterton chose art school. After a brief flirtation with an art school in St. John's Woods, Chesterton enrolled in and attended the Slade School of Art from 1892 through 1895. While at Slade he also attended other classes and lectures at University College, where he studied Latin under A. E. Housman and literature with W. P. Ker. His study in the fine arts under Frederick Brown and drawing master Henry Tonks exposed him to the currently popular technique of Impressionism. Given Chesterton's love of form, edges, and limits, it is easy to see why he disliked the fragmented haziness so important to the Impressionists. Impressionism challenged the reality of external objects and fueled young Chesterton's fear that nothing existed outside of the mind. Critic John Coates, however, calls this analysis reductionist. Instead, Coates sees Chesterton as disliking Impressionism because "it destroyed imagination by fixing it to one aspect of one object seen at one time unrelated to anything else."

Compounding Chesterton's solipsistic conflict at Slade School were his explorations in spiritualism that confirmed his belief in–even an acquaintance with–the devil. Although Chesterton would never act on them, he could imagine the worst possible crimes and sins. At Slade he felt surrounded by people who might actually be that evil and came face to face with inner devils and the presence of sin in the world. The nihilism and solipsism that Chesterton faced at Slade were later recounted in many of his works but especially in his essay "The Diabolist" in *Tremendous Trifles* (1909). In this essay Chesterton relates an encounter he had with another student, the diabolist, who constantly pushed all limits. What Chesterton considered evil, this diabolist might consider good. The essay's narrator reports that he heard the diabolist confide to another colleague, "If I do that I shan't know the difference between right and wrong." The narrator believed that the diabolist later committed suicide, and the narrator never dreamed of considering the terrible act or the point where the diabolist would hesitate. Some scholars believe that the diabolist was an actual classmate of Chesterton; some suggest that this essay is an example of the dialogue of the mind with itself.

Upon leaving Slade, Chesterton, at twenty-one, needed to construct a system that protected him from solipsism and nihilism, from thinking back to thought itself. He wrote in his *Autobiography,* "With little help from philosophy and no real help from religion, I invented a rudimentary mystical theory of my own. It was substantially this; that even mere existence, reduced to its most primary limits, was extraordinary enough to be exciting. Anything was magnificent next to nothing." Armed with this defense, he began writing seriously in 1895, all the while working in literary publishing houses. First, he was charged with running the entire press department at the Redway House, which published mostly books on the occult. There Chesterton read manuscripts for acceptance and wrote reviews of Redway books to appear in periodicals. After a few months he left to edit for the larger house of Fisher Unwin. After three years at Fisher Unwin, Chesterton quit working in the publishing business in late 1899 to attempt to make a living solely from his pen.

Chesterton's overwhelming drive to make money in order to marry and support a wife drove him to write much and to write often. He was able to double his pay on essays by first selling them to a newspaper and later collecting them to be sold as books. Finally, Chesterton, who was never known to be especially careful with his money, could afford to marry Frances Blogg on 28 June 1901. Frances brought order and stability to Chesterton's life, keeping track of his accounts and appointments. She fashioned his famous hat, cape, and swordstick style as a practical remedy to his careless dress and large body. She was also a devoutly religious Anglican who served as a model for Chesterton of a practicing Christian. She converted to Roman Catholicism in 1926, some four years after her husband.

During the early years of his marriage Chesterton was extremely productive. Through journalism, the favorite mass medium in England during the first third of the twentieth century, Chesterton had a great and immediate impact on his society. His newspaper columns also provided him with an incubator for ideas and a pulpit for his witty, humorous, commonsensical voice. Chesterton was convinced that his society needed change; he was also convinced that as a young journalist he could play a role in effecting positive change. Coates suggests that Chesterton understood the Edwardian cultural crisis to be an imbalance brought about by a narrowing of English culture through an abundance of skepticism. To Chesterton the most damaging of these skepticisms was the doubting of and disappearance of God, a legacy of the Romantic and Victorian periods.

Although he did not convert to Roman Catholicism until 1922, Chesterton defended God and religion from the outset of his writing career and was purposefully moral in all he wrote. In his early journalistic and expository writing, Chesterton the critic often defended the fairy tale against charges of being frivolous and corrupt. In an *Illustrated London News* article dated 2 December 1905 he writes of the duchess of Somerset's outrage over a school teaching fairy tales. The duchess reasoned that, since fairy tales are "nonsense," it would be better to teach the children serious, "real" subjects. Chesterton defends fairy stories by pointing out the mutability of the duchess and the schools versus the immutability of fairy tales. The fairy tale "is simply the ancient and enduring system of human education"–a system that has at its core "the spirit of folk-lore; and folk-lore is, in strict translation, the German for common-sense."

Against the charge of fairy tales promoting immoral, escapist utopias, places where savage freedom is allowed and prized, Chesterton argues that fairy tales are in fact moral and not so free. In an *Illustrated London News* article of 29 February 1908 Chesterton writes, "If you really read the fairy-tales, you will observe that one idea runs from one end of them to the other–the idea that peace and happiness can only exist on some condition. This idea, which is the core of ethics, is the core of the nursery-tales. The whole happiness of fairyland hangs upon a thread, upon

All rights reserved

The Man Who Was Thursday : A Nightmare. by G. K. Chesterton.

BRISTOL
J. W. ARROWSMITH, 11 QUAY STREET
LONDON
SIMPKIN, MARSHALL, HAMILTON, KENT & COMPANY LIMITED

Title page for Chesterton's 1908 novel, in which the protagonist's fantastic daydream deepens his faith in God

one thread." Chesterton's examples of this single thread include Cinderella not staying out after midnight, Bluebeard's wife not opening a specific door, and Pandora not opening the box. These examples stress personal responsibility and free will, but they also speak of external consequences. "This great idea, then, is the backbone of all folk-lore–the idea that happiness hangs on a thin veto; all positive joy depends on one negative." Hence fairy tales, with their morals, truths, and codes of ethics, imply the presence of an external power.

The tenets offered in these newspaper essays are further and definitively articulated and illustrated in Chesterton's well-known essay "The Ethics of Elfland," which appears in *Orthodoxy*. In the essay Chesterton terms his fantasy ethic "The Doctrine of Conditional Joy." Conditional joy depends on the existence of some other. This recognition of an external power is an important element in considering Chesterton as a fantasy writer.

In her *G. K. Chesterton: Explorations in Allegory* (1979), the most exhaustive study of Chesterton and fantasy, Lynette Hunter argues that Chesterton's fantasy writing was merely a dangerous stage that he passed through on his way to becoming an allegorist. Because the fantasist has the power to create from nothing, as God did, Hunter believes that Chesterton "doubted that any man could create responsibly without belief in the standard of an external authority. And he doubted further, that responsibility could be attained unless this external were included in the invention." Thus, Hunter asserts that ultimately Chesterton rejected fantasy because of the solipsistic danger it posed. Yet near the end of her book she rightfully claims that Chesterton's definition of fantasy would be similar to the one articulated by C. N. Manlove in *Modern Fantasy: Five Studies* (1975). According to Manlove fantasy is "a fiction evoking wonder and containing a substantial and irreducible element of the supernatural with which the mortal characters in the story or the readers become on at least partly familiar terms."

Hunter posits that Chesterton rejects what is now known as high fantasy, as distinguished from low fantasy. Each type of fantasy features a supernatural presence. The distinguishing feature is setting. As Robert Boyer and Kenneth Zahorski, in their introduction to *Fantasists on Fantasy* (1984), explain it, a created "secondary world, with its discernible though nonrational causality, is what characterizes high fantasy. Low fantasy, on the contrary, features nonrational happenings that are without causality or explanation because they occur in the rational world where such things are not supposed to occur." As a creative artist Chesterton refrained from creating the unique otherworlds of high fantasy; thus Hunter is correct. However, Chesterton did create low fantasy, or the seemingly nonrational in the rational world, by merely presenting a common object from a different perspective, by reawakening a forgotten memory, by presenting allegorical truth, or by including a believable or explainable external cause (usually nature or God) in his work.

At first the external power for Chesterton was the pure wonder of existence in nature. But as he matured he realized that God was present in nature and ultimately responsible for this wonder. Hence God as present in nature and the world becomes the supernatural and external to Chesterton. Possibly he is best described in regard to fantasy as a Christian fantasist who had a stated respect for earlier Christian fantasists such as George MacDonald, a description not in opposition to Hunter's "allegorist."

Chesterton would further agree with Manlove that fantasy celebrates "things as they are"; and if one believes in a Christian world order, then fantasy also celebrates "things as they were meant to be." Chesterton was fascinated by exploring and celebrating the self, and as Manlove states, "Many fantasies describe a spiritual development on the part of the hero at the same time as delighting in the constant nature of the world to which he belongs." Thus the hero "becomes more fully himself through a process of spiritual evolution"; and "such change is only the dynamic aspect of 'things as they are': the self that is finally produced is the true self, the-self-as-it-ought-to-be." This self is the hero of Chesterton's best fiction.

But as Hunter asserts, Chesterton believed that deep introspection of the self is dangerous. So in order to make such exploration safe, Chesterton's fantasy fiction includes some sort of frame that provides the reader with an "out." By acknowledging that his fiction's fantastic events occur in an admittedly unreal future or past or in a nightmare or a daydream, he provides his readers with the same sanity-preserving escape that he had with his father's toy theater: he allows the reader to reach into the fantasy and touch it and, by doing so, realize that it is only a story, only "cardboard." Chesterton's fantasy is a fairy tale for adults that serves as a safe educational tool while providing an escape or release. In a review of *Manalive* (1912) for the April 1912 issue of the *Bookman,* C. M. Francis commented that, since adults are not made wondrous by physical impossibilities, Chesterton's fairy tales "turn only on what is socially improbable, which to a grown up person is a far wilder thing. They represent men as acting logically upon some natural impulse which though innocent and even praiseworthy, it is customary to repress." Chesterton's fantasies are full of humor and dance and sword fights; they please and instruct.

Another of Chesterton's conventions was the use of stock characters. Hunter suggests that Chesterton borrowed this trait from Dickens: "Chesterton attributes the brilliance of [Dickens's] characterisations to an ability to find the vast yet unchanging essence of a person; and then create the physical limits of a human being around it. The reader knows that the physical appearance of the character is connected to essence by the transcendental truth of his soul." Thus, the character allegorically represents such ideas as religion or science and, once created and identified, is placed in situations that reveal to both character and reader that a person cannot function in society if polarized and acting monomaniacally. Garry Wills describes these character conflicts as "sustained encounters between opposed principles, each correct in its own sphere, yet each in need of the tempering influence of its opponents." When left alone, the characters and their views narrow and polarize; when brought into conflict, Wills writes, "they achieve that flying balance and poise of stress which is praised in *Orthodoxy.*"

A third standard convention of Chesterton is reliance on the past. Critical of futurists for believing in nothing, Chesterton strongly believed that one can base nothing on what is about to happen or what might happen; for him, present and future are built on the past, as evidenced in *The Napoleon of Notting Hill* (1904), when he wrote that he "must slip into a prophetic past." Indeed, the past is a common source for fantasy. Manlove notes, "Fantasy often draws spiritual nourishment from the past . . . particularly from a medieval and/or Christian world-order." Wills believes that Chesterton infuses a poetry of being in all that he writes, and often this poetry is also expressed in tales in which "the past is a symbol of one's half-forgotten self." As Chesterton expresses it in *Orthodoxy,* "We are all under the same mental calamity; we have all forgotten our names. We have all forgotten what we really are." Or, possibly, we have not yet found out who we are or who we ought to be.

Some have criticized Chesterton's use of the past, accusing him of glorifying bygone days and attempting to return England to a utopian Middle Ages that never existed. In one of the first full-length critiques of Chesterton, Julius West characterizes the Middle Ages as a "servile state, stricken with plague, and afflicted with death in all its forms," then summarizes Chesterton's societal solutions: "His political principles, roughly speaking, are England was decent once–let us apply the same recipe to the England of to-day." But this oversimplification is only partially correct. Chesterton did believe in the reality and lessons of history. He was not, however, so naive as to think the imperfect medieval social structure, for example, could be applied to his day or eighty years in the future. Rather, he believed that an adapted version of the past could be applied and that one should evaluate and synthesize past and present in order to go forward. Wills writes, "He finds in Romance the higher dialectic of paradox the quest for things at once ancient and new, the need for exploration and return, for Troy and Ithaca."

Besides understanding and defending fantasy, Chesterton began producing literary texts designed to provoke wonder and decrease skepticism about the presence of God. Although his earliest works were not great financial successes, they succeeded in increasing his recognition among the reading public, which broadened his newspaper readership and made it easier to publish what would become more popular works. To the overabundance of prophecies littering the early twentieth century Chesterton felt compelled to add his own with *The Napoleon of Notting Hill,* his first novel. Although the story is set eighty years in the future in the London of 1984 the city resembles the London of 1904 and, especially in the sword-and-ax battle scenes, has the look and feel of an even earlier England. The novel describes an English society in which "Democracy was dead; for no one minded the governing class." There is still a monarchy, but no royal family; the king is chosen randomly from the usually bureaucratic automatons that make up the civil servant class. This seemingly senseless practice is not as dangerous as it might seem, since the king is only allowed to make laws and decrees that do not interfere with the true rulers of England, the merchant class. When the time comes to select a new king, the lot falls to a curious little artist named Auberon Quin. Quin, one of Chesterton's stock characters, is the dangerous aesthete–dangerous because he is singular in focus, caring only for a joke. Quin's ultimate practical joke is to turn the modern suburbs back into medieval cities. Each city becomes autonomous, building walls, employing guards, flying banners, and adopting coats of arms. For ten years the joke runs smoothly.

A problem finally arises when big business wants to tear down Notting Hill's Pump Street in order to make way for a modern road. But the newly elected provost of Notting Hill, Adam Wayne, will not allow such destructive construction. The fanatic Wayne, another stock character, views Pump Street, with its grocery, toy store, curiosity shop, and barbershop, as the heart of Notting Hill. The business-minded provosts then try to buy Pump Street, but Wayne refuses to sell; Wayne and Notting Hill will fight. What follows is a series of fantastic battles in which the Notting Hill loyalists defeat the larger, probusiness mercenary armies. It seems impossible that massive armies could be defeated by such a smaller foe. But the England of 1984 has lost the ideas of democracy and revolution; men who had become soldiers had lost the ability, the knowledge, and the desire to fight.

Wayne promises that Notting Hill will not interfere with the defeated cities as they had interfered with Notting Hill. But he cannot prevent Notting Hill from becoming an empire, and twenty years after the Battle of the Waterworks, Notting Hill is interfering with its neighbors, who now have hopes, dreams, and the patriotism to fight. In Wayne's own words, "it is the glory of our achievement that we have infected the other cities with the idealism of

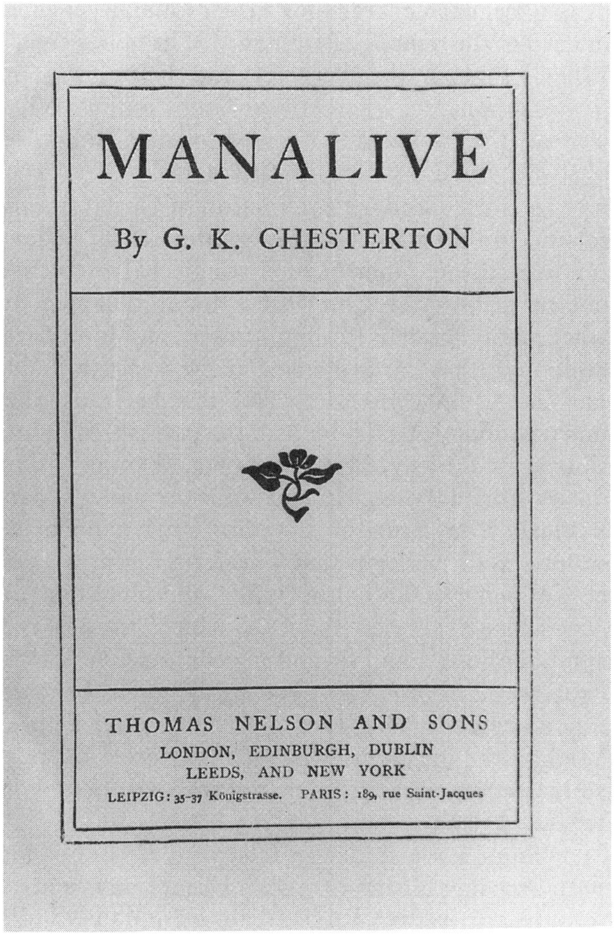

MANALIVE

By G. K. CHESTERTON

THOMAS NELSON AND SONS
LONDON, EDINBURGH, DUBLIN
LEEDS, AND NEW YORK
LEIPZIG: 35-37 Königstrasse. PARIS: 189, rue Saint-Jacques

Frontispiece and title page for Chesterton's 1912 book in which the boarders of Beacon House are reawakened to the wondrous nature of the commonplace

Notting Hill." Personal autonomy and a sense of democracy transform the stale English society of the future with the idea that the empowered individual makes a better citizen.

In the novel's last chapter Quin and Wayne comment on all that has happened and essentially sum up the importance of the book. Throughout the novel Quin and Wayne have been called mad, and Wayne believes that they are mad "because we are two lobes of the same brain, and that brain has been cloven in two." Quin counters that nothing can alter their polarized antagonism, but Wayne sees an external authority in the common man, "whom mere geniuses like you and me can only worship like a god." The healthy brain is found in the common ploughman, whose single brain contains both humor and gravity and who is the half-forgotten self of the past. Quin and Wayne embark together into the new world re-created after the war annihilated the old. *The Napoleon of Notting Hill* thus attacks the narrow, skeptical individual of Chesterton's day who would make a mess of England's future big enough to warrant a revolution.

The Club of Queer Trades (1905) is a collection of short "philosophical mysteries" that are only explained when interpreted by Basil Grant. Rather than drawn-out logical conclusions, Basil relies on his intuition and his ability to distinguish between good and evil. He is described as a poet, a stargazer, a mystic, and a hermit. Professionally a jurist, he supposedly went mad and began dispensing personal and moral advice from the bench. He sentenced men to the seaside instead of to prison; in summation of a high-profile libel case, Basil sang a nonsense verse and then retired to a small garret filled with fantastic, romantic objects.

Basil is pried from these lodgings by his private-detective brother, Rupert. Since Rupert is always attempting to deduce things, to be logical and factual, and then jumps to the obviously wrong conclusions, he is useless in solving philosophical mysteries. Thus Basil is called upon to solve "crimes" in six cases. But they are not crimes. Instead, they are the actions of entrepreneurs who end up joining the Club of Queer Trades, "an eccentric and Bohemian Club, of which the absolute condition of membership lies in this, that the candidate must have invented the method by which he earns his living. It must be an entirely new trade."

In the first story a respected major believes that he is the target of a murder plot. After moving into a new house the major strolls through the neighborhood and comes across a bed of pansies that spell out "DEATH TO MAJOR BROWN." As he tries to learn who is behind the threat, the major is told not to mention "jackals," converses with a mysterious lady, gets yelled at by a seemingly decapitated head, and battles a beast in a coal cellar. Doubting the ill will behind these events, Basil leads his brother and client to the Adventure and Romance Agency, LTD., where he discovers that the plot against Major Brown was paid for by the previous owner of his villa, a Mr. Gurney-Brown. So what appeared fantastic was in the end perfectly explainable.

This type of humorous plot introduces readers to a variety of practitioners–moral or immoral–of new and fantastic trades. Basil exposes an immoral "Organizer of Repartee" in chapter 2, "Professional Detainers" in chapter 3, an "Arboreal House-Agent" in chapter 4, and a "Dancing Language Inventor" in chapter 5. In chapter 6 Basil reveals that he is the president of the Club of Queer Trades, in which he earns membership as a "purely moral judge to settle purely moral differences." Basil explains that as a jurist he had little impact on justice.

The Club of Queer Trades is not high fantasy; it might not be fantasy at all, for people may actually pay for adventure, pay for a stooge, live in a tree house, communicate by dancing, or seek a moral judge. But the fictions work as fantasy for Chesterton–they restore proper vision to characters and readers by challenging their assumptions that the world is as it appears.

The Man Who Was Thursday (1908) is Chesterton's most popular and most critically acclaimed book. It, too, makes a variety of comments on politics, philosophy, and religion. The dedicatory verses to E. C. Bentley indicate that the novel will deal with the fears and doubts of their boyhood days. Kingsley Amis, in an introduction to a reprint edition, writes, "Chesterton's story is indeed a fantasy, unrealistic in outline and, towards the end, in what happens or seems to happen. Or, better, call it a nightmare, as he did himself in its subtitle, saying afterwards that it was the key to understanding the book." Biographer Dudley Barker describes this nightmare element as a "strand of increasing unreality in the narrative itself, yet woven with the dialogue of the most natural and commonplace kind, in which the comparison with Kafka's surrealist nightmares enters the mind. But there is a difference that, where Kafka's are sombre and bitter, Chesterton's is clothed with comedy and good humour, as though, as once was said of the book, he had been commissioned by a publisher to write *Pilgrim's Progress* in the style of *Pickwick Papers*."

Ultimately, the reason this novel may be Chesterton's most successful is that it focuses on one hero, and the reader is able to follow closely his evolution to the self-that-ought-to-be. The hero, Gabriel Syme, is the poet of law, order, and respectability; he is viewed in opposition to Lucian Gregory, the poet of anarchy. Strolling in a park, the two poets agree on next to nothing. But they do agree on one thing, that "a paradox might wake men up to a neglected truth." One truth attached to Syme's character is the transformative, dutiful, patriotic spirit so present in *The Napoleon of Notting Hill*. As he measures Gregory, Syme is struck by hearing "very faintly in some distant street a barrel-organ begin to play, and it seemed to them that his heroic words were moving to a tiny tune from under or beyond the world."

Irritated by Syme's dismissal of his anarchy, Gregory takes Syme to a serious anarchist meeting, making him swear that he will not reveal the group to the police. But unbeknownst to Gregory, Syme is a detective in Scotland Yard's special division, The Last Crusade. This entertaining evening takes Syme to a meeting of Gregory's local chapter of anarchists, who are electing a representative to the seven-member supreme council of anarchists, who all have code names corresponding to the days of the week. Because Syme is bound by silence not to reveal him, Gregory smugly confides his imminent election as "Thursday." At this point Syme asks if Gregory would make the same promise of secrecy: if Gregory would swear not to reveal Syme's secret, especially to the anarchists, then he would explain his own secret identity. Gregory promises, and Syme tells him that he is a detective. With Gregory shocked and subdued, Syme roars anarchic rhetoric and influences the chapter to elect him as Thursday instead of Gregory.

As Syme travels up the Thames to his first weekly meeting of the supreme council, he reflects on his recruitment to Scotland Yard by a man who sat in a dark room and never turned to face Syme and on his trip to meet Sunday, the mysterious, intimidating leader of the supreme council of anarchists. But even with the title of "The Last Crusade," a mysterious leader, a terrible opponent, and an eerie night, Syme envisions his adventure a

Cover for Chesterton's 1913 play, in which fairyland is said to be "rather different from other places. It's either nowhere or it's wherever you are."

healthy one. He takes comfort in common objects, in his nationalistic spirit, and in his mission. The narrator comments, "For even the most dehumanised modern fantasies depend on some older and simpler figure; the adventures may be mad, but the adventurer must be sane."

At the council meeting the strangely great and terrible president, Sunday, stares at Syme and demands that the council retire to a private room. Once alone, Sunday announces that there is a spy in their midst. Syme, sure that he has been found out, faces the choice of staying or running. Here the symbolic barrel-organ returns and gives Syme strength: "A barrel organ in the street suddenly sprang with a jerk into a jovial tune. Syme stood up taut, as if it had been a bugle before battle. He found himself with a supernatural courage that came from nowhere." Syme keeps silent and stands steady as another member, Tuesday, is exposed as a policeman and is ordered out of the meeting. This begins a series of wild adventures that reveal all six members of the supreme anarchist council as Last Crusade detectives. Ultimately, the only thing left for the detectives to do is track down Sunday.

When the six detectives assemble for their weekly breakfast meeting with Sunday, the question of identity and being, of both the detectives and Sunday, is articulated in dialogue. Dr. Bull claims that "we are six men going to ask one man what he means," but Syme, who has a clearer notion of what is occurring, responds, "I think it is six men going to ask one man what they mean." Sunday, telling them that they are all jackasses and that they will never understand all of him, leads the detectives on one last great adventure. As he begins to bound away, Sunday taunts them, "There's one thing I'll tell you though about who I am. I am the man in the dark room, who made you all policemen." What follows is a wild chase involving cabs, elephants, and hot-air balloons, and all the while Sunday toys with his pursuers by throwing them nonsensical messages.

The detectives chase Sunday to his country estate, which feels like a fairyland to Syme. Once there the detectives change into costumes decorated with symbols of their days of the Creation and sit in the garden with Sunday, who reveals, "I am the peace of God." Just then the scene is interrupted by the real anarchist, Gregory, who accuses the seven of having never suffered. Syme furiously replies that each, in his isolated, mysterious uncertainty, has indeed suffered. Realizing that he might not speak for Sunday, Syme turns and asks, as Job did, if God had suffered. With his huge, terrible face expanding and blocking out the universe, Sunday asks, "Can ye drink of the cup that I drink of?"

Syme is propelled out of this daydream and finds himself still strolling in Saffron Park with Gregory but transformed: "He felt he was in possession of some impossible good news." Through the strange adventures Syme had wrestled with God; and, like Job, Syme is comforted by his existence, even if he allows evil and suffering in the world.

First published serially in 1905, *The Ball and the Cross* (1909) has the most fantastic beginning of any of Chesterton's fiction. The novel opens with Professor Lucifer streaking through the sky in a flying ship. Lucifer had been engaged in debate with Brother Michael, a cloistered Bulgarian monk Lucifer had kidnapped. These two stock characters represent science/evolution and religion/creation. The debate heats up as the two argue about the ball and the cross that adorn the top of St. Paul's Cathedral. Lucifer argues that the ball is the perfect symbol because of its complete unity. The monk argues for the cross's ability to stand on its own and, if the lines of the cross are continued, its transcendence of all circular limits. The cross is the perfect mirror of humanity as it also has at its essence a contradiction. This frame ends with Lucifer pushing the monk out of his ship and onto the cross atop St. Paul's. Fighting off terror, the monk realizes that he must fearlessly work his way down. Knowing that "whoever shall lose his life the same shall save it" and with faith intact, the monk successfully makes his way to safety. This brush with death destroys and re-creates his world; he sees everything with joy. But, upon telling his story, he is immediately taken into custody as a lunatic.

He is arrested during a street disturbance caused by two other antithetical characters. Evan MacIan, a devoutly Catholic Highlander, is offended by an editorial taped to the office window of an atheist press. The paper's editor, a fiery Scot named Turnbull, is always trying to pick a fight with religion, but no one had argued with him until MacIan broke the window because of the editorial insulting the Virgin Mary. Both characters are polarized in their staunch belief or unbelief; the only answer is to fight a duel.

What follows is a madcap attempt at fighting a duel, but in this modern day of relativism there are no duels, as convictions are not held strong enough to fight over. Their willingness to fight makes them wanted by the authorities; thus they flee together, searching for a place and time to duel. Yet every time they begin to fight, characters representing all the modern philosophies intervene. Turnbull and MacIan soon begin to see how similar they are compared to everyone else. As the mad adventure draws out, these similarities are intensified. They slowly become friends. The duelists, running from the law, climb a tree and drop into a garden belonging to the lunatic asylum that houses the monk. They are soon joined by the characters who assisted them in fleeing the authorities in their attempt to fight the duel. Chesterton suggests that modern science and its psychologists will soon either have the public in the prisons of their own minds or in the asylum.

Turnbull and MacIan join the monk in high-security isolation cells. There, separated for the first time since the adventure began, MacIan and Turnbull dream of life without the balancing tension of their antithetical characters. The characters' dreams represent the distortion of the truth that each represents. Without a balancing tension science and religion would lead to a totalitarian regime that would abuse the English people. Because of these premonitions, the two escape from their cells into the asylum's garden. After their long isolation, the garden of the asylum appears wondrous. The narrator comments, "Only once or twice in life is it permitted to a man thus to see the very universe from outside, and feel existence itself as an adorable adventure not yet begun." Their new adventure is to prevent what they realize has already started to happen: science narrowing and limiting people by not allowing humanity contradictions or conflict. The duel represents the freedom to fight, to revolt.

After all participants are imprisoned the doctors proclaim that the duelists and their challenge never existed. Science denies their being. Thus the imprisoned have to risk their lives in order to gain them. The patients revolt, set the asylum on fire, and escape to remake the world. As the duelists leave they drop their swords, which fall in the shape of a cross, and the two friends go off together.

Manalive, by dealing directly with the recovery of childlike joy, wonder, and purity of vision, is the most perfect example of Chestertonian fantasy. The supernatural, symbolized by wind, is present in the first sentence of the book. The west wind blows through England variously affecting the people and landscape. The narrator suggests, "Perhaps this wind was a kind of fairy wand to test men and women." The people present to be tested reside at a boardinghouse called Beacon House and include an accomplished young doctor named Warner; his student, Arthur Inglewood; an Irish journalist named Michael Moon; Diana Duke; Rosamund Hunt; and a Miss Gray. Standing in the garden, the men begin to discuss friendship, and Inglewood mentions his greatest school friend, Innocent Smith. Inglewood laments that Smith went to Oxford and went mad. A year ago he had received a telegram from Smith that apparently confirmed the insanity: "Man found alive with two legs." A man is expected to have two legs, but Smith's point is that most people do not think of legs as wonderful things.

While discussing Smith's madness the group is interrupted by his appearance at Beacon House. Smith's large presence changes the attitude of the residents of that sterile boardinghouse. They begin to see common objects and everyday life with an awakened vigor. They take up old and new hobbies, and they rediscover their feelings for each other. The one common object of Smith's that causes some confusion is a large revolver, which he explains is not an agent of death but of life.

Before Rosamund begins to understand Smith, she sends an urgent message to Dr. Warner asking him to come and save the house from the lunatic. Warner arrives with an associate and goes in search of Smith. All are shocked when Smith shoots the doctor's hat off. The shooting surprises the tenants of Beacon House, but it does not surprise Warner, who has gathered evidence that Smith is a dangerous criminal. Smith is suspected of kidnapping women, committing burglary, and attempting to kill the warden of Brakespeare College.

Warner wants to take Smith before an official magistrate, but Moon prevents this. Moon, the women, and Inglewood believe that he is somehow innocent. Moon argues for the right for Beacon House to hold court in this case. The mock court is convened, and it clearly appears that Smith is guilty: all heard the shots, and all saw Smith holding the smoking gun. But appearances do not explain or represent intent. In his defense of Smith, Moon uncovers a letter written by the warden of Brakespeare College revealing that the warden–like Warner, a calculating intellectual–acted vacuously. Smith recognized that the warden needed to be reawakened to the wonder of life, and to do so he drew his revolver and began shooting. An excellent shot, Smith was intentionally missing the warden; only by reminding the cynic of death was Smith able to save him. The same was true for the incident with Warner. In the letter Smith reportedly vowed, "I am going to hold a pistol to the head of the Modern Man. But I shall not use it to kill him. Only to bring him to life." The same situation exists in all of the other charges against him: he continually remarried his own wife and repeatedly burglarized his own house in order to remember the preciousness of his relationships and possessions.

Magic: A Fantastic Comedy (1913), a play set at the house and on the grounds of an English duke, presents the usual Chestertonian themes and comments on fairies and fairyland, but absent is his humor. In the play's prelude the audience is introduced to a mysterious stranger and the duke's niece, Patricia. While walking the grounds Patricia is startled by the presence of a stranger who claims to be a fairy. She is surprised that a fairy can be as large as a human, but the stranger claims that fairies are larger than the world. The play also includes Chesterton's usual two antithetical characters, in this case the Reverend Smith and Dr. Grimthorpe. Smith and Grimthorpe each acknowledge that the other is sane, although they are on different sides of the issue. Grimthorpe becomes quite concerned upon discovering that the niece has seen fairies on the grounds. As the family physician to the duke's brother in Ireland, the doctor knows that the family has a history of seeing fairies. While the doctor will not allow Patricia to bring those fairies with her from Ireland to England, the duke will bring a local conjurer to his estate to provide a night of entertainment.

On the night Morris, the duke's nephew, is to arrive from America, Patricia is walking on the cool, damp grounds. Questioned upon her return, Patricia replies that she has been in fairyland. Queried as to where fairyland is, Patricia replies, "It's rather different from other places. It's either nowhere or it's wherever you are." Asked who resides in fairyland, Patricia responds, "Generally only two. Oneself and one's shadow. But whether he is my shadow or I am his shadow is never found out." Patricia describes the shadow as a trick-performing wizard, but the men think that this wizard is probably a swindler. The local conjurer arrives, and indeed, to Patricia's regret, he is the fairy wizard she had met on the grounds. Turning out to be a mortal, the conjurer has stolen Patricia's fairy tale.

The extremely practical Morris begins to catalogue the magical props of the conjurer. In a gleeful tone Morris identifies the mechanisms of each trick. But the conjurer makes an important distinction: more important than merely knowing how something is done is knowing how to do it. As the group argues, Morris's temper flares during the discussions of old and new, real and false magic, and the Reverend Smith, defending religion, further agitates him. A series of strange events ensues–a portrait sways; a plank slides; and Morris accuses the conjurer of rigging the room with wires. The conjurer admits that he could have done so. But just then something happens that drives Morris mad: looking out over the garden and toward the doctor's home, he sees the doctor's red door lamp turn blue.

Morris, bedridden and babbling, can only be understood to ask how the lamp trick was done. The doctor believes that the conjurer is the only one who can save Morris. The conjurer explains that the trick was simple: it was done by a supernatural spirit. But only Reverend Smith and Patricia seem to believe him. In revealing the secret, the conjurer professes his love for Patricia, and the two become engaged. He tells her that he not only believes in fairies but in devils too, and he even admits to having some connection with devils but has always been able to suppress and control them. During the argument with Morris the conjurer admits that he became so upset that he called upon real devils to change the light. The conjurer turns to thoughts of God after he sends the demons back to hell and tells Morris a sanity-saving lie. But he will not tell the others what the lie is, lest they believe it too. He does not want them to disbelieve the "Immortal Mystery" that had been played out.

By the end of 1914 Chesterton had produced, in addition to his journalism and fantasy, two volumes of Father Brown mystery stories and was in the process of selecting poems for a collection. Barker reports that producing such a vast amount of work, combined with the threat of the Boer War, his brother's libel trial, his own danger of being accused of libel, a split with the *Daily News,* and excessive drinking, drove Chesterton to mental and physical collapse. Chesterton's almost completely comatose state lasted from approximately Christmas Eve 1914 to Easter 1915. He recovered and continued with his literary career, but it took a different turn. He no longer wrote as much philosophical fantasy but continued to write philosophical detective stories. He expended a great deal of time and energy in journalism, taking temporary editorship of his brother's *New Witness* when Cecil went off to fight in World War I. When Cecil died in 1918 Chesterton was forced to assume the editorship as well as to write a good portion of the paper. The *New Witness* folded in 1922, thereby freeing Chesterton for other writing. He did some traveling and wrote some travelogues of his adventures. He also produced some texts of social criticism, several short stories, and his well-known biographies of Saint Francis of Assisi and Saint Thomas Aquinas.

Chesterton died on 14 June 1936. That same year his exceptional *Autobiography* was published. He left a large literary estate that included unpublished plays, notebooks, short stories, and fragments of verse. Some of these writings made it to print after his death–for example, *The Coloured Lands* (1938), a children's book of nonsense verse, fairy tales, and fantastic short stories.

Also published posthumously was *The Surprise* (1952), a second drama that emphasizes the presence of God while promoting recovery. Some critics believe it is Chesterton's most significant work of the last half of his career because of its strong theme and open-endedness. Written in 1932, *The Surprise* makes a poignant statement about creator and created. The creator is an author of plays and a maker of puppets; he creates life-size, windup puppets that move and speak in order to act out his plays. In the foothills of the Pyrenees the author encounters a wandering friar. In explaining himself to the friar, the author says that he is king to his created court–he has created it all. The friar reminds the author that he should not forget God and the real world. The author replies that the real world is mean and grievous, and in his plays he has made a world where virtue is always rewarded and characters are never mean. The discussion of God leads the two to a discussion of miracles, and the author asks the friar to hear his confession as spoken in his puppets' play. The friar is really being asked to judge whether or not the author is solipsistically mad.

The first play-within-a-play involves the activities of the Princess of Garfagna. A wandering poet,

The Chestertons' gravestone, designed by Eric Gill

who cannot find the final four notes for his masterpiece, is in the princess's courtyard. He is so inspired to serenade the kindly princess that he successfully completes his song. The princess hears the completed song but, although enchanted, does not reveal her presence. Beneath the princess's balcony the poet is accosted by guards. Battling the guards for the opportunity to serenade the princess, the poet is knocked unconscious, thus losing the crucial four notes. The princess rushes down to intercede on his behalf, reprimanding the guards and ordering the poet to be taken to an apartment.

The poet awakes in darkness. At first he does not know if he is dead, blind, or drunk, but then he remembers his battle with the guards. In what must be a dark jail cell, he finds vessels that he is sure will hold stale bread and tepid water. Instead, he is greatly surprised to find the jug filled with wine and the plate full of cake. The poet believes that he has stumbled onto a new philosophy: all people want to and should be continually surprised by life. "If ever, one fine morning, they found the pig in the ale-cask and could drink ale out of the pig-trough–they would think they were in a fairy-tale." When she visits, the princess carries on a conversation with the poet about his newfound philosophy and adopts it.

The princess is that day to marry King Phoebus, who as a young man was protected by the princess's father. But Phoebus is in love with the princess's faithful servant, Maria. The princess enlists the poet to act as her ambassador. He is to greet the young king and make him reaffirm his vow to marry the princess. Meanwhile, the princess will meet with Maria, who will bring the heavy, opaque wedding veil and robes for the princess to try on. The princess asks Maria to model the clothes and, after making Maria reaffirm her faithfulness, orders her to remain in the wedding dress and take her place at the ceremony.

While the surprise wedding is taking place, the poet runs into the princess, who is wearing Maria's veil, and he confesses his love for the princess. When the princess unmasks herself, the poet is concerned that his declaration has offended the princess on her wedding day. She explains her deed and assures the poet that someday, possibly beyond this world, someone will sound his four notes and ask to be serenaded. The princess then ascends to her balcony, plays the forgotten four notes, and asks to be serenaded.

After the play is over the author asks the friar if there was anything peculiar about the play. The friar responds positively but wants to hear the author's intent. The author tells him that he was tired of most plays being populated by villains and set out to create a play without a villain. After an intense discussion of free will and being, the author, who earlier had asked the friar to give his puppets wills of their own, insists that his puppets should be alive. The friar replies that "no man can work a miracle. But to some men there has been granted a sign." As a sign to the author, the puppets suddenly begin to move without being wound; they begin to act out their play, but now, with free wills, they are changed.

The new play's opening is less noble. There is no honorable fight on behalf of a lady or of poetry. There is only a drunk guard and a drunk poet. The poet never gets around to caring about his missing four notes, for he passes out first. He awakens in a cell and finds only water and bread. There are no surprises. The interview in the cell between princess and poet occurs, and the philosophy of surprise is brought up. However, the discussion is not nearly as virtuous as in the first play. The princess and Maria meet and switch roles; but instead of letting the surprise be carried out, the princess meets the young king dressed as Maria. The young king sees who he thinks is Maria and confesses his love for her and pledges marriage. The princess accepts his proposal and unmasks herself. The young king cannot back out of this public proposal. The princess accepts her unhappiness gladly, as King Phoebus, the poet, and Maria would not honorably accept her direct, good intentions. All the puppet characters end up fighting over love, hate, and intentions but not for honor or virtue.

The author bursts in to stop his puppets from acting anymore. This was not what he had intended. With this realization by the author, *The Surprise* ends: God makes a play with the author's puppets, and the author realizes that he does not rival God. He answers his own confession.

As such a powerful voice at the beginning of the twentieth century, Chesterton exerted an influence on society still felt today. With the current popularity of fantasy literature, a proper understanding of Chesterton is necessary to understand the genre. Hunter admits that although Chesterton chose not to create high fantasy, "There is absolutely no reason why fantasy should not be used to maintain a beneficial effect." That beneficial effect is wonder-inspired recovery. In *Orthodoxy* he explained how fantastic fairy tales remind readers of normality and allow them to revel in it. "These tales say that apples are golden only to refresh that forgotten moment when we found that they were green. They make rivers run with wine only to make us remember for one wild moment that they run with water."

The recovery of wonder and vision includes for Chesterton a rediscovery of the banished, unprovable God. In an *Illustrated London News* article dated 15 May 1909 he explains fantasy's role:

> It is this power of recovery after extravagance that I urgently recommend. Indulge in all the most decadent or futile fantasies, as long as you can curb the indulgence, like that of alcohol. . . . Learn to be nonsensical, and then sensical again; to create strange things and still to be independent of them. Learn to suggest a thing, to urge it, to prove it, and still to disbelieve it. For the very few things that are really worth believing are not worth proving.

Bibliographies:

John Sullivan, *G. K. Chesterton: A Bibliography* (London: University of London Press, 1958);

Sullivan, *Chesterton Continued: A Bibliographical Supplement* (London: University of London Press, 1968).

Biographies:

Maisie Ward, *Gilbert Keith Chesterton* (New York: Sheed & Ward, 1943);

Garry Wills, *Chesterton: Man and Mask* (New York: Sheed & Ward, 1952);

Dudley Barker, *G. K. Chesterton: A Biography* (London: Constable, 1973);

Alzina Stone Dale, *The Outline of Sanity: A Life of G. K. Chesterton* (Grand Rapids, Mich.: Eerdmans, 1982);

Michael Ffinch, *G. K. Chesterton: A Biography* (San Francisco: Harper & Row, 1986);

Michael Coren, *Gilbert: The Man Who Was G. K. Chesterton* (New York: Paragon House, 1990).

References:

Kingsley Amis, Introduction to Chesterton's *The Man Who Was Thursday: A Nightmare* (New York: Penguin, 1986);

Ian Boyd, *The Novels of G. K. Chesterton: A Study in Art and Propaganda* (New York: Barnes & Noble, 1975);

Robert Boyer and Kenneth Zahorski, eds., *Fantasists on Fantasy: A Collection of Critical Reflections* (New York: Avon, 1984);

Lawrence J. Clipper, *G. K. Chesterton* (New York: Twayne, 1974);

John Coates, *Chesterton and the Edwardian Cultural Crises* (Hull: Hull University Press, 1974);

C. M. Francis, "G. K. Chesterton's *Manalive*," *Bookman* (London) (April 1912);

Christopher Hollis, *The Mind of Chesterton* (London: Hollis & Carter, 1970);

Lynette Hunter, *G. K. Chesterton: Explorations in Allegory* (New York: St. Martin's Press, 1979);

C. N. Manlove, *Modern Fantasy: Five Studies* (New York: Cambridge University Press, 1975);

J. R. R. Tolkien, *Tree and Leaf* (Boston: Houghton, Mifflin, 1965);

Julius West, *G. K. Chesterton: A Critical Study* (London: Secker, 1915).

Papers:

The Robert John Bayer Collection held at the Grasselli Library of John Carroll University, Cleveland, Ohio, has a complete holding of Chesterton's first editions and includes many supplemental and manuscript items. Another complete holding is the John Bennett Shaw Collection on G. K. Chesterton located in the Theodore M. Hesbrough, C.S.C. Library at Notre Dame University, South Bend, Indiana. The Marquette University, Columbia University, and Indiana University libraries all have special Chesterton holdings, as does the British Library.

Sir Arthur Conan Doyle

(22 May 1859 – 7 July 1930)

Terry Heller
Coe College

See also the Doyle entries in *DLB 18: Victorian Novelists After 1885; DLB 70: British Mystery Writers, 1860–1919;* and *DLB 156: British Short-Fiction Writers, 1880–1914: The Romantic Tradition.*

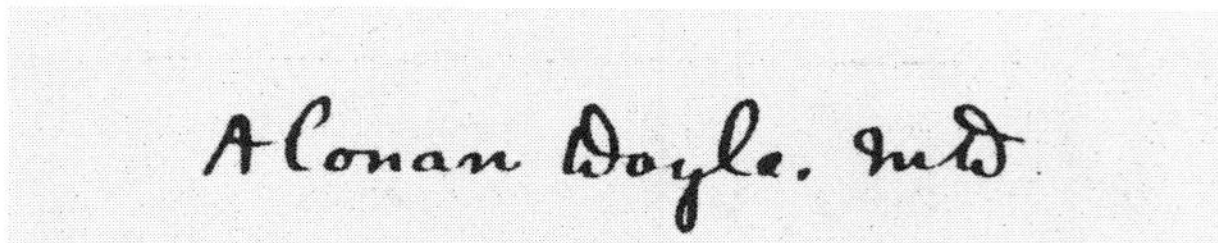

BOOKS: *A Study in Scarlet* (London: Ward, Lock, 1888; Philadelphia: Lippincott, 1890);

The Mystery of Cloomber (London: Ward & Downey, 1889; New York: Fenno, 1896);

Micah Clarke (London: Longmans, Green, 1889; New York: Harper, 1889);

Mysteries and Adventures (London: Scott, 1890); republished as *My Friend the Murderer and Other Mysteries and Adventures* (New York: Lovell, Coryell, 1893);

The Captain of the Polestar and Other Tales (London & New York: Longmans, Green, 1890);

The Firm of Girdlestone (London: Chatto & Windus, 1890; New York: Lovell, 1890);

The Sign of Four (London: Blackett, 1890; New York: Collier, 1891); republished as *The Sign of the Four* (Philadelphia: Lippincott, 1890);

The White Company (3 volumes, London: Smith, Elder, 1891; 1 volume, New York: Lovell, 1891);

The Doings of Raffles Haw (New York: Lovell, 1891; London: Cassell, 1892);

The Adventures of Sherlock Holmes (London: Newnes, 1892; New York: Harper, 1892);

The Great Shadow (Bristol: Arrowsmith, 1892; New York: Harper, 1893);

The Refugees (3 volumes, London: Longmans, Green, 1893; 1 volume, New York: Harper, 1893);

Jane Annie; or, The Good Conduct Prize: A New and Original English Comic Opera, by Doyle and J. M. Barrie (London: Chappell, 1893);

The Great Shadow and Beyond the City (Bristol: Arrowsmith, 1893);

The Memoirs of Sherlock Holmes (London: Newnes, 1894; New York: Harper, 1894);

An Actor's Duel and The Winning Shot (London: Dicks, 1894);

Round the Red Lamp: Being Facts and Fancies of Medical Life (London: Methuen, 1894; New York: Appleton, 1894);

The Parasite (London: Constable, 1894; New York: Harper, 1895);

The Stark Munro Letters (London: Longmans, Green, 1895; New York: Appleton, 1895);

The Exploits of Brigadier Gerard (London: Newnes, 1896; New York: Appleton, 1896);
Rodney Stone (London: Smith, Elder, 1896; New York: Appleton, 1896);
Uncle Bernac: A Memory of the Empire (London: Smith, Elder, 1897; New York: Appleton, 1897);
The Tragedy of the Korosko (London: Smith, Elder, 1898); republished as *A Desert Drama* (Philadelphia: Lippincott, 1898);
Songs of Action (London: Smith, Elder, 1898; New York: Doubleday & McClure, 1898);
A Duet with an Occasional Chorus (London: Richards, 1899; New York: Appleton, 1899);
Hilda Wade: A Woman with Tenacity of Purpose, by Doyle and Grant Allen (London: Richards, 1900; New York: Putnam, 1900);
The Green Flag and Other Stories of War and Sport (London: Smith, Elder, 1900; New York: McClure, Phillips, 1900);
The Great Boer War (London: Smith, Elder, 1900; New York: McClure, Phillips, 1900);
The Immortal Memory (Edinburgh: Mitchell, 1901);
The Hound of the Baskervilles (London: Newnes, 1902; New York: McClure, Phillips, 1902);
The War in South Africa: Its Cause and Conduct (London: Smith, Elder, 1902; New York: McClure, Phillips, 1902);
Adventures of Gerard (London: Newnes, 1903; New York: McClure, Phillips, 1903);
A Duet (A Duologue) (London: French, 1903);
The Return of Sherlock Holmes (London: Newnes, 1905; New York: McClure, Phillips, 1905);
The Fiscal Question: Treated in a Series of Three Speeches (Hawick, Scotland: Henderson, 1905);
An Incursion into Diplomacy (London: Smith, Elder, 1906);
Sir Nigel (London: Smith, Elder, 1906; New York: McClure, Phillips, 1906);
The Story of Mr. George Edalji (London: Roberts, 1907); republished as *The Case of Mr. George Edalji* (Putney: Blake, 1907);
Through the Magic Door (London: Smith, Elder, 1907; New York: Doubleday, Page, 1908);
The Croxley Master: A Great Tale of the Prize Ring (New York: McClure, Phillips, 1907);
Waterloo (London & New York: French, 1907);
Round the Fire Stories (London: Smith, Elder, 1908; New York: McClure, Phillips, 1908);
The Crime of the Congo (London: Hutchinson, 1909; New York: Doubleday, Page, 1909);
Divorce Law Reform: An Essay (London: Divorce Law Reform Union, 1909);
Why He Is Now in Favour of Home Rule (London: Liberal Publication Department, 1911);
Songs of the Road (London: Smith, Elder, 1911; Garden City, N.Y.: Doubleday, Page, 1911);
The Last Galley: Impressions and Tales (London: Smith, Elder, 1911; Garden City, N.Y.: Doubleday, Page, 1911);
The Speckled Band: An Adventure of Sherlock Holmes (New York & London: French, 1912);
The Case of Oscar Slater (London: Hodder & Stoughton, 1912; New York: Doran, 1913);
The Lost World (London: Hodder & Stoughton, 1912; New York: Doran, 1912);
The Poison Belt (London: Hodder & Stoughton, 1913; New York: Doran, 1913);
Great Britain and the Next War (Boston: Small, Maynard, 1914);
To Arms! (London: Hodder & Stoughton, 1914);
The German War (London: Hodder & Stoughton, 1914);
The Valley of Fear (London: Smith, Elder, 1915; New York: Doran, 1915);
A Visit to Three Fronts (London: Hodder & Stoughton, 1916; New York: Doran, 1916);
The British Campaign in France and Flanders, 6 volumes (London: Hodder & Stoughton, 1916–1919; New York: Doran, 1916–1920);
His Last Bow: Some Reminiscences of Sherlock Holmes (London: John Murray, 1917); republished as *His Last Bow: A Reminiscence of Sherlock Holmes* (New York: Doran, 1917);
The New Revelation (London: Hodder & Stoughton, 1918; New York: Doran, 1918);
Danger! and Other Stories (London: John Murray, 1918; New York: Doran, 1919);
The Vital Message (London: Hodder & Stoughton, 1919; New York: Doran, 1919);
Spiritualism and Rationalism (London: Hodder & Stoughton, 1920);
The Wanderings of a Spiritualist (London: Hodder & Stoughton, 1921; New York: Doran, 1921);
The Coming of the Fairies (London: Hodder & Stoughton, 1922; New York: Doran, 1922);
The Poems of Sir Arthur Conan Doyle: Collected Edition (London: John Murray, 1922);
Tales of the Ring and Camp (London: John Murray, 1922); republished as *The Croxley Master and Other Tales of the Ring and Camp* (New York: Doran, 1922);
Tales of Terror and Mystery (London: John Murray, 1922); republished as *The Black Doctor and Other Tales of Terror and Mystery* (New York: Doran, 1922);
Tales of Twilight and the Unseen (London: John Murray, 1922); republished as *The Great Keinplatz Experiment and Other Tales of Twilight and the Unseen* (New York: Doran, 1922);

The Case for Spirit Photography (London: Hutchinson, 1922; New York: Doran, 1923);
Our American Adventure (London: Hodder & Stoughton, 1923; New York: Doran, 1923);
Three of Them: A Reminiscence (London: John Murray, 1923);
Our Second American Adventure (London: Hodder & Stoughton, 1924; Boston: Little, Brown, 1924);
Memories and Adventures (London: Hodder & Stoughton, 1924; Boston: Little, Brown, 1924; revised edition, London: Hodder & Stoughton, 1930);
Psychic Experiences (London & New York: Putnam, 1925);
The Land of Mist (London: Hutchinson, 1926; New York: Doran, 1926);
The History of Spiritualism, 2 volumes (London: Cassell, 1926; New York: Doran, 1926);
Pheneas Speaks (London: Psychic Press & Bookshop, 1927; New York: Doran, 1927);
The Case-Book of Sherlock Holmes (London: John Murray, 1927; New York: Doran, 1927);
The Maracot Deep and Other Stories (London: John Murray, 1929; Garden City, N.Y.: Doubleday, Doran, 1929);
Our African Winter (London: John Murray, 1929);
The Edge of the Unknown (London: John Murray, 1930; New York: Putnam, 1930);
The Field Bazaar (London: Atheneum, 1934; Summit, N.J.: Pamphlet House, 1947);
The Professor Challenger Stories (London: John Murray, 1952);
The Crown Diamond: An Evening with Sherlock Holmes, a Play in One Act (New York: Privately printed, 1958);
Strange Studies from Life: Containing Three Hitherto Uncollected Tales (New York & Copenhagen: Candlelight Press, 1963);
Tales of Terror and Mystery (New York: Penguin, 1977);
The Best Supernatural Tales of Arthur Conan Doyle, edited by E. F. Bleiler (New York: Dover, 1979);
The Best Science Fiction of Arthur Conan Doyle, edited by Charles G. Waugh and Martin H. Greenberg (Carbondale: Southern Illinois University Press, 1981);
The Unknown Conan Doyle: Uncollected Stories, edited by John Michael Gibson and Richard Lancelyn Green (London: Secker & Warburg, 1982; Garden City, N.Y.: Doubleday, 1984);
The Unknown Conan Doyle: Essays on Photography, edited by Gibson and Green (London: Secker & Warburg, 1982; North Pomfret, Vt.: David & Charles, 1983);
Masterworks of Crime and Mystery, edited by Jack Tracy (New York: Dial, 1982);
The Uncollected Sherlock Holmes, edited by Green (London: Penguin, 1983);
The Secret Conan Doyle Correspondence, edited by Leslie Harper (Provo, Utah: Hascom Publishers, 1986).

PLAY PRODUCTIONS: *Jane Annie,* by Doyle and J. M. Barrie, London, Savoy Theatre, 13 May 1893;
Foreign Policy, London, Terry's Theatre, 3 June 1893;
A Story of Waterloo, Bristol, Princess's Theatre, 21 September 1894; London, Lyceum Theatre, 4 May 1895;
Halves, Aberdeen, Scotland, Haymarket Theatre, 10 April 1899; London, Garrick Theatre, 10 June 1899;
Brigadier Gerard, London, Imperial Theatre, 3 March 1906;
The Fires of Fate, London, Lyric Theatre, 15 June 1909;
The House of Temperley, London, Adelphi Theatre, 27 December 1909;
A Pot of Caviare, London, Adelphi Theatre, 19 April 1910;
The Speckled Band, London, Adelphi Theatre, 4 June 1910;
The Crown Diamond, London, London Coliseum, 16 May 1921.

Sir Arthur Conan Doyle is better known for his detective, Sherlock Holmes, than for his stories and novels of fantasy and science fiction, but his contributions to these other genres were formidable. His enormous popularity as a result of the Sherlock Holmes stories allowed him to command a large audience for nearly anything he wrote, and those who read his fantasy and science fiction almost always were rewarded with skillfully written, entertaining, and sometimes thought-provoking tales. Most scholars of Doyle's work agree that in the Sherlock Holmes stories, he advanced plot formulas, narrative conventions, and characters inherited from writers he admired, such as Edgar Allan Poe and Émile Gaboriau, to realize a truly new and powerful literary form, what has come to be known as the classical detective story. Few would argue that his work in fantasy and science fiction advanced those genres in the same way, but his fantasy is influential, and his science fiction, mainly *The Lost World* (1912), inspired future writers and, along with the novels of H. G. Wells, helped to expand the audience that had been created to a large extent by the French novelist Jules Verne.

During Doyle's career, fantasy blossomed into a large and varied field ranging from the Gothic romances of Bram Stoker and Robert Louis Stevenson through the exotic fantasy adventures of H. Rider Haggard and Edgar Rice Burroughs to the technical science adventures of Jules Verne to the light and whimsical tales and poems of J. M. Barrie and Lewis Carroll as well as to the satiric works of Oscar Wilde, to mention some of the main currents. If fantasy is seen as those narratives that present story elements that violate audience consensus concerning what is real in its world, then Verne's science adventures are as much fantasy as Alice's excursion behind the looking glass. At the turn of the twentieth century, Verne and Wells between them divided science fantasy into two branches; Verne emphasized possible technology within a fairly narrow range, and Wells emphasized exotic adventure in invented worlds and speculation about the future, devising whatever technology was necessary to make such stories possible. Both of these branches have come to be seen as generally distinct from other types of fantasy in part because they rely on technology as opposed to magic or the supernatural to produce those elements of the story that most readers at the time would agree to be fantastic or outside the realm of currently possible experience.

While Doyle's fantasies follow the convention of depending upon magic or the apparently supernatural, those of his stories that have been called science fiction show relatively little interest in technology. Rather, they tend to have scientists as characters and to deal with scientific topics. Still, in each of the major works, Doyle develops ideas that have come to be important in science fiction: the journey to another world; confronting a cosmic catastrophe such as the destruction of humanity; and the discovery and exploitation of hitherto untapped human mental powers. Especially when Doyle works with "spiritualist phenomena" his stories seem to occupy a border between science fiction and fantasy. This occurs in his early comic story, "The Great Keinplatz Experiment" (1885), first published in *Belgravia* and collected in *The Captain of the Polestar and Other Tales* (1890), in which a scientist experimenting with mesmerism unintentionally exchanges bodies with the student who wishes to marry his daughter. Whether mesmerism in this case is to be seen as a technique or magic is not easily decided.

Arthur Conan Doyle was born on 22 May 1859 in Edinburgh, Scotland. In many ways his family were outsiders who were made at least sometimes to feel their differences. They were Roman Catholic in a center of Protestantism. They were artistic in a practical, puritanical society. His grandfather and his uncle were successful illustrators. His uncle Richard Doyle gained fame drawing for *Punch*. Though Arthur's father, Charles, became clerk of the Board of Works in Edinburgh, he also had artistic ambitions, and he illustrated the first edition of his son's *A Study in Scarlet* (1888), the first tale of Sherlock Holmes, but Doyle grew up without a strong father. Charles suffered from depression, alcoholism, and epilepsy and was institutionalized at various locations from 1879 until his death by epileptic seizure in 1893. Michael Baker's 1978 publication of one of Charles's notebooks, *The Doyle Diary,* suggests that Charles was kept in institutions more because of the family's needs than because of his own.

Separated from the mainstream by religion, artistic temperament, an unsuccessful and mentally ill father, and resulting comparative poverty, Doyle was also separated racially by the Irish origins of his family. His Irish-born mother, Mary Foley, acted as the head of the family, though Dr. Bryan Charles Waller, who came to board with the Doyles and eventually shared his home with them, took on some of these functions. Arthur was the second of Mary's seven children, the older of two boys. Seeing talent in him, Mary carefully arranged his education, sending him to Jesuit schools at Stoneyhurst and at Feldkirch, Austria, at considerable financial sacrifice. She fostered his medical studies at the University of Edinburgh and remained his close friend and consultant until her death in 1921. Doyle moved himself further from the mainstream of popular thought when he broke with his Roman Catholic faith and became an agnostic in the years before his marriage. Biographical study has yet to work through the influence upon his works of these many ways in which Doyle inhabited his world as an outsider.

Doyle began his writing career while at the University of Edinburgh, where he started medical studies in 1876. At the university Doyle met two professors who became models for his most famous literary creations, though he drew upon himself and others in his characterizations: Dr. Joseph Bell was the original model for Sherlock Holmes, and William Rutherford, the model for Professor George Edward Challenger, the scientist hero of several of his science-fiction stories. While at the university Doyle in 1879 published his first story in *Chambers's Journal,* "The Mystery of Sasassa Valley," which was collected in *Mysteries and Adventures* (1890).

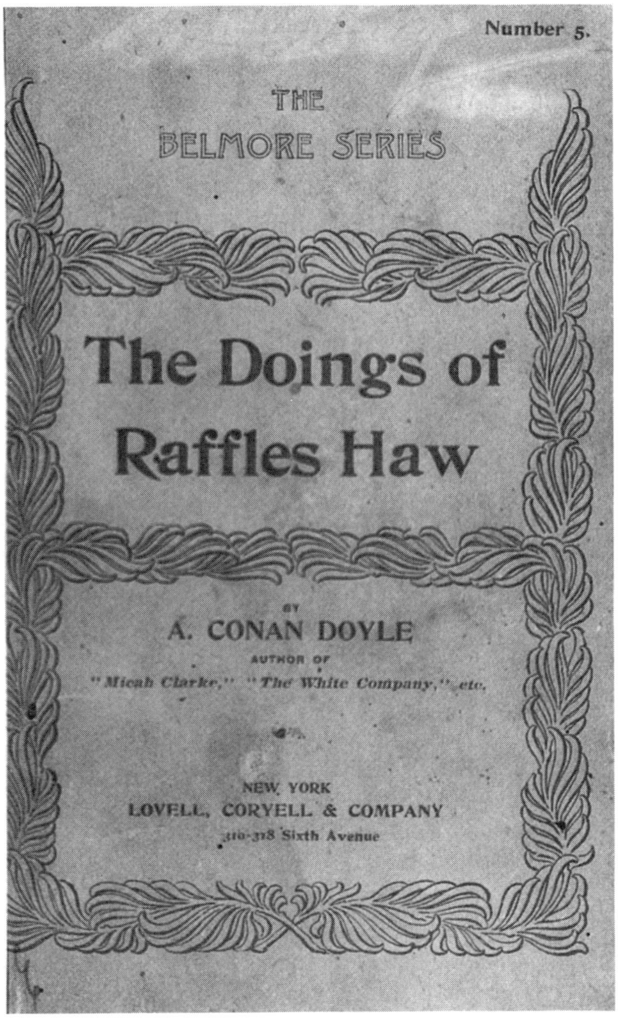

Cover for Doyle's first science-fiction novel, in which an idealistic scientist discovers that material comforts corrupt spiritually

Large, healthy, and energetic, Doyle enjoyed sports and physical activities. This interest led him to seek a vigorous and adventurous physical as well as mental life. Before completing his bachelor's degree in medicine, he served briefly in 1880 as surgeon on the arctic whaler *Hope,* gathering some of the material for his later fantasy story "The Captain of the Polestar," first published in *Temple Bar* magazine in 1883. In this story a whaler captain, grieving over the loss of his fiancée, comes to believe he sees her in the blowing snow on the ice floes that surround his ship. When he leaves the ship to follow this vision, he eventually freezes to death. Those who find his body also think they may see this mysterious snow-woman.

After finishing his bachelor of medicine degree in 1881, Doyle was medical officer on a ship traveling to Africa's west coast. This trip provided materials for one of his more-notorious early stories, "J. Habakuk Jephson's Statement," a hoax in the manner of Poe that appeared in the prestigious *Cornhill Magazine* in 1884 and was collected in *The Captain of the Polestar.* Jephson claims to be the only survivor of the *Mary Celeste*–Doyle changes this to *Marie Celeste*–a ship found abandoned but in excellent condition off the coast of Portugal in 1872. He explains how the ship was taken over by Septimus Goring, an American mulatto who has tired of his secret campaign of revenge against American whites for enslaving Africans and has decided to settle in Northwest Africa after eliminating the passengers of this ship. Goring spares Jephson because he possesses a good-luck piece given him by a grateful black man whom he helped during the Civil War. The story achieved notoriety because the British official at Gibraltar who oversaw the salvaging of the *Mary Celeste* treated the anonymously published story as an intentional hoax and went to considerable lengths to deny its truth in public documents.

Though Doyle worked earnestly at establishing a medical practice, first in an ill-fated partnership with a university friend–treated fictionally in *The Stark Munro Letters* (1895)–and later on his own in Plymouth, then in a Portsmouth suburb, and finally in London, he found earning a living quite difficult. He continued writing during his attempt at a medical career but with limited success. Among these early stories was a parody of Gothic fantasies told in a style of an author he admired, Tobias Smollett. "Selecting a Ghost," first published in 1883 in *London Society* and collected in *Mysteries and Adventures,* is narrated by Silas D'Odd, who achieved wealth selling groceries and retired into the nobility, changing his given name to Argentine. Deciding he needs a ghost to haunt his feudal mansion, he sends up to London for one that will guarantee his respectability. A notorious burglar pretends to be an agent for ghosts needing haunts and drugs Silas to gain time to steal his newly acquired silver. During the burglary Silas has visions of likely ghosts for his mansion. Doyle arranges this bagatelle so that it appears to be an actual ghost story until the drug wears off.

Doyle completed his M.D. in 1885 and then married Louisa Hawkins. More desperate for a comfortable income and still gaining little from medicine, he put more of his energy into writing, completing two novels within the first year of his marriage, one of which was eventually published, *The Firm of Girdlestone* (1890), a business novel. Also among these early works is a little-regarded novel of supernatural retribution, *The Mystery of Cloomber* (1889), which reveals his early interest in spiritualism and the occult. General Heatherstone retires from the Indian service and takes his family to Cloomber Castle in Scotland. He hopes to avoid the vengeance of Buddhist magicians for killing one of their adepts, but they pursue him, using their magical powers to enter his heavily guarded castle.

A decisive point in Doyle's medical and writing careers came in 1887, when he published *A Study in Scarlet,* his first tale of Sherlock Holmes, in *Beeton's Christmas Annual.* Though this publication did not immediately make him famous, it was successful enough to be republished as a book in 1888. Thus encouraged, he completed *Micah Clarke* (1889), a historical novel inspired by the work of Sir Walter Scott. Indeed, Scott was his literary model, and Doyle hoped during most of his life to achieve the kind of fame Scott achieved by writing the kinds of novels Scott wrote. While he did achieve Scott's fame, it was not for writing historical romances, though *Micah Clarke* and several of Doyle's other historical novels were popular successes. He now found it easier to publish his work, and so more and more of his energy went into writing.

Doyle concentrated his efforts on historical novels such as *The White Company* (1891), but he found that he needed to publish magazine short stories to sustain his income. Among these were some of his first publications in fantasy and science fiction: "The American's Tale" and "The Silver Hatchet," first published in *London Society* in 1880 and 1883, respectively, and collected in *Mysteries and Adventures;* "John Barrington Cowles" (*Cassell's Saturday Journal,* 1886), "A Literary Mosaic" (*The Boy's Own Paper,* 1886), and "The Ring of Thoth" (*Cornhill Magazine,* 1890), all collected in *The Captain of the Polestar and Other Tales;* and "The Winning Shot," collected in *An Actor's Duel and The Winning Shot* (1894). In these stories, as in his development of the detective story, the influence of Poe is often detectable.

"The American's Tale" is a tall tale in the manner of Bret Harte and Mark Twain. A westerner, who claims first to be from Arizona and then from Montana, narrates an adventure in which the villain is killed by a giant Venus's-flytrap. "The Silver Hatchet" is set in Budapest. A set of mysterious murders in which people are killed by their best friends is solved when investigators discover a written curse hidden in the silver handle of the antique hatchet that is the murder weapon. Anyone who grasps the weapon will kill his best friend with it. "A Literary Mosaic" is the playful narration of a failing writer who dreams that several of the novelists he admires meet to help him write a story.

Doyle in his study, 1894

Two of the stories of this period deal with a favorite Doyle theme, mesmerism. The little known "The Winning Shot" presents a group of documents, narratives by various people directly and indirectly involved in the events. Doyle often used this device, and his contemporary, Bram Stoker, used it with great dexterity in *Dracula* (1897). This story opens with Lottie Underwood explaining why she has placed an advertisement in the *Times,* warning the public against Octavius Gaster. She tells how she and her fiancé, Charley Pillar, first met Gaster during a walk on the moor in Devon. They befriended the charming traveler, but then he fell in love with Lottie. When he tried to rape her and then accused her of sexual infidelity, Charley expelled him from the estate. Like the other villainous mesmerists of Doyle's earlier stories, Gaster has cultivated the power of projecting his will on others. Gaster later appears at a local shooting match in which Charley is participating. When Charley is called upon to make the last shot upon which the match depends, he is bothered by a vision of a man who looks much like himself between himself and the target. Lottie and others notice Gaster on a nearby hill seemingly in great concentration. Charley's shot wins the match, but it also kills him in a way not explained. Finally, as Lottie's mother reports in a separate document, Gaster apparently carries Lottie off in a mysterious way. In "John Barrington Cowles" the beautiful

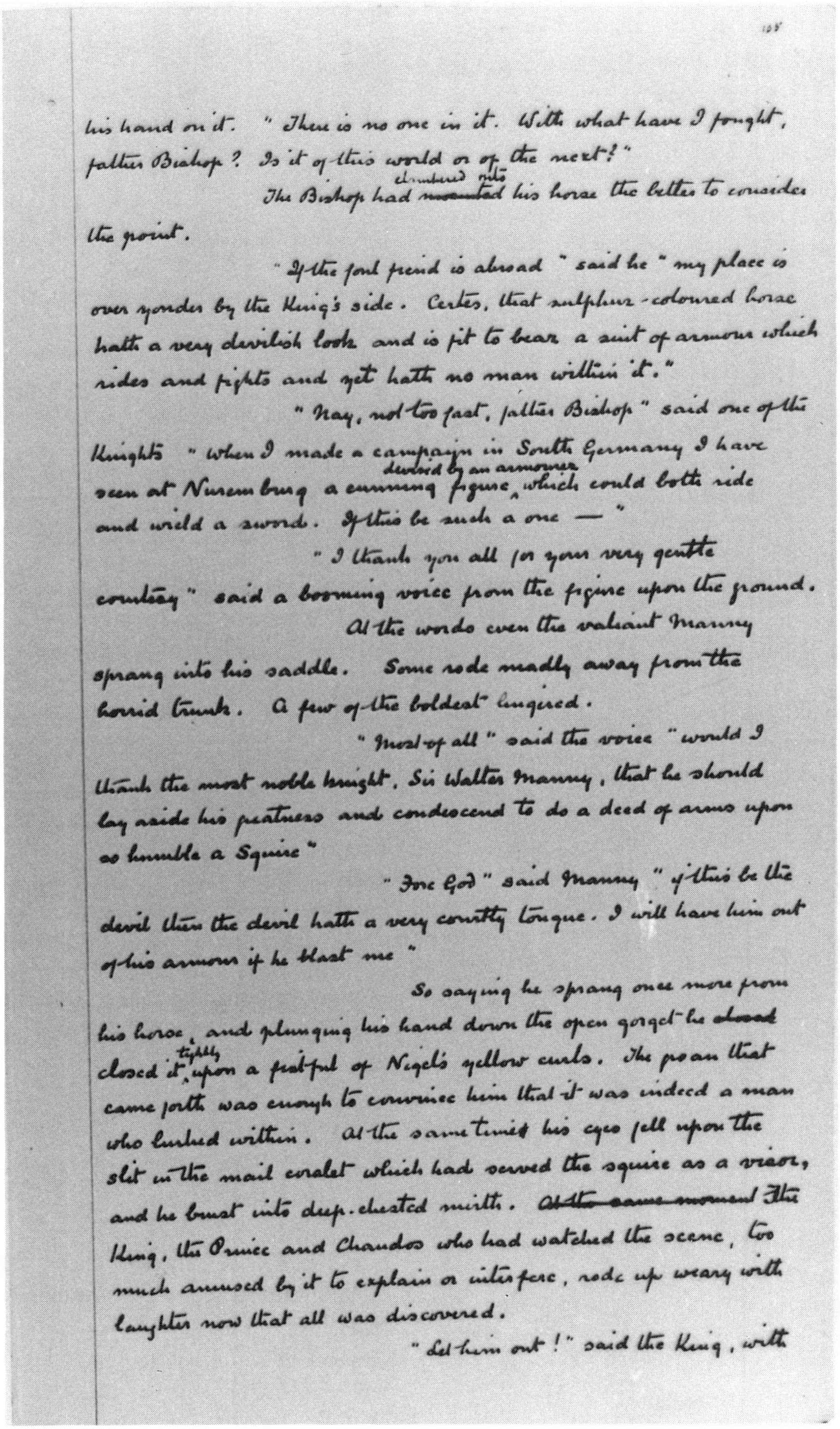
108

his hand on it. "There is no one in it. With what have I fought, father Bishop? Is it of this world or of the next?"

The Bishop had ~~mounted~~ clambered onto his horse the better to consider the point.

"If the foul fiend is abroad" said he "my place is over yonder by the King's side. Certes, that sulphur-coloured horse hath a very devilish look and is fit to bear a suit of armour which rides and fights and yet hath no man within it."

"Nay, not too fast, father Bishop" said one of the Knights "when I made a campaign in South Germany I have seen at Nuremburg a cunning figure devised by an armourer which could both ride and wield a sword. If this be such a one —"

"I thank you all for your very gentle courtesy" said a booming voice from the figure upon the ground.

At the words even the valiant Manny sprang into his saddle. Some rode madly away from the horrid trunk. A few of the boldest lingered.

"Most of all" said the voice "would I thank the most noble knight, Sir Walter Manny, that he should lay aside his greatness and condescend to do a deed of arms upon so humble a Squire"

"Fore God" said Manny "if this be the devil then the devil hath a very courtly tongue. I will have him out of his armour if he blast me"

So saying he sprang once more from his horse, and plunging his hand down the open gorget he ~~closed~~ closed it tightly upon a fistful of Nigel's yellow curls. The groan that came forth was enough to convince him that it was indeed a man who lurked within. At the same time his eyes fell upon the slit in the mail corslet which had served the squire as a visor, and he burst into deep-chested mirth. ~~At the same moment~~ The King, the Prince and Chandos who had watched the scene, too much amused by it to explain or interfere, rode up weary with laughter now that all was discovered.

"Let him out!" said the King, with

Page from a manuscript for Doyle's Sir Nigel, *dated 1905 (MA 3217, Pierpont Morgan Library)*

and sinister Kate Northcott draws men to fall in love with her, then reveals a dark secret about what they must do if they marry her. Each one then refuses to marry her, and she takes revenge upon them. Like the title character in Poe's "Ligeia" (1838), Kate has a powerful will and fascinating eyes. Her revenge consists of exerting the power of her will over these men, mesmerizing them at a distance, and driving them to suicide. Doyle later developed these ideas at greater length in *The Parasite* (1894).

Another favorite source of materials for Doyle is ancient Egypt. "The Ring of Thoth" is a sophisticated, ironic story with two levels of plot. John Vansittart Smith, an aspiring Egyptologist, is humbled at the Louvre in Paris when he accidentally meets Sosra, a genuine four-thousand-year-old Egyptian priest-scientist who developed an elixir of life. Because he can never die, he cannot rejoin his beloved Princess Atma in the next life. After centuries of searching, Sosra has found in the Louvre the Ring of Thoth, which contains the antidote to immortality. Sosra's story shows the narrator how little he knows about Egypt, how little he has mastered of his own profession, and also how shallow his passions seem next to those of this accomplished Egyptian.

Though Doyle, at the request of *Lippincott's Magazine,* had provided *The Sign of Four,* another Holmes novel, in 1890, it was "A Scandal in Bohemia," appearing in the new *Strand Magazine* early in 1891, that finally launched him toward stardom as a popular writer. *The Strand Magazine* bought twelve tales of Sherlock Holmes over the next year, and though Doyle wrote them quickly, the public demand seemed insatiable. Wanting to turn away from what he considered potboilers after the first series of six in order to work on projects closer to his heart, he at first demanded "ridiculous" prices, £50 per story for the second series, and then £1,000 for the next series of twelve. It seemed *The Strand Magazine* would pay any price for them. To escape this remunerative trap, he decided to kill Holmes. The last story in the series for 1892–1893, collected as *The Memoirs of Sherlock Holmes* (1894), was "The Final Problem," in which Holmes dies, falling down the Swiss Reichenbach Falls in the grip of the sinister and ruthless Moriarty.

While this "final" series was appearing monthly in *The Strand Magazine,* Doyle worked energetically, writing *The Refugees* (1893), another historical novel, and publishing two other novels and several stories. During this period his career flourished, and all seemed to be going well at home. Louisa bore a daughter, Mary, in 1889 and a son, Kingsley, in 1891. During 1890 and 1891 the family lived in Vienna, where Doyle prepared for another attempt at professional medicine, studying eye surgery. The practice he set up after returning to London was a failure. He reportedly had no patients at all and used his office hours to write. Doyle gave up his medical practice in 1891 and made writing fiction his profession.

The Doings of Raffles Haw (1891) is Doyle's first science-fiction novel. Raffles Haw is a brilliant young chemist who has discovered a means of converting lead into gold. This technique reduces heavier metals to lighter ones, making it possible to reduce any metal on the periodic table of the elements to the next lighter metal. This process can be repeated until one arrives at a primal dust. Haw is also an idealist who wishes to use his power to become a benefactor to humanity, dedicating the magnificent wealth he can so easily create to alleviate suffering and to bring people closer together. He finds, however, that his wealth corrupts almost everyone he meets, isolating him from humanity, and that, ultimately, making people materially comfortable destroys them spiritually. He comes to see that human refinement comes through sorrow. This theme eventually dominated Doyle's work and his life. He believed strongly that Western civilization was declining because of a loss of appreciation for spiritual values in the face of an overwhelming materialism. *The Doings of Raffles Haw* makes a case that the human spirit is strengthened by adversity and weakened by material security. When Haw learns that he cannot use his wealth to benefit humanity, he reduces all his remaining gold to dust and destroys his record of the process. Then he dies, perhaps of a broken heart.

Among Doyle's short stories of the early 1890s, several belong to the fantasy and science-fiction genres. First published in the *Idler* in 1892 and collected in *The Last Galley: Impressions and Tales* (1911), "De Profundis," like "Selecting a Ghost," belongs to the "explained" Gothic, a genre in which fantastic events appear to occur, only to be explained as natural events at the end of the story. Emily Lawson and the narrator witness an apparition at sea that she interprets as a final message from her dying fiancé. It turns out, however, that he died several days before the apparition and was buried at sea at about the spot where they saw the apparition. The narrator concludes that they saw the actual body rise, released from its restraining weights. Their vision was a coincidence, not a spirit communication.

"Lot No. 249," published in *Harper's* in 1892 and collected in *Round the Red Lamp: Being Facts and Fancies of Medical Life* (1894), is another story of

"SUDDENLY, OUT OF THE DARKNESS, OUT OF THE NIGHT, THERE SWOOPED SOMETHING WITH A SWISH LIKE AN AEROPLANE."

(See page 616.)

"PROFESSOR CHALLENGER."

"HE WAS AT MY VERY HEELS. I WAS LOST."

(See page 134.)

Illustrations by Harry Rountree for the serialization of The Lost World *in* The Strand Magazine

Egyptology. Edward Bellingham discovers an ancient papyrus containing instructions for reanimating a mummy and making it serve him. He uses the mummy he has purchased, as part of Lot Number 249, to attack those people who have opposed his desires, including the story's hero, Abercrombie Smith, a medical student who threatens to expose him after learning what he is doing. Smith eventually forces Bellingham at gunpoint to destroy the mummy and the papyrus.

"The Los Amigos Fiasco," published in the *Idler* in 1892 and collected in *Round the Red Lamp,* is a humorous story of the American West. Los Amigos is a town so rich in electrical power that the authorities decide to electrocute rather than to hang murderers. Sure that more is better, they decide to use the maximum power on their first convict, despite warnings from Peter Stupnagel. As he predicts, however, the effect of massive electricity on the human body is to strengthen it, so that after several attempts, they have made their murderer virtually indestructible.

Sherlock Holmes met his doom in the December 1893 *Strand Magazine.* Doyle's pleasure at his demise was small, for 1893 had proven a very difficult year. About a month after the death of his father, Doyle learned that Louisa had tuberculosis, usually fatal then. In November he and Louisa went to Switzerland, where they hoped the climate would help her, so he was not at home to hear the public outcry when Holmes died. During this stay in Switzerland Doyle introduced to that country the Norwegian sport of cross-country skiing. Also in November Doyle joined the Society for Psychic Research, where he heard Arthur Balfour lecture on mesmerism. After completing *The Stark Munro Letters* early in 1894, Doyle began work on *The Parasite,* a novel of mesmerism.

Doyle's interest in spiritualism and psychic research was virtually lifelong. He attended his first séance in 1879, before he renounced his Roman Catholicism. Joining the Society for Psychic Research, then, was not merely a reaction to his wife's illness but the continuation of his persistent search for religious truth in response to his sense of civilization's decline into self-destructive materialism. *The Parasite* reveals some of the reasons for Doyle's interest in spiritualism, but it may reveal more about his response to Louisa's disease. Critics and biographers who have attempted to explain Doyle's decision to suppress the work after it was serialized in *Harper's* and then issued as a novella have produced provocative and persuasive explanations. It is quite possible that in his anguish, Doyle wrote a book that inadvertently revealed more about his inner self than he wished, especially about his sexual struggles, for Helen Penelosa seems quite clearly to be an external manifestation of Professor Austin Gilroy's sexual desire.

Gilroy is a physiologist, quite skeptical of a colleague's psychological research into mesmerism until he sees Helen demonstrate her ability on his fiancée, Agatha. He begins a study of mesmerism, with Helen performing hypnosis on him as subject. Helen, who is much older than he and neither physically nor spiritually attractive to him, falls in love with him and uses her growing power over him to force him to behave as if he loves her. When he realizes this and breaks with her, she refuses to let him go. Her power over him goes far beyond the known powers of hypnotism. She is able to project her soul out of her body from a distance and to occupy his, overpowering his soul. She takes control of him during his lectures, turning them into nonsense. She leads him to attempt bank robbery and finally to attempt to scar Agatha with acid. However, her efforts seem finally to destroy her health, and she dies before she can force him to harm Agatha. The story is interesting especially for its similarity to other contemporary stories of persecution, such as Guy de Maupassant's "The Horla" (1887), in which possession by an invisible other is a main element, and Henry James's *The Turn of the Screw* (1898), in which the governess who believes she sees ghosts is, like Gilroy, trapped alone in her vision, constrained from sharing it with others and so getting the kind of help she desperately needs.

Between 1894 and 1908 Doyle continued to publish vigorously in several genres, but he produced little in fantasy and science fiction. He collected just three notable short stories in *Round the Fire Stories* (1908)–"The Brown Hand" and "Playing with Fire," originally published in *The Strand Magazine* in 1899 and 1900, and "The Leather Funnel," first published in *McClure's* in 1900. His personal life changed dramatically when in 1897 he met and fell in love with twenty-four-year-old Jean Leckie. Though later commentators continue to look upon this relationship with amazement, it seems indisputable that Doyle maintained a platonic relationship with her until after Louisa's death in 1906. He married Jean in 1907. They had three children: Denis (1909), Adrian (1910), and Lena Jean (1912).

In "The Brown Hand," Dr. Hardacre explains how he became heir to Sir Dominick Holden by exorcising a ghost. Holden, while in India, amputated the hand of a man who required that he preserve it because his religion teaches that his

"IT IS, IN MY OPINION, THE END OF THE WORLD."

366 THE STRAND MAGAZINE.

"CATHEDRAL AND CHURCHES ARE FULL TO OVERFLOWING. THE DEAD OUTNUMBER THE LIVING."

he had sat while he announced the doom of the planet. "Come," said he. "If there is little time left, there is the more need that we should spend it in sober and reasonable enjoyment."

And, indeed, it proved to be a very merry meal. Do not think us flippant, my good reader. The full solemnity of the event loomed ever at the back of our minds and tempered our thoughts. But surely it is the

THE POISON BELT. 373

at the little birds under the trees!"

We drew four chairs up to the long, low window, the lady still resting with closed eyes upon the settee. I remember that the monstrous and grotesque idea crossed my mind—the illusion may have been heightened by the heavy stuffiness of the air which we were breathing—that we were in four front seats of the stalls at the last act of the drama of the world.

"THE CHAUFFEUR DOWN IN THE YARD HAD MADE HIS LAST JOURNEY."

(To be continued.)

THE POISON BELT. 489

hip-hurrah about it," said Lord John. "What does it matter?"

"It just matters this, that the world is a living instead of a dead one. If you had the scientific imagination, you would cast your mind forward from this one fact, and you would see some few millions of years hence—a mere passing moment in the enormous flux of the ages—the whole world teeming once more with the animal and human life which will spring from this tiny root. You have seen a prairie fire, where the flames have swept every trace of grass or plant from the surface of the earth and left only a blackened waste. You would think that it must be for ever desert. Yet the roots of growth have been left behind, and when you pass the place a few years hence you can no longer tell where the black scars used to be. Here in this tiny creature are the roots of growth of the animal world, and by its inherent development, and evolution, it will surely in time remove every trace of this incomparable crisis in which we are now involved."

"Dooced interestin'!" said Lord John, lounging across and looking through the microscope. "Funny little chap to hang number one among the family portraits. Got a fine big shirt-stud on him!"

"The dark object is his nucleus," said

"'YOU CAN SEE IT FOR YOURSELVES,' HE KEPT REPEATING."

Vol. xlv.—50.

Illustrations by Harry Rountree for the serialization of The Poison Belt *in* The Strand Magazine

body must be whole in order to enter the afterlife. The preserved hand is later lost in a fire, and after the man's death he visits Holden each night to demand its return. Dr. Hardacre obtains a substitute brown hand from a corpse through his medical contacts, but fails on the first attempt because he provides the wrong hand. The second attempt gives the ghost a pair, and it is content.

In "Playing with Fire," a group of dabblers in spiritualism of varying sorts is joined by a French guest who likes to have fun. He helps them use their medium better to make contact with a dead spirit. That spirit tells them about the serious side of their activity, the opportunity for spiritual knowledge. The Frenchman, however, likes dramatic effects, and so he helps them to materialize a unicorn. They frighten the unicorn, and it causes damage and minor injuries before it dissipates. This story implicitly criticizes those who play with spiritual forces for entertainment, and its most authoritative voice argues for the serious study and use of spiritual phenomena.

"The Leather Funnel" also deals with an aspect of the spirit world, the tendency of places and objects to retain spiritual echoes of traumatic events associated with them. The narrator learns the history and use of a mysterious antique leather funnel by sleeping next to it and dreaming the events in which it was involved. The funnel was used to punish a notorious poisoner by forcing gallons of water into her stomach.

Between 1894 and 1908 Doyle wrote many popular stories for *The Strand Magazine,* including several about Etienne Gerard, a comic soldier in Napoleon's army; these stories were collected in *The Exploits of Brigadier Gerard* (1896) and *Adventures of Gerard* (1903). His prizefighting novel *Rodney Stone* (1896) was a popular success, serialized with illustrations in *Strand Magazine*. He added to his less successful series of historical novels about the Napoleonic era in France: *The Great Shadow* (1892), *The Refugees,* and *Uncle Bernac* (1897). *Songs of Action* (1898) collected his poems. *Sir Nigel* (1906) was a companion volume to *The White Company,* a historical adventure novel set in medieval England. *The Green Flag and Other Stories of War and Sport* (1900) brought together some of his recent stories. Among these was "The New Catacomb" (1898), which is rather indirectly about sport and more clearly a redoing of Poe's "The Cask of Amontillado." Julius Berger takes revenge upon Mr. Kennedy, a fellow researcher of Roman antiquities, for making a game of seducing Berger's fiancée. He leads Kennedy into an unexplored labyrinthine catacomb and abandons him. He is later found dead.

With *The Hound of the Baskervilles* (1902) Doyle resurrected Sherlock Holmes, though this story takes place before the detective's death in Switzerland. While vacationing after an unsuccessful bid for a seat in Parliament in 1900, Doyle heard the legends that became the inspiration for what is probably the most famous of his tales of Sherlock Holmes, and one of the Holmes stories that draws heavily upon Gothic conventions. Though the story usually is not characterized as a fantasy, still it strikingly resembles stories, such as those of Ann Radcliffe, that are sometimes called "explained" Gothic. Dr. Watson and Holmes explain the mysterious death of a gentleman and save his heir, both of whom are threatened by a "hellhound." The villain, Stapleton, is a distant relative. Should he eliminate the more immediate heirs without being detected, he could inherit the valuable Baskerville estate. To accomplish this, he terrorizes the men with a trained hound kept hidden in the treacherous moor. The first heir dies of fear, and the second is nearly killed before Holmes can reveal that the hound is really an ordinary dog rather than a spirit animal carrying out a curse that evil members of the family have incurred in the distant past. Until the end of the novella Doyle maintains the fantastic effect of keeping the characters, and, to some extent, the reader uncertain about whether the hound is an evil specter or an actual dog.

In 1901, while *The Hound of the Baskervilles* was appearing in *The Strand Magazine,* William Gillette's popular play, *Sherlock Holmes,* opened in London, and the clamor for more Holmes stories increased. Doyle was offered about $7,500 per story to write another series. He began this new series with "The Adventure of the Empty House," in which Holmes returns, announcing that he did not really fall over the falls and that he has had to hide for three years from surviving members of Moriarity's gang. This series was collected in *The Return of Sherlock Holmes* (1905). Once the ice was broken Doyle found more occasions to write stories about his much-loved detective hero. *The Valley of Fear* (1915), a novel, and two more collections were published during Doyle's lifetime: *His Last Bow* (1917) and *The Case-Book of Sherlock Holmes* (1927).

After the turn of the century Doyle increasingly played the role of a public figure and wrote nonfiction. In 1899 Doyle volunteered as a medical officer in the Boer War. He published *The Great Boer War* in 1900, a respected history of the war. *The War in South Africa: Its Cause and Conduct* (1902), a defense of the British role in the war, brought him his knighthood in 1902. In *The Story of Mr. George Edalji* (1907) he recounts his mostly successful effort to clear Edalji after the Straffordshire solicitor was mistakenly convicted for animal mutilation. This case and *The Case of Oscar Slater* (1912) were notable examples of Doyle's willingness to stake his enormous popular reputation to help those he thought unjustly accused. They also show that he could match his great detective's methods of investigation and argument in the pursuit of justice. An important result of his efforts was the establishment in England of a court of criminal appeals. In 1906 he again stood for Parliament but failed to be elected.

After Louisa's death and his remarriage, he continued to take interest in public events, writing for divorce-law reform and Irish home rule and against Belgian colonialism and woman's suffrage. He also traveled widely in Europe and America, publishing accounts of these trips. As World War I approached, he wrote about the threat of Germany and especially about England's lack of readiness for modern warfare. During the war he served as a correspondent and published a six-volume collection of his observations, *The British Campaign in France and Flanders* (1916–1919).

The next milestone in Doyle's career was his conversion to spiritualist beliefs, which his biographers, though they acknowledge that his drift in this direction was long and gradual, place in 1916. Apparently, in that year he became convinced that he had received a spirit message. Thereafter, he threw himself into this new religion that he believed promised to restore the spiritual balance against materialism that had worried him at least since he wrote *The Doings of Raffles Haw.* He wrote many volumes about spiritualism during the remainder of his life, including his respected two-volume *The History of Spiritualism* (1926). His great fame and considerable financial support helped to propel him rapidly to a position of leadership in the movement. After 1920 he was persuaded of the reality of fairies and wrote *The Coming of the Fairies* (1922) to present a case that later was admitted to have been a fraud perpetrated upon him.

Many of his contemporary readers as well as most of his later readers found his conversion to a religion that few people take seriously and that was, as he well knew, so pervaded by charlatanism and wishful thinking to be troubling. How could the creator of the great rationalist detective fall for such obvious silliness? The answer to this question is complex. One must take into account the spirit of the times and Doyle's position in society as a public man. Regardless of the actual merit of Doyle's beliefs, his best and most justly famous works of fantasy and science fiction date from this period, and they reveal clearly that, especially in the early years of his conversion, Doyle was no simpleminded dupe.

To Doyle's age spiritualism as a religion did not appear especially silly. His career coincides with a crisis in modernist thought, when Western intellectuals were coming to the conclusion that the materialist scientific view of the universe tended to erase meaning and human value from the cosmos without providing any adequate substitute. To practical-minded rationalists, spiritualism seemed to offer scientifically verifiable proof of the existence of the soul and of the continuation of the soul's life and identity after physical death. The active Society for Psychical Research, which attracted leading artists and thinkers such as William James, really did attempt to bring scientific rigor to the study of widely reported spirit phenomena. Such an organization could only increase the attractiveness of the associated religion of spiritualism for Doyle.

During the last two decades of his life, Doyle wrote his most notable science fiction and fantasy, the Professor Challenger stories. The character was featured in the novels *The Lost World, The Poison Belt* (1913), and *The Land of Mist* (1926) as well as in two stories collected in *The Maracot Deep and Other Stories* (1929), "When the World Screamed" (*Liberty,* 1928) and "The Disintegration Machine" (*The Strand Magazine,* 1929). Challenger became one of Doyle's favorite characters; he enjoyed occasionally disguising himself as Professor Challenger. During this period Doyle also published several relatively minor short stories usually grouped with his science fiction and fantasy, including "How it Happened" (*The Strand Magazine,* 1913), collected in *Danger! and Other Stories* (1918), and one of his better fantasy stories, "The Bully of Brocas Court" (*The Strand Magazine,* 1921), collected in *Tales of the Ring and Camp* (1922), in which an accomplished young boxer meets his match in a ghostly fighter from the previous century. Of special interest during World War I was "Danger!," a prophetic science-fiction story that appeared in *The Strand Magazine* in July 1914, just before the outbreak of the war. Doyle's fictional warning about the dangers to Britain of a submarine blockade caused public controversy because some people thought the story aided the Germans.

In *The Lost World,* Ned (Edward) Malone, an Irish-born London newspaper reporter and a sportsman, finds that Gladys Hungerton will not marry him because he is not a romantic hero like Richard Burton, the famed Victorian explorer of Africa. Emboldened by Gladys, Ned seeks

Illustrations by Tom Peddie for the serialization of The Maracot Deep *in* The Strand Magazine

more-challenging assignments. His quest leads him to Professor George Edward Challenger. In the prime of his life, the egotistical, arrogant Challenger is so angered by the scientific world's skepticism regarding his recent discoveries in the Amazon that he has virtually secluded himself, but Ned earns his confidence. When Challenger successfully goads the Zoological Institute into sending another expedition to prove him wrong, Ned persuades the sponsors to include him.

The original party includes Ned, Professor Summerlee, and Lord John Roxton. Summerlee is a sixty-six-year-old zoologist, chosen because of his utter disbelief of Challenger's assertions. Like Challenger, he is arrogant and self-assured. Though forced to change his mind in the course of the adventures, Summerlee is always Challenger's antagonist. The two scientists argue tirelessly and tiresomely, and it falls to Roxton and Ned to invent strategies to divert these childish intellectuals from their bickering. Lord Roxton is a sportsman adventurer who travels the world, looking for the riskiest hunting and exploration.

Challenger leads the group to the virtually inaccessible secret plateau he has discovered, where Jurassic flora and fauna still flourish. They construct a precarious log bridge to reach it and then are betrayed by two native helpers who bear an old grudge against Lord Roxton. Marooned on the plateau, they must contend with the impracticality of Challenger; with prehistoric monsters and apemen, the almost-human, "missing link" inhabitants; and finally with the too friendly natives who desire the protection of advanced European weapons.

After adventures and close calls, the men find an escape route and return to England with their evidence. The climax is a second meeting of the Zoological Institute. Somewhat foolishly but effectively, Challenger defeats the remaining doubters by exhibiting a young, live pterodactyl, which escapes and terrorizes the crowd before disappearing on its attempted return home. The novel ends after Ned revisits Gladys, only to learn that she has married a solicitor's clerk in his absence. So disappointed, he returns to the warmth of his newfound masculine friendships, where he learns that John discovered diamonds on the plateau and they are now all independently wealthy. The novel closes with Ned and John clasping hands, agreeing to return to the lost world.

Though *The Lost World* develops themes of manhood and human evolution, it is notable mainly for its adventure and humor. Doyle carries modern readers through exotic landscapes to Earth's dim past to witness the elemental forces that have determined the shape of their world and of their species. In his next Professor Challenger novel, Doyle shows much more interest in thematic issues.

The Poison Belt takes place three years after the adventure in Brazil. The four comrades come together again to witness the end of the world. On its journey through the universal ether, Earth passes through a current that Challenger compares to the Gulf Stream in the ocean. According to his physics, this ether rapidly interpenetrates all matter passing through it, and its main effect appears to be to make the atmosphere poisonous to all the higher forms of animal life. Plants remain, and the group observes a surviving amoeba during their confinement in an airtight room, where a supply of bottled oxygen preserves their lives.

Ned Malone, again the narrator, is called to Challenger's country home, along with Summerlee and Roxton, after Challenger connects recent astronomical observations with reports of massive plagues in equatorial regions. Upon their arrival Challenger explains his theories to them, demonstrates that their behavior has already been affected by atmospheric changes, and lays out his plans. He believes no one can be saved, but that they can prolong their lives long enough to observe the end of this phase of natural history. From the window of their sealed room, they witness such death and destruction as can be seen from a country hilltop near London. When their oxygen runs out they choose to die in fresh though poisoned air and open their window. To their surprise and horror, they find that the crisis is over and that they are probably among the very few and widely scattered survivors. Challenger persuades them that their duty is to continue to observe and learn, so they travel to London by car to observe how the city died. After the five adventurers return home in the evening, humans and animals return to life. Apparently the noxious atmosphere produced a sort of universal coma rather than actual death. Still, many people have died in the fires, wrecks, and other disasters that occurred when everyone lost consciousness.

Unlike *The Lost World,* this novel offers considerable moralizing. Challenger takes a spiritualist point of view against Summerlee's materialism and asserts his conviction that human life continues on a higher plane after it ends in this world. This becomes the foundation for his repeated assertions that they should never willingly surrender their lives, for whatever they gain in knowledge in this life will benefit themselves and others in the next. Challenger several times asserts that material human life in the universe is not nearly so important as humans habitually believe. Individual lives and cultures and even the largest concerns of the present are insignificant in the grand scheme of creation. This is demonstrated in what they all believe is happening, the end of the world. On the other hand, human life is so important that the earth, at least, is programmed to produce it. Reversing the usual understanding of natural selection–that species propose and environment selects–Challenger asserts that the human species is latent in the simplest living cells and even in the environment itself; therefore, as long as the earth continues to exist, even if all humanity is destroyed at one time, over the aeons that follow, human life will inevitably appear again. These ideas lead directly to the moral of the story given in the novel's last paragraph. Humanity has a destiny imposed upon it by a higher power, a destiny to be taken seriously, but not one that justifies an exalted view of the value of humanity in its present state.

Of considerable thematic interest as well is the implicit popular anthropology of the novel. Challenger hypothesizes that the poisoned atmosphere's effects take hold spatially and racially. The symptoms are first observed at sea level and near the equator, from which they spread upward and northward. Little information is given about its operations in the Southern Hemisphere. Racially, the poison first affects the least complex and last reaches the most complex. This allows him to lay out a map of racial complexity, which follows the familiar patterns of European racism: Africans, Australian aborigines, and Southern Asians succumb first, having both geographical and racial disadvantages. Arabs and Slavs follow, while white Northern Europeans are the last to succumb, though the Spanish, French, and Welsh go before the British. Complexity in Challenger's mind is clearly connected with intellectual development as recognized by Western civilization. He offers himself as the most highly developed of European minds. Hence, after hiding under the breakfast table and biting his maid on the leg, he recognizes that his behavior is being adversely influenced and sets about exerting his mind to keep his less sane impulses under control. In this way he reveals a psychological level to this popular anthropology.

As Challenger's behavior shows, there is a good deal of comedy in *The Poison Belt* just as there is in *The Lost World.* As Roxton, Ned, and Summerlee travel by train to Challenger's, they also behave strangely and somewhat comically, and like his oddity, their behavior shows potentially dangerous passions bubbling to the surface, apparently freed of their usual conscious control. Similarly, the development of the poison's effects has a class dimension, the lower classes more often giving way to mass violence–for example, riots among Welsh coal miners and among African Americans in the southern United States. Implicit in the novel, then, are many aspects of the popular Victorian social theory that saw the key dangers to social order in the "colored" races, the laboring classes, the southern and non-European, and in the physical passions.

Doyle's remaining Professor Challenger stories reflect the interests of his final years. *The Land of Mists* dramatizes the conversion of Challenger, now a materialist, to spiritualism. Technically, this story is neither fantasy nor science fiction, yet for nonbelievers it has many of the characteristics of a fantasy novel, consisting largely of episodes in which skeptics confront persuasive evidence of the truth of the spiritualist worldview. Set soon after World War I and after the deaths of Summerlee and of Challenger's wife, it shows Ned Malone and the professor's daughter, Enid, gradually coming to accept spiritualist beliefs. At the same time they fall in love. Challenger refuses to accept their marrying because of their heretical beliefs, and this act moves them to try to convince him. In the climax Enid for the first time goes into a medium's trance and enables the materialization of Challenger's beloved wife, who admonishes him to become a believer. Challenger also receives a message like the one Doyle himself received, containing information about his past that only he could remember. Though marred occasionally by sermonizing, *The Land of Mists* is an entertaining introduction to spiritualism, its beliefs, the associated phenomena, and social attitudes toward it after the Great War. The book helps to make clear why Doyle, himself inclined to skepticism, could come to take this religion seriously.

"The Disintegration Machine" prefigures the science fiction of the next generation. Professor Challenger meets his dark double, Theodore Nemor, the Latvian inventor of a machine that can reduce any object of any size to ethereal parts and then, if the user chooses, restore its physical existence. The clever and absolutely unscrupulous Nemor sells himself to the first nation to bid high enough, the Soviet Union. He will not reveal how to build the machine, but he will build models exclusively for the Russians. The skeptical Challenger, along with Malone, meets Nemor and experiences a comic demonstration. Challenger soon realizes that Nemor takes delight in the prospects of destruction and of holding absolute power over a groveling humanity. Challenger tricks Nemor into the right position and disintegrates him. When Malone protests, Challenger defends his action with the argument that morality requires him to prevent murder and that his act prevents the murder of millions without, technically, killing Nemor.

"When the World Screamed" is probably the most complex of the Professor Challenger stories. In *The Lost World* and in "The Disintegration Machine," Challenger is the target of considerable comic irony, but he always remains a much-admired man and usually far ahead of the rest of humanity in his scientific knowledge. In "When the World Screamed" Challenger appears to be severely undercut, offered perhaps as a parody of the arrogance of Western scientific materialism. Having received a large bequest for research, Challenger uses it to sink a shaft through the earth's crust. Eight miles below the surface, he encounters what he believes is living tissue. He decides to drop an artesian drill into this tissue in order to let the living planet know that there are living and intelligent beings on its surface. When he does this the world screams through the shaft he has made, then closes it up from within. Witnesses of his success honor Challenger, and he feels he deserves it; he stands before them "his eyes half closed, a smile of conscious merit upon his face, his left hand upon his hip, his right buried in the breast of his frock-coat." Not everyone is pleased, however, especially not those victimized by consequent volcanic eruptions.

Challenger demonstrates that the planet is a conscious organism and he has communicated with it, but Ned Malone and Peerless Jones, among his helpers, offer quite a different view. Challenger covers the living surface when the shaft reaches it and forbids the workers to look at it. Malone and Jones remove this "veil" and gaze upon their mother earth. What they see evokes their reverence and a feeling that to pierce the planet would be a violation, a kind of rape of the great mother. Again, just before they actually drop the drill, they feel as if the living surface is aware of them and attempting to communicate with them. This view suggests a criticism of the masculine, scientific, imperialistic approach to contacting what is probably an alien intelligence, and it suggests further that Challenger's crude communication by attack earns the enmity of the organism that provides humanity its congenial home.

This story seems to prefigure such profound later stories of alien contact as Stanislaw Lem's *Solaris* (1961; translated, 1970) and *Fiasco* (1986; translated, 1987). More than any of the other stories in the series, "When the World Screamed" seems to put Challenger in his place as a representation of British imperialism, raping the planet with a kind of maddeningly cheerful innocence, convinced that this is his natural right as shown by his having the imagination to conceive it and the power to carry it out. This side of Challenger is present in all of the stories, but readers may remain unsure whether Doyle saw this side so negatively before writing this story.

Of Doyle's non-Challenger science fiction published in the last two decades of his life, the best stories are probably "The Terror of Blue John Gap" (*The Strand Magazine,* 1910), collected in *The Last Galley,* and "The Horror of the Heights" (*The Strand Magazine,* 1913), collected in *Danger! and Other Stories.* In the earlier story, Dr. John Hardcastle, while recuperating in the country, encounters a kind of giant opossum that on dark nights issues from Blue John Gap, an abandoned mine, to hunt farm animals and, eventually, a man. He concludes that when miners dug into the great limestone caves in the area, they opened another gap into a subterranean world, out of which this one monster found its way. "The Horror of the Heights" also explores the idea of worlds of terror nearby and parallel to the normal human world. Mr. Joyce-Armstrong, a pilot seeking altitude records and adventure, finds another world of floating beings in the upper atmosphere, some beautiful and benign, others aggressively carnivorous. His second visit to this realm to bring back proof of his discoveries ends with his death, when he is, presumably, enveloped by a great diaphanous purple beast of the sort H. P. Lovecraft would soon invent.

The novella *The Maracot Deep* is a wonderfully imaginative adventure that most critics wish was more fully developed. Dr. Maracot, a marine biologist with universal knowledge; Cyrus Headley, an American biologist; and Bill Scanlon, an American technician, descend in an advanced bathysphere to explore the Maracot Deep, a rift in the Atlantic near the Canary Islands. Attacked by a giant crayfish, they are cut loose and drift twenty-five thousand feet to the bottom. There they are rescued by survivors of Atlantis, people who built a kind of stationary ark in which they could continue to live after their city sank beneath the sea ten thousand years before. Their adventures in this lost world include many wonders, but Doyle focuses on the history of Atlantis, which proves to be like contemporary Europe. As scientific knowledge advanced, Atlantis fell into a materialism that blinded people to the spiritual and led inevitably to their corruption. The sinking of Atlantis was a direct expression of the spirit world's wish to purify the planet. The survivors were those who heeded the prophets and followed the one who designed their ark.

In the second half of the narrative, Headley describes their encounter with a human immortal, Baal-seepa, a satanic being first evoked by the spiritual blindness of the Atlanteans. The nonbelieving explorers reevoke this being, and he threatens now to destroy the remnant of Atlantis. But Maracot becomes a medium for Warda, the more powerful and good antithesis of Baal-seepa. Under this influence Maracot drives the demon from the material realm forever. As a result the materialist Maracot undergoes a partial conversion to a spiritualist point of view. Doyle creates a vivid and interesting world in this novella, then uses it to further what were then his most pressing concerns, to warn his readers that contemporary materialism was self-destructive and to encourage them to take seriously the worldview of the spiritualist movement.

Doyle fell ill with heart disease in 1929 and died in 1930 at his home, Windlesham, where he is buried. At his death he was one of the best-known Englishmen in the world, his tales of Sherlock Holmes translated into many languages, read and loved almost everywhere. The detective stories are reinterpreted and adapted to film and television anew with each generation. Among his works of science fiction and fantasy, only *The Lost World* has had comparable influence. The novel has been adapted to film at least twice (1925, 1960), and the historically related film, *King Kong* (1933), became a classic with multiple sequels and a reinterpretation in 1976. Furthermore, the novel itself inspired many related stories and films, notably *The Land that Time Forgot* (1918, film 1974) by Edgar Rice Burroughs. Doyle's literary reputation tended to remain rather low in comparison to contemporaries such as Thomas Hardy and Joseph Conrad, perhaps in part because of his association with spiritualism and because of critics' not always helpful distinction between "high" literature and popular entertainment. When scholars and critics in the later years of the twentieth century undertook the serious study of Doyle, they found him more complex than expected. Many critics, scholars, and especially biographers have turned their attention to his works. Though his reputation rests mainly on his achievement in the stories of Sherlock Holmes, collections of his fantasy and science-fiction stories continue to appear, attesting to an enduring lively interest among his readers.

Letters:

Letters to the Press, edited by John Michael Gibson and Richard Lancelyn Green (London: Secker & Warburg, 1986; Iowa City: University of Iowa Press, 1986).

Interviews:

R. Blathwayt, "Talk with Dr Conan Doyle," *Bookman* (London), 2 (May 1892): 50–51;

Harry How, "A Day with Dr. Conan Doyle," *Strand Magazine,* 4 (August 1892): 182–188;

Robert Barr, "A Chat with Conan Doyle," *Idler,* 6 (October 1894): 340–349; republished in *McClure's Magazine,* 3 (November 1894): 503–513;

Sir Arthur Conan Doyle: Interviews and Recollections, edited by Harold Orel (New York: St. Martin's Press, 1991).

Bibliographies:

Harold Locke, *A Bibliographical Catalogue of the Writings of Sir Arthur Conan Doyle, M.D., LL.D., 1879–1928* (Tunbridge Wells: Webster, 1928);

Edgar W. Smith, *Baker Street Inventory: A Sherlockian Bibliography* (Summit, N.J., 1945);

Jay Finley Christ, *The Fiction of Sir Arthur Conan Doyle* (N.p.: Privately printed, 1959);

Ronald Burt De Waal, *The World Bibliography of Sherlock Holmes and Dr. Watson* (Boston: New York Graphic Society, 1974);

De Waal, *The International Sherlock Holmes: A Companion Volume to the World Bibliography of Sherlock Holmes and Dr. Watson* (Hamden, Conn.: Archon, 1980; London: Mansell, 1980);

Richard Lancelyn Green and John Michael Gibson, *A Bibliography of A. Conan Doyle* (Oxford: Clarendon Press, 1983, reprinted with corrections, 1984).

Biographies:

John Lamond, *Arthur Conan Doyle: A Memoir* (London: John Murray, 1931; Port Washington, N.Y.: Kennikat Press, 1972);

Hesketh Pearson, *Conan Doyle: His Life and Art* (London: Methuen, 1943; New York: Walker, 1961);

Adrian Conan Doyle, *The True Conan Doyle* (London: John Murray, 1945; New York: Coward-McCann, 1946);

John Dickson Carr, *The Life of Sir Arthur Conan Doyle* (London: John Murray, 1949; New York: Harper, 1949; New York: Carroll & Graf, 1987);

Sir Arthur Conan Doyle Centenary, 1859–1959 (London: John Murray, 1959; Garden City, N.Y.: Doubleday, 1959);

Michael Hardwick and Mollie Hardwick, *The Man Who Was Sherlock Holmes* (London: John Murray, 1964; Garden City, N.Y.: Doubleday, 1964);

Pierre Nordon, *Sir Arthur Conan Doyle: L'Homme et L'Oeuvre* (Paris: Didier, 1964): translated as *Conan Doyle: A Biography* (London: John Murray, 1966; New York: Holt, Rinehart & Winston, 1967);

Ivor Brown, *Conan Doyle, a Biography of the Creator of Sherlock Holmes* (London: Hamish Hamilton, 1972);

Charles Higham, *The Adventures of Conan Doyle: The Life of the Creator of Sherlock Holmes* (New York: Norton, 1976; London: Hamish Hamilton, 1976);

Ronald Pearsall, *Conan Doyle: A Biographical Solution* (London: Weidenfeld & Nicolson, 1977; New York: St. Martin's Press, 1977);

Julian Symons, *Portrait of an Artist–Conan Doyle* (London: Whizzard Press, 1979; New York: Mysterious Press, 1987);

Owen Dudley Edwards, *The Quest for Sherlock Holmes: A Biographical Study of Arthur Conan Doyle* (Edinburgh: Mainstream, 1983; Totowa, N.J.: Barnes & Noble, 1983);

Jon L. Lellenberg, ed., *The Quest for Sir Arthur Conan Doyle: Thirteen Biographies in Search of a Life* (Carbondale: Southern Illinois University Press, 1987).

References:

Dana Martin Batory, "The Climax of 'When the World Screamed,'" *Riverside Quarterly,* 8 (March 1988): 124–128;

Batory, "A Look Behind Conan Doyle's *Lost World,*" *Riverside Quarterly,* 6 (December 1977): 268–271;

Batory, "*The Poison Belt* as a Morality Tale," *Riverside Quarterly,* 7 (March 1982): 97–100;

Batory, "The Rime of the Polestar," *Riverside Quarterly,* 7 (December 1985): 222–227;

James L. Campbell, "Sir Arthur Conan Doyle," in *Science Fiction Writers,* edited by E. F. Bleiler (New York: Scribners, 1982), pp. 45–50;

Don R. Cox, *Arthur Conan Doyle* (New York: Ungar, 1985);

Anne Cranny-Francis, "Arthur Conan Doyle's *The Parasite*: The Case of the Anguished Author," in *Nineteenth-Century Suspense: From Poe to Conan Doyle,* edited by Clive Bloom and others (New York: St. Martin's Press, 1988), pp. 93–106;

Howard Davies, "*The Lost World*: Conan Doyle and the Suspense of Evolution," in *Nineteenth-Century Suspense: From Poe to Conan Doyle,* edited by Bloom and others (New York: St. Martin's Press, 1988), pp. 107–119;

Jacqueline A. Jaffe, *Arthur Conan Doyle* (Boston: Twayne, 1987);

Sam Moskowitz, "Arthur Conan Doyle: A Study in Science Fiction," in his *Explorers of the Infinite* (Cleveland: World, 1963), pp. 157–171;

Harold Orel, ed., *Critical Essays on Sir Arthur Conan Doyle* (New York: Macmillan, 1992);

Audrey Peterson, "Arthur Conan Doyle and the Great Detective," in her *Victorian Masters of Mystery: From Wilkie Collins to Conan Doyle* (New York: Ungar, 1984), pp. 197–217;

Alvin E. Rodin and Jack D. Key, *The Medical Casebook of Dr. Arthur Conan Doyle: From Practitioner to Sherlock Holmes and Beyond* (Malabar, Fla.: Krieger, 1984).

Papers:

Doyle manuscripts may be found in the Berg Collection of the New York Public Library; the Harry Ransom Humanities Research Center at the University of Texas, Austin; and the Lilly Library of Indiana University, Bloomington. Letters and a variety of other materials are available in the Arthur Conan Doyle Collection of the Metropolitan Toronto Library, Toronto, Ontario, and in the Philip S. and Mary Kahler Hench Collection in the O. Meredith Wilson Library, University of Minnesota, Minneapolis.

George Du Maurier

(6 March 1834 – 8 October 1896)

Elizabeth S. Davidson
University of South Carolina–Spartanburg

See also the Du Maurier entry in *DLB 153: Late-Victorian and Edwardian British Novelists, First Series.*

BOOKS: *Peter Ibbetson* (New York: Harper, 1891; London: Osgood, McIlvaine, 1891);
Trilby (London: Osgood, McIlvaine, 1894; New York: Harper, 1894);
The Martian (3 volumes, London: Bell, 1897; 1 volume, New York: Harper, 1897).

During the last seven years of his life George Du Maurier wrote and illustrated three fantasy novels with the barest touch of science in each of them. Though each of the novels is firmly rooted in the London and Paris of the middle 1800s, each also includes elements that in some way set the main characters apart, letting them rise above the difficult circumstances connected with their daily existence. In two of the novels this ability to transcend normal patterns of life gains for them also an authority to deliver a message of hope to a world that has lost faith in traditional religion and in the goodness of society. In a third he created the villain Svengali, a name long associated with those who misuse mesmerism or hypnotism for power and profit or with manipulators in general.

The name Du Maurier comes from a fantasy of George's grandfather Robert Mathurin Busson, who after moving to England adopted the name Du Maurier for its aristocratic sound. One of his sons, Louis-Mathurin Busson Du Maurier, married Ellen Clarke in 1831. On 6 March 1834 their first son, George Louis Palmella Busson du Maurier, nicknamed "Kicky," was born in Paris. All of his life Du Maurier was bilingual in French and English, so comfortable with both that he used them in all of his novels, even occasionally satirizing the pronunciation and word choice of both French and English speakers who fumble their way through the other language. Du Maurier's childhood and youth were spent in France. There his parents sent him to the Pension Froussard, a boarding school. Because he preferred to read the romantic adventures of Alexander Dumas *père,* Victor Hugo, Pierre de Ronsard, François Villon, and Sir Walter Scott rather than the Latin classics that were part of the curriculum, Du Maurier failed the Latin paper on the exam required for graduation. His young adulthood was spent in England as a student in chemistry at University College, London, then in Paris and Antwerp as an art student. In Antwerp in 1857 he first experimented with mesmerism in the company of his friend Felix Moscheles, later a model for Svengali. After two months in Antwerp he suddenly became blind in the left eye: "I was drawing

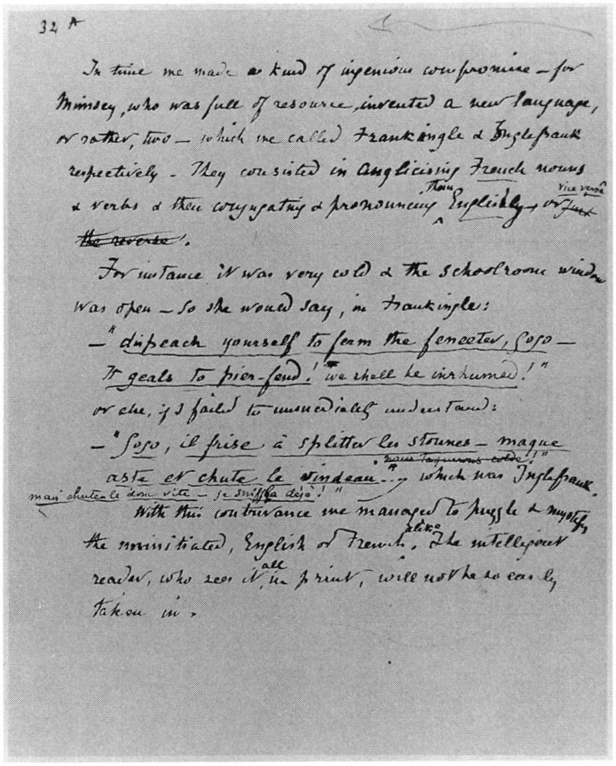

32 A

In time we made a kind of ingenious compromise – for Mimsey, who was full of resource, invented a new language, or rather, two – which we called Frankingle & Englefrank respectively. They consisted in anglicising French nouns & verbs & then conjugating & pronouncing them Englishly, or vice versa.

For instance it was very cold & the schoolroom window was open – so she would say, in Frankingle:

– "dispeach yourself to ferm the fenester, Gogo – it geals to pier-fend! we shall be enrhumed!"

or else, if I failed to immediately understand:

– "Gogo, il frise à splitter les stounes – mais [illegible] la windeau ..." which was Englefrank.

With this contrivance we managed to puzzle & mystify the uninitiated, English or French. The intelligent reader, who sees it all in print, will not be so easily taken in.

Page from a draft for Peter Ibbetson *(MA 115, Pierpont Morgan Library)*

from a model," he said in an October 1895 interview in *McClure's,* "when suddenly the girl's head seemed to me to dwindle to the size of a walnut. . . . My left eye had failed me; it might be altogether lost. It was so sudden a blow that I was as thunderstruck." He moved to the quiet town of Malines for a rest cure and sought medical advice from doctors in Antwerp, Ghent, and Düsseldorf during the next two years, but all attempts to restore his vision failed. Back in London in the summer of 1860, he began a long career as an illustrator for magazines such as *Punch* and for novels, including *Washington Square* by Henry James, with whom he formed a friendship that lasted the rest of his life. Marrying Emma Wightwick in 1863, Du Maurier fathered five children: Beatrix (1864), Guy Louis (1865), Sylvia Jocelyn (1866), Marie Louise (1868), and Gerald (1873). His most famous descendant is a granddaughter, Daphne, best known for her novel *Rebecca.* When Du Maurier was fifty-five he began writing his first fantasy novel, *Peter Ibbetson* (1891). In 1892 he began *Trilby* (1894), and in 1894 he began *The Martian* (1897). Each first appeared in serial form in *Harper's Monthly Magazine* before appearing in hardcover. *Trilby* was also adapted for the stage and produced in 1895. George Du Maurier died on 8 October 1896.

Du Maurier's most fantastic novel, *Peter Ibbetson,* is also the one in which he demonstrates the most skill in presentation. The frame narrator, Madge Plunkett, is the literary executrix for her cousin Peter Ibbetson, a prisoner all his adult life–first in prison itself and then in an asylum for the criminally insane. The frame narrator is not intrusive but serves to remind readers from time to time that the life and experiences of Peter Ibbetson are a complete thing, creating a distance from the murderer so that an audience may see the hidden dreamer within. Except for an extremely wonderful childhood, the main character, Peter, has little good happen to him as an adult, and he *is* guilty of homicide. Having killed his guardian in a fit of rage over that man's slander of his mother, the waking Peter loses his freedom for life. The sleeping Peter, however, is another matter altogether. His dream life becomes his true life–literally: Peter states, "Yet of all that ever lived on this earth I have been, perhaps, the happiest and most privileged" and "A cell in a criminal lunatic asylum! That does not sound like a bower in the Elysian Fields! It is, and has been for me." The second main character, Mary, the Duchess of Towers, who is involved in a failing marriage, has been taught the secret of "dreaming true." After she teaches it to Peter, in a dream as she rescues him from a nightmare, she and Peter for the rest of their lives enter each other's dreams nightly and explore interests common to both–a shared childhood, a shared interest in the beauties of music and art–through these interests finding the secret to a kind of mental time travel. They go back in time as observers, even seeing and sketching the great mammoths. The death of the Duchess of Towers ends one stage of their dreaming true, precipitating the crisis that sends Peter from the prison to the asylum, but it also makes possible the last lesson of Peter's dream life. His beloved Mary is not dead but has moved on to another plane of existence, from which she manages to visit him seven times. Mary speaks of the genius, physically incapacitated in some way, who, like Ludwig van Beethoven, manages to convey his message with the help of those who love him. Death then becomes for Peter a door to an ultimate freedom. His cousin Madge, though she does not understand, views him in his coffin, where "he appeared gigantic–the most magnificent human being I ever beheld; and the splendor of his dead face will haunt my memory till I die."

Trilby involves mesmerism, in which Du Maurier had been interested since 1857. A poor artist's model with no ear for music whatsoever but with the most artistically appealing foot in the world and

Illustrations by Du Maurier for Peter Ibbetson *(top) and* Trilby *(bottom)*

a wonderfully cavernous mouth becomes a world-famous concert singer–so long as Svengali lives and keeps her mesmerized. Once he has mesmerized Trilby, she has no other life. Removed from her captor, she cannot live the life he has created for her, cannot return to her former life, goes into a decline, and dies. The people who are her friends before she falls into Svengali's control are English art students in Paris: Taffy, the Laird, and Little Billee. Because Little Billee, who loves Trilby, cannot overcome his middle-class prejudices about Trilby's having worked as a nude model and because his middle-class mother does not think an artist's model of any kind good enough to marry her son, he misses the chance to marry the one woman he will ever love and to keep her safe from a musician they both consider disreputable, Svengali. Desperately seeking relief from an eye pain sounding rather like migraine headaches, Trilby first accepts his offer to mesmerize the pain away. Svengali, who has long noticed the potential that the structure of Trilby's mouth offers for resonant tones, then trains her, mesmerized, to produce the sounds necessary to become a great singer. After they have great success on the Continent, they perform in London. Svengali, seeing Taffy, the Laird, and Little Billee in the audience, is overcome by his hatred for them so severely that he has a heart attack and dies. Though her English friends rescue her from the stage on which she no longer knows what to do, this reunion is happy only in that she is cherished in her last days and does not die alone.

The title of Du Maurier's last novel, *The Martian,* is a kind of promise of an interplanetary contact to come that is not fulfilled until about 255 pages into the book. There are hints from time to time that this promise will be kept, but the narrator, Robert Maurice, a prosperous politician and wine merchant supposedly writing a biography of the novelist Barty Josselin, delays the reader a long time in coming to that point. One overwhelming darkness in Barty Josselin's otherwise bright life is his illegitimate birth. In their shared days as schoolboys in Paris, Robert Maurice and Barty Josselin form a friendship that lasts a lifetime. Another darkness in Barty's life is both literal and figurative: like Du Maurier, he loses his sight in one eye. Barty's attempted suicide finally brings Martia, the Martian, into the story. Sharing Barty's body off and on throughout his life, at this moment of crisis she finds she can take control when he is sleeping, make his body throw away the poison he was planning to take, and write him a letter. He now learns that a talent he has had as a schoolboy, always being able to sense the North Pole, is made possible because of the magnetism of this disembodied consciousness. From this moment on he knows that when he feels the north again he is to leave writing materials by the bed so that Martia can communicate with him. It is a kind of automatic writing, but he is asleep, not in a spiritualist's trance.

Martia has been reincarnated many times in the ancient civilization of Mars, becoming improved each time, until she (and others like her) begin the disembodied journey toward the Sun. She wants to be born again on Earth and has selected Barty and the beautiful Julia, a wealthy young aristocrat, as her prospective parents. Barty, however, prefers Leah, not only a tradesman's daughter but a bookkeeper and then manager of her father's house of business. Though Martia is offended, through many nights of possession of his body and automatic writing, she turns Barty into a world-famous novelist in both French and English. Finally, having seen and accepted Leah's excellence, Martia tells Barty she will be reincarnated again as their ninth child. This child, a girl called Marty, lives the life of a tomboy until an accident makes her an invalid. When she dies, she calls Barty and Leah to come to her. Barty dies almost immediately and Leah within twenty-four hours. Together, Robert Maurice suggests, they continue the next step in Martia's immortal journey as disembodied spirits.

Du Maurier's last book thus is similar to his first in an emphasis on a life beyond that of the body. When the body is alive, that life beyond may be touched only through sleep, differing in means in the two books between the altered consciousness of the dream state and the possession demonstrated by the automatic writing. When the body dies the consciousness goes on living, not in a Judeo-Christian heaven or hell but in a state perhaps more Neo-Platonic. Best described near the end of *Peter Ibbetson,* the senses are not terminated by death but are freed to become more intense, even to blend. The potential is unlimited.

Peter Ibbetson was well received because of Du Maurier's reputation as an illustrator and because of warm praise by literary friends such as James. *Trilby* was an international publishing phenomenon, the first modern best-seller after the passage of a new copyright law in 1891 put the cheap reprint houses out of business. The use of a new marketing system led to a "Trilbymania" so widespread that three hundred thousand copies were sold by the end of 1894 and interest in Du Maurier's first novel was revived and increased. "Trilby worship" became a fad. There is even a town in south Florida named Trilby, beside which is a lake named Little Billee. Though the three novels are rather obscure one hundred years after their first publication, Du Maurier was one of the early writers to deal with the fundamental duality in personality. Du Maurier was an important contributor to the development of the novel, using the forms of the nineteenth century but subjects foreshadowing the interests of the twentieth, the freedom of the subconscious mind but also mind control. Though the dream relationship of Peter and the duchess and the deathbed relationship of Trilby to her circle of rescuing English artists may seem too sentimental, and though the linking of Svengali's Jewishness to his villainous behavior is excessive, *Trilby* and *Peter Ibbetson* still have much to say to a modern reader about the underlying psychological causes of human behavior and the value of fantasy for developing greater insights into the survival of the human spirit.

Letters:

The Young George Du Maurier: A Selection of His Letters, 1860–67, edited by Daphne Du Maurier (London: Davies, 1951).

Biographies:

Daphne Du Maurier, *The Du Mauriers* (London: Gollancz, 1937);

C. C. Hoyer Millar, *George Du Maurier and Others* (London: Cassell, 1937);

Leonée Ormond, *George Du Maurier* (Pittsburgh: University of Pittsburgh Press, 1969).

References:

Richard J. Allen, "Documentary Realism and Artistic Licence: A Note on an Emblematic Prison Gate in *The Nether World* and *Peter Ibbetson*," *Gissing Newsletter*, 17 (April 1981): 20–25;

Nina Auerbach, "Magi and Maidens: The Romance of the Victorian Freud," *Critical Inquiry*, 8 (Winter 1981): 281–300;

Martha Banta, "Artists, Models, Real Things, and Recognizable Types," *Studies in the Literary Imagination*, 16 (Fall 1983): 7–34;

Rebecca A.Carey, "J. M. Barrie and the Du Mauriers," *Mythlore: A Journal of the J. R. R. Tolkien, C. S. Lewis, Charles Williams, General Fantasy and Mythic Studies*, 15 (Summer 1989): 40–42;

Elizabeth S. Davidson, "*Peter Ibbetson*," in *Survey of Modern Fantasy Literature*, 5 volumes, edited by Frank N. Magill (Englewood Cliffs, N.J.: Salem Press, 1983), III: 1227–1229;

J. B. Gilder and J. L. Gilder, *Trilbyana: The Rise and Progress of a Popular Novel* (New York: Critic, 1895);

Henry James, "George Du Maurier," *Harper's New Monthly Magazine*, 95 (September 1897): 594–609;

James, "George Du Maurier," *Harper's Weekly Magazine*, 38 (14 April 1894): 341–342;

Richard Kelly, *George Du Maurier* (Boston: Twayne, 1983);

Isaac Hull Platt, *The Ethics of Trilby: With a Supplemental Note on Spiritual Affinity* (Philadelphia: Conservator, 1895);

L. Edward Purcell, "*Trilby* and Trilby-Mania: The Beginning of the Best-Seller System," *Journal of Popular Culture*, 11 (Summer 1977): 62–76;

Edgar Rosenberg, *From Shylock to Svengali: Jewish Stereotypes in English Fiction* (Stanford, Cal.: Stanford University Press, 1960);

Eve Kosofsky Sedgwick, "The Beast in the Closet: James and the Writing of Homosexual Panic," in *Speaking of Gender*, edited by Elaine Showalter (New York: Routledge, 1989), pp. 243–268;

George Bernard Shaw, "*Trilby* and 'L'Ami des Femmes,'" in his *Dramatic Opinions and Essays*, volume 1 (New York: Brentano's, 1922), pp. 228–236;

Robert H. Sherard, "The Author of Trilby," *McClure's Magazine*, 4 (October 1895): 391–400;

Lionel Stevenson, "George Du Maurier and the Romantic Novel," in *Essays by Divers Hands*, edited by Hardy Wallace (London: Oxford University Press, 1960), pp. 36–54;

Edmund Wilson, "The Jews," in his *A Piece of My Mind: Reflections at Sixty* (London: W. H. Allen, 1956), pp. 64–103;

T. Martin Wood, *George Du Maurier–The Satirist of the Victorians: A Review of His Art and Personality* (London: Chatto & Windus, 1913);

Jules Zanger, "A Sympathetic Vibration: Dracula and the Jews," *English Literature in Transition (1880–1920)*, 34, no. 1 (1991): 33–44.

E. M. Forster

(1 January 1879 – 7 June 1970)

Johan Heje

See also the Forster entries in *DLB 34: British Novelists, 1890–1929: Traditionalists; DLB 98: Modern British Essayists, First Series; DLB 162: British Short-Fiction Writers, 1915–1945;* and *DLB Documentary Series 10: The Bloomsbury Group.*

E. M. Forster, circa 1915

BOOKS: *Where Angels Fear to Tread* (Edinburgh & London: Blackwood, 1905; New York: Knopf, 1920);

The Longest Journey (Edinburgh & London: Blackwood, 1907; New York: Knopf, 1922);

A Room with a View (London: Arnold, 1908; New York & London: Putnam, 1911);

Howards End (London: Arnold, 1910; New York & London: Putnam, 1910);

The Celestial Omnibus and Other Stories (London: Sidgwick & Jackson, 1911; New York: Knopf, 1923);

The Story of the Siren (Richmond: Hogarth Press, 1920);

Alexandria: A History and a Guide (Alexandria: Whitehead Morris, 1922; Garden City, N.Y.: Doubleday, 1961; London: Michael Haag, 1982);

Pharos and Pharillon (Richmond: Hogarth Press, 1923; New York: Knopf, 1923);

A Passage to India (London: Arnold, 1924; New York: Harcourt, Brace, 1924);

Anonymity: An Enquiry (London: Hogarth Press, 1925);

Aspects of the Novel (London: Arnold, 1927; New York: Harcourt, Brace, 1927);

The Eternal Moment and Other Stories (London: Sidgwick & Jackson, 1928; New York: Harcourt, Brace, 1928);

A Letter to Madam Blanchard (London: Hogarth Press, 1931; New York: Harcourt, Brace, 1932);

Sinclair Lewis Interprets America (Cambridge, Mass.: Harvard University Press, 1932);

Goldsworthy Lowes Dickinson (London: Arnold, 1934; New York: Harcourt, Brace, 1934);

Abinger Harvest (London: Arnold, 1936; New York: Harcourt, Brace, 1936);

What I Believe (London: Hogarth Press, 1939);

England's Pleasant Land, a Pageant Play (London: Hogarth Press, 1940);

Nordic Twilight, Macmillan War Pamphlet, no. 3 (London: Macmillan, 1940);

Virginia Woolf: The Rede Lecture (Cambridge: Cambridge University Press, 1942; New York: Harcourt, Brace, 1942);

The Collected Tales of E. M. Forster (New York: Knopf, 1947); republished as *Collected Short Stories of E. M. Forster* (London: Sidgwick & Jackson, 1948);

Two Cheers for Democracy (London: Arnold, 1951; New York: Harcourt, Brace, 1951);

Billy Budd: An Opera in Four Acts. Libretto by E. M. Forster and Eric Crozier, Adapted from the Story by Herman Melville (London, New York, Toronto,

Sydney, Cape Town, Buenos Aires, Paris & Bonn: Boosey & Hawkes, 1951);

The Hill of Devi (London: Arnold, 1953; New York: Harcourt, Brace, 1953);

Marianne Thornton (1797–1887): A Domestic Biography (London: Arnold, 1956; New York: Harcourt, Brace, 1956);

Maurice (London: Arnold, 1971; New York: Norton, 1971);

"Albergo Empedocle" and Other Writings by E. M. Forster, edited by George H. Thomson (New York: Liveright, 1971);

The Life to Come and Other Stories (London: Arnold, 1972); republished as *The Life to Come and Other Short Stories* (New York: Norton, 1972);

The Manuscripts of "Howards End," edited by Oliver Stallybrass (London: Arnold, 1973; New York: Holmes & Meier, 1973);

Goldsworthy Lowes Dickinson and Related Writings, edited by Stallybrass (London: Arnold, 1973; New York: Holmes & Meier, 1973);

Aspects of the Novel and Related Writings, edited by Stallybrass (London: Arnold, 1974; New York: Holmes & Meier, 1974);

The Lucy Novels: Early Sketches for "A Room with a View," edited by Stallybrass (London: Arnold, 1977; New York: Holmes & Meier, 1978);

The Manuscripts of "A Passage to India," edited by Stallybrass (London: Arnold, 1978; New York: Holmes & Meier, 1978);

Commonplace Book, facsimile edition (London: Scolar Press, 1978);

Arctic Summer and Other Fiction, edited by Elizabeth Heine and Stallybrass (London: Arnold, 1980; New York: Holmes & Meier, 1981);

The Hill of Devi and Other Indian Writings, edited by Heine (London: Arnold, 1983; New York: Holmes & Meier, 1983);

Commonplace Book, edited by Philip Gardner (London: Scolar Press, 1985; Stanford: Stanford University Press, 1985).

Collection: *The Abinger Edition of E. M. Forster*, 14 volumes, edited by Oliver Stallybrass and Elizabeth Heine (London: Arnold, 1972–1984; New York: Holmes & Meier, 1972–1984).

E. M. Forster holds a rather unusual position in English literature. By the age of thirty-two he had gained recognition for four out of the five novels that were to appear in his lifetime. After that he published little for fourteen years, then achieved an international reputation with his prize-winning, best-selling, and much-debated novel about India and the British in India, *A Passage to India* (1924). Except for a slim volume of short stories written many years before, he did not publish any new fiction during the remaining years of his life, but he was far from forgotten. Though never a prolific writer, he was esteemed for his criticism, a few biographies, books dealing with his travels in Egypt and India, and particularly for his essays on more general topics, some of which have become classics. His unfaltering advocacy for his distinctive and personal brand of liberal humanism in an age of crumbling values and advancing totalitarianism earned him much attention and respect and, in his later years, the status of a sage. His novels continued to be reprinted, read, and studied while other literary figures of the Edwardian period seemed to recede into the past. The years after his death at age ninety-one gave rise to renewed interest in Forster, nourished by the publication of a novel, *Maurice* (1971), and some short stories, which he had not wanted to be published in his lifetime, in most cases because of their homosexual themes. In recent years James Ivory's film versions of four of his novels have contributed to keeping interest in Forster's works alive.

Edward Morgan Forster was born in London and grew up in moderately affluent middle-class homes in what was then still the rural southeast of England. His father, architect Edward Morgan Llewellyn Forster, died when Forster was less than two years old, and he was brought up by his mother, Alice Clara Forster, who outlived her husband by sixty-five years, and his maternal grandmother. A timid and sensitive child, Forster was not happy in the type of private boarding school that boys of his class were expected to attend–in his case as a day student, an outsider. Later, in "Notes on the English Character" (1924; collected in *Abinger Harvest*, 1936) he criticized the British public-school system for producing youths "with well-developed bodies, fairly developed minds, and undeveloped hearts." In his novels and short stories the British middle class is generally the target of satire, and often outright condemnation, because of its "undeveloped hearts," its narrow-mindedness and stunted emotional life.

More formative influences on Forster were his childhood experience of a vanishing rural way of life, his close communion with nature, and his four years as a student of classics and, later, history at King's College, Cambridge. He felt at home with the freedom of thought and discussion he found in Cambridge and formed lasting friendships, always important to him, both with fellow students and some of his teachers. One of these teachers–Goldsworthy Lowes Dickinson, whose biography Forster later wrote (1934)–was an enthusiastic advocate of what he saw as "The Greek View of Life," the title of his best-known book. The ancient Greeks, he said, recognized "the whole range of our complex human nature," as opposed to the medieval "antithesis of soul and body which . . . dominates still our current ethical conceptions." A less direct but perceptible influence was that of philosopher G. E. Moore, who in his *Principia Ethica* (1903) stated that "the most valuable things, which we know or can imagine, are certain states of consciousness, which can roughly be described as the pleasures of human intercourse and the enjoyment of beautiful objects." These values were shared by the artists and intellectuals who formed the famed Bloomsbury Group, which later counted Forster among its affiliates though he did not belong to its inner circle. They were clearly echoed in his novel *Howards End* (1910), where the paramount value of "personal relationships" is held up against the businessman's world of "telegrams and anger."

After Cambridge, Forster spent 1901–1902 traveling in Italy, which provided the subject for his earliest short stories and two of his novels, *Where Angels Fear to Tread* (1905) and *A Room with a View* (1908), with their depiction of dramatic clashes between British middle-class respectability and a more intuitive and sensual response to life represented by Italy and the Italians. In 1903 he visited Greece, where he found the setting for one of his most admired stories, "The Road from Colonus" (1904). Forster's novels established him in the solid realistic tradition of English social comedy and satire. In his short stories, however, realism often merges with fantasy, and the supernatural enters the scene of social comedy. On closer examination his themes are the same in both genres, and the fantastic is never far off in the novels. It does not appear in the form of supernatural or seemingly supernatural incidents, but it is present in hints given by the ever-present narrator, in certain unconventional characters who instinctively perceive what seems hidden from others, and in the novels' pervasive symbolism, their recurrent and expanding imagery that opens vistas beyond scenes of specific, mundane conflict.

In his *Aspects of the Novel* (1927) Forster makes a distinction between "fantasy" and "prophesy." His definition of fantasy does not rest merely on the inclusion of the supernatural but on what he terms "either the supernatural or its absence." Laurence Sterne's *Tristram Shandy* (1759–1767) falls within his category of fantasy because "Muddle . . . almost incarnate" is its ruling principle, so does James Joyce's *Ulysses* (1922), because it is patterned on the *Odyssey*. Parody or adaptation of mythology are types of fantasy. Fantasy may illuminate aspects of life not explicable in terms of "time or people or logic or any of their derivatives," but when a novelist succeeds in implying fundamental laws of life the realm of "prophesy" is entered. Forster cites examples from Fyodor Dostoyevsky, whose characters are at certain moments representative of all humanity, as well as Herman Melville, D. H. Lawrence, and Emily Brontë.

Forster's distinction bears a striking resemblance to a development in his own fiction. The short stories that include a supernatural element are nearly all among his earliest works; some of them even preceded his first novels. In his two major novels, *Howards End* and *A Passage to India,* the supernatural is ostensibly absent, but the author obviously endeavors to invest his plot and characters with a degree of universal significance. In *Howards End* a visualization of Ludwig van Beethoven's Fifth Symphony as a battle between good and evil forces develops into a symbolic pattern underlying the novel, and in *A Passage to India* the echo in the Marabar Caves, where a young Englishwoman imagines she has been sexually assaulted by an Indian, becomes a recurrent theme integrated in the novel as a whole, suggestive of both the immensity and the emptiness of the universe, but apprehended only by an elderly woman visitor from England and a Hindu mystic.

In Forster's early short stories he is most clearly a fantasy writer by any definition, wide or narrow. These stories have generally been admired for their originality and lucid style, but they have seldom been granted the same attention as his novels. "The Story of a Panic" (*Independent Review,* 1904), the first story he wrote, and "Albergo Empedocle," the first of his stories that achieved publication (in the magazine *Temple Bar,* December 1903), were both products of the revelatory experience Italy had been to him. They both deal with British middle-class tourists in Italy faced with phenomena that beggar their understanding. In both an apparently dim-witted youth undergoes a transformation that is alarming and incomprehensible to his travel companions. In "Albergo Empedocle" the main character, Harold, traveling with his fiancée and prospective in-laws in Sicily, discovers that he had been a Greek in what seems to be an earlier life. When he is not believed he loses all interest in the present and ends up in an asylum where he, in the narrator's words, "is living the life he knew to be greater than the life he lived with us." In "The Story of a Panic" a group of British picnickers in the mountains above Ravello are discussing the death of "the great God Pan" when they are suddenly seized with terror and flee. Only Eustace, an apathetic adolescent, is left behind, asleep. When he wakes up he seems to have become a different person. His uncontrollable ecstasies cause so much embarrassment that he is locked up in his bedroom. A young Italian waiter, who understands what has happened to Eustace, sets him free at the sacrifice of his own life.

The theme uniting these and most stories by Forster is that of another life not ruled by conventions that cripple natural impulses and potential for self-realization. In accordance with the thinking of his mentor, Dickinson, the other life is "Greek," and Greek mythology is often its frame of reference. It is hinted that Eustace is not just saved for the moment but that he comes to live a fuller life. This type of happy ending is found only in one other story by Forster, "The Curate's Friend" (*Pall Mall Magazine,* 1907), in which a Wiltshire curate is helped by a faun, invisible to others, to become aware of who he is and so become a better curate. In "Other Kingdom" (*English Review,* 1909), one of Forster's best executed stories, a young woman, unstudied in the conventional life, escapes her rich, possessive husband-to-be by metamorphosing into a dryad. He believes of course that she has eloped with a lover. In "The Celestial Omnibus" (*Albany Review,* 1908), a boy is transported to a heavenly pantheon of great poets and mythical heroes in a mysterious bus. The boy with the still-unspoiled and inquisitive mind is lifted high on the shield of Achilles, whereas a horrible end awaits Mr. Bons (his name a reversal of *snob*), president of the local literary society and owner of "Seven Shelleys," when he accompanies the boy on a trip in the bus.

Forster's view of modern life as contrasted with the "other" life is demonstrated in "The Other Side of the Hedge" (*Independent Review,* 1904), a fairly simple and explicit allegory. The narrator, for a moment tired of walking on the brown, dusty road of a busy, competitive life, is tempted to take a look at what is behind the hedge that lines the road. It turns out to be an Arcadian paradise, which "fills up slowly, though it was meant for all mankind." There he meets Death, a harvester with a scythe, who turns out to be his long-vanished brother. Death is also the theme of *The Story of the Siren* (1920). It opens with a view of a "sea change": the narrator, in a boat off the coast of Italy, inadvertently drops his notebook into the water where it is transformed into a vibrant thing of beauty. The Italian boatman who dives for the book then tells the story of how his brother, Giuseppe, nearly drowned when diving for money to amuse a party of tourists. He had faced "the Siren," Death, and could no longer bear to have contact with the living because he knew that they would all die. When he married a girl, Maria, who suffered from the same "madness," and she became pregnant a rumor was spread that the child would be the Antichrist. Eventually she was pushed into the sea from a cliff, presumably by a priest. The boatman, however, has a final twist to his story: a child will one day be born "who will fetch up the Siren from the sea, and destroy silence, and save the world!" The story expresses a belief that is also voiced in *Howards End:* when the reality of death is no longer shrouded by fear, silence, or religion people need no longer be lonely.

Forster wearing a cloak he bought in Italy, circa 1902–1903

Forster ventured into other types of fantasy but without much success, as in "The Purple Envelope" (first published in *The Life to Come and Other Stories,* 1972), his halfhearted 1905 attempt at writing a ghost story. "The Road from Colonus" may be viewed as a more mature treatment of the theme of

"Albergo Empedocle," but without the supernatural element. In a relatively late allegory, "The Point of It" (*English Review,* 1911), the other life is identified with the betrayed dreams of youth, with which the meritorious citizens condemned to an afterlife buried in a sandy desert fear to be confronted.

The overt romanticism of Forster's fantastic stories may make them seem more dated than his novels. But one of his stories is still disturbing: "The Machine Stops," written in 1908, his only science-fiction story. Forster viewed the rapid advance of industrialization and urbanization with apprehension. In *Howards End* the transformation of London is seen as "a foretaste of a nomadic civilization" cut off from its roots. Early in 1908 Henry Farman, a Frenchman, flew an airplane over a kilometer circuit in a minute and a half. Forster, in his diary, did not share the enthusiasm this event called forth: ". . . if I live to be old I shall see the sky as pestilential as the roads. . . . Science, instead of freeing man–the Greeks nearly freed him by right feeling–is enslaving him to machines. Nationality will go, but the brotherhood of man will not come. . . ." In these reflections one may see the nucleus of "The Machine Stops" (*Oxford and Cambridge Review,* 1909), which he wrote shortly after. Another incentive was H. G. Wells's utopianism. The story was "a reaction to one of the earlier heavens of H. G. Wells," he explained many years later in his introduction to the 1947 edition of his *Collected Tales.* "The Machine Stops" is a dystopia showing humanity's growing dependence on machinery in its ultimate consequence.

In "The Machine Stops" the future is envisioned as a bleak negation of all that Forster believed in. Personal relationships have gone, and the surface of the earth is barren and presumably uninhabited. Humanity has retreated into a hivelike existence underground, reduced to complete dependence on the all-powerful Machine for survival. Each individual has been provided with a cell, where all needs are provided for by the Machine. Time is spent on exchanging inconsequential "ideas" through the telescreens that have replaced nearly all direct human contact. Physical strength as well as daring thinking are discouraged. The Machine is an object of religious worship.

The story has only two characters, Vashti and Kuno, mother and son. She is worried by her son's unorthodoxy. He has managed to visit the surface of the earth without using an airship, a crime punishable by "Homelessness," which means certain death, since humans are no longer able to survive in the "outer air." His Forsterian speculations on what humanity could be seem absurd to Vashti. But Kuno is the first to detect that the Machine is breaking down. Slowly, imperceptibly, its functions are deteriorating, and apparently no one alive knows how to mend it. The perfection of technology has become the end of human life: "People were crawling about, people were screaming, whimpering, gasping for breath, touching each other, vanishing in the dark and ever and anon being pushed off the platform . . . at last the final horror approached–light began to ebb, and she knew that civilization's long day was closing." Though Forster's dystopia is hardly likely to literally happen it is not easily forgotten because it poses some still-pertinent questions about the goals of science.

Forster acknowledged his debt to Samuel Butler, the author of *Erewhon* (1872). Butler's belief in the instinctual life, his agnosticism, his distrust of conventions, and his skepticism toward technology are clearly present in Forster, as well. The boy in "The Celestial Omnibus" is a Butlerian character. In *Erewhon* Forster saw "the human race re made," as he put it in a 1907 letter to his friend Bob Trevelyan. Still, Forster's fantastic short stories may justly be rated as original contributions to fantasy literature. "The Machine Stops," also reminiscent of Butler, stands out among them as the first of the many antitechnological, and often anti-Wellsian, dystopias that were to figure prominently in the literature of the following decades. After Forster came Yevgeny Zamyatin, Aldous Huxley, C. S. Lewis, C. M. Kornbluth and Frederik Pohl, Anthony Burgess, and many other writers of mainstream as well as science fiction.

Letters:

Selected Letters of E. M. Forster, 2 volumes, edited by Mary Lago and P. N. Furbank (London: Collins, 1983, 1985; Cambridge, Mass.: Harvard University Press, 1983, 1985);

A Calendar of the Letters of E. M. Forster, edited by Lago (London: Mansell, 1985).

Bibliographies:

Frederick P. W. McDowell, ed., *E. M. Forster: An Annotated Bibliography of Writings about Him* (DeKalb: Northern Illinois University Press, 1977);

B. J. Kirkpatrick, *A Bibliography of E. M. Forster,* second edition (Oxford: Clarendon Press, 1985).

Biographies:

P. N. Furbank, *E. M. Forster: A Life, Volume One: The Growth of a Novelist (1879–1914)* (London: Secker & Warburg, 1977); *E. M. Forster: A Life, Volume Two: Polycrates' Ring (1914–1970)* (London: Secker & Warburg, 1978), republished as *E. M. Forster: A Life*, 1 volume (New York: Harcourt Brace Jovanovich, 1978; London: Oxford University Press, 1979);

Francis E. King, *E. M. Forster and His World* (London: Thames & Hudson, 1978; New York: Scribners, 1978);

Nicola Beauman, *Morgan: A Biography of E. M. Forster* (London: Hodder & Stoughton, 1993).

References:

Glen Cavaliero, *A Reading of E. M. Forster* (London: Macmillan, 1979);

John Colmer, *E. M. Forster: The Personal Voice* (London & Boston: Routledge & Kegan Paul, 1975);

Philip Gardner, ed., *E. M. Forster: The Critical Heritage* (London & Boston: Routledge & Kegan Paul, 1974);

Denis Godfrey, *E. M. Forster's Other Kingdom* (Edinburgh & London: Oliver & Boyd, 1968);

K. W. Gransden, *E. M. Forster* (Edinburgh & London: Oliver & Boyd, 1962; New York: Grove, 1962);

J. K. Johnstone, *The Bloomsbury Group* (London: Secker & Warburg, 1954);

Frederick P. W. McDowell, *E. M. Forster* (New York: Twayne, 1969; revised, 1982);

Harry T. Moore, *E. M. Forster* (New York & London: Columbia University Press, 1965);

Norman Page, *E. M. Forster* (London: Macmillan, 1987);

Oliver Stallybrass, ed., *Aspects of E. M. Forster* (London: Arnold, 1969; New York: Harcourt, Brace & World; 1969);

George H. Thomson, *The Fiction of E. M. Forster* (Detroit: Wayne State University Press, 1967).

Papers:

Forster's papers are held at King's College Library, Cambridge University. Scripts for radio talks and radio and television interviews and productions are in the BBC Written Archives Centre, London.

Kenneth Grahame

(8 March 1859 – 6 July 1932)

Walter Albert
University of Pittsburgh

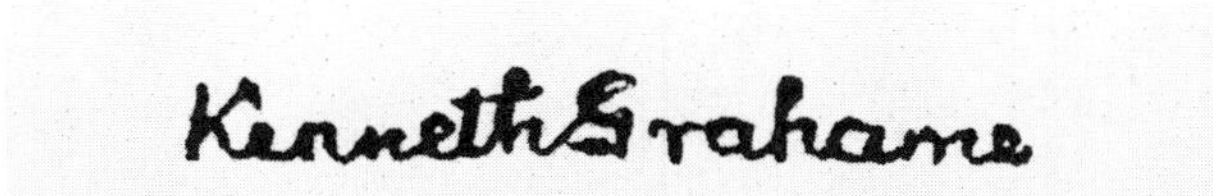

See also the Grahame entries in *DLB 34: British Novelists, 1890–1929: Traditionalists* and *DLB 141: British Children's Writers, 1880–1914.*

BOOKS: *Pagan Papers* (London: Matthews & John Lane, 1894; Chicago: Stone & Kimball, 1894);
The Golden Age (London: John Lane, 1895; Chicago: Stone & Kimball, 1895);
The Headswoman (New York & London: John Lane, 1898);
Dream Days (New York & London: John Lane, 1898);
The Wind in the Willows (London: Methuen, 1908; New York: Scribners, 1908);
First Whisper of "The Wind in the Willows," edited by Elspeth Grahame (London: Methuen, 1944; Philadelphia & New York: Lippincott, 1945);
Paths to the River Bank: The Origins of "The Wind in the Willows" from the Writings of Kenneth Grahame, edited by Peter Haining (London: Souvenir, 1983).

OTHER: Eugene Field, *Lullaby-Land,* edited by Grahame (New York: Scribners, 1897; London: John Lane, 1898);
A Hundred Fables of Aesop from the English Version of Sir Roger L'Estrange, edited by P. J. Billinghurst, introduction by Grahame (London & New York: John Lane, 1899);
The Cambridge Book of Poetry for Children, 2 volumes, edited by Grahame (Cambridge: Cambridge University Press, 1916; New York: Putnam, 1916; revised edition, Cambridge: Cambridge University Press, 1932; New York: Putnam, 1933);
George Sanger, *Seventy Years a Showman,* preface by Grahame (New York: Dutton, 1926; London & Toronto: Dent, 1926).

Although it has often been pointed out that *The Wind in the Willows* (1908), Kenneth Grahame's most enduring work, presents an idealized portrait of rural nineteenth-century England, Edmund Little in *The Fantasts: Studies in J. R. R. Tolkien, Lewis Carroll, Mervyn Peake, Nicolai Gogol, and Kenneth Grahame* (1984) includes him among a group of writers who have created "unreal worlds." According to Lois R. Kuznets, Grahame's novel is an "animal fantasy" in which animals and humanity live quietly in close

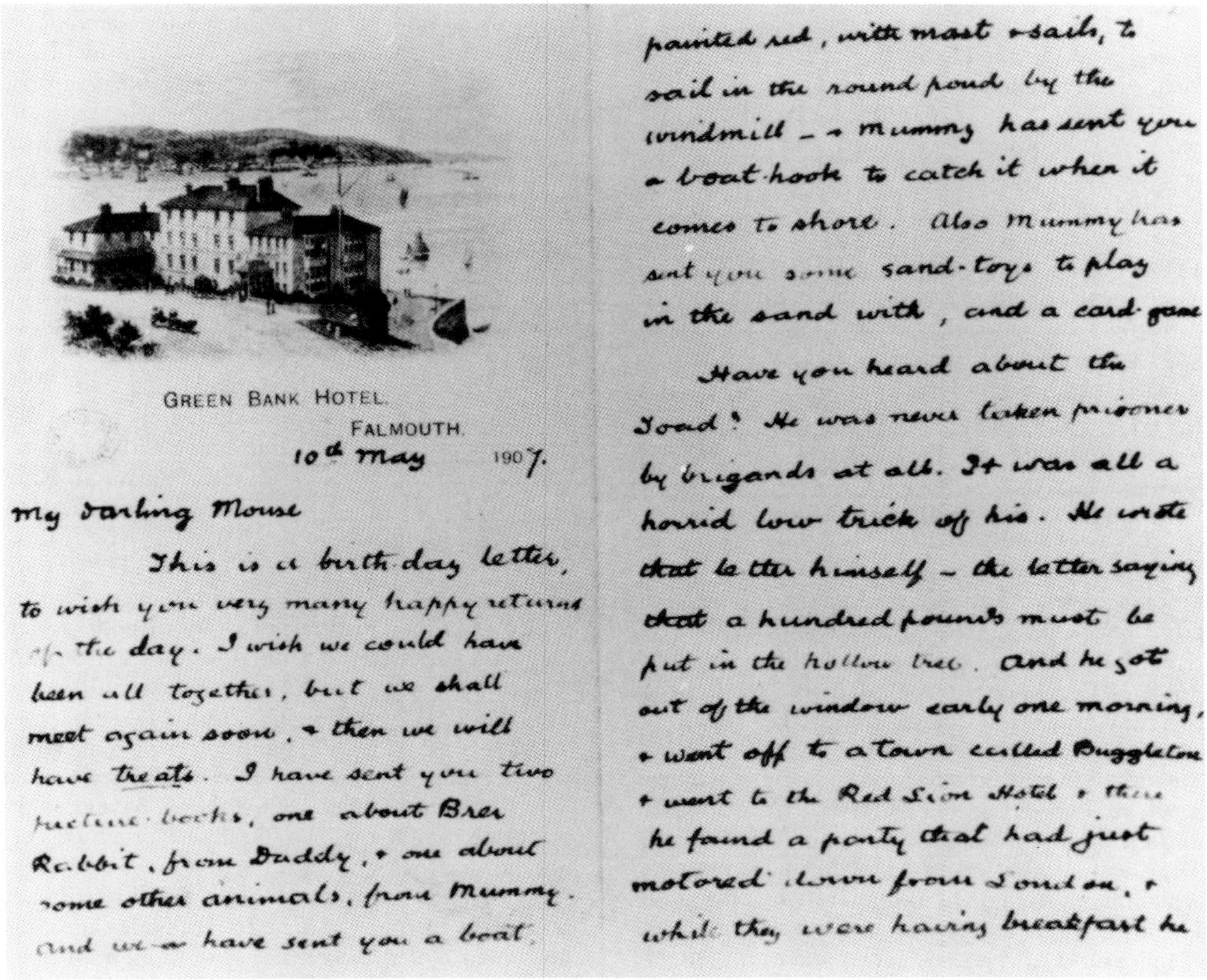

GREEN BANK HOTEL.
FALMOUTH.
10th May 1907.

My darling Mouse

This is a birth-day letter, to wish you very many happy returns of the day. I wish we could have been all together, but we shall meet again soon, & then we will have treats. I have sent you two picture-books, one about Brer Rabbit, from Daddy, & one about some other animals, from Mummy. And we have sent you a boat painted red, with mast & sails, to sail in the round pond by the windmill – & Mummy has sent you a boat-hook to catch it when it comes to shore. Also Mummy has sent you some sand-toys to play in the sand with, and a card-game.

Have you heard about the Toad? He was never taken prisoner by brigands at all. It was all a horrid low trick of his. He wrote that letter himself – the letter saying that a hundred pounds must be put in the hollow tree. And he got out of the window early one morning, & went off to a town called Buggleton & went to the Red Lion Hotel & there he found a party that had just motored down from London, & while they were having breakfast he

Grahame's letter to his son, Alastair, that includes the beginning of the earliest written version of The Wind in the Willows *(Ms. Misc d. 281, fol. 1^v–2^r; Bodleian Library, Oxford)*

proximity, with the animals conversing, dressing, and living like humans in a seamless conjunction of the real and the unreal. It also develops out of his earlier work and from a life that may seem at first glance too prosaic to support it.

Kenneth Grahame, the second son and third child of Bessie Ingles and James Cunningham Grahame, was born on 8 March 1859 in Edinburgh, Scotland. When he was about a year old the family moved to Argyllshire, where his father had been appointed to the post of sheriff substitute. There, on 16 March 1864, his mother died from scarlet fever following the birth of her third son, Roland. Kenneth also caught the infection but recovered under the care of his maternal grandmother. Shortly after this the four children went to live with her at Cookham Dene in Berkshire. Their father stayed behind to mourn his wife and began to develop an increasing dependency on alcohol.

Grahame would later recall these few years at Cookham Dene with affection in his two collections of reminiscences, *The Golden Age* (1895) and *Dream Days* (1898). "Granny" Ingles may not have been the stereotypical doting grandmother–straitened financial circumstances and a stern Presbyterian nature undoubtedly mitigated against that–but the happy memories of those years were also, in part, the foundation for *The Wind in the Willows*. He and his brothers and sister first moved with their grandmother in 1866 to a smaller cottage after repairs became necessary to Cookham Dene, then back to their father's house when he unexpectedly summoned them home. The stay was less than a year. In the spring of 1867 their father resigned his post and went abroad, and the children were sent back to their grandmother. It was the last time they were to live with their father; he died in France in 1887, and of the three surviving children only Kenneth was present at the funeral in Le Havre.

In 1868 both Kenneth and his older brother, Willie, were enrolled in St. Edwards School in Oxford, where Kenneth excelled during the next seven years. However, in 1875, at the start of Kenneth's last year at St. Edwards, Willie Grahame died from a severe inflammation of the lungs. The following year the family, refusing to support Kenneth's application for Oxford, insisted instead that he apply for a clerkship in the Bank of England. He was nominated for the position and worked in his uncle's office until a vacancy opened in 1879.

From outward appearances Grahame had a successful career at the Bank of England–he became a secretary of the bank, one of its highest positions, in 1898–but it was soon apparent that his real interests lay elsewhere. Now living in London, he made lasting literary friendships and became a contributor to various publications–including *The Yellow Book,* an important periodical whose reputation for outrageous attacks against the literary and social establishment was more extreme than its actual content. Many of his essays were published in *Pagan Papers* (1894), which also included six sketches that form the nucleus of *The Golden Age.* His topics, including rail travel, smoking, books, and traveling, are often treated and developed fancifully, with elements that border on the fantastic. The memorable episode of the encounter with Pan in *The Wind in the Willows* is prefigured in "The Lost Centaur," in which the god is linked to humanity by "his gift of melody," while in other essays Grahame celebrates the pleasures of river and forest and unforgettably evokes in animals "the mute and stunted human embryo that gazes appealing from out the depths of their eyes."

Later editions of *Pagan Papers* omit the sketches that were incorporated into *The Golden Age,* and the modest literary success of the first book was eclipsed by the reception of *The Golden Age* and *Dream Days.* These pieces, which depict an imaginative child's retreat from his distant, repressive family, whom he calls the "Olympians," were so successful that the initial reception of *The Wind in the Willows* was colored by the disappointment contemporary reviewers felt when Grahame apparently abandoned his realistic, if poetic, evocations of childhood for a fantasy involving animal characters. Some thought Grahame had forfeited his laurels as a serious writer for children's literature.

However, fantasy was not absent from *The Golden Age* and *Dream Days.* Both works are rich in the fantasizing of childhood with its games peopled by fierce pirates, intrepid adventurers, and bold sea captains, while the tale of "The Reluctant Dragon" in *Dream Days,* recounted by an adult friend of the narrator, is informed with the warm camaraderie among the characters that is a notable feature of *The Wind in the Willows.* In this medieval fantasy, a revisionist version of Saint George and the Dragon, there is a public adherence to the rules of chivalric romance, but the dragon-poet puts up with the external trappings of the legend only long enough to "get it over with, and then . . . go on with the sonnets." This may reflect Grahame's willingness to do what was necessary in his professional life until he was able to retire to the country–like the dragon to his cave–and "go on" with his writing.

This success brought Grahame to the attention of a much larger audience, among them Elspeth Thomson, who, like Grahame, was a native of Edinburgh. After the death of her father, inventor Robert William Thomson, Elspeth's mother married Fletcher Moulton, a poor barrister who subsequently became a successful lawyer. The family moved to London, where Moulton became accustomed to having his house run by Elspeth, which may have influenced his initial dislike of Grahame. However, strong-willed Elspeth was not deterred by Moulton's reservations. The couple were married on 22 July 1899, with Elspeth rejecting her expensive wedding gown for a tattered muslin dress and a daisy chain around her neck. She may have thought this outfit would appeal to Grahame's rustic tastes, but her clothes and extravagant behavior were symptomatic of a profound incompatibility with her husband's reserved nature. In Grahame there was no evidence of any eccentricity of thought or manner. They had a son, Alastair, in May 1900, and he did not write another book until his final one, *The Wind in the Willows.*

In a letter to a friend in May 1904, Grahame wrote that he had told "Mouse" (the Grahames' pet name for their son) "stories about moles, giraffes & water-rats" to quiet the boy after a crying spell. In May 1906 the Grahame family moved from London to Cookham Dene, where Grahame had spent the childhood years that inspired his two books of reminiscences. By 1907 the giraffe had been replaced by Toad, whose adventures were the subject of letters written by Grahame to his "dearest Mouse" when the boy was on holiday with his governess. Grahame spent much of the summer in London, and the letters continued into the autumn. The completed manuscript was submitted to publishers but was only reluctantly accepted by Methuen after John Lane, the publisher of Grahame's earlier books, had turned it down.

Critics generally consider the main story to be that of the ebullient, manic Toad, even though his character is not introduced until chapter 6. The book is tripartite in its structure: the friendship of Mole and Ratty (the Water Rat) and their domestic arrangements take up the first third of the book; the second involves the flight and adventures of Toad until he returns to Toad Hall, which has been taken over by a motley, unpleasant crew of weasels and ferrets; finally, Toad, Mole, and Rat band together under the leadership of their mentor, the wise and intrepid Badger, to eject the intruders and restore Toad to his rightful place. The book, with its roots in the places that Grahame loved, is one of the most intimate and realistic of childhood masterpieces. It is also a poetic fantasy, imbued with personal feelings that resonate in the memory after the exact nature of Toad's adventures may have faded with time.

Grahame was at best an indifferent poet in verse; in prose he was a master stylist. Chapter 7, "The Piper at the Gates of Dawn," placed almost at the center of the interlocking narratives, is the thematic heart of the book. This vision of *The Yellow Book* coterie's favorite pagan divinity, Pan, ignores the erotic implications to focus on his piping, music that arouses a "longing that . . . is pain" and that is a call "stronger even than the music is sweet." In the context of the narrative the call both draws Water Rat and Mole to the demigod, whom they "worship" with heads bowed, and leads them to Little Portly, the lost otter child, who has wandered off as he is accustomed to do.

This wanderlust, which afflicts almost all the major characters in the book and was a call to which Grahame himself responded many times in his life, is even more explicit in "Wayfarers All," chapter 9, where the inland Rat meets up with the seafaring Rat, who tempts him with stories of marvelous voyages and wonderful adventures in distant lands. The seafaring Rat's narrative exercises a hypnotic spell on Ratty from which he is recalled only by Mole, who wrestles him to the ground and holds him down until the fit passes. This is also the fit that intermittently takes hold of Toad, capturing his imagination and drawing him away from older pleasures to new interests and landscapes. An "altered" Toad finally discovers peace and some measure of security only back at Toad Hall, among his friends.

Toad's reunion with his friends and the portrait of a male fraternity dominate Grahame's domestic fantasy. Furthermore, although humans and animals seem to live in some variation of a peaceable kingdom, only Toad ventures fearlessly out of the Riverbank world to move freely among the humans, and he is portrayed as a transgressor whose wild ways bring him into conflict with their laws.

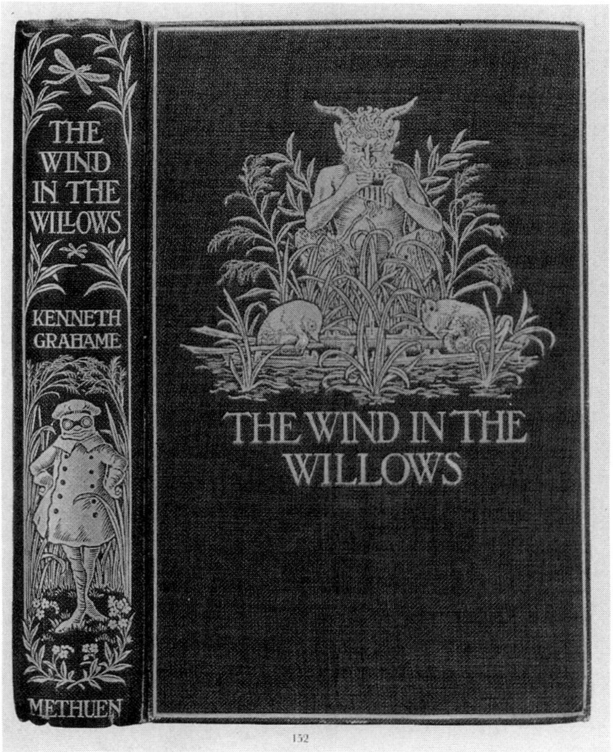

Cover for Grahame's 1908 book, in which Toad, Rat, Mole, and Badger work together to remove a band of invading weasels and ferrets from Toad Hall

Grahame's animals seem as naturally reclusive as he was, but while Rat and Mole succeeded in establishing a cozy housekeeping arrangement, Grahame's relationship with his wife is thought to have lacked the security and freedom from worry that Rat and Mole knew, so the emotional bonds in the novel may be another of its fantasy elements.

Grahame resigned from the Bank of England in June 1908, three months after his forty-ninth birthday and four months before the publication of *The Wind in the Willows.* Some writers have suggested that this early retirement was the result of frequent absences for bronchial disorders (the same weakness that had killed his brother) or conflicts with William Campbell Middleton, the bank's governor; some say he left at the request of his employers because of insufficient attention to his duties. Although Grahame may not have been the most devoted employee of the bank, he was promoted at least once, and there is sufficient evidence to suggest that he was a competent secretary. Biographer Peter Green points out that his annual pension did not exceed £400, but Grahame may have seen retirement as preferable to continuing in a job that he had not chosen for himself. With the royalties that accrued from the unexpected and continuing success of *The Wind in the Willows,* the pension allowed him to live in comfort and to travel extensively.

In 1910 the Grahames moved from Cookham Dene to Boham's, a farmhouse in the Berkshire Downs. Alastair left in 1911 for school at the Old Malthouse on the Dorset coast, where he seems to have spent a fairly uneventful two years. He entered Rugby in September but, deeply unhappy there, stayed only six weeks. He was admitted to Eton in January 1915 but left in the middle of the Lent Term in 1916 and worked with a private tutor until he was admitted to Christ Church, Oxford, in 1918. His academic and personal problems, aggravated by a religious crisis, worsened, and on 7 May 1920 his decapitated body was found on railroad tracks near the university. An inquest ruled the decapitation an accidental death, but the circumstances make it more likely that Alastair committed suicide, strained, as Green puts it, "almost beyond belief [to live up to the] myth that Elspeth wove around him." The myth of his "special" qualities and great promise was undoubtedly constructed by both parents. There is no record of the extent of Elspeth's bereavement, but a friend commented in a letter that Grahame's "suffering must have been intense."

Biographers have generally made Elspeth out to be the villain in the ill-matched pair's marriage. She is described by Green in a caption accompanying a 1930 photograph as "a pallid, haggard ghost whose meanness was proverbial." Yet the photograph, taken when Elspeth was in her late sixties, seems little more than a realistic portrait of a woman of that age, and it may be presumptuous to assign to her all the blame in this unhappy couple's marriage.

Grahame was not well during his last years at Pangbourne, where the couple moved on their 1924 return from a long stay in Italy following Alastair's death. He suffered from circulatory problems and died on 6 July 1932 of a cerebral hemorrhage. He was buried in Holywell Churchyard in Oxford.

To some extent, the perception that Grahame was a man whose reserve prevented him from having many close friends has its counterpart in the final lines of *The Wind in the Willows.* The animals in the forest, when they want to quiet their children, threaten them with stories of the "terrible grey Badger." This, the author adds, was a "base libel on Badger, who, though he cared little about Society, was rather fond of children; but it never failed to have its full effect." Grahame was clearly a man who cared little for society and largely lived apart from it in his final years, but he, too, was surely "rather fond" of children. The book that began as a story for his "dearest Mouse" is his imperishable legacy to them and their parents.

Biographies:

Patrick R. Chalmers, *Kenneth Grahame: Life, Letters and Unpublished Work* (London: Methuen, 1933);

Peter Green, *Kenneth Grahame 1959–1932: A Study of His Life, Work and Times* (London: John Murray, 1959); abridged as *Beyond the Wild Wood: The World of Kenneth Grahame, Author of "The Wind in the Willows"* (Exeter: Webb & Bower, 1982);

Eleanor Graham, *Kenneth Grahame* (London: Bodley Head, 1963);

Alison Prince, *Kenneth Grahame: An Innocent in the Wild Wood* (London: Allison & Busby, 1994).

References:

Humphrey Carpenter, *Secret Garden: A Study of the Golden Age of Children's Literature* (Boston: Houghton Mifflin, 1985), pp. 115–125, 151–169;

Peter Hunt, *"The Wind in the Willows:" A Fragmented Arcadia* (New York: Twayne, 1994);

Lois R. Kuznets, *Kenneth Grahame* (Boston: Twayne, 1987);

Edmund Little, *The Fantasts: Studies in J. R. R. Tolkien, Lewis Carroll, Mervyn Peake, Nicolai Gogol, and Kenneth Grahame* (Amersham: Avebury, 1984);

W. W. Robson, "On *The Wind in the Willows,*" in his *The Definition of Literature and Other Essays* (Cambridge: Cambridge University Press, 1982), pp. 119–144;

Roger Sale, *Fairy Tales and After: From Snow White to E. B. White* (Cambridge, Mass.: Harvard University Press, 1978), pp. 165–193.

George Griffith

(20 August 1857 - 4 June 1906)

Darren Harris-Fain
Shawnee State University

BOOKS: *Poems,* as Lara (London: Stewart, 1883);
The Dying Faith, as Lara (London: Stewart, 1884);
The Angel of the Revolution (London: Tower, 1893; Westport, Conn.: Hyperion, 1974);
Olga Romanoff; or, The Syren of the Skies (London: Tower, 1894; Westport, Conn.: Hyperion, 1974);
A Heroine of the Slums (London: Tower, 1894);
The Outlaws of the Air (London: Tower, 1895);
Valdar the Oft-Born (London: Pearson, 1895; Naperville, Ill.: FAX, 1972);
Briton or Boer? (London: F. V. White, 1897);
The Knights of the White Rose (London: F. V. White, 1897);
The Romance of Golden Star (London: F. V. White, 1897; New York: Arno, 1978);
Men Who Have Made the Empire (London: Pearson, 1897);
The Virgin of the Sun (London: Pearson, 1898);
The Destined Maid (London: F. V. White, 1898);
The Gold-Finder (London: F. V. White, 1898);
Gambles with Destiny (London: F. V. White, 1899);
The Great Pirate Syndicate (London: F. V. White, 1899);
Knaves of Diamonds (London: Pearson, 1899);
The Rose of Judah (London: Pearson, 1899);
"Thou Shalt Not–," as Stanton Morich (London: Pearson, 1900);
Brothers of the Chain (London: F. V. White, 1900);
Denver's Double (London: F. V. White, 1901);
The Justice of Revenge (London: F. V. White, 1901);
Captain Ishmael (London: Hutchinson, 1901);
In an Unknown Prison Land (London: Hutchinson, 1901);
A Honeymoon in Space (London: Pearson, 1901; New York: Arno, 1975);
The Missionary (London: F. V. White, 1902);
The White Witch of Mayfair (London: F. V. White, 1902);
"To the Rescue" (London, 1903);
The Lake of Gold (London: F. V. White, 1903);
A Woman against the World (London: F. V. White, 1903);

George Griffith

With Chamberlain through South Africa (London: Routledge, 1903);
Sidelights on Convict Life (London: John Long, 1903);
The World Masters (London: John Long, 1903);
A Criminal Croesus (London: John Long, 1904);
The Stolen Submarine: A Tale of the Russo-Japanese War (London: F. V. White, 1904);
His Better Half (London: F. V. White, 1905);
An Island Love Story (London: F. V. White, 1905);
A Mayfair Magician (London: F. V. White, 1905);

A Conquest of Fortune (London: F. V. White, 1906);
The Great Weather Syndicate (London: F. V. White, 1906);
His Beautiful Client (London: F. V. White, 1906);
The Mummy and Miss Nitocris: A Phantasy of the Fourth Dimension (London: Laurie, 1906);
The World Peril of 1910 (London: F. V. White, 1907);
John Brown, Buccaneer (London: F. V. White, 1908);
The Sacred Skull (London: Everett, 1908);
The Lord of Labour (London: F. V. White, 1911);
The Diamond Dog (London: Pearson, 1913).
Collection: *The Raid of 'Le Vengeur' and Other Stories* (London: Ferret Fantasy, 1974).

OTHER: Alphonse Daudet, *The Hope of the Family,* adapted by Griffith as Levin Carnac (London: Pearson, 1898).

SELECTED PERIODICAL PUBLICATIONS–UNCOLLECTED: "The Curse of Ham," as Levin Carnac, *Pick-Me-Up* (30 July 1898);
"The Searcher of Souls," *Pearson's Weekly Extra,* Christmas Number (November 1899).

Though not as well known as Sir Arthur Conan Doyle or H. G. Wells, George Griffith was one of the most popular writers of fantasy and science fiction in England in the closing years of the nineteenth century. Astonishingly prolific, Griffith serialized dozens of works in late-Victorian newspapers and magazines, which found a thriving audience with the rise in literacy and increases in income and leisure time. In his stories and novels Griffith contributed significantly to the development of the "scientific romance" in Britain.

George Griffith was born George Chetwynd Griffith-Jones on 20 August 1857 in Plymouth to George Alfred Jones, a clergyman, and Jeanette Henry Capinster Jones. He was the second of two sons. His family moved as his father changed positions–first in 1860 to Tring in Buckinghamshire, then in 1861 to Ashton-under-Lyne near Manchester. The family moved one more time, to nearby Mossley, where his father was appointed vicar in 1864. In 1872, when Griffith was fourteen, his father died of a stomach hemorrhage.

Before his father's death Griffith was educated at home. Although his father rose slowly in his changing positions, he did not earn enough to pay for extensive educations for his sons. However, Griffith profited not only from his instruction at home but also from his father's library, from which he read writers such as Edward Bulwer-Lytton, Charles Dickens, Sir Walter Scott, and Jules Verne–all of them formative influences on Griffith's later development as a writer. After his father died Griffith was sent to a private institution in Southport, where he stayed for fifteen months before going to sea as an apprentice.

He was also unhappy with life as a sailor, so when the ship arrived in Melbourne, he deserted it and worked as a manual laborer in Australia until he had earned enough money to take to sea again. He traveled the world and later claimed that he turned down an offer of marriage to the daughter of a Polynesian king. He returned to England, much to his mother's relief, before his twentieth birthday.

Surprisingly, Griffith's next job was as a master at a prep school, a position he detested but one he endured for seven years at Worthing College and elsewhere. During this time he began to contribute short pieces to local newspapers. By the early 1880s he was publishing poetry under the pen name Lara, which he used for his two book-length collections, *Poems* (1883) and *The Dying Faith* (1884).

While serving as a master at Bolton he met Elizabeth Brierly, and they married in February 1887. Their family eventually included two sons and a daughter. That same year he passed the College of Preceptors Examination and immediately quit teaching to try to earn his living as a writer. Griffith and his wife moved to London, where his work for a newspaper led to his becoming its editor. As editor he was charged with libel, and the paper ceased publication when Griffith, defending himself in court, lost the case. For a time he was jobless and almost homeless, but then a friend introduced him to the well-known newspaper and magazine publisher C. Arthur Pearson. Griffith was put to work for *Pearson's Weekly* addressing envelopes. His wide range of experiences, however, soon made him useful in other ways, and before long Griffith was writing for the magazine.

Griffith's formative experiences as a writer of fiction came from his work as a professional journalist and columnist, which encouraged his talent for writing clearly and quickly and for keeping readers' attention. Also, when he turned from journalism to fiction, he had already established important publishing contacts–in particular Pearson, in whose periodicals most of Griffith's novels were serialized before their publication in book form.

Griffith's first successful book was his novel *The Angel of the Revolution* (1893), which was serialized in *Pearson's Weekly* from January to October 1893. The novel is part of the British tradition known as the scientific romance, a combination of modern science with the nineteenth-century mode of adventurous or exotic fiction called romance. In the case of *The Angel of the Revolution,* Griffith writes

A hideous shape rose out of the water behind them.

Whole mountain ranges of glowing lava were hurled up miles high.

"The Earth, the Earth—thank God, the Earth!"

Illustrations by Stanley Wood and Harold Piffard for Griffith's A Honeymoon in Space

of a future war (1903–1905) fought with airships, submarines, and poison gas; what would be nightmarish in the world wars to come instead delighted readers, who also enjoyed Griffith's depictions of mass destruction, their realism no doubt aided by his journalistic experience. In addition, Griffith interpolated his ideas about the viability of anarchism and socialism into the novel, which provided him–as has often been the case with science-fiction writers–a relatively nonconfrontational means of discussing concepts that otherwise might alienate readers.

Pearson had wanted to capitalize on two popular trends in the early 1890s: future-war stories and serialized novels. He succeeded admirably at both with *The Angel of the Revolution,* which was serialized anonymously. The book publication in October 1893, however, did feature Griffith's name. In book form it was equally successful, making Griffith an instantly popular author hailed as a second Jules Verne, and Pearson wasted no time in gaining exclusive access to his serialized fiction. Griffith was even given a secretary, to whom he dictated his work; this no doubt helps to account for part of his prolific output. However, the anti-Americanism expressed in this novel likely hurt Griffith's career, for none of his books and only a few of his stories were published in the United States during his lifetime.

Olga Romanoff; or, The Syren of the Skies (1894), serialized in *Pearson's Weekly* (December 1893–August 1894) simply as *The Syren of the Skies,* is a sequel to *The Angel of the Revolution.* The story has advanced five generations to the year 2030, when the peace established in the previous book is challenged. The later setting enables Griffith to give freer rein to his imagination, which helps to compensate for the novel's weaknesses in other areas–such as the deus ex machina use of a comet (a common device in Griffith's fiction) that foils the title character's plans to reclaim the power once held by her czarist ancestors.

Also in 1894 Griffith published a collection of his earlier stories called *A Heroine of the Slums* and engaged in a publicity stunt for *Pearson's Weekly.* A few years earlier an American writer named Elizabeth Bisland had made headlines by traveling around the world in seventy-six days–four days less than the number in Jules Verne's well-known novel. Given Griffith's background as a world traveler, Pearson thought he would be ideal to try to beat that number. Frequently reporting to *Pearson's Weekly* via telegraph, Griffith set out on 12 March through France, Italy, and Germany, where he boarded a steamer for Japan. From Japan he sailed to Canada, crossed the country by train, and then sailed from New York back to London, arriving on 16 May.

Pearson not only vigorously promoted Griffith's trip and his articles about it, but he also used Griffith's popularity to try to support a magazine he had acquired in 1893, *Short Stories.* Despite the title, Pearson intended to do this by serializing Griffith's next novel in its pages. Thus *The Outlaws of the Air* (published in book form in 1895) appeared in *Short Stories* beginning in September 1894. The story, densely plotted like all of Griffith's novels, is similar to that of *The Angel of the Revolution* and includes aircraft, anarchists, and heroes who naturally save the day.

Griffith's next serialized novel, *Valdar the Oft-Born* (1895), began appearing in *Pearson's Weekly* in January 1895 and ended in August of that year. It belongs to a type of fantasy popular in the late Victorian era, the historical romance with reincarnation as a major plot element. One of the most popular of these works was Edwin Arnold's *The Wonderful Adventures of Phra the Phoenician* (1890), which probably served as the model for *Valdar the Oft-Born.* (Such imitation was common in the popular magazines of the day, and Griffith's novels were imitated often.) Griffith's novel, unlike Arnold's, provides no other explanation for its story than the idea of reincarnation, making *Valdar the Oft-Born* not a scientific romance but a work of fantasy–and thus a departure of sorts for the author.

Valdar, a demigod in the otherworldly Norse realm of Asgard, attempts to pry into the secrets of the gods and is sentenced by Odin to become mortal. His lover, not wanting for them to be separated, asks to become mortal as well. The lovers are then incarnated and reincarnated over thousands of years, living through key moments in world history and rediscovering each other in many of their new incarnations. The pattern is still not broken by the end of the novel, when Valdar anticipates future incarnations and reunions with the woman he will love for all eternity. Audiences responded well not only to the love story but also to the historical sweep of the novel, but it failed to surpass Arnold's in reputation.

Griffith's reputation was also threatened in the mid to late 1890s by another writer of scientific romance. H. G. Wells had obviously read Griffith and was probably influenced by him in some of his fiction, but then Wells began to supersede Griffith and other writers of scientific romance with such highly imaginative and well-wrought novels as *The Time Machine* (1895), *The Island of Doctor Moreau* (1896), *The Invisible Man* (1897), *The War of the Worlds* (1898), and *The First Men in the Moon* (1901). Writers such as Arnold, Griffith, Wells, Doyle, H. Rider Haggard, William Le Queux, M. P. Shiel, and others were often published in the same magazines during this period, but obviously posterity has granted a few of these writers a higher standing than others. Although he remained popular at this time, Griffith would not be one of these.

Griffith began his next fantastic novel, *The Romance of Golden Star* (1897), on the return trip to England after an assignment for *Pearson's Weekly* in South America. Basing it on the history of the Incas in Peru, Griffith tells a story of an Incan prince and his sister, who, anticipating the destruction of their civilization by the Spanish conquistadors, place themselves in suspended animation so that in the future they may restore their lost empire. Instead the prince's mummified body is taken to England, and he is revived, initiating a typically convoluted plot of intrigue and amazing adventures.

Not all of Griffith's work for Pearson's magazines was fantasy or science fiction; in fact, in the nineteenth century few commercial writers were so specialized. Griffith also wrote historical adventures, detective fiction (Doyle's Sherlock Holmes was extremely popular at this time), and other types of fiction as well as occasional nonfiction pieces, especially travel writing. However, by the second half of the 1890s Griffith was no longer Pearson's favorite writer of science fiction (it was now Wells, until he left for higher-paying markets), let alone his favorite writer in general; in addition, Pearson stopped publishing Griffith's novels in book form, forcing him to find another publisher. Insult was added to injury when, after Pearson announced that Griffith would be the editor of one of the four new magazines he planned to launch in 1897–*The Passport,* which would reprint articles from around the world; the magazine was never published.

Despite such disappointments, Griffith continued to produce material for Pearson's magazines at a prodigious rate, such as "The Gold Magnet," published in book form as *The Gold-Finder* (1898). Inspired by a contemporary interest in gold rushes, the novel is about the invention of a device that is attracted to gold and the consequences that result. He also wrote another future-war story, *The Great Pirate Syndicate* (1899), but it was neither as imaginative nor as successful as *The Angel of the Revolution* and *Olga Romanoff,* although it does include some interesting battle craft. Also interesting in terms of Griffith's writing of fantasy is his inclusion of a scene in his 1898 novel *The Destined Maid* in which the villain is transformed into a skeleton after being shot in a duel.

Gambles with Destiny (1899) collects several of Griffith's short stories. These cover a wide range of subjects. "Hellville, U.S.A." is about a penal colony established in Arizona as a result of radical political reforms. The colony is so violent that politicians consider shelling it into oblivion, but conveniently a meteor shower does the job for them. A massive object from outer space also appears in "The Great Crellin Comet." Here a wealthy amateur astronomer learns that a comet will collide with the Earth in little more than a year and spearheads the construction of an enormous gun to fire an explosive device at the comet in order to save the world. Some historians also believe this story contains the first use of a countdown from ten to zero. More fantastic in nature is "Genius for a Year," in which a character experiments with hashish to bolster his creativity. It works for a year, but then the side effects are disastrous. Tragic consequences also are part of "A Corner in Lightning," in which a man's efforts to become the richest person on earth by controlling the world's supply of electricity has some unexpected results.

Just as Wells published his best scientific romances in the 1890s (along with the 1901 *The First Men in the Moon*), Griffith's work of this period is also his best. Though he continued to publish extensively into the twentieth century, his later efforts are generally less interesting than his earlier novels, with the exception of *A Honeymoon in Space* (1901), an entertaining romance in both senses of the term.

A Honeymoon in Space marked a return to the scientific romance for Griffith, whose novels toward the end of the nineteenth century tended toward historical adventure and works about contemporary issues. In this novel an inventor of an antigravity spacecraft and his new wife take the first honeymoon in space as they visit the Moon, Venus, Mars, an asteroid, Jupiter, and Saturn.

Two other novels published in 1901 include the fantastic. In *Denver's Double* one finds spiritual possession as part of a story of hypnotism and identical twins, while a character in *Captain Ishmael* has several encounters with the legendary Wandering Jew; he also creates powerful weapons in a "world of if."

The fantastic is also present in two of Griffith's later short stories. In "The Lost Elixir" (*Pall Mall Magazine,* October 1903) a mummy searches for his ancient love, whose chemically treated kiss will free him from his life-after-death. "From Pole to Pole" (*Windsor Magazine,* October 1904) is about the discovery of a tunnel from the North Pole to the South. Both are collected in *The Raid of 'Le Vengeur' and Other Stories* (1974).

Toward the end of his career Griffith turned again to the future-war novel. *The Lake of Gold* (1903), *The World Masters* (1903), and *The Stolen Sub-*

marine: A Tale of the Russo-Japanese War (1904) are all thickly plotted thrillers about political ends attempted through fantastic means–through advanced aircraft and submarines in *The Lake of Gold,* an invention that makes metal brittle in *The World Masters,* and high-powered explosives and submarines in *The Stolen Submarine.* The story is similar in *The Great Weather Syndicate* (1906), in which a cartel attempts to achieve world peace by controlling the weather, and in *The World Peril of 1910* (1907), which combines Griffith's typical future-war scenarios with the threatened impact of a comet.

A Criminal Croesus (1904) belongs to another popular category of fantasy in this period, the lost-race novel. Here the lost race, the descendants of a party of Norsemen trapped below the earth years ago, is found underground by a group searching for precious metals. The novel also includes the creation of a United States of South America.

Another late Griffith novel, *A Mayfair Magician* (1905), concerns a science-fiction invention typical of the period: a machine that can read minds. As in many such works, the emphasis is on adventure and intrigue rather than such niceties as characterization and ideas about the implications of science and technology. Griffith included many of his fantasy and science-fiction ideas in *The Mummy and Miss Nitocris: A Phantasy of the Fourth Dimension* (1906), which is similar in plot to *The Romance of Golden Star.* Here the reader can find the threat of world war, amazing aircraft, and the fourth dimension. It is as if Griffith had merged his scientific romances, future-war novels, and supernatural fantasies into one package. A more typical fantasy is *The Sacred Skull* (1908), which includes a special set of jewels capable of discovering guilt and innocence.

Griffith's last novel, *The Lord of Labour,* was dictated on his deathbed and published posthumously in 1911. The main invention here is a radium rifle that kills at close range by stopping the heart and disables at a distance by causing a nervous breakdown. (Griffith fails to mention what happens to the person who fires the gun.) He also revives the metal-weakening weapon here. These weapons are developed respectively by Britain and Germany, whose real war in the not-too-distant future was carried out more prosaically but just as lethally.

Griffith's later career was likely hurt by three problems. First, his once-high reputation suffered in the wake of the success of Wells and other new writers, and by the early years of the twentieth century Griffith had lost his connections with Pearson's magazines. Second, though he was a fast and prolific writer, he probably became too productive for his own good. The quality of his writing suffered, as did his health. Third, his health was not aided by his apparent alcoholism, and when he died on 4 June 1906, a little over a year before his fiftieth birthday, it was of cirrhosis of the liver.

Though Griffith and Wells mined similar territory, Griffith's reputation in science fiction is secondary and in general literature is almost completely nonexistent. While Wells possessed the ability to combine imaginative ideas and exciting stories with serious themes and philosophical speculations, the chief interest in Griffith's work lies mainly in his imaginative ideas and exciting stories. Even an appreciation of these, however, is often hampered by an uninspired, if not clichéd, style, poor characterization, weak ideas, and repetition. Still, Griffith is noteworthy in the history of science fiction for his contributions to the scientific romance and for what his work reveals not of the future but of his own times.

Bibliography:

George Locke, "Bibliography," in *The Raid of 'Le Vengeur' and Other Stories,* by Griffith (London: Ferret Fantasy, 1974), pp. 51–58.

Biography:

Sam Moskowitz with George Locke, "George Griffith–The Warrior of If," in *The Raid of 'Le Vengeur' and Other Stories,* by Griffith (London: Ferret Fantasy, 1974), pp. 6–50.

Reference:

Brian M. Stableford, *Scientific Romance in Britain 1890–1950* (London: Fourth Estate, 1985; New York: St. Martin's Press, 1985).

H. Rider Haggard

(22 June 1856 - 14 May 1925)

Elizabeth S. Davidson
University of South Carolina-Spartanburg

See also the Haggard entries in *DLB 70: British Mystery Writers, 1860–1919; DLB 156: British Short-Fiction Writers, 1880–1914: The Romantic Tradition;* and *DLB 174: British Travel Writers, 1876–1909.*

BOOKS: *Cetywayo and His White Neighbours; or, Remarks on Recent Events in Zululand, Natal, and the Transvaal* (London: Trübner, 1882; enlarged, 1888; revised, 1889);

Dawn (3 volumes, London: Hurst & Blackett, 1884; 1 volume, New York: Harper, 1887);

The Witch's Head (3 volumes, London: Hurst & Blackett, 1885; 1 volume, New York: Appleton, 1885);

King Solomon's Mines (London, Paris, New York & Melbourne: Cassell, 1885; New York: Harper, 1886);

She: A History of Adventure (New York: Harper, 1886; London: Longmans, Green, 1887);

Jess (London: Smith, Elder, 1887; New York: Harper, 1887);

Allan Quatermain, Being an Account of His Further Adventures and Discoveries in Company with Sir Henry Curtis, Bart., Commander John Good, R. N., and One Umslopogaas (London: Longmans, Green, 1887; New York: Harper, 1887);

A Tale of Three Lions (New York: Lovell, 1887);

Maiwa's Revenge; or, The War of the Little Hand (London & New York: Longmans, Green, 1888; New York: Harper, 1888);

Mr. Meeson's Will (London: Blackett, 1888; New York: Harper, 1888);

My Fellow Laborer (New York: Munro, 1888);

Colonel Quaritch, V. C.: A Tale of Country Life (3 volumes, London: Longmans, Green, 1888; 1 volume, New York: Lovell, 1888);

Allan's Wife and Other Tales (London: Blackett, 1889);

Cleopatra, Being an Account of the Fall and Vengeance of Harmachis, the Royal Egyptian, As Set Forth by His Own Hand (New York: Munro, 1889; London: Longmans, Green, 1889);

Beatrice: A Novel (London: Longmans, Green, 1890; New York: Munro, 1890);

H. Rider Haggard

The World's Desire, by Haggard and Andrew Lang (London: Longmans, Green, 1890; New York: Harper, 1890);

Eric Brighteyes (London: Longmans, Green, 1891; New York: Harper, 1890);

Nada the Lily (London & New York: Longmans, Green, 1892);

An Heroic Effort (London: Butler & Tanner, 1893);

Montezuma's Daughter (London: Longmans, Green, 1893; New York: Longmans, Green, 1894);

The People of the Mist (London & New York: Longmans, Green, 1894);

Church and State (New Style), An Appeal to the Laity (London: McCorquodale, 1895);

Heart of the World (New York: Longmans, Green, 1895; New York & Bombay: Longmans, Green, 1896);

Joan Haste (London & New York: Longmans, Green, 1895);

The Wizard (London: Simpkin, Marshall, Hamilton, Kent, 1896; New York, London & Bombay: Longmans, Green, 1896);

Dr. Thorne (London, New York & Bombay: Longmans, Green, 1898);

A Farmer's Year, Being His Commonplace Book for 1898 (London, New York & Bombay: Longmans, Green, 1899);

The Last Boer War (London: Kegan Paul, Trench, Trübner, 1899); republished as *A History of the Transvaal* (New York: New Amsterdam / London: Kegan Paul, Trench, Trübner, 1899);

Swallow: A Tale of the Great Trek (London, New York & Bombay: Longmans, Green, 1899);

Black Heart and White Heart and Other Stories (London, New York & Bombay: Longmans, Green, 1900); republished in part as *Elissa* (New York: Longmans, Green, 1900);

Lysbeth: A Tale of the Dutch (London, New York & Bombay: Longmans, Green, 1901);

A Winter Pilgrimage, Being an Account of Travels through Palestine, Italy, and the Island of Cyprus, Accomplished in the Year 1900 (London, New York & Bombay: Longmans, Green, 1901);

Rural England, Being an Account of Agricultural and Social Researches Carried Out in the Years 1901 and 1902, 2 volumes (London, New York & Bombay: Longmans, Green, 1902);

Pearl-Maiden: A Tale of the Fall of Jerusalem (London, New York & Bombay: Longmans, Green, 1903);

Stella Fregelius: A Tale of Three Destinies (New York: Longmans, Green, 1903; London, New York & Bombay: Longmans, Green, 1904);

The Brethren (London, Paris, New York & Melbourne: Cassell, 1904; New York: McClure, Phillips, 1904);

Ayesha: The Return of She (London: Ward, Lock, 1905; New York: Doubleday, Page, 1905); republished as *The Return of She: Ayesha* (New York: Lancer, 1967);

A Gardener's Year (London, New York & Bombay: Longmans, Green, 1905);

Report on the Salvation Army Colonies in the United States and at Hadleigh, England, with Scheme of National Land Settlement (London: Printed for His Majesty's Stationery Office, 1905); enlarged as *The Poor and the Land, Being a Report on the Salvation Army Colonies in the United States and at Hadleigh, England, with a Scheme of National Settlement and an Introduction* (London, New York & Bombay: Longmans, Green, 1905);

Benita: An African Romance (London, Paris, New York & Melbourne: Cassell, 1906); republished as *The Spirit of Bambatse: A Romance* (New York: Longmans, Green, 1906);

The Way of the Spirit (London: Hutchinson, 1906);

Fair Margaret (London: Hutchinson, 1907); republished as *Margaret* (New York: Longmans, Green, 1907);

The Wizard and Black Heart and White Heart (London: Newnes, 1907); republished as *Black Heart and White Heart and The Wizard* (London: Hodder & Stoughton, 1924);

Reports of the Royal Commission on Coast Erosion, 3 volumes, by Haggard and others (N.p., 1907–1911);

The Ghost Kings (London, Paris, New York, Toronto & Melbourne: Cassell, 1908); republished as *The Lady of the Heavens* (New York: Lovell, 1908);

The Real Wealth of England (London: Ward, Lock, 1908);

The Yellow God: An Idol of Africa (New York: Cupples & Leon, 1908; London, New York, Toronto & Melbourne: Cassell, 1909);

Morning Star (London, New York, Toronto & Melbourne: Cassell, 1910; New York: Longmans, Green, 1910);

Queen Sheba's Ring (London: Eveleigh Nash, 1910; Garden City, N.Y.: Doubleday, Page, 1910);

Regeneration, Being an Account of the Social Work of the Salvation Army in Great Britain (London & New York: Longmans, Green, 1910);

Red Eve (London, New York & Toronto: Hodder & Stoughton, 1911; Garden City, N.Y.: Doubleday, Page, 1911);

The Mahatma and the Hare: A Dream Story (London, New York, Bombay & Calcutta: Longmans, Green, 1911; New York: Holt, 1911);

Rural Denmark and Its Lessons (London, New York, Bombay & Calcutta: Longmans, Green, 1911);

Marie (London, New York, Toronto & Melbourne: Cassell, 1912; New York: Longmans, Green, 1913);

Reports of the Dominions Royal Commission, 24 volumes, by Haggard and others (N.p., 1912–1917);

Child of Storm (London, New York, Toronto & Melbourne: Cassell, 1913; New York: Longmans, Green, 1913);

A Call to Arms to the Men of East Anglia (London & Bungay: Privately printed, 1914);

The Wanderer's Necklace (London, New York, Toronto & Melbourne: Cassell, 1914; New York: Longmans, Green, 1914);

The Holy Flower (London, Melbourne & Toronto: Ward, Lock, 1915); republished as *Allan and the Holy Flower* (New York: Longmans, Green, 1915);

The After-War Settlement and Employment of Ex-Service Men in the Overseas Dominions: Report to Royal Colonial Institute (London: Published for the Royal Colonial Institute by the Saint Catherine Press, 1916);

The Ivory Child (London, New York, Toronto & Melbourne: Cassell, 1916; New York: Longmans, Green, 1916);

Finished (London, Melbourne & Toronto: Ward, Lock, 1917; New York: Longmans, Green, 1917);

Elissa, or, The Doom of Zimbabwe (London: Hodder & Stoughton, 1917);

Love Eternal (London, New York, Toronto & Melbourne: Cassell, 1918; New York: Longmans, Green, 1918);

Moon of Israel: A Tale of the Exodus (London: John Murray, 1918; New York: Longmans, Green, 1919);

When the World Shook, Being an Account of the Great Adventure of Bastin, Bickley and Arbuthnot (London, New York, Toronto & Melbourne: Cassell, 1919; New York: Longmans, Green, 1919);

The Ancient Allan (London, New York, Toronto & Melbourne: Cassell, 1920; New York: Longmans, Green, 1920);

Smith and the Pharaohs and Other Tales (London: Simpkin, Marshall, Hamilton, Kent, 1920; New York: Longmans, Green, 1921);

She and Allan (London: Hutchinson, 1921; New York: Longmans, Green, 1921);

The Virgin of the Sun (London, New York, Toronto & Melbourne: Cassell, 1922; Garden City, N.Y.: Doubleday, Page, 1922);

Wisdom's Daughter: The Life and Love Story of She-Who-Must-Be-Obeyed (London: Hutchinson, 1923; Garden City, N.Y.: Doubleday, Page, 1923);

Heu-Heu, or The Monster (London: Hutchinson, 1924; Garden City, N.Y.: Doubleday, Page, 1924);

Queen of the Dawn: A Love Tale of Old Egypt (Garden City, N.Y.: Doubleday, Page, 1925);

The Days of My Life: An Autobiography, edited by C. J. Longman, 2 volumes (London, New York, Toronto, Bombay, Calcutta & Madras: Longmans, Green, 1926); republished in part as *A Note on Religion* (London, New York, Toronto, Bombay, Calcutta & Madras: Longmans, Green, 1927);

The Treasure of the Lake (London: Hutchinson, 1926); republished as *Treasure of the Lake* (Garden City, N.Y.: Doubleday, Page, 1926);

Allan and the Ice-gods: A Tale of Beginnings (London: Hutchinson, 1927; Garden City, N.Y.: Doubleday, Page, 1927);

The Works of H. Rider Haggard (New York: Black, 1928); republished as *The Favorite Novels of H. Rider Haggard* (New York: Blue Ribbon, 1928);

Mary of Marion Isle (London: Hutchinson, 1929); republished as *Marion Isle* (Garden City, N.Y.: Doubleday, Doran, 1929);

Belshazzar (London: Paul, 1930; Garden City, N.Y.: Doubleday, Doran, 1930);

Five Adventure Novels of H. Rider Haggard (New York: Dover, 1951);

Lost Civilizations: Three Adventure Novels (New York: Dover, 1953);

She and King Solomon's Mines (New York: Modern Library, 1957);

Three Adventure Novels of H. Rider Haggard: She, King Solomon's Mines, Allan Quatermain (New York: Dover, 1960);

She and The Return of She (New York: Lancer, 1972);

The Private Diaries of Sir H. Rider Haggard, edited by D. S. Higgins (New York: Stein & Day, 1980).

MOTION PICTURES: *She,* dialogue and narration by Haggard, Reciprocity Films, 1925.

A man of diverse talents and a writer eager to try his skills at many different kinds of writing projects, both nonfiction and fiction, H. Rider Haggard is best remembered by general readers of fantasy for his fourth and fifth books, *King Solomon's Mines* (1885) and *She: A History of Adventure* (1886), action adventures in which something long hidden is found: in one, a lost treasure; in the other, a veiled woman of incredible power and beauty awaiting the reincarnation of the one love of her long lifetime amid the ruins of a lost civilization that had ceased to be even before the flourishing of the civilization of ancient Egypt. The continued popularity of these works alone would secure Haggard a place in reference books dedicated to fantasy and science fiction. Yet these works were just the start of his lucrative thirty-nine-year writing career, which includes his development of further adventures for the main characters of *King Solomon's Mines* and *She,* his exploration of historical fantasies, his other lost-race adventures, his look at the last survivors of Atlantis, and a kind of time travel through dreams of past

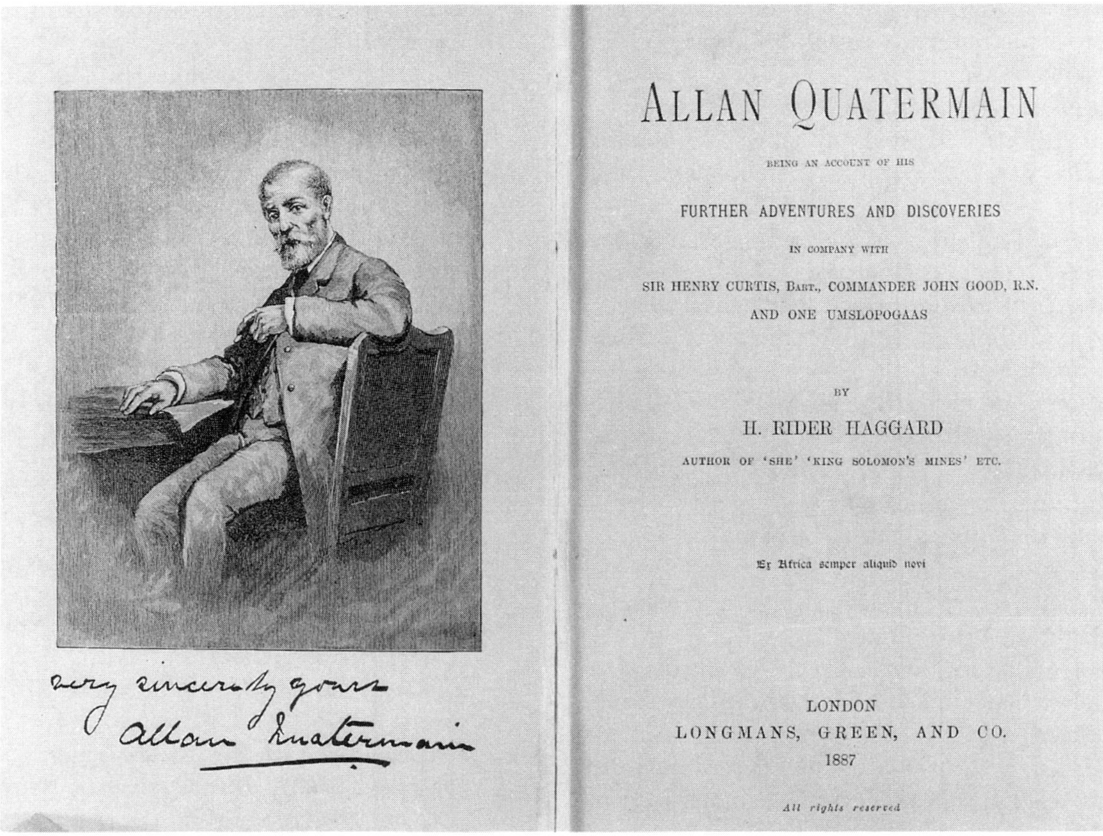

Frontispiece and title page for the second of Haggard's sixteen novels recounting Quatermain's adventures

lives. A good friend of Andrew Lang and Rudyard Kipling, Haggard collaborated with them on a couple of his works.

Influenced by the writings of Robert Louis Stevenson and Olive Schreiner, Haggard himself influenced many writers, such as mainstream author Graham Greene, and, perhaps more noteworthy for fantasy, Edgar Rice Burroughs and Philip José Farmer. Haggard's fantasies have influenced not only writers of novels but also writers of screenplays. Starting in his lifetime, filmmakers have returned repeatedly to his work, especially to *King Solomon's Mines* and *She,* not only in English but in other languages. Many children were introduced to these two works in *Classics Illustrated* comics in the 1950s, grew up to look for them in hardcover and paperback editions, and enjoyed reading the entire novels as adults. They are books that children may enjoy but that have even more to offer adults. Some of them are still in print in the popular fantasy and science-fiction market, more than one hundred years after Haggard's first fantasies were put on the market.

Born 22 June 1856 in Norfolk, Henry Rider Haggard was the son of William Meybohm Rider Haggard and Ella Doveton Haggard. His schooling included education at home by private tutors and some time at Ipswich Grammar School. At age nineteen Haggard left England to join the staff of Sir Henry Bulwer, lieutenant governor of Natal, South Africa. A capable young man, Haggard was chosen by Sir Theophilus Shepstone to join in a special mission to the Transvaal, helping to raise the British flag in Pretoria following annexation of the Transvaal by the British government and later serving as master and registrar of the high court of the Transvaal. When English policy on South Africa changed, Haggard returned to England.

During his years in Africa his first love had married someone else. Though most scholars believe this disappointment influenced the lost-love theme in his fiction, on his return to England he met Norfolk heiress Mariana Louisa Margitson, married her, and took her to South Africa to farm. He returned to England with her and their newborn son when conditions in South Africa became especially dangerous a little more than a year later. Though Haggard traveled extensively during the remaining forty-four years of his life, including trips to Egypt, Iceland, Mexico, the Middle East, and Canada, they lived in England.

Although Haggard began to study law in 1881 and joined the bar in 1885, his career as a lawyer was almost immediately overshadowed and displaced by his career as a novelist. In these years he wrote the nonfiction *Cetywayo and His White Neighbours* (1882) and the nonfantasy novels *Dawn* (1884) and *The Witch's Head* (1885). After the 1885 publication and stunning commercial success of *King Solomon's Mines,* followed quickly by *She* and other successful fantasies, Haggard was an established author with new material released every year or so, sometimes with more than one work released in a year. In addition to writing fiction, Haggard became an expert on scientific farming and on the Salvation Army, writing thoughtful nonfiction on both. In the last years of his life he was also interested in an anti-Bolshevik movement in England. In 1912 he was knighted. He died on 14 May 1925. His autobiography, *The Days of My Life,* was published in 1926.

King Solomon's Mines, written in about six weeks on a dare from one of his brothers after Haggard had read Stevenson's *Treasure Island* (1883), began his African fantasies and introduced the hero, Allan Quatermain the hunter, whom Haggard's readers would not let die even after Haggard had killed him in *Allan Quatermain* (1887), the second book published in a series that finally ended after Haggard's own death with the publication of *Allan and the Ice-gods* (1927), one of the two books in which Haggard was influenced by Kipling. Even then the public was not tired of Allan Quatermain, especially when he was finding lost treasures and civilizations. Not only have there been many reprints, but the film industry, even as recently as 1987 with *Allan Quatermain and the Lost City of Gold,* has also borrowed and adapted the tales of this hero, especially the adventures from *King Solomon's Mines.*

The Allan Quatermain series includes sixteen novels and some short stories. Though there have been some attempts to establish an internal chronology for the series, for instance that in *The Science Fiction Encyclopedia* (1979), it is obvious that such a chronology can be only an estimate. The series was not planned with such an internal chronology in mind from the beginning but grew as Haggard backtracked to create adventures for a younger Allan and twice introduced Quatermain's consciousness to a kind of drug-induced time travel to prehistoric times in *The Ancient Allan* (1920) and *Allan and the Ice-gods.* Also, Allan was twice married, both times to beautiful women who died young; his memories of them protect Allan from the allure of the beautiful temptresses in some of his many adventures. The Allan Quatermain stories, in the order of publication, are : *King Solomon's Mines, Allan Quatermain, Maiwa's Revenge* (1888), *Allan's Wife* (1889), *Marie* (1912), *Child of Storm* (1913), *The Holy Flower* (1915), *The Ivory Child* (1916), *Finished* (1917), *The Ancient Allan, Smith and the Pharaohs and Other Tales* (1920), *She and Allan* (1921), *Heu-Heu* (1924), *The Treasure of the Lake* (1926), and *Allan and the Ice-gods.* Many of the adventures take Allan away from the known into the unknown, from the modern places colonized by Europeans along the coast of Africa to the sites of ancient, vanished civilizations in its dangerous, unexplored interior. Usually the companions of his journey include both other white explorers such as Sir Henry Curtis and mysterious blacks such as Umbopa, who is later discovered to be a king in exile. This group has to cross formidable terrain, fight formidable battles, and come to know themselves in a way those never removed from their own civilizations cannot understand. As Samuel H. Vasbinder has pointed out, Allan Quatermain's strength of character is a match for the dangers, natural and supernatural, that "tax [his] courage and strength."

The most memorable of all of the Allan Quatermain material is the novel that set the pattern for this series, *King Solomon's Mines.* It involves a lost brother, a map of the fabled diamond mines, a king in exile, and a malignant witch. The handsome young Sir Henry Curtis, with the aid of Quatermain and an English naval officer, set off to find Sir Henry's missing brother and the mines. The darkly magnificent African, Umbopa, seems from the first to be more than a hired servant. After a long, perilous, adventure-filled journey they reach their destination, Kukuanaland, a pastoral, Zulu-like African culture not yet spoiled by contact with white traders, missionaries, and soldiers. There also are stone ruins as evidence of a vanished civilization. Yet all is not well in this seemingly Edenic setting. The rulers are evil tyrants who must be challenged and replaced.

One challenge involves the explorers' knowledge of an imminent eclipse, solar in the first edition but later changed to lunar, which they use to awe the people into sparing a beautiful young woman from her role as sacrifice in a ceremony. The other challenge is that of Umbopa, who reveals his identity to a people ready to join him in reclaiming his heritage. His white companions fight so fiercely at his side that he rewards them by forcing the ancient witch, Gagool, to lead them to the hidden store of diamonds. Gagool, who plans the men's death in the mines, is foiled by the grateful maiden, and both women lose their lives. The white men accept some diamonds, refuse a life of privilege in Kukuanaland, and return to England. The new king of Kukuanaland intends to seal off his land from further contact

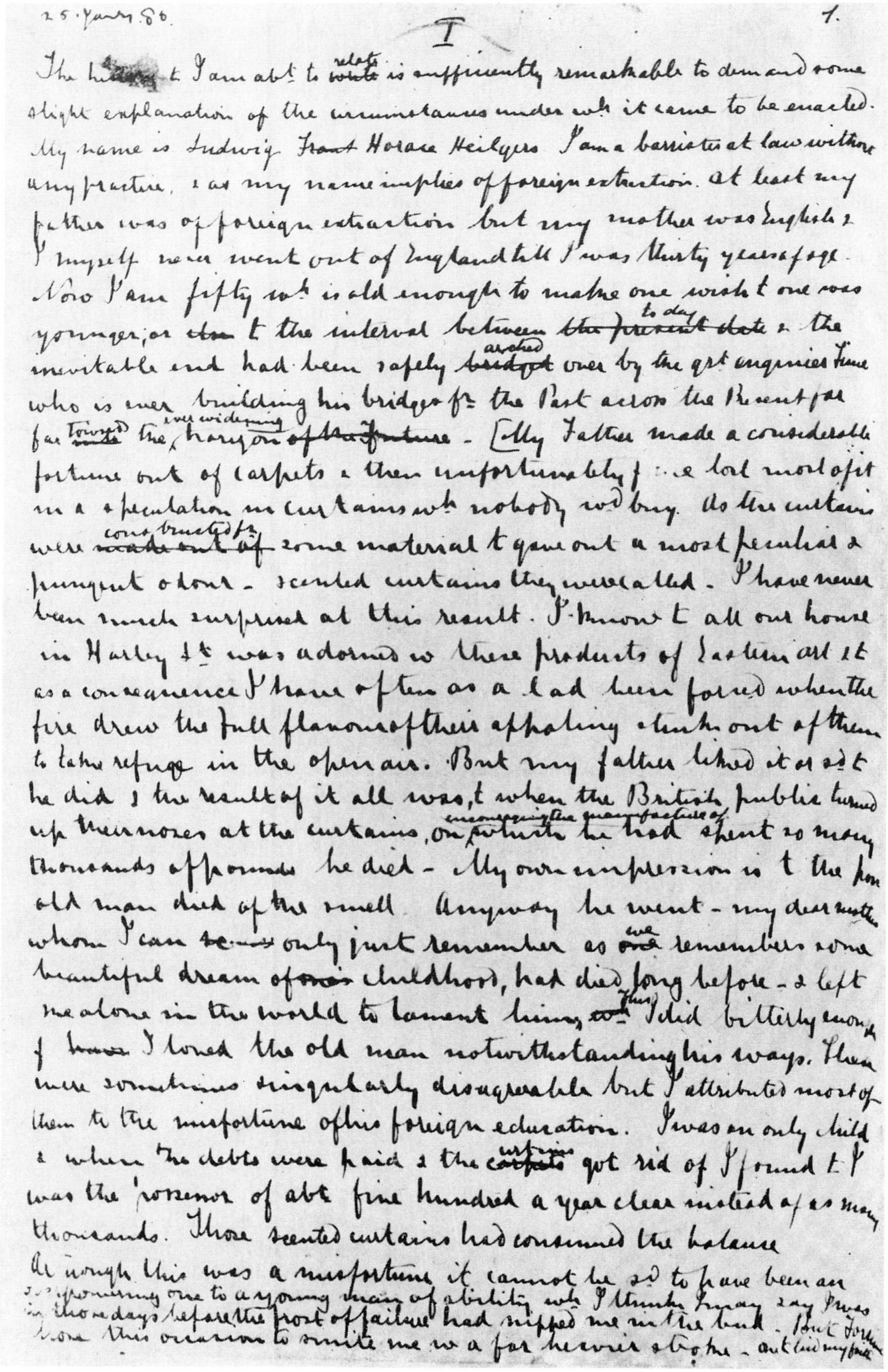

25 · Jany 86. I 1.

The history that I am about to relate is sufficiently remarkable to demand some slight explanation of the circumstances under which it came to be enacted. My name is Ludwig Horace Heilgers I am a barrister at law without any practice, and as my name implies of foreign extraction. At least my father was of foreign extraction but my mother was English and I myself never went out of England till I was thirty years of age. Now I am fifty which is old enough to make one wish that one was younger, or that the interval between today and the inevitable end had been safely arched over by the great engineer Time who is ever building his bridges from the Past across the Present far towards the ever widening horizon. My Father made a considerable fortune out of carpets and then unfortunately he lost most of it in a speculation in curtains which nobody would buy. As the curtains were constructed of some material that gave out a most peculiar and pungent odour – scented curtains they were called – I have never been much surprised at this result. I know that all our house in Harley Street was adorned with these products of Eastern art and as a consequence I have often as a lad been forced when the fire drew the full flavour of their appalling stink out of them to take refuge in the open air. But my father liked it or said that he did and the result of it all was, that when the British public turned up their noses at the curtains, on encouraging the manufacture of which he had spent so many thousands of pounds he died – My own impression is that the poor old man died of the smell. Anyway he went – my dear mother whom I can only just remember as we remember some beautiful dream of childhood, had died long before – and left me alone in the world to lament him, which I did bitterly enough for I loved the old man notwithstanding his ways. These were sometimes singularly disagreeable but I attributed most of them to the misfortune of his foreign education. I was an only child and when the debts were paid and the curtains got rid of I found that I was the possessor of about five hundred a year clear instead of as many thousands. Those scented curtains had consumed the balance. Although this was a misfortune it cannot be said to have been an overpowering one to a young man of ability which I think I may say I was in those days before the frost of failure had nipped me in the bud. But Fortune chose this occasion to smite me with a far heavier stroke – and [illegible]

Page from the first draft of She *(J.E. Scott,* Bibliography of the Works of Sir Henry Rider Haggard, 1856–1925, *1947)*

with Europeans. As Norman Etherington puts it in his *Rider Haggard* (1984) "daily life in Kukuanaland is healthy, tidy, and chaste–soap, civilization, and missionaries are not wanted here."

In *Allan Quatermain* the lost civilization hidden in the interior of Africa is a white one, the people of Zu-Vendis living in an agricultural and feudal society. Though accompanied there by the marvelous Zulu warrior Umslopogaas, Allan Quatermain, Sir Henry Curtis, and others are tested once again to know themselves; as Etherington explains, "white men discover their own interior savage selves–Sir Henry Curtis puts on skins and feathers, and becomes African; he picks up a battle-ax and becomes a Viking; he puts on chain-mail and becomes a medieval knight. . . . [Haggard's] purpose was to use Africans to lay bare the inner core of European man. . . ."

Haggard quickly followed his success with *King Solomon's Mines* with a novel that became the cornerstone of his second important series, *She*. Again he had not intended to create a series, but again he had created a character whom his readers would not be satisfied to let him leave dead–Ayesha, the She of the title. The series includes four novels, and one of these, the action adventure *She and Allan,* belongs more in the spirit of the Allan Quatermain series. The characters Leo Vincey and L. Horace Holly appear in only two, *She* and *Ayesha: The Return of She* (1905). The fourth, *Wisdom's Daughter* (1923), about Ayesha's beginnings, was published only two years before Haggard's death.

Like *King Solomon's Mines, She* has remained the most popular work in its series. Filmmakers were drawn to *She* before they were drawn to *King Solomon's Mines* and continue to find the material worth reviving, usually with some revision for modern audiences. The material in *She* held a distinct appeal for Sigmund Freud, who recommended Haggard's works to one of his patients eager to follow up on the theories that Freud was then discussing before he developed those theories in writings of his own.

Ayesha, the mysterious, veiled, omniscient, omnipotent, many-thousand-year-old beauty, is at the heart of the novel. Her seekers are at least two-thirds misogynists and one-third wide-eyed youthful inexperience. This novel, too, has hints of not just one lost civilization, but two or perhaps three. (Other non-Quatermain, non-Ayesha tales about finding lost races are *The People of the Mist* [1894], *Heart of the World* [1895], *Elissa* [1900], *Benita* [1906], *The Yellow God* [1908], and *Queen Sheba's Ring* [1910].) There is also a family with an incredibly ancient pedigree and lineal curse.

Through this curse the infant Leo Vincey became the ward of the ape-featured Cambridge don L. Horace Holly, whom Billali, an Amahagger leader in Africa, later refers to as "my son, the baboon." Holly is a misogynist because of his former jilting by a beautiful flirt. Holly will hire no female servants to steal the boy's affections, hiring instead the corpulent Job, "the pig" to the Amahagger and a misogynist on general principle. When Leo, having attained his twenty-fifth birthday and having finished the course of study laid out by his dying father, examines the artifacts bequeathed to him by that father, the three men decide to test the truth of the artifacts. Written by the beautiful former priestess of Isis, the Egyptian Amenartas, Leo's many-times great-grandmother, they tell of a love triangle in which an Arabian woman destroys Kallikrates, Leo's many-times great-grandfather, for preferring the love of Amenartas. Even from the grave Amenartas wants revenge and charges her descendants to find and destroy this murderess. Holly and Leo undertake the journey out of curiosity and a desire to travel; Job, with no curiosity and no desire to travel, as a good and faithful servant goes with his gentlemen. Off the east coast of Africa they and the Arab captain of their ship survive a shipwreck, coincidentally at the one place they need to be to follow the river inland to the lands ruled by the mysterious, legendary white queen. Rescued from a swamp by the Amahagger people, a neither black nor white matriarchal race, Leo and Job are claimed in marriage by Amahagger women. Though Job ungraciously rejects his offer, Leo accepts that of Ustane, unknowingly setting the stage for history to repeat itself in a love triangle.

The Amahagger people and Ayesha, the white woman who has ruled them for thousands of years, live in the burial caves of Kor, a civilization in ruins at the time of ancient Egypt. Because Leo is in a deathly fever, Ayesha first befriends Holly, captivating him by the sight of her unveiled face. After seeing Leo, identifying him as the reincarnation of her murdered love, Kallikrates, and using her knowledge of science to bring him back from the brink of death, Ayesha executes Ustane with a glance. Taking Leo, Holly, and Job on a further journey through the ruins of the city of Kor and beyond, Ayesha offers Leo and Holly the secret of eternal youth, which Holly wisely declines. Because Leo is fearful, Ayesha steps into the flame for the second time, but this overdose is deadly, aging her thousands of years in minutes. The sight frightens Job to death. Leo and Holly escape, return to England, leave Holly's manuscript with a stranger, and go on further mysterious travels.

Though *Ayesha: The Return of She* takes Holly and Leo in search of Leo's now-lost love to another

Illustration by E. K. Johnson of Ayesha entering the Pillar of Fire in She

lost civilization somewhere in the wilds of central Asia, Leo and Ayesha are doomed not to be together in this life. The aged Ayesha is now priestess at a sacred mountain and can defend them from the forces of the vengeful queen whose husband they have been forced to kill. Here Ayesha's powers seem more magical than based on a science no longer understood in the modern world. Though they marry before Leo dies, there is no consummation in life. This time, to be with him outside the bounds of mortal life, Ayesha follows him in death. Donald L. Lawler judges *She* a more powerful book than *Ayesha,* citing "Haggard's preference for ambiguity" over "an obvious or gratuitous moral" and a theme "great enough . . . to leave its imprint on the reader's sensibilities."

The last novel in the She series, *Wisdom's Daughter,* finally addresses separately the story told so indirectly as an undercurrent in *She.* Born in Arabia, and grown to a dangerous beauty, Ayesha flees with Noot, a great magician, to Egypt, becomes a priestess of Isis, and first sees and loves Kallikrates, a slightly different version of their first meeting than that in *She.* When the Persians conquer Egypt, Ayesha uses her magic against them and is forced to flee to Kor. There Noot (later to be reincarnated as Horace Holly, not really hinted at in *She*) refuses to bathe in the elemental Flame of Life. Ayesha does, and when Kallikrates turns from her increased gloriousness to her mortal rival, Ayesha kills him in a jealous rage and lives with no equal for any of her passions though she rules the people of that country for thousands of years. Though surrounded by various sets of servants genetically bred to meet her requirements at any given time, Ayesha is always alone, a great solitary superwoman, not exactly above the law but beyond it so far that it has no relevance to her, a kind of female Byronic hero. A truly alien being, she does not aspire to a greater humanity, but to raise just one human, the love of her ageless life, above his origins and environment to be her equal.

Soon after his success with *She* Haggard created fantastic tales of two well-remembered women, Cleopatra from history and Helen of Troy from myth. Haggard's Cleopatra is not just about a beautiful ruler who refuses to be made subordinate in marriage. One dominant element of fantasy in it is its link to the theories of Charles Darwin and C. G. Jung, through subconscious memories of the evolution of humankind, which are made apparent to the hero in significant dreams. According to Etherington, Haggard's "notion that modern 'civilized' people harbor within themselves elements of previous stages of evolution . . . gives the primitive, violent, animalistic side of man a time dimension." Haggard later returned to ancient Egyptian materials in *Morning Star* (1910), *Moon of Israel* (1918), and *Queen of the Dawn* (1925) and wrote a Jewish/Roman tale, *Pearl-Maiden* (1903).

The World's Desire (1890), about Helen of Troy, is the fantasy that Haggard wrote in collaboration with Lang. It is a kind of sequel to Homer's epic *The Odyssey,* one that lovers of the clever Penelope of the original will find quite different in spirit. In order to free Odysseus for further wanderings and another love, Haggard and Lang destroy the residents of Ithaca with a plague, changing Odysseus's homecoming beyond recognition. Having received a vision in the temple of Aphrodite, Odysseus, an erstwhile worshipper of Athena, searches for Helen in Egypt. A powerful sorceress, Meriamun, queen of Egypt, desires him for her own and, magically disguising herself as Helen, diverts him from his goal until he discovers the deception. Fighting with the Egyptians in battle against Achaean raiders, Odysseus drives them back but seeks and finds his own death in battle. George P. Landow observes that "the resulting story suggests a very non-Homeric and nonheroic vision of man and his fate . . . presenting [Odysseus] as an isolated, alienated hero" and "convey[s] a vision of human possibilities more characteristic of the aesthetes and decadents, who were obsessed with visions of predatory women, than of Homer or the classical tradition." The shift from Penelope to Helen, like that from Athena to Aphrodite, as the object of Odysseus's longing certainly brings into play a different set of symbols.

In *King Solomon's Mines* Sir Henry's resemblance at one point to an ancient Viking warrior and in *Allan Quatermain* the existence of a feudal culture in lost Zu-Vendis suggest that Haggard might later write fantasy actually set in the Middle Ages. *Eric Brighteyes* (1891) and *Red Eve* (1911) are set, respectively, in tenth-century Iceland and fourteenth-century England.

Growing up together, Eric, Gudruda the Fair, and Swanhild the witch form another love triangle that cannot be resolved in life. Gudruda is the legitimate daughter of Asmund Asmundson, whose wife dies at her birth. Though not acknowledged as a daughter until she is given in marriage to Earl Atli the Good, Swanhild is also Asmundson's daughter by the Finnish woman Groa, who came to him as the sea's gift, an ill-omened survivor of a shipwreck that claimed the life of her husband. Because of a deathbed promise to his wife, Asmundson does not marry Groa but keeps her as a mistress until after their daughter is married and he arranges his own second marriage to Unna, a kinswoman of Eric Brighteyes. Groa uses her knowledge of herbal lore to poison Asmundson and Unna at their wedding feast, the first somewhat magical action the reader sees her do. Though both Groa and Swanhild are called witches, there are no actions in the first half of the book demonstrating any extraordinary powers. From childhood Swanhild is jealous that Eric prefers the blond, brown-eyed Gudruda, but her first attempt to kill Gudruda, though apparently magically inspired by a wolf familiar, is an ordinary push over a cliff. After she has been married and sent into exile for attempted murder, Swanhild acquires the ability, though not in her own person, to walk the waves on two occasions. First she saves Eric from ambush with the aid of a rat, a less noble familiar than the wolf. Next she wrecks his ship and destroys all of his crew except Eric and his best friend, Skallagrim Lambstail, stranding them in Earl Atli's court until she can destroy Earl Atli and make the blame appear to be Eric's rather than her own. Her familiar for this adventure is a loathesome toad.

Eric fights foes mightily with the aid of his berserker companion Lambstail, a warrior possessed when the heat of battle is on him by the spirit of a mighty animal. Eric, of course, cannot be united with Gudruda in life. Though they finally are wed, Swanhild persuades a Viking noble to break in and slay Gudruda as she sleeps, using Eric's own magic sword. When Eric and Lambstail are killed in the ensuing battle, Swanhild takes his body to her ship and sails away to an unknown destination, probably to death at sea.

Red Eve begins in England with a competition for the hand of Eve between the heroic Hugh de Cressi and the villainous Sir Edmund Acour but continues in France and ends in Venice. In this work there is a mysterious immortal who influences individual and European human destiny. Murgh, the Lord of Death from the East, not only takes a hand in uniting Eve and de Cressi but also brings the Black Death to Europe.

Haggard later wrote another Viking novel, *The Wanderer's Necklace* (1914), and a novel of the Crusades, *The Brethren* (1904). Other historical tales include two set in Mexico, *Montezuma's Daughter* (1893) and *The Virgin of the Sun* (1922). There are also two historical novels involving Spain, *Lysbeth* (1901), which also deals with the Dutch, and *Fair Margaret* (1907).

Etherington notes that the plot of *Nada the Lily* (1892) is in most ways parallel to that of *Eric Brighteyes* but that these similarities are more than simply writing to a successful formula. "They arise from consistent identification of the European past with the African present. His contemporaries regarded Africans as primitive, mired in a savage

He ran in upon her and smote her on the head.

He lifted the spear . . . and drove it down between the shoulders.

Illustrations by Charles Kerr for Haggard's Nada the Lily

stage of development through which Europeans had passed long ago. Treating the hypothesis as fact, Haggard uses both European past and African present to provide examples of repressible but ineradicable human desires." The characters in *Eric Brighteyes* are all white, Icelandic, tenth-century; the characters in *Nada the Lily* are all black, Zulu, nineteenth-century. There are also witchcraft and political perversions of witchcraft in *Nada the Lily*. The point-of-view character is Mopo, not only a witch doctor but one of the chief associates of Chaka, the Zulu military genius and king. The aged Mopo, in disguise as Zweete, an uncannily accurate but not so notorious witch doctor, tells the tale to a white man who has come for Zweete's magic to help him find his stray animals. Though Nada, Mopo's beautiful daughter, is the title character and the great Umslopogaas, her supposed twin brother, the hero in the making, the first hundred pages are really the story of how Mopo and his real sister, Baleka, come to Chaka. Because as a child Mopo has had the insight to give a drink to the exiled child Chaka, the adult Mopo becomes Chaka's most trusted witch doctor, and his sister Baleka becomes Chaka's most favored wife. Chaka fears having a son who might someday oust him from the throne and therefore orders all his newborn children slain. Mopo, Baleka, and Unandi, Chaka's mother, deceive him by having Mopo raise as his own son, Umslopogaas, the son of Chaka and Baleka. When Chaka discovers the imposture and destroys all of Mopo's family he can, Mopo is prevented from taking immediate revenge by the divine intervention of the queen of heaven herself. She later reappears to Mopo when the time is right for his revenge on Chaka, who has not been able to destroy Umslopogaas because of divine intervention. A lioness has carried him away from camp, making it possible for him to be rescued by Galazi, a warrior with a magic spear and a pack of wolf spirits at his command. Using two magical wolf skins, Galazi and Umslopogaas win for the latter a magical club. The two become known as the Wolf Brethren, whom no one can defeat until the last ghost wolf is finally destroyed. Though Mopo does destroy Chaka, Umslopogaas never inherits Chaka's kingdom. Instead, he becomes a wanderer destined to a life full of action and adventure, becoming better known through his being a companion of Allan Quatermain. Other non-Quatermain African tales are *The Wizard* (1896), *Swallow: A Tale of the Great Trek* (1899), *Black Heart and White Heart* (1900), and *The Ghost Kings* (1908).

The use of the new technology of radio has a place in the otherwise realistic *Stella Fregelius* (1903), where it is used in an attempt to contact the dead. Some other tales with modern mystical or spiritualist elements are *The Way of the Spirit* (1906), *The Mahatma and the Hare* (1911), and *Love Eternal* (1918).

In *The Mahatma and the Hare* the main character, the dead hare, enlightens readers from beyond the grave about the feelings of the prey and makes a harsh judgment of hunting as a sport. The hare does not enjoy the hunt at all. Though skilled in eluding both dogs that hunt by sight and by scent, the hare is doomed to a short life of almost constant terror, merely to alleviate the hunter's boredom. This particular hare eventually dies for such an unworthy cause but gets an unusual chance to contribute his experiences of hell on earth to the wisdom of the ages.

Following some suggestions by Kipling, Haggard also tried his hand at an Atlantis fantasy, bringing two miraculously preserved Atlanteans out of their 250,000-year suspended animation in *When the World Shook* (1919). Three English travelers–the rich Arbuthnot, the narrow-minded preacher Bastin, and the atheist doctor Bickley–rather unlike the Allan Quatermain type of rugged adventurer come ashore after a shipwreck on an uncharted island in the South Pacific. Though warned by the initially friendly people to beware the shrine of the god Oro, they are forced to ignore the warning after Bastin burns the natives' idol and the three have to escape the angry worshippers by fleeing to the mysterious island in the middle of the lake. With few other options left to them, they enter the cave of wonders, finding airships and suspended-animation chambers, which they activate, restoring the last king of Atlantis, Oro, and his daughter, Yva, to life. Oro has destroyed his own culture because of its contentiousness and desires to destroy that of the madly warring modern world. Yva, having fallen in love with Arbuthnot, foils her father's plans at the cost of her own life. Oro lets the three live to return home, and Bastin and Bickley become heroes in World War I. John Scarborough finds this novel disappointing, "a work of the third rank" when compared to Haggard's other novels: "There is little of the crisp narrative in *When the World Shook* that one would expect from the author of *Nada.* There are few passages of sheer beauty like those that festoon . . . *She.* There is none of the exuberance that delights the reader of *Mines.* . . . Arbuthnot seems to represent Haggard's bitterness at the end of a long and productive career." Nevertheless it has some powerful "graphic descriptions of the horrors of World War I," which influence Oro's decision once again to bring civilization to an end.

Although his many fantasies from 1885 through 1925 were a commercial success, few people now come across any but *King Solomon's Mines* and *She,* and even those two titles are no longer well known to the casual reader. Yet lovers of fantasy and science fiction can still feel the force of Haggard's discovery of places for adventures of the body, mind, and spirit. In Allan Quatermain and Ayesha he created two characters who, once encountered, are not easily forgotten. Though Kukuanaland may not give rise to as much speculation as Kor, lost in time as well as in the heart of a then-unknown Africa, it offers worldly wealth to be gained by the practical adventurer. The experience of Kor–the burial caves, the ruined city with the statue of a veiled Truth, the cave of the Pillar of the Fire of Life–certainly gives rise to speculation about the place of humanity in the plan of science or the supernatural. It asks the age-old questions about truth and life and gives the age-old evasions, which may itself be an answer–a veiled mystery.

Bibliographies:

J. E. Scott, *Bibliography of the Works of Sir Henry Rider Haggard 1856–1925* (Bishop's Stortford: Elkin Mathews, 1947);

D. E. Whatmore, *H. Rider Haggard: A Bibliography* (London: Mansell, 1987).

Biographies:

Lilias Rider Haggard, *The Cloak That I Left* (London: Hodder & Stoughton, 1951);

Morton Cohen, *Rider Haggard, His Life and Works* (London: Hutchinson, 1960);

Cohen, *Rudyard Kipling to Rider Haggard* (London: Hutchinson, 1965);

Peter B. Ellis, *Rider Haggard, A Voice from the Infinite* (London: Routledge & Kegan Paul, 1978);

D. S. Higgins, ed., *The Private Diaries of Rider Haggard* (London: Cassell, 1980);

Higgins, *Rider Haggard, The Great Storyteller* (London: Cassell, 1981).

References:

Jeff D. Bass, "The Romance as Rhetorical Dissociation: The Purification of Imperialism in *King*

Solomon's Mines," *Quarterly Journal of Speech,* 67 (August 1981): 259–269;

David Bunn, "Embodying Africa: Woman and Romance in Colonial Fiction," *English in Africa,* 15 (May 1988): 1–28;

Claudia Crawford, "She," *Sub-stance,* 29 (1981): 83–96;

Maria Pia De Angelis, "She di Rider Haggard: Il fascino degli archetipi," *Quaderni di Filologia Germanica della Facolta di Lettere e Filosofia dell'Universita di Bologna,* 2 (1982): 75–95;

Marysa Demoor, "Andrew Lang's Letters to H. Rider Haggard: The Record of a Harmonious Friendship," *Etudes Anglaises: Grand-Bretagne, Etats-Unis,* 40 (July–September 1987): 313–322;

Norman Etherington, *The Annotated "She": A Critical Edition of H. Rider Haggard's Victorian Romance* (Bloomington: Indiana University Press, 1991);

Etherington, *Rider Haggard* (Boston: Twayne, 1984);

Sandra M. Gilbert, "Rider Haggard's Heart of Darkness," in *Coordinates: Placing Science Fiction and Fantasy,* edited by George E. Slusser, Eric S. Rabkin, and Robert Scholes (Carbondale: Southern Illinois University Press, 1983), pp. 124–138;

Susan Gubar, "*She* and *Herland*: Feminism as Fantasy," in *Coordinates: Placing Science Fiction and Fantasy,* pp. 139–149;

Wendy R. Katz, *Rider Haggard and the Fiction of Empire: A Critical Study of British Imperial Fiction* (Cambridge: Cambridge University Press, 1987);

George P. Landow, "The World's Desire," in *Survey of Modern Fantasy Literature,* 5 volumes, edited by Frank N. Magill (Englewood Cliffs, N.J.: Salem Press, 1983), V: 2176–2179;

Donald L. Lawler, "The She Series," in *Survey of Modern Fantasy Literature,* 5 volumes, III: 1396–1401;

C. S. Lewis, "The Mythopoeic Gift of Rider Haggard," in his *Of This and Other Worlds,* edited by Walter Hooper (London: Collins, 1982);

Harold Orel, "Adapting the Conventions of the Historical Romance: Rider Haggard's *Eric Brighteyes,*" *English Literature in Transition (1880–1920),* 36 (1993): 40–59;

Orel, "Hardy, Kipling, and Haggard," *English Literature in Transition (1880–1920),* 25 (1982): 232–248;

Peter Pierce, " 'Weary with Traveling through Realms of Air . . .': Romance Fiction of 'Boldrewood,' Haggard, Wells and Praed," *Westerly: A Quarterly Review,* 32 (June 1987): 79–90;

Murray Pittock, "Rider Haggard and *Heart of Darkness,*" *Conradiana: A Journal of Joseph Conrad Studies,* 19 (Autumn 1987): 206–208;

John Scarborough, "H. Rider Haggard, 1856–1925," in *Science Fiction Writers: Critical Studies of the Major Authors from the Early Nineteenth Century to the Present Day,* edited by E. F. Bleiler (New York: Scribners, 1982), pp. 19–24;

Samuel H. Vasbinder, "The Allan Quatermain Saga," in *Survey of Modern Fantasy Literature,* 5 volumes, I: 22–25;

Vasbinder, "Aspects of Fantasy in Literary Myths about Lost Civilizations," in *The Aesthetics of Fantasy Literature and Art,* edited by Roger C. Schlobin (Notre Dame: University of Notre Dame Press: 1982), 192–210;

Hartwig A. Vogelsberger, *"King Romance": Rider Haggard's Achievement* (Salzburg: Universität Salzburg, 1984).

Papers:

There is a collection of Haggard papers in the Rare Books and Manuscript Library of Columbia University. Haggard's diaries from 1914 until 1925 are in the Norfolk Records Office.

William Hope Hodgson

(15 November 1877 – 17 April 1918)

Samuel W. Bruce

See also the Hodgson entries in *DLB 70: British Mystery Writers, 1860–1919; DLB 153: Late-Victorian and Edwardian British Novelists, First Series;* and *DLB 156: British Short-Fiction Writers, 1880–1914: The Romantic Tradition.*

BOOKS: *The Boats of the "Glen Carrig"* (London: Chapman & Hall, 1907); republished in *The House on the Borderland and Other Novels* (Sauk City, Wis.: Arkham House, 1946);

The House on the Borderland (London: Chapman & Hall, 1908); republished in *The House on the Borderland and Other Novels;*

The Ghost Pirates (London: Stanley Paul, 1909); republished in *The House on the Borderland and Other Novel;* abridged as *The Ghost Pirates, A Chaunty and Another Story* (New York: Reynolds, 1909);

Carnacki, the Ghost Finder, and a Poem (New York: Reynolds, 1910);

The Night Land (London: Nash, 1912); republished in *The House on the Borderland and Other Novels;* condensed in *Poems and the Dream of X* (London: Watt, 1912; New York: Paget, 1912);

Carnacki, the Ghost Finder (London: Nash, 1913; enlarged edition, Sauk City, Wis.: Mycroft & Moran, 1947);

Men of the Deep Waters (London: Nash, 1914);

Cargunka and Poems and Anecdotes (London: Watt, 1914; New York: Paget, 1914);

The Luck of the Strong (London: Nash, 1916);

Captain Gault, Being the Exceedingly Private Log of a Sea-Captain (London: Nash, 1917; New York: MacBride, 1918);

The Calling of the Sea (London: Selwyn & Blount, 1920);

The Voice of the Ocean (London: Selwyn & Blount, 1921);

Out of the Storm: Uncollected Fantasies by William Hope Hodgson, edited by Sam Moskowitz (West Kingston, R.I.: Grant, 1975);

The Haunted Pampero, edited by Moskowitz (West Kingston, R.I.: Grant, 1991);

William Hope Hodgson (Hulton Getty Picture Collection Limited)

Terrors of the Sea, edited by Moskowitz (West Kingston, R.I.: Grant, 1996).

Editions: *Deep Waters* (Sauk City, Wis.: Arkham House, 1967);

Poems of the Sea (London: Ferret Fantasy, 1977).

The influence of the works of William Hope Hodgson on the literature of fantasy and science fiction has yet to be fully charted. Writers as diverse as H. P. Lovecraft, Olaf Stapledon, and Dennis Yates Wheatley have clearly been influenced by Hodgson's work. Although he won critical acclaim during his lifetime, Hodgson never gained a popular audience until after his death. His posthumous reputation, however, has grown formidable among critics of fantasy fiction, and while praise of Hodgson's work is often qualified (frequently regarding questions of style), critics almost invariably use such superlatives as *classic* or *masterpiece* to describe the author's best efforts. His reputation rests for the most part on two novels, *The House on the Borderland* (1908) and *The Night Land* (1912), and a group of stories set in the Sargasso Sea, a region in the Atlantic where seaweed is so thick it appears to be a landmass. Though generally considered of less importance, Hodgson's short stories featuring an occult detective, which were collected as *Carnacki, the Ghost Finder* (1913), have also influenced other writers in the genre.

William Hope Hodgson was born 15 November 1877 in Blackmore End, Wethersfield, Finchingfield, Essex. His parents were Samuel Hodgson, an Anglican priest, and Lissie Hodgson, née Lissie Sarah Brown. He was the second of twelve children, though three did not survive beyond infancy. Hodgson's father was reportedly a powerful speaker who could deliver inspiring sermons extemporaneously, but he was also unorthodox in his beliefs. As a result, he was frequently shifted to new parishes–eleven in a period of twenty-one years–and the family was often in financial straits.

Lissie Hodgson hoped that her son would follow in his father's footsteps and become a clergyman. At the age of thirteen, however, lured by romantic notions of a sailor's life, Hodgson ran away from Margaret's Boarding School in Margate, intending to go to sea. He was caught and returned to his family, but they were unable to dissuade him from his desire. With the help of an uncle, Hodgson convinced his family to let him attempt his chosen vocation, and he began a four-year apprenticeship as a cabin boy on 28 August 1891. Of small build, Hodgson quickly found that he was ill equipped to deal with the brutal treatment afforded to apprentices by their superiors. He began to lift weights and became obsessed with bodybuilding; he also learned judo. Before long, when sailors picked the diminutive Hodgson as a target for bullying, they found themselves facing a powerful opponent. According to Sam Moskowitz, whose introductory critical biography for *Out of the Storm: Uncollected Fantasies by William Hope Hodgson* (1975) is the essential source of biographical information on the author, "there is strong evidence that throughout [Hodgson's] life one of his most delightful diversions was to pound seamen to jelly at the slightest provocation."

While Hodgson was serving as an apprentice, his father died of throat cancer on 11 November 1892. Hodgson completed his apprenticeship in 1895 and spent two years at school in Liverpool preparing to take the test for a mate's certificate. He passed in 1897 and began his career as a professional seaman. When not engaged in his duties, Hodgson spent much of his time at sea working out with a punching bag, lifting weights, and boxing with other crew members. He also picked up other hobbies, including photography. In fact, Hodgson gained some renown as an amateur photographer and was allowed to set up a darkroom onboard his ship on at least one voyage, during which he took photographs of sharks caught at sea and of the maggot-filled food served to the sailors; he also took some impressive action shots of typhoons. Sometime around 1898 or 1899 the ship on which Hodgson was serving was caught up in a cyclone, and the amateur photographer set up his equipment to make a record of the experience, which included stalk lightning, the cyclone's funnel, huge waves ("as high as a seven-story building," according to Moskowitz), and a view of the near capsizing of the ship. His pictures of stalk lightning (which rises from the sea rather than descending from the clouds) are apparently the earliest photographs of the phenomenon.

Despite such shows of courage (on another occasion he dived into shark-infested waters off the coast of New Zealand to rescue a crewman who had fallen overboard), Hodgson chafed against the harshness of sea life. He finally concluded that only a masochist would remain in such a career, and he embarked on a new enterprise in the fall of 1899, opening W. H. Hodgson's School of Physical Culture in Blackburn, an industrial city to the north of Liverpool.

Hodgson printed and sent out postcards bearing a front and back view of his physique, which undoubtedly convinced many prospects to sign up. Members of the Blackburn police force were regular customers, a connection that led to Hodgson's involvement in an embarrassing episode. The renowned escape artist Harry Houdini appeared at the Blackburn Palace Theatre in 1902, and as a publicity stunt he was locked in the Blackburn jail, from which he easily escaped. The police apparently complained to Hodgson over their lost honor, and on 24 October the bodybuilder attended one of Houdini's performances and accepted the escape artist's standing challenge: anyone who could bind him so as to prevent escape would receive a cash award. Hodgson "scientifically" bound Houdini to allow minimal play of muscle, even causing the performer to cry out in pain. After thirty-five minutes, Houdini asked that the chains be loosened for a few minutes–they were so tight they were cutting off his circulation–but Hodgson refused. Finally, after two

THE GODDESS OF DEATH. 565

I assented, and invited him to join me at my wine. He did so, and we passed the time away very pleasantly in reminiscences of old times.

"What about weapons?" I asked presently. "I suppose it will be advisable to take something in that line?"

For answer Will unbuttoned his coat, and I saw the gleam of a brace of pistols. I nodded, and, going to my trunk, opened it and showed him a couple of beautiful little pistols I often carried. Having loaded them, I put them in my side pockets. Shortly afterwards eleven chimed, and getting into our cloaks, we left the house.

It was very cold, and a wintry wind moaned through the night. As we entered the park, we involuntarily kept closer together.

Somehow, my desire for adventure seemed to be ebbing away, and I wanted to get out from the place, and into the lighted streets.

"We'll just have a look at the statue," said Will; "then home and to bed."

A few minutes later we reached a little clearing among the bushes.

"Here we are," Will whispered. "I wish the moon would come out a moment: it would enable us to get a glance at the thing." He peered into the gloom on our right. "I'm hanged," he muttered, "if I can see it at all!"

Glancing to our left, I noticed that the path now led along the edge of a steep slope, at the bottom of which, some considerable distance below us, I caught the gleam of water.

"The park lake," Will explained in answer to a short query on my part. "Beastly deep too!"

He turned away, and we both gazed into the dark gap among the bushes.

A moment afterwards the clouds cleared for an instant, and a ray of light struck down full upon us, lighting up the little circle of bushes, and showing the clearing plainly. It was only a momentary gleam, but quite sufficient. There stood a *pedestal great and black*; but *there was no statue upon it!*

Will gave a quick gasp, and for a minute we stood stupidly; then we commenced to retrace our steps hurriedly. Neither of us spoke. As we moved we glanced fearfully from side to side. Nearly half the return journey was accomplished, when, happening to look behind me, I saw in the dim shadows to my left the bushes part, and a huge, white, carven face, crowned with black, suddenly protrude.

I gave a sharp cry and reeled backwards. Will turned. "Oh, mercy on us!" I heard him shout, and he started to run.

The Thing came out of the shadow. It looked like a giant. I stood rooted; then it came towards me, and I turned and ran. In the

The Thing came out of the shadow. . . . and I turned and ran.

THE GODDESS OF DEATH

A Weird Story of an Indian Deity Stolen by a Colonel, and the Horrible Revenge Wrought by its High Priest.

BY WILLIAM HOPE HODGSON.

IT was in the latter end of November when I reached T——worth to find the little town almost in a state of panic. In answer to my half-jesting inquiry as to whether the French were attempting to land, I was told a harrowing tale of some restless statue that had formed a nightly habit of running amuck amongst the worthy townspeople. Nearly a dozen had already fallen victims, the first having been pretty Sally Morgan, the town belle.

These and other matters I learnt. Wherever I went it was the same story. "Good Heavens! what ignorance, what superstition!" This I thought, imagining that they were the dupes of some murderous rogue. Afterwards I was to change my mind. I gathered that the tragedies had all happened in some park near by, where, during the day, this Walking Marble rested innocently enough upon its pedestal.

Though I scouted the story of the walking statue, I was greatly interested in the matter. Already it had come to me to look into it and show these benighted people how mistaken they had been; besides, the thing promised some excitement. As I strolled through the town I laughed, picturing to myself the absurdity of some people believing in a walking marble statue. Pooh! What fools there are! Arriving at my hotel, I was pleased to learn from the landlord that my old friend and schoolmate, William Turner, had been staying there for some time.

That evening while I was at dinner he burst into my room, and was delighted at seeing me.

"I suppose you've heard about the town bogey by now?" he said presently, dropping his voice. "It's a dangerous enough bogey, and we're all puzzled to explain how on earth it has escaped detection so long. Of course," he went on, "this story about the walking statue is all rubbish, though it's surprising what a number of people believe it."

"What do you say to trying our hands at catching it?" I said. "There would be a little excitement, and we should be doing the town a public benefit."

Will smiled. "I'm game if you are, Herton—we could take a stroll in the park to-night, if you like; perhaps we might see something.

"Right," I answered heartily. "What time do you propose going?"

Will pulled out his watch. "It's half-past eight now; shall we say eleven o'clock? It ought to be late enough then."

Copyright, 1904, by C. Arthur Pearson Ltd., in the United States of America.

The first two pages of Hodgson's story appearing in the April 1904 issue of the Royal Magazine, *illustrated by Cyrus Cuneo*

hours, Houdini, his arms and wrists bleeding and his clothes torn, managed to escape for a cheering audience. He proclaimed Hodgson's treatment of him to be the cruelest he had ever received. The *Blackburn Star* reported that Hodgson sensed the hostility of the audience and quickly departed for the sanctuary of the police station. Despite this episode, Hodgson seems to have been generally liked and respected in the community.

His business proved less profitable than he had hoped, however, particularly because of its seasonal nature. In the summer, enrollment at the school dropped drastically, but rent and other expenses continued. Hodgson closed the school and turned to a new outlet for his interest in physical culture: he began to write articles for magazine publication. His first was titled "Physical Culture versus Recreative Exercises," published in the February 1903 issue of *Sandow's Magazine;* the article included a muscle-pose photograph of the author. Hodgson subsequently sold other photo-illustrated articles, including "Health from Scientific Exercise," which appeared in *Cassell's Magazine* for October 1903.

During this period Hodgson joined the Society of Authors, which included among its members H. G. Wells, Arthur Conan Doyle, and George Meredith. He learned valuable information about the business of writing from its monthly publication, *The Author,* and he eventually met Wells through his society connection.

Hodgson's first professional fiction also dates from this period. His short story "The Goddess of Death" appeared in the April 1904 issue of the *Royal Magazine.* Although the story is resolved with a supposedly rational explanation, it reveals Hodgson's attraction to fantastic events. The plot concerns an eight-foot-tall statue of Kali, the Hindu goddess of death, which has been shipped to England by a colonel and erected in the center of a pond in a small park. Several murders ensue, and the culprit is reported to resemble the statue. The narrator investigates and is terrified when the statue emerges from hiding and pursues him. He escapes, and, after more murders, a suspicious turbaned figure is seen run-

Illustration by E. F. Skinner for Hodgson's "From the Tideless Sea," which was published in the May 1907 issue of the London Magazine

ning through the park. The townspeople pursue and shoot the fugitive as he jumps into the pond, where he dies. A tunnel and a secret room are discovered under the statue, and the man in the turban is revealed to be a Hindu assassin who was avenging the removal of the idol from its native land.

Hodgson's second published story reveals more of the characteristics for which his best fiction would be remembered. "A Tropical Horror" appeared in the June 1905 issue of the *Grand Magazine* and is a horror story set at sea. The narrator recounts how his ship is boarded by a sea monster with "a vast slobbering mouth a fathom across" that proceeds to eat the ship's crew. They battle the creature to little avail, and soon the narrator and one other crewman are the only survivors. They barricade themselves in a steel cabin, but the monster breaks a window and probes inside the cabin with its tongue, killing the crewman. The narrator, who has found an ax, wields it against the invader, chopping off a portion of the huge tongue. Rescuers find only the narrator alive; in the cabin with him is "a huge curled-up mass of whitish flesh, weighing about half a ton, one end of which appeared to have been hacked through with a sharp instrument."

Although an early work, the story is characteristic of Hodgson's fiction-writing technique. The plot, although exciting, is subordinate to the author's evocation of a dark, brooding atmosphere and his use of bizarre and horrific imagery. Hodgson seems to take a fiendish delight in describing the creature, which is a Freudian's dream (or nightmare) come true: "From the huge dripping lips hang great tentacles," the narrator relates, and, "There are no eyes visible; only that fearful slobbering mouth set on the tremendous trunklike neck." Furthermore, each time the monster snares a sailor with its tongue, it makes the sound, "Glut! Glut!" The story's dark mood and its descriptions of the monster anticipate the work of Lovecraft, who read and admired Hodgson's fiction.

Hodgson's next published fiction was a short fantasy story titled "The Valley of Lost Children," a sentimental tale about the ghost of a lost child, published in the February 1906 issue of the *Cornhill Magazine.* Apart from its possible autobiographical elements, stemming from Hodgson's feelings about his siblings who died in infancy, the story is not particularly noteworthy.

Meanwhile, Hodgson had learned from the pages of *The Author* about *Monthly Story Magazine,* a new fiction pulp being published in America. Encouraged by its policy of including fantasy and weird fiction in every issue, he submitted "From the Tideless Sea," which was published in the April 1906 issue. Besides being Hodgson's first sale to the American market, the tale also introduced the Sargasso Sea setting that the author would develop in his subsequent fiction. This "Sargasso Sea Mythos," as it has become known, would become a central motif of Hodgson's oeuvre, and some readers feel that it is his crowning achievement. As Brian M. Stableford points out in his *Scientific Romance in Britain 1890–1950* (1985), Thomas Janvier's novel *In the Sargasso Sea* (1898) preceded Hodgson's use of the locale as a trap for unwary ships, but Hodgson's use of the setting "developed the legend into its most fantastic form."

"From the Tideless Sea" utilizes one of Hodgson's favorite plot devices: a frame narrative describing the finding of a manuscript. In this case the manuscript is a diary found floating in a keg by a captain and his cabin boy. It tells of the ship *Homebird,* which is trapped in the Sargasso Sea along with countless other vessels, none of which shows any signs of life. The crew of the *Homebird* soon learns that giant octopi and other horrible creatures live underneath the mass of seaweed and driftwood. These creatures attack and kill the sailors, until only the injured captain, his daughter, and the narrator are left alive. The narrator encloses the ship's deck with a frame covered by canvas and tar, and the captain, who is dying, presides over the couple's marriage. The captain dies, but the narrator and his wife forge a new life amid their strange surroundings. They have a child, and they plan to send the diary up in a keg tied to a fire balloon in hopes that they will be rescued–the narrator estimates their provisions should last for seventeen years. After the story is finished, the captain sadly muses over the fate of the castaways, noting that the diary was written "sumthin' like twenty-nine years ago!"

Meanwhile, Hodgson had been working on book-length fiction, and he signed a contract in August 1907 with Chapman and Hall for the publication of his first novel, *The Boats of the "Glen Carrig."* Until recently, most critics (including Moskowitz) have assumed that Hodgson wrote this novel first, with the others following in the order of their publication, but in "Writing Backwards: The Novels of William Hope Hodgson" (1992) Sam Gafford cites evidence from letters Hodgson wrote in 1905 that suggests he had finished all four novels before the first was published. Although *The Night Land* is not specifically mentioned in the letters Gafford examined, he concludes that the novels were actually written in the reverse order of their publication; while this is not certain, it is clear that *Boats of the "Glen Carrig"* was not the first novel Hodgson wrote; he proclaims in a letter to Coulson Kernahan dated 25 September 1905 that "I've just finished my fourth book," and he continues, "The title of this book is *The Boats of the "Glen Carrig."*

Hodgson's first published novel is written in an antiquated, eighteenth-century style. It is the tale of a shipwreck, purportedly "told by John Winterstraw, Gent., to his son James Winterstraw, in the year 1757, and by him committed very properly and legibly to manuscript." After the *Glen Carrig* strikes a hidden rock and sinks, the crew escapes in lifeboats and sights land after five days. The men journey inland via a creek, from which they see mudflats and strange-looking trees and hear eerie cries, compared to "the weeping of lost souls."

Farther up the creek, they find an abandoned ship with provisions aboard. They spend the night in the cabin, where they hear strange noises, and the next night they sight a bizarre "many-flapped" creature. Exploring the land, they find that the odd-looking trees are spongy to the touch, and they find what appear to be the faces of a male and a female in one of the trunks. When the bosun makes a cut in the trunk, the tree screams and blood flows from the wound. The crew makes a hasty retreat to the boats and rows out to sea, where they read a manuscript found aboard the abandoned ship: its female author writes of her intention to search for her lover, who has gone ashore and whose voice she hears "wailing in the night."

After a storm, the crew members aboard the lifeboats find themselves in the Sargasso Sea, drifting among the abandoned hulks of trapped ships. After battles with a monstrous crab, a giant octopus, malevolent mushrooms, and intelligent but hostile slug creatures, the surviving crewmen join some survivors on one of the trapped ships, and all manage to break free of the weed after a final battle with the slug creatures.

Moskowitz quotes several of the laudatory contemporary reviews of the novel and claims that "literally no unfavorable reviews have yet been discovered." Despite critical success, however, the novel did not capture a popular audience and thus failed to provide Hodgson with much-needed income. The work would establish the pattern of Hodgson's writing career.

Meanwhile, although the American publication of "From the Tideless Sea" had drawn little notice, Hodgson sold the British serial rights to the *London Magazine,* where it appeared in May 1907, amid a flurry of editorial praise that ranked the story as "a real masterpiece," among other superlatives. The second story in the Sargasso Sea Mythos, "The Mystery of the Derelict," was published in the July 1907 issue of the *Story Teller,* after which Hodgson wrote a sequel to "From the Tideless Sea." As with the first story, the author submitted the sequel to an American market first, and "More News from the Homebird" appeared in the August 1907 *Blue Book Magazine* (a retitling of *Monthly Story Magazine*). Its first British publication was in the May 1911 *London Magazine,* under the title "The Fifth Message from the Tideless Sea."

Hodgson's first story combining conventions of the detective story and the weird tale was published the following month in the September 1907 *Blue Book Magazine.* Titled "The Terror of the Water Tank," the story presents a series of murders that have taken place in the vicinity of the town's water tank. The tank caretaker is charged with the crimes, but the story's climax reveals that a mutated monster living in the tank has been strangling passersby. Another weird sea tale appeared in the November 1907 *Blue Book Magazine.* "The Voice in the Night" tells the story of a young man and his fiancée who are marooned on a fungus-covered island. The fungus begins to grow on them until they are covered with it and become fungus creatures.

Hodgson's second novel publication, *The House on the Borderland,* was probably originally titled "The House of Mysteries," judging by remarks in Hodgson's 25 September 1905 letter to Kernahan. The novel again employs the narrative device of the found manuscript, with an added dimension: in addition to a framing sequence, Hodgson adds an introduction in which he presents himself as the editor of the text to follow. The frame-narrator, camping with a friend in western Ireland, discovers a manuscript titled "The House on the Borderland" in the ruins of a strange building on the brink of a chasm. The manuscript is "a kind of diary" kept by a self-described old man who has moved into the countryside with his sister as housekeeper and his "one friend," a dog named Pepper. He has purchased an odd old house with an evil reputation–its design is "curious and fantastic to the last degree. Little curved towers and pinnacles, with outlines suggestive of leaping flames, predominate; while the body of the building is in the form of a circle." (Hodgson reportedly modeled his description after a house that he had once rented in Ireland.)

One evening in his study, the narrator has a visionary experience, in which he streaks through space "beyond the fixed stars, and . . . into the huge blackness that waits beyond." He arrives upon an immense plain lit by a black sun ringed with red flame. He passes through enormous mountains, upon which he sees the colossal forms of ancient gods staring down into the valley. The mountains form a natural arena, at the center of which is an oversized version of his house in Ireland. He sees a giant, swine-faced humanoid trying to gain entry into the structure. The creature sees the narrator and lopes toward him, as the horrified onlooker helplessly floats nearer the monster. But he suddenly rises above the scene and streaks through space back toward earth; he is awakened from his trance with Pepper licking his hand.

After the narrator's return, swine creatures emerge from the gorge near his house. He barricades himself and his sister (who has not seen the creatures and believes her brother to be mad) in the structure. Through the night he fights them off, and after the following night proves uneventful the narrator decides to explore the cellars, where he finds an ancient-looking trapdoor. He opens it and stares down into what appears to be a bottomless pit. When he hears "a soft titter, that grew into a hideous chuckling, faint and distant," he slams the trapdoor in horror.

The narrator gradually explores the environs of the house. He descends into a cave or tunnel inside the chasm, from which he had seen the swine creatures emerge. He finds that it leads to a bottomless pit, which he later discovers is directly beneath the trapdoor in the cellars. Heavy rains have swollen a lake in the ravine, however, and water pours into the pit. The narrator wonders if the creatures have been drowned, and hopes at least that their means of egress into the world has been closed.

The narrator considers leaving the house, which he feels is evil, but after some time he has another vision–this time he enters a misty realm called the Sea of Sleep, where he sees his lost beloved. At this point the "editor" intrudes to inform the reader that a subsequent section of the manuscript is damaged and indecipherable–he includes fragments that further describe this visionary episode, which includes another trip through outer space, before picking up with the intact remainder of the document.

The manuscript resumes with a recounting of the central visionary sequence of the book, in which the narrator perceives time quicken to such a pace that the Sun and Moon alternately race through the heavens. He watches Pepper crumble and turn to dust as ages pass. He witnesses the Sun's death and the descent of a new Ice Age upon the Earth, although the stone structure of his house remains standing. The burned-out solar system drifts through space for aeons, until it is pulled into the orbit of an enormous Green Sun. After countless further ages, even the house begins to decay. The narrator sees eyes around him, and after hearing "a hideous swine-noise," he leaps out of a window. He discovers that the house is crawling with the swine creatures he had battled. After further ages he sees the house burst into red flames and hears the squeals of the creatures. He rises from the Earth, which begins to ignite from proximity to the Green Sun. The house and the swine creatures, who have not been harmed by the flames, fall into the pit below.

Floating in space, the narrator watches the solar system plunge toward the Green Sun–he suddenly perceives it as "the vast *Central* Sun–the great sun, round which our universe and countless others revolve." The vision continues, and the narrator witnesses "celestial globes" that seem to originate in this sun; he is drawn inside one of the spheres and finds that he is on the shores of the Sea of Sleep, where he is reunited with his loved one. After ages of happiness, a shadow falls, brought by the Sun of Darkness, and the narrator's beloved leaves. He discovers that there are two Central Suns: the Green Sun and the Dead Sun. After further travels through

THE
GHOST PIRATES

BY
WILLIAM HOPE HODGSON
AUTHOR OF
"THE BOATS OF THE 'GLEN CARRIG'"
"THE HOUSE ON THE BORDERLAND"

WITH A FRONTISPIECE BY
SIDNEY H. SIME

LONDON
STANLEY PAUL & CO.
1 CLIFFORD'S INN, TEMPLE BAR
1909

Frontispiece and title page for Hodgson's third novel, in which the barrier between the material and immaterial dimensions breaks down (courtesy of the Lilly Library, Indiana University)

dimensions–including a return to the immense plain and the giant version of his house–the narrator returns to his starting point to find all unchanged, with one exception: where Pepper was lying there is only "an elongated, little heap of grey, ash-like dust."

The narrator gets another dog but keeps it outside. The narrator hears footsteps ("pad, pad, pad") outside one night, after which the dog howls. He finds that it has been infected with a glowing, green fungus. Later the narrator discovers that he, too, has been infected. The swine creatures return, and the manuscript closes with another of Hodgson's favorite devices, breaking off in the middle of a fevered passage: "O God, help me now! Jesus– The door is opening–slowly. Somethi––"

The novel has been praised by Lovecraft (who called it "perhaps the greatest of all Mr. Hodgson's works"), Stableford, and Brian W. Aldiss and David Wingrove in *Trillion Year Spree* (1986), with special emphasis on the ultimate visionary sequence (which, along with *The Night Land,* probably influenced Olaf Stapledon's *Last and First Men*, 1930, and *Star Maker,* 1937). Although Moskowitz criticizes the sequence, calling it "the great weakness of the novel, since it is extraneous," he may be missing an important point: just as the main story is embedded within other narrative strata (the "editorial" and frame-narrative levels), so are deeper realities embedded within the "reality" described in the diary; they are symbolically linked by the reappearance of the house, the swine creatures, and the narrator's lost love. Somehow, these eternal realities (both beautiful and horrible) have found a passage into the temporal world through the strange house and the pit beneath. Thus, what might have been simply, as Aldiss puts it, "a tawdry horror story" takes on an unsettling, cosmic dimension. Stableford, taking as his cue some of "editor" Hodgson's remarks in the introduction, offers an insightful allegorical reading of the novel: "The house is the home of consciousness–the psyche–which exists both in the mundane world and (in a grander form) in the world of Ideas. The 'two central suns of creation,' the Green and the Dark, are . . . Growth and Decay, the former associated with love and ecstasy and the latter with emotional pain"; he sums up Hodgson's ultimately pessimistic vision: "In the end, man–individually and collectively–is bound to be consumed by the forces of decay, and so are the earth and sun."

Hodgson's "Date 1965; Modern Warfare: Extract from the Phono-Graphic," a futuristic antiwar satire, was published in the 24 December 1908 issue of the *New Age,* a socialist weekly. "Out of the Storm," a story purportedly sent via wireless from a sinking ship to a scientist in his laboratory, appeared in the February 1909 *Putnam's Monthly.* Of interest in the tale is the narrator's loathing for the sea, which he calls "the *Thing.*" He dutifully transmits until the end, sending the words, "My God! I am drowning! I–am–dr."

Another novel, *The Ghost Pirates,* was published in 1909. Hodgson claims in a prefatory note that the book completes a trilogy with *The Boats of the "Glen Carrig"* and *The House on the Borderland;* their connection resides not in their subject matter but in "conceptions that have an elemental kinship." The book relates the story of the ship *Mortzestus.* The crew begins to notice extra shadows aboard, and they soon see one of them walk overboard and into the sea. A crewman falls from the rigging to his death, and a ghostly figure seems to board the ship and then pass back into the sea. The narrator puts forward the theory that the barrier that normally separates the material and immaterial dimensions has been broken down. More ghostly phenomena appear in the night, and the next day the crew sees shadowy vessels following them beneath the waves. An "undulating greyness" appears along the port rails and then resolves itself into "hundreds of strange men." The bow of the *Mortzestus* plunges into the sea, and the ship sinks, although the narrator escapes to tell the tale. The theme of an alien reality breaking into the material, quotidian world bears a marked resemblance to *The House on the Borderland.*

Despite growing literary recognition, Hodgson's income from his writing was meager. He decided to boost his magazine sales by writing a series of stories around a continuing character. Detective characters had proved successful since Doyle had introduced Sherlock Holmes in the 1890s, and Hodgson followed the lead of Algernon Henry Blackwood's *John Silence* stories by creating a detective character specializing in supernatural cases–an occult detective named Carnacki. Robert Barr, the editor of the *Idler* magazine, agreed to publish the series, and the first Carnacki story, "The Gateway of the Monster," appeared in the January 1910 issue. Carnacki was more than just an imitation Sherlock Holmes or John Silence–he was, in fact, one of the earliest examples of the "scientific detective," that is, he employs new gadgets (often of his own invention) made possible by scientific technology. This first story introduces one of Carnacki's most potent gadgets, or "apparatus," as he refers to them–the electric pentacle, a defense drawing on the detective-inventor's knowledge of the *Sigsand MS.* (a fictional fourteenth-century text mentioned frequently in these stories) and Professor Gardner's *Experiments with a Medium.* The device consists of vacuum tubes formed in the shape of a mystic pentacle, and it protects whoever is within its borders from "the Immaterial"; when connected to its battery, it emits a "pale blue glare." Carnacki also frequently employs a camera in his cases, which sometimes are supernatural in nature and sometimes are liable to rational explanation.

Six Carnacki stories were collected in book form in 1913, but after the author's death three further tales featuring the "Ghost-Finder" were discovered and included in an expanded edition published by Arkham House in 1947. One of these additional stories, "The Hog," is also the longest Carnacki tale and offers echoes of the porcine horror Hodgson invoked in *The House on the Borderland.* In it the detective introduces his refinement of the electric pentacle, "the new spectrum 'defense,'" which "consists of seven glass vacuum circles with the red on the outside, and the colour circles lying inside it, in the order of orange, yellow, green, blue, indigo, and violet." He also employs "a specially made camera, a modified form of phonograph with ear-pieces instead of a horn, and a glass disk composed of many fathoms of glass vacuum tubes arranged in a special way." All of this to discover what is causing a client's nightmares, in which he hears pigs grunting, through which "every now and then there comes a gargantuan GRUNT, breaking through the million pig-voiced roaring–a stupendous GRUNT that comes in with a beat." Carnacki diagnoses the man's trouble as a "failure to be what I might term efficiently insulated–spiritually–from the outer monstrosities" and identifies the evil force as the "Hogge . . . of ye outer Monstrous Ones" discussed in the *Sigsand MS.* Although Carnacki defeats the creature, he almost loses his life in the process. Stableford asserts that the story "is clearly associated with the allegory of *The House on the Borderland.*"

The Carnacki stories have several devoted followers, especially Dennis Wheatley, who wrote an introduction for a 1975 republication of the series and whose acclaimed horror novel *The Devil Rides Out* (1935) includes scenes that are obviously inspired by episodes from "The Gateway of the Monster" and "The Hog." Moskowitz is less enamored of the tales, describing the first as "an ill-concealed pastiche on the genre," and finding Hodgson's use of humor in the stories to be ill advised. The stories definitely have moments of comedy, as in Carnacki's refreshingly honest reportage of his unheroic response to the suspected supernatural in "The Thing Invisible": while awaiting a ghostly manifestation in a chapel at night, the detective "sat and sweated coldly . . . with the 'creep' busy at my spine," he admits; later, after he has apparently been attacked, he reacts in a decidedly un-Holmesian manner: "I bowed my head, and made a scrambling run in the complete darkness and dashed into a pew. I jumped back, staggering, got my bearings a little, and raced down the centre of the aisle, putting my mailed arms over my face. I plunged into my camera, hurling it among the pews. I crashed into the font, and reeled back. Then I was at the exit." This rather slapstick passage is more subtle when taken in context; in fact, Wheatley treats the series as straight horror tales: "All of them have that quality of built-up suspense and terror of the unknown, which so greatly exceeds terror of the known, and makes Hope Hodgson's tales real masterpieces of storytelling."

Hodgson produced a variety of work for magazine sales in 1910–1912, including essays, various types of formula fiction, and a three-part "exposé" of sea life titled "Prentices' Mutiny." One horror story, "The Albatross," appeared in the July 1911 issue of the American pulp magazine *Adventure,* although it had been written four years earlier. Another story set in the Sargasso Sea, titled "The Thing in the Weeds," appeared in the January 1912 *Storyteller.*

More important, in 1911 Hodgson managed to sell his longest (two hundred thousand words), and in his opinion best, novel, *The Night Land.* The story, written in an eccentric, antiquated style, is ostensibly a dream message received by a seventeenth-century man from a telepathic youth living in the earth's final days. This telepathic "narrator" lives in the Great or Last Redoubt, a seven-mile-high metal pyramid that serves as a home to the last remnants of humanity. This world of millions of years in the future is a "Night Land," a cold, eternally dark place (the sun has died) except for the heat and light provided in some areas by volcanic "fire-holes." The redoubt, protected by a force field powered by the "Earth-Current," is surrounded by mysterious and frightening creatures, including the Silent Ones, the Strange Things, the monstrous Ab-Humans, and, most fearsome of all, the four mountainous Watchers (in the northeast, southeast, northwest, and southwest) and a fifth "Watching Thing of the South," whose threatening progression toward the redoubt had been halted in the distant past by a glowing dome that had inexplicably risen from the earth. The Watchers remain immobile but apparently alert for the failure of the technology that keeps them at bay (and, the narrator suggests, which probably allowed them and the other monstrous

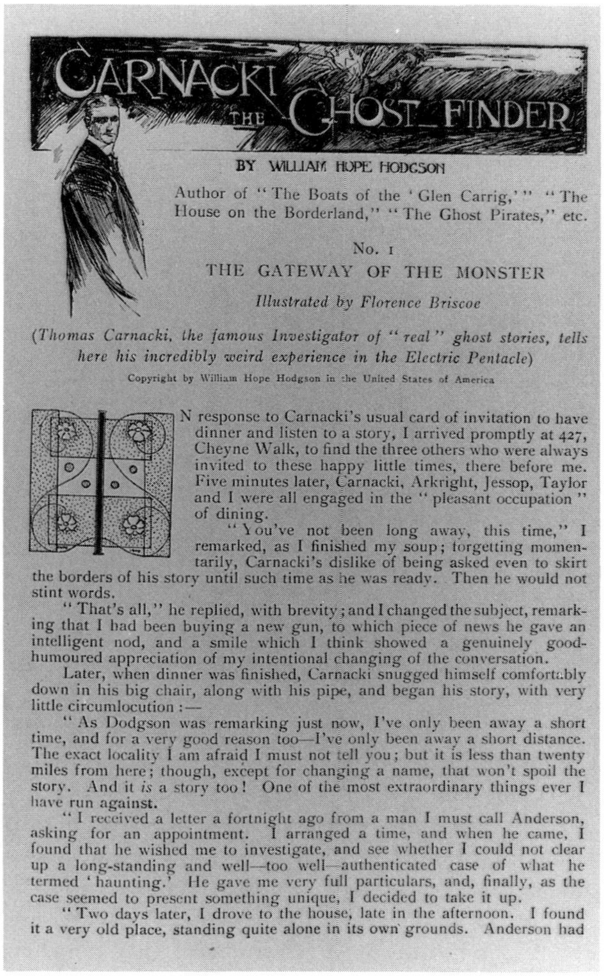

CARNACKI THE GHOST FINDER

BY WILLIAM HOPE HODGSON

Author of "The Boats of the 'Glen Carrig,'" "The House on the Borderland," "The Ghost Pirates," etc.

No. I

THE GATEWAY OF THE MONSTER

Illustrated by Florence Briscoe

(Thomas Carnacki, the famous Investigator of "real" ghost stories, tells here his incredibly weird experience in the Electric Pentacle)

Copyright by William Hope Hodgson in the United States of America

IN response to Carnacki's usual card of invitation to have dinner and listen to a story, I arrived promptly at 427, Cheyne Walk, to find the three others who were always invited to these happy little times, there before me. Five minutes later, Carnacki, Arkright, Jessop, Taylor and I were all engaged in the "pleasant occupation" of dining.

"You've not been long away, this time," I remarked, as I finished my soup; forgetting momentarily, Carnacki's dislike of being asked even to skirt the borders of his story until such time as he was ready. Then he would not stint words.

"That's all," he replied, with brevity; and I changed the subject, remarking that I had been buying a new gun, to which piece of news he gave an intelligent nod, and a smile which I think showed a genuinely good-humoured appreciation of my intentional changing of the conversation.

Later, when dinner was finished, Carnacki snugged himself comfortably down in his big chair, along with his pipe, and began his story, with very little circumlocution:—

"As Dodgson was remarking just now, I've only been away a short time, and for a very good reason too—I've only been away a short distance. The exact locality I am afraid I must not tell you; but it is less than twenty miles from here; though, except for changing a name, that won't spoil the story. And it *is* a story too! One of the most extraordinary things ever I have run against.

"I received a letter a fortnight ago from a man I must call Anderson, asking for an appointment. I arranged a time, and when he came, I found that he wished me to investigate, and see whether I could not clear up a long-standing and well—too well—authenticated case of what he termed 'haunting.' He gave me very full particulars, and, finally, as the case seemed to present something unique, I decided to take it up.

"Two days later, I drove to the house, late in the afternoon. I found it a very old place, standing quite alone in its own grounds. Anderson had

First page of Hodgson's story appearing in the January 1910 issue of the Idler, *illustrated by Florence Brisco*

creatures access to the material world in the first place).

The protagonist of the novel is a reincarnation of the seventeenth-century frame narrator, and he becomes aware telepathically of the existence of a Lesser Redoubt to the North, in which resides the female incarnation of his earlier self's beloved wife. The inhabitants are under attack by the forces of evil and are calling for help from the Great Redoubt. After a heroic attempt by five hundred youths to reach the Lesser Redoubt fails, the narrator ventures forth into the Night Land to reach his lost love and her people.

After passing such dangers as the Grey Man, the Place Where the Silent Ones Kill, Mist Men, and the forest of the Humped Men, the narrator finally reaches the Lesser Redoubt, only to find that it has fallen to the forces of evil. One of the few to survive, however, is his love, Naani; the narrator and his love are reunited and begin their journey back across the Night Land to the Great Redoubt. They must surmount further obstacles on the return trip, as when they pass through one of the many lands of total darkness in this future world, in which horrible monsters are discernible only by sounds and smells:

> And we only to be able to make our best speed, and to be weary with hope that we rush not blind into death; and alway to be chilled in the back with the belief that the thing that made chase did be very close in that utter dark, and to gain upon us. . . . And we came nigh to the fire-pit, and lo! the smell did be grown very drear and foul; but there did be no Slug-Beast near the fire. And by this, we perceived that we did be truly in an utter danger, in that the Pursuer did be upon us, and the stinking to tell this to be truth, and our spirits to have warned us very strong and proper.

The pair arrive within sight of the Great Redoubt, only to have Naani struck by a "Force which did have in it an utter Silence and bleakness of Desolation," sent from the ominous House of Silence. The narrator carries her unconscious form to the redoubt and manages to get her inside after a final battle at the gates with the amassed forces of evil, only to learn that she has died. The millions of inhabitants gather to honor the narrator and grieve for his lost love. They descend to the lowest underground level of the pyramid, the Country of Silence, where Naani's body is placed on the Last Road, which carries the departed to eternity. As she is about to pass out of sight, however, Naani stirs and revives; the narrator later explains: "the great life-force of the Earth-Current to have waked her spirit, and her body then to live and her blood to flow proper again. And the Doctors had talkt much and searched much of late in the olden Records of their Work; and they to have found somewhat of one such happening in the olden time; but truly, naught such to have been ever through a mighty age of years." The novel closes with the couple married and living happily ever after.

No plot summary can adequately convey the essence of *The Night Land;* its plot structure is merely that of the epic quest, with the hero surmounting a series of episodic obstacles, culminating in seeming defeat and then final triumph. The novel's originality lies in its strange use of language, bizarre imagery, and setting. Although such a technique is now commonplace, Hodgson was one of the first fantasy writers to set such a quest in a world of the far future rather than the remote past or a magical land, and the strangeness of the world he created has yet to be surpassed.

Modern critical consensus on the work praises the sweep of its imaginative vision while voicing impatience with its pseudoarchaic style. Aldiss proclaims that the novel "flares into magnificence and dies into unreadability," while Stableford writes, "such are its idiosyncracies that it is virtually unreadable–in terms of literary style it must be one of the worst books ever published–but it also has an imaginative intensity that could belong only to an extremely personal book." Even Moskowitz, who regards the book as Hodgson's masterpiece, admits the shortcomings of its style but argues that the language, which is "tedious and absurd" in the lengthy frame-narrative sequences, becomes more effective when describing the dark world of the future, where it "succeeds in painting a canvas of that milieu with awesome literary strokes, inspired by imaginative genius." A less qualified dissent is provided by James Cawthorn and Michael Moorcock, who assert in their *Fantasy: The 100 Best Books* (1988) that the novel's "difficulties have been greatly exaggerated" and conclude that "anyone willing to make the effort will find that the manner of the telling heightens the atmosphere of a remote and alien Earth."

Hodgson continued submitting stories to periodicals as well, and in 1912 the *Red Magazine* published two humorous weird tales chronicling the adventures of Captain Jat and his cabin boy, Pibby Tawles. A serious supernatural tale, "The Derelict," appeared in the 1 December 1912 issue of the same magazine. The year 1913 marked a turning point in Hodgson's life: on 26 February he married Betty Farnworth, whom he had met through his connection with the *Red Magazine;* Farnworth had been a staffer on *Home Notes,* a women's magazine published by the same company. The couple honeymooned in the south of France and were so taken with the area that they decided to move there, settling in Chalet Mathilda, Sanary.

Following his marriage, Hodgson turned his efforts to producing short fiction in practically every genre imaginable. His next fantasy story was another tale set in the Sargasso Sea, "The Finding of the Graiken," which appeared in the 15 February 1913 issue of the *Red Magazine.* "Sea Horses," a sentimental fantasy utilizing the lost-child motif, appeared in the March 1913 *London Magazine.* In the same month, *Carnacki, the Ghost Finder,* a collection of Hodgson's stories featuring his occult detective, was published by Eveleigh Nash to positive reviews.

The last important fantasy tale by Hodgson appeared the following year, in the 1 July 1914 issue of the *Red Magazine.* "The Stone Ship" relates how the crew of the becalmed *Alfred Jessop* witnesses strange events onboard a petrified ship that has risen from the depths of the sea. When the story was included in Hodgson's collection of short stories titled *The Luck of the Strong* (1916), the reviewer for the *Bookman* praised it as "an excellent example of the author's skill." In the meantime another collection of Hodgson's fiction had appeared, *Men of the Deep Waters* (1914), which includes some of the author's finest horror and fantasy fiction.

After the outbreak of World War I Hodgson and his wife returned to England, where he joined the Officer's Training Corps at the University of London. He received his commission as a lieutenant in the Royal Field Artillery in 1915, but in 1916 he was thrown from a horse and recieved a serious head injury; as a result he was discharged from the service. Not content to sit out the remainder of the war, however, Hodgson fought his way back to health and reenlisted in the Royal Field Artillery. He was reported killed by an exploding shell at Ypres on 17 April 1918.

Hodgson's works, which had never attained commercial success, fell into relative obscurity after his death. In 1920–1921 inexpensive editions of his eight books were published by Holden and Hardingham (and two posthumous volumes of his verse were published by Selwyn and Blount), but Hodgson's fiction remained generally ignored, eclipsed by the works of better-known writers of fantasy fiction such as Joseph Sheridan Le Fanu, Arthur Machen, M. R. James, and Blackwood. Hodgson's reputation began to be established in the United States almost three decades after his death, when a devotee and collector of fantasy fiction named H. C. Koenig discovered Hodgson's story "The Voices in the Night" in Colin de la Mare's anthology *They Walk Again* (1931). Impressed, Koenig began a strident search for more of Hodgson's work. He discovered a general neglect of the author but eventually managed to obtain first editions of his books.

Convinced of Hodgson's importance, Koenig began to champion the author's cause among fantasy readers; he eventually persuaded Mary Gnaedinger, the editor of the pulp *Famous Fantastic Mysteries,* to republish some of Hodgson's work. "The Derelict" appeared in the December 1943 issue, and the following issue (March 1944) presented *The Ghost Pirates.* Hodgson proved popular with the magazine's readers, and the June 1945 issue featured *The Boats of the "Glen Carrig."* These publications of Hodgson's fiction generated much interest in his work among readers of the genre.

This interest, combined with Koenig's continued campaigning, convinced August Derleth of the influential specialty imprint, Arkham House, to publish an omnibus edition of Hodgson's novels. *The House on the Borderland and Other Novels* (1946) introduced many American readers to Hodgson's book-length fiction, including *The Night Land.* His reputation was quickly established as a major writer of horror and fantasy fiction, and his books (including the Arkham House omnibus) began to command premium prices on the collector's market. The attention accorded to Hodgson's work is still largely that of critics and historians of the genre, but the author's reputation continues to grow and could be poised for more general critical success.

References:

Brian W. Aldiss and David Wingrove, *Trillion Year Spree: The History of Science Fiction* (New York: Atheneum, 1986), pp. 167–169;

Lin Carter, "Strange Odyssey," in *The Boats of the "Glen Carrig,"* by Hodgson (New York: Ballantine, 1971), pp. vii–xii;

James Cawthorn and Michael Moorcock, *Fantasy: The 100 Best Books* (New York: Carroll & Graf, 1988), pp. 53–54, 65–66;

Richard Dalby, "William Hope Hodgson," *Book and Magazine Collector,* 35 (February 1987): 30–38;

Sam Gafford, "Writing Backwards: The Novels of William Hope Hodgson," *Studies in Weird Fiction,* 11 (Spring 1992): 12–15;

H. P. Lovecraft, *Supernatural Horror in Literature* (New York: Abramson, 1945), pp. 82–85;

Sam Moskowitz, "William Hope Hodgson," in *Out of the Storm: Uncollected Fantasies by William Hope Hodgson,* by Hodgson (West Kingston, R.I.: Grant, 1975), pp. 9–117;

Brian M. Stableford, *Scientific Romance in Britain 1890–1950* (New York: St. Martin's Press, 1985), pp. 91–102.

Papers:

Some of William Hope Hodgson's letters are at the Harry Ransom Humanities Research Center, University of Texas at Austin.

Charles Kingsley

(12 June 1819 - 23 January 1875)

Corbin Scott Carnell
University of Florida

See also the Kingsley entry in *DLB 21: Victorian Novelists Before 1885; DLB 32: Victorian Poets Before 1850;* and *DLB 163: British Children's Writers, 1800–1880.*

BOOKS: *The Saint's Tragedy; or, The True Story of Elizabeth of Hungary* (London: Parker, 1848; New York: International Book Company, 1855);

Twenty-Five Village Sermons (London: Parker, 1849; Philadelphia: Hooker, 1855);

Introductory Lectures, Delivered at Queen's College, London (London: Parker, 1849);

Cheap Clothes and Nasty, as Parson Lot (Cambridge: Macmillan, 1850);

Alton Locke, Tailor and Poet: An Autobiography (2 volumes, London: Chapman & Hall, 1850; 1 volume, New York: Harper, 1850);

The Application of Associative Principles and Methods to Agriculture (London: Bezer, 1851);

Yeast: A Problem (London: Parker, 1851; New York: Harper, 1851);

The Message of the Church to Labouring Men: A Sermon (London: Parker, 1851);

Who Are the Friends of Order? (London: Lumley, 1852);

Phaeton; or, Loose Thoughts for Loose Thinkers (Cambridge: Macmillan, 1852; Philadelphia: Hooker, 1854);

Sermons on National Subjects (London: Griffin, 1852); republished as *The King of the Earth, and Other Sermons Preached in a Village Church* (London: Macmillan, 1872);

Hypatia; or, New Foes with an Old Face (2 volumes, London: Parker, 1853; 1 volume, New York: Lowell, 1853);

Alexandria and Her Schools (Cambridge: Macmillan, 1854);

Sermons on National Subjects: Second Series (London & Glasgow: Griffin, 1854);

Who Causes Pestilence? (London & Glasgow: Griffin, 1854);

Brave Words for Brave Soldiers and Sailors, anonymous (Cambridge: Macmillan, 1855); republished as *True Words for Brave Men* (London: Kegan Paul, 1878; New York: Whittaker, 1886);

Westward Ho! or, The Voyages and Adventures of Sir Amyas Leigh, Knight, of Burrough, in the County of Devon, in the Reign of Her Most Glorious Majesty Queen Elizabeth (3 volumes, Cambridge: Macmillan, 1855; 1 volume, Boston: Ticknor & Fields, 1855);

Glaucus; or, The Wonders of the Shore (Cambridge: Macmillan, 1855; Boston: Ticknor & Fields, 1855);

Sermons for the Times (London: Parker, 1855; New York: Dana, 1856);

Sermons for Sailors (London, 1855?); republished as *Sea Sermons* (London: Kegan Paul, 1885);

The Heroes; or, Greek Fairy Tales for My Children (Cambridge: Macmillan, 1856; Boston: Ticknor & Fields, 1856);

Two Years Ago: A Novel (3 volumes, Cambridge: Macmillan, 1857; 1 volume, Boston: Ticknor & Fields, 1857);

Andromeda and Other Poems (London: Parker, 1858; Boston: Ticknor & Fields, 1858);

Miscellanies, 2 volumes (London: Parker, 1859);

The Good News of God: Sermons (London: Parker, 1859; New York: Burt, Hutchinson & Abbey, 1859);

The Massacre of the Innocents: An Address (London: Jarrold, 1859?);

The Limits of Exact Science as Applied to History: An Inaugural Lecture (Cambridge: Macmillan, 1860);

The Example of the Early Navigators: A Sermon (London: Parker, 1860);

Why Should We Pray for Fair Weather? A Sermon (London: Parker, 1860);

New Miscellanies (Boston: Ticknor & Fields, 1860);

Ode Performed in the Senate-House Cambridge, Composed for the Installation of His Grace the Duke of Devonshire, Chancellor of the University (Cambridge & London: Macmillan, 1862);

Speech of Lord Dundreary in Section D on Friday Last: On the Great Hippocampus Question, anonymous (Cambridge & London: Macmillan, 1862);

A Sermon on the Death of His Royal Highness the Prince Consort (London: Parker, 1862);

The Water-Babies: A Fairy Tale for a Land-Baby (London & Cambridge: Macmillan, 1863; Boston: Burnham, 1864);

The Gospel of the Pentateuch: A Set of Parish Sermons (London: Parker, Son & Bourn, 1863);

Mr. Kingsley and Dr. Newman: A Correspondence on the Question Whether Dr. Newman Teaches That Truth Is No Virtue (London: Longmans, Green, 1864);

"What, Then, Does Dr. Newman Mean?": A Reply (London & Cambridge: Macmillan, 1864);

The Roman and the Teuton (Cambridge & London: Macmillan, 1864);

The Irrationale of Speech, by a Minute Philosopher (London: Longmans, 1864); also published as *Hints to Stammerers, by a Minute Philosopher* (London: Longmans, 1864);

David: Four Sermons Preached before the University of Cambridge (London: Macmillan, 1865);

Hereward the Wake: "Last of the English" (2 volumes, London: Macmillan, 1866; 1 volume, Boston: Ticknor & Fields, 1866);

The Temple of Wisdom: A Sermon (London: Macmillan, 1866);

Three Lectures Delivered at the Royal Institution on the Ancien Régime before the French Revolution (London: Macmillan, 1867);

The Water of Life and Other Sermons (London: Macmillan, 1867; Philadelphia: Lippincott, 1868);

The Hermits (London: Macmillan, 1868; Philadelphia: Lippincott, 1868);

Discipline and Other Sermons (London: Macmillan, 1868; Philadelphia: Lippincott, 1868);

Women and Politics (London: National Society for Women's Suffrage, 1869);

God's Feast: A Sermon (London & Cambridge: Macmillan, 1869);

The Address on Education, Read before the National Association for the Promotion of Social Science (London: Head, 1869);

Madam How and Lady Why; or, First Lessons in Earth Lore for Children (London: Bell & Daldy, 1870; New York: Macmillan, 1885);

At Last: A Christmas in the West Indies, 2 volumes (London: Macmillan, 1871; New York: Harper, 1871);

Letter to a Public School Boy, on Betting and Gambling (London: Society for Promoting Christian Knowledge, 1871?);

Town Geology (London: Strahan, 1872; New York: Appleton, 1873);

Poems: Collected Edition (London: Macmillan, 1872; expanded, 1873; New York: Hurst, 1880);

Plays and Puritans, and Other Historical Essays (London: Macmillan, 1873);

Prose Idylls, New and Old (London: Macmillan, 1873);

Frederick Denison Maurice: A Sermon (London: Macmillan, 1873);

Health and Education (London: Isbister, 1874; New York: Appleton, 1874);

Westminster Sermons (London: Macmillan, 1874);

Lectures Delivered in America in 1874 (London: Longmans, Green, 1875; Philadelphia: Coates, 1875);

Letters to Young Men on Betting and Gambling (London: King, 1877);

All Saints' Day and Other Sermons, edited by W. Harrison (London: Kegan Paul, 1878; New York: Scribner, Armstrong, 1878);

Out of the Deep: Words for the Sorrowful from the Writings of C.K., edited by Fanny Kingsley (London & New York: Macmillan, 1880);

From Death to Life: Fragments of Teaching to a Village Congregation, with Letters on the Life after Death, Edited by His Wife (London & New York: Macmillan, 1887);

Words of Advice to Schoolboys, Collected from Hitherto Unpublished Notes and Letters, edited by E. F. Johns (London: Simpkin, 1912; Winchester: Warren, 1912);

The Tutor's Story: An Unpublished Novel, revised and completed by Kingsley's daughter, Lucas Malet (Mrs. Mary St. Leger Harrison) (London: Smith, Elder, 1916; New York: Dodd, Mead, 1916).

Collection: *The Works of Charles Kingsley*, 28 volumes (London & New York: Macmillan, 1885–1902).

OTHER: *Most Glorious Majesty Queen Elizabeth*, 3 volumes, rendered into modern English by Kingsley (Cambridge: Macmillan, 1855);

Charles Blackford Mansfield, *Paraguay, Brazil and the Plate, Letters Written in 1852–1853*, biographical sketch by Kingsley (Cambridge: Macmillan, 1856);

The History and Life of J. Tauler with Twenty-Five of His Sermons, preface by Kingsley (London: Smith, Elder, 1857);

H. Brooke, *The Fool of Quality; or, The History of Henry, Earl of Moreland*, preface by Kingsley (London: Smith, Elder, 1859);

John Bunyan, *The Pilgrim's Progress*, preface by Kingsley (London: Longman, 1860).

Charles Kingsley was a clergyman, social reformer, novelist, poet, polemicist, university professor, and chaplain to the queen. He was prolific, publishing thirty-five works in addition to a volume of sermons that was published after his death. By 1850, when he was only thirty-one, he was an established author whom Alfred Tennyson visited and whom foreign visitors endeavored to see. His younger brother, Henry, was also a novelist, though he is now forgotten. Though Kingsley's works are not widely read today, his life and thought continue to be of interest.

In the history of fantasy, Kingsley is remembered primarily for his novel *The Water-Babies: A Fairy Tale for a Land-Baby* (1863), a tale about a boy who dies and is transformed as a fantastic creature. Popular in its own time, the book is historically significant not only for its use of the fantastic but also for its Victorian blend of evolutionary theory and religious concerns, as well as for its anticipations of the psychological ideas of Sigmund Freud.

Kingsley was born on 12 June 1819 in Holne in Devonshire, the son of the Reverend Charles Kingsley and his wife, née Mary Lucas. The elder Kingsley had been bred as a country gentleman, but finances had forced him to adopt a profession, and he took orders after age thirty. Young Charles was a precocious child, writing sermons and poems at age four. Living in the fen country, he early learned boating, horseback riding, and shell collecting.

In 1836, when he was seventeen, Kingsley went with his family to London and became a student at King's College, walking back and forth from Chelsea. He found London life dismal, and he was bored with his parents' parish work. In 1838 he entered Magdalene College at Cambridge, and at the end of his first year he won a scholarship. In July 1839 he met his future wife, Frances "Fanny" Grenfell.

Kingsley neither thought nor studied systematically, and his beliefs were more matters of instinct and feeling than based on reason. He had a great many interests and an enormous capacity for enjoyment. He considered himself a Platonist and also a disciple of Frederick Denison Maurice's "social gospel" Christianity. He greatly admired Charles Darwin, T. H. Huxley, and Charles Lyell, and he corresponded with John Stuart Mill, sharing especially Mill's interest in the emancipation of women.

During his college years Kingsley was plagued by religious doubt, and he avidly read Samuel Taylor Coleridge, Thomas Carlyle, and Maurice. He was popular at college and played several sports, including horseback riding, rowing, and boxing. He was a good walker and once hiked from Cambridge to London in a day. He was a somewhat desultory student up until his last six months at Cambridge, but he managed to get a first in classics. In 1842 he was ordained to the curacy of Eversley, Hampshire. He was what was known as a Broad Churchman: fairly liberal, with an emphasis on Christian social concern. Kingsley was an energetic parson. He loved and respected the poor. Matthew Arnold said in a letter of condolence to his family that he was "the most generous man I have ever known; the most forward to praise what he thought good, the most willing to admire, the most free from all thought of himself . . . and the most incapable of being made ill-natured."

As a young man Kingsley tended to be somewhat restless and high-strung. His love for Frances Grenfell stabilized him and helped to guide him toward ordination in the Church of England. The same solution–the love of God and a good woman–he grants to several of the heroes of his novels.

The year 1844 was important for Kingsley. He was married and pursued his pastoral work in Eversley parish, trying to repair the dilapidated church and manse. He established a variety of clubs, an adult evening school, a lending library, a lecture series, and a loan fund. He began daily visits among his parishioners and experimented in his garden with scientific agricultural methods. Also in this year he met Maurice, the only one of his friends he ever called his "master." They had corresponded before the meeting and continued to do so until Kingsley's death.

In 1848 Kingsley was appointed professor of English literature at Queen's College in London, but ill health forced him to give up the post a year later. Also in 1848 he contributed papers to *Politics for the People* and was known for his passionate concern for the poor; he especially called attention to the bad sanitation that the cities afforded the poor. He also contributed to a journal called *Christian Socialist* and was associated with the Christian Socialist movement for several years. In 1848 he had a complete breakdown of health and went for a rest in Bournemouth. After a second collapse he spent the winter in north Devonshire.

In spite of bouts of ill health, Kingsley wrote throughout these years. He published a poetic drama based on the life of a thirteenth-century saint called *The Saint's Tragedy; or, The True Story of Elizabeth of Hungary* (1848). Reviewers gave the play both praise and disapproval. They identified what were to become Kingsley's characteristic flaws: exaggeration, some bitterness, and occasional outbursts of fine talk. *The Saint's Tragedy* also points up the themes that interested Kingsley most: the glorification of marriage and the importance of the family, the ennobling influence of women, the importance of serving the poor, and a Christianity of practical good works. This latter attitude was later referred to as "muscular Christianity," and this term is still closely associated with Kingsley.

Kingsley also anonymously published "Why Should We Fear the Romish Priests?" (1848) in *Fraser's Magazine.* Here he argues for the superiority of Christian marriage over priestly celibacy, for the authority of the Protestant tradition, and against sacerdotalism. He published his first novel, *Yeast: A Problem* (1851), anonymously in *Fraser's* in 1848, although his second serialized novel of social purpose, *Alton Locke, Tailor and Poet: An Autobiography,* was published first in book form in 1850. Both novels have held up fairly well with the passage of time. Kingsley certainly reads as well as Sir Walter Scott

A drawing by Kingsley for his unpublished life of Saint Elizabeth of Hungary (Susan Chitty, The Beast and the Monk*)*

or George Eliot. He is not as strong as they in creating memorable characters, but his plots keep moving and the thematic ideas are clear.

In 1849 Kingsley published a collection of twenty-five sermons, and in 1851 he preached a sermon in London in which he said it was un-Christian for the rich to continue to amass wealth at the expense of the suffering masses. This claim was in line with his Christian Socialist thinking, which many in England viewed with suspicion. Although initially forbidding Kingsley to preach in his diocese, the bishop withdrew his prohibition after reading the published version of the sermon.

Early in his career Kingsley had been interested in Chartism, which took its name from the People's Charter, published in 1838. The Chartists sought equal electoral areas, universal suffrage, payment for members of Parliament, no property qualifications for voting, vote by ballot, and annual parliaments. The movement failed in 1839 and was revived in 1842, when it failed a second time. From then until 1852 Kingsley was active in the Christian Socialist movement. He wrote articles for the move-

An illustration by Linley Sambourne for Kingsley's The Water-Babies

ment and began *Yeast* in the middle of Christian Socialist activities.

Exhausted after the political agitation of 1848 and the strain of completing *Yeast,* Kingsley placed his parish duties in the care of his father and took his family to Ilfracombe in the West Country for the winter. There he recuperated by walking along the seashore, gathering specimens and shells, and talking with friends. He returned to Eversley in June, but his health began to fail again and he returned to the West Country, where he fished and read François Rabelais and John Ruskin. In September he returned to his parish.

Kingsley contined to be interested in the agitation for better sanitation, especially in view of the appearance of cholera among the poor, and in other Christian Socialist causes. However, he was not specifically aligned with the Christian Socialists after the early 1850s, though he submitted many poems and articles to their publication, *Christian Socialist.* With the demise of the magazine in June 1852, he withdrew from active participation. He remained attuned to Christian Socialist thinking, but he believed also that society is changed as individuals are changed and that all cannot be accomplshed by political means alone. In later life he inclined toward the Tory way of thinking (with views close to those of Queen Victoria) and believed fervently in the House of Lords. He finally looked to Christianity rather than politics to bring about the deepest changes.

In 1853 Kingsley published the novel *Hypatia; or, New Foes with an Old Face*, set in ancient Alexandria. Some critics regard *Hypatia* as the best of his novels. In it, according to Larry K. Uffelman, the novelist "admirably portrays the doubts and dangers of life in an age of transition. He does so on a canvas packed with incidents, themes, and characters . . . [and] develops his conviction that the story of history is the story of God's developing plan."

In 1855 Kingsley published another novel, *Westward Ho! or, The Voyages and Adventures of Sir Amyas Leigh, Knight, of Burrough, in the County of Devon, in the Reign of Her Most Glorious Majesty Queen Elizabeth.* Although it was an immediate success, George Eliot complained that Kingsley's preaching spoiled the book. Also in 1855 he anonymously published *Brave Words for Brave Soldiers and Sailors*, inspired by his interest in the Crimean War. In 1856 he brought out *The Heroes; or, Greek Fairy Tales for My Children.* In 1857 another novel appeared, this one with a more contemporary setting: *Two Years Ago,* a tale of the cholera epidemic of 1854.

In 1859 Kingsley was appointed chaplain to Queen Victoria, and Prince Albert in particular admired him, later recommending him for an honorary degree. Kingsley was appointed to a professorship of modern history at Cambridge, though he was not really trained as a historian. He sensed that he was not qualified to be a professor of history in a great university and in 1869 resigned his post. He continued, however, to lecture widely. He sometimes spoke out in support of Darwinism, holding that it was compatible with Christian theology. In 1873 he was appointed canon of Westminster Abbey.

Kingsley published his best-known work, *The Water-Babies,* in 1863. It stands apart from his other novels, belonging instead with the fantasies of George MacDonald and Lewis Carroll though probably not of their stature. It is an absorbing story for children, though there are words and details that only adults would understand. It tells of the adventures of a forlorn young chimney sweep named Tom who travels as a tiny "water-baby" through a strange and unpredictable world.

On occasion in *The Water-Babies* Kingsley has a good time stringing big words together or creating unpronounceable names, evoking a child's sense of play, and some of the creatures he describes would also appeal to a child's whimsy. The story includes an ugly fairy named Mrs. Bedonebyasyoudid and a beautiful fairy named Mrs. Doasyouwouldbedoneby. They both give Tom moral instruction.

The Water-Babies reflects Kingsley's keen interest in nature and his ability to describe, with some precision, creatures and scenes from nature. The narrative is a bit loose, and there is no climax to the tale, but it holds one's interest. It lacks the magical beauty of favorite fairy stories, but it does transport the reader to another world. C. S. Lewis said that no children's story that does not also appeal to adults can last. The fact that this story continues to be published and republished suggests that adults can enjoy it, too.

In 1870 Kingsley's son Maurice went to the United States, where he worked as a railway engineer. Kingsley had long been interested in America and had favored the abolition of slavery, though before the Civil War he came to believe that slavery would die once the South were free of the North and that under English influence blacks would be better treated. In 1874 Kingsley sailed for America, where he traveled and lectured. He became ill in Colorado with a respiratory ailment and returned to England in August. He was preaching again at Westminster by November. He became ill again in December and died on 23 January 1875 at age fifty-five. A pension was granted to his widow, but she declined the queen's offer of rooms in Hampton Court Palace. She died in 1891, sixteen years after her husband. They were survived by their four children. Kingsley's posthumous reputation was somewhat tainted by his attack on Catholic convert John Henry Newman, who certainly won the argument and gained tremendous prestige with his autobiography, *Apologia pro Vita Sua* (1864), which was written in part in response to Kingsley's attack.

Kingsley has been remembered as an apostle of "muscular Christianity," for he believed that a Christianity that does not minister to the poor and the unfortunate is not a true Christianity. He–not Karl Marx, who took the metaphor from Kingsley–first spoke of religion as an opiate of the people, and he fought against a Christianity that led to complacency and inactivity. He opposed Neoplatonism with its gnostic tendencies, and he raised serious questions about Ralph Waldo Emerson's "pseudo-spiritualism." Kingsley wanted a faith with substance, and he felt such a robust faith could be reconciled with the findings of science.

Charles Kingsley can be considered a competent novelist, an engaging writer of sermons, and the author of a significant work of fantasy. His lively engagement with the issues of his day will make his life and ideas of interest even as his writings are read with declining frequency.

Letters:

Charles Kingsley: Letters and Memories of His Life, 2 volumes, edited by Frances Kingsley (London: King, 1877);

American Notes: Letters from a Lecture Tour, 1874, edited by R. B. Martin (Princeton, N.J.: Princeton University Library, 1958).

Bibliographies:

Robert A. Campbell, "Charles Kingsley: A Bibliography of Secondary Studies," *Bulletin of Bibliography,* 33 (1976): 78–91, 104, 127–130;

Styron Harris, *Charles Kingsley: A Reference Guide* (Boston: G. K. Hall, 1981).

Biographies:

Stanley E. Baldwin, *Charles Kingsley* (Ithaca, N.Y.: Cornell University Press, 1934; London: Humphry Milford, 1934);

Margaret F. Thorp, *Charles Kingsley, 1819–1875* (Princeton, N.J.: Princeton University Press, 1937; London: Humphry Milford, 1937);

Una Pope-Hennessy, *Canon Charles Kingsley: A Biography* (London: Chatto & Windus, 1948; New York: Macmillan, 1949);

Robert B. Martin, *The Dust of Combat: A Life of Charles Kingsley* (London: Faber & Faber, 1959; New York: Norton, 1960);

Susan Chitty, *The Beast and the Monk: A Life of Charles Kingsley* (London: Hodder & Stoughton, 1975; New York: Mason, Charter, 1975);

Brenda Colloms, *Charles Kingsley* (London: Constable, 1975; New York: Barnes & Noble, 1975).

References:

Dorothy Coleman, "Rabelais and 'The Water Babies,'" *Modern Language Review,* 66 (1971): 511–521;

Allan J. Hartley, *The Novels of Charles Kingsley: A Christian Social Interpretation* (Folkestone: Hour-Glass Press, 1977);

Arthur Johnson, "'The Water-Babies': Kingsley's Debt to Darwin," *English,* 12 (1959): 215–219;

Q. D. Leavis, "The Water-Babies," *Children's Literature in Education,* 23 (Winter 1976): 155–163;

Colin N. Manlove, "Charles Kingsley and 'The Water-Babies,'" in his *Modern Fantasy: Five Studies* (Cambridge & New York: Cambridge University Press, 1975), pp. 13–54;

Stephen Prickett, "Adults in Allegory Land," in his *Victorian Fantasy* (Hassocks: Harvester, 1979; Bloomington: Indiana University Press, 1979), pp. 150–173;

Larry K. Uffelman, *Charles Kingsley* (Boston: Twayne, 1979).

Papers:

Kingsley's papers are located at the British Library; the Cambridge University Library; the Princeton University Library; the Imperial College, London; the Pierpont Morgan Library, New York; the Berg Collection, New York Public Library; the Huntington Library, San Marino, California; and the University of Illinois, Urbana-Champaign.

Vernon Lee
(Violet Paget)

(14 October 1856 – 13 February 1935)

Amelia A. Rutledge
George Mason University

See also the Lee entries in *DLB 57: Victorian Prose Writers After 1867; DLB 153: Late-Victorian and Edwardian British Novelists, First Series; DLB 156: British Short-Fiction Writers, 1880–1914: The Romantic Tradition;* and *DLB 174: British Travel Writers, 1876–1909.*

Vernon Lee

BOOKS: *Studies of the Eighteenth Century in Italy* (London: Satchell, 1880; Chicago: McClurg, 1908);

Tuscan Fairy Tales, Taken Down from the Mouths of the People, anonymous (London: Satchell, 1880);

Belcaro: Being Essays on Sundry Aesthetical Questions (London: Satchell, 1881);

The Prince of the Hundred Soups: A Puppet-Show in Narrative (London: Unwin, 1883; New York: Lovell, 1886);

Ottilie: An Eighteenth-Century Idyll (London: Unwin, 1883); republished with *The Prince of the Hundred Soups* (New York: Harper, 1886);

The Countess of Albany (London: W. H. Allen, 1884; Boston: Roberts, 1884);

Euphorion: Being Studies of the Antique and the Medieval in the Renaissance (2 volumes, London: Unwin, 1884; Boston: Roberts, 1884; revised, 1 volume, London: Unwin, 1885; Boston: Roberts, n.d.);

Miss Brown: A Novel (3 volumes, Edinburgh & London: Blackwood, 1884; 1 volume, New York: Harper, 1885);

A Phantom Lover: A Fantastic Story (Edinburgh: Blackwood, 1886; Boston: Roberts, 1886);

Baldwin: Being Dialogues on Views and Aspirations (London: Unwin, 1886; Boston: Roberts, 1886);

Juvenilia: Being a Second Series of Essays on Sundry Aesthetical Questions (2 volumes, London: Unwin, 1887; 1 volume, Boston: Roberts, 1887);

Hauntings: Fantastic Stories (London: Heinemann, 1890; New York: Lovell, 1890);

Vanitas: Polite Stories (London: Heinemann, 1892; New York: Lovell, Coryell, 1892); republished and expanded as *Vanitas: Polite Stories, Including the Hitherto Unpublished Story Entitled "A Frivolous Conversation"* (London: John Lane, Bodley Head, 1911; New York: John Lane, 1911);

Althea: A Second Book of Dialogues on Aspirations and Duties (London: Osgood, McIlvaine, 1894); republished as *Althea: Dialogues in Aspirations and Duties* (London & New York: John Lane, 1910);

Renaissance Fancies and Studies: A Sequel to "Euphorion" (London: Smith, Elder, 1895; New York: Putnam / London: Smith, Elder, 1896);

Art and Life (East Aurora, N.Y.: Roycroft, 1896);
Limbo and Other Essays (London: Richards, 1897); republished and enlarged as *Limbo and Other Essays, to Which Is Now Added "Ariadne in Mantua"* (London: John Lane, Bodley Head / New York: John Lane, 1908);
Genius Loci: Notes on Places (London: Richards, 1899; London: John Lane, Bodley Head / New York: John Lane, 1908);
The Child in the Vatican (Portland, Maine: Mosher, 1900);
Chapelmaster Kriesler: A Study of Musical Romanticists (Portland, Maine: Mosher, 1901);
In Umbria (Portland, Maine: Mosher, 1901);
Ariadne in Mantua: A Romance in Five Acts (Oxford: Blackwell, 1903; Portland, Maine: Mosher, 1906);
Penelope Brandling: A Tale of the Welsh Coast in the Eighteenth Century (London: Unwin, 1903);
Hortus Vitae: Essays on the Gardening of Life (London: John Lane, Bodley Head / New York: John Lane, 1903);
Pope Jacynth and Other Fantastic Tales (London: Richards, 1904; London: John Lane, Bodley Head / New York: John Lane, 1907);
The Enchanted Woods, and Other Essays on the Genius of Places (London: John Lane, Bodley Head / New York: John Lane, 1905);
Sister Benvenuta and the Christ Child: An Eighteenth Century Legend (New York: Kennerley, 1905; London: Richards, 1906);
The Spirit of Rome: Leaves from a Diary (London: John Lane, Bodley Head / New York: John Lane, 1906);
The Sentimental Traveller: Notes on Places (London: John Lane, Bodley Head / New York: John Lane, 1908);
Gospels of Anarchy and Other Contemporary Studies (London & Leipzig: Unwin, 1908; New York: Brentano's / London: Unwin, 1909);
Laurus Nobilis: Chapters on Art and Life (London: John Lane, Bodley Head / New York: John Lane, 1909);
Beauty and Ugliness and Other Studies in Psychological Aesthetics, by Lee and Clementina Anstruther-Thomson (London: John Lane, Bodley Head / New York: John Lane, 1912);
Vital Lies: Studies of Some Varieties of Recent Obscurantism, 2 volumes (London: John Lane, Bodley Head / New York: John Lane / Toronto: Bell & Cockburn, 1912);
The Beautiful: An Introduction to Psychological Aesthetics (Cambridge: Cambridge University Press / New York: Putnam, 1913);
The Tower of the Mirrors, and Other Essays on the Spirit of Places (London: John Lane, Bodley Head / New York: John Lane / Toronto: Bell & Cockburn, 1914);
Louis Norbert: A Two-Fold Romance (London: John Lane, Bodley Head / New York: John Lane, 1914);
The Ballet of the Nations: A Present-day Morality (London: Chatto & Windus, 1915; New York: Putnam, 1915); enlarged as *Satan the Waster: A Philosophic War Trilogy* (London: John Lane, Bodley Head / New York: John Lane, 1920);
Peace with Honour: Controversial Notes on the Settlement (London: Union of Democratic Control, 1915);
The Handling of Words and Other Studies in Literary Psychology (London: John Lane, Bodley Head, 1923; New York: Dodd, Mead, 1923);
The Golden Keys and Other Essays on the Genius Loci (London: John Lane, Bodley Head, 1925; New York: Dodd, Mead, 1925);
Proteus; or, The Future of Intelligence (London: Kegan Paul, Trench, Trübner / New York: Dutton, 1925);
The Poet's Eye: Notes on Some Differences between Verse and Prose (London: Hogarth Press, 1926; Folcroft, Pa.: Folcroft Library, 1974);
For Maurice: Five Unlikely Stories (London: John Lane, Bodley Head, 1927; New York: Arno, 1976);
Music and Its Lovers: An Empirical Study of Emotional and Imaginative Responses to Music (London: Allen & Unwin, 1932; New York: Dutton, 1933).

Collections: *A Vernon Lee Anthology: Selections from the Earlier Works Made by Irene Cooper Willis,* with an explanatory note by Lee (London: John Lane, Bodley Head, 1929; Folcroft, Pa.: Folcroft Library, 1977);
Etudes et Réflexions sur l'Art, edited by Berthe Noufflard (Paris: Editions Corrêa, 1938);
The Snake Lady, and Other Stories, edited by Horace Gregory (New York: Grove, 1954);
Supernatural Tales: Excursions into Fantasy, introduction by Irene Cooper Willis (London: Owen, 1955); republished as *The Virgin of the Seven Daggers* (London: Transworld, 1962);
Pope Jacynth and More Supernatural Tales: Excursions into Fantasy (London: Owen, 1956).

OTHER: "Le rôle de l'élément moteur dans la perception esthétique visuelle," by Lee and Clementina Anstruther-Thomson, *Papers of the IV International Congress of Psychology, Paris, 1900* (Paris: Alcan, 1901), pp. 468–477;
The Life of Saint Mary Magdalen, translated from the Italian of an unknown fourteenth-century

writer by Valentina Hawtrey, introduction by Lee (London: John Lane, 1904; New York: John Lane, 1904), pp. i-xxi;

R. W. Sémon, *Mnemic Psychology,* translated by Bella Duffy, introduction by Lee (London: Allen & Unwin, 1923), pp. 11–53;

Anstruther-Thomson, *Art and Man: Essays and Fragments,* edited, with an introduction, by Lee (London: John Lane, 1924);

"John Singer Sargent. In Memoriam," in *John Sargent,* by Evan Charteris (London: Heinemann, 1927), pp. 233–255;

Irene Forbes-Mosse, *Don Juan's Daughters,* preface by Lee (London: John Lane, 1930), pp. vii–xiv.

SELECTED PERIODICAL PUBLICATIONS–UNCOLLECTED: "Les aventures d'une pièce de monnaie," as Mlle. V. P., *La Famille* (Lausanne), no. 10 (May 1870): 233–237; no. 12 (June 1870): 268–271; no. 14 (July 1870): 327–334;

"Tuscan Peasant Plays," *Fraser's Magazine,* old series 95/new series 15 (February 1877): 224–234;

"The Art of Singing," *British Quarterly Review,* 72 (October 1880): 318–342;

"The Influence of the Italian Renaissance on the Elizabethan Stage," *British Quarterly,* 75 (1882): 295–323;

"Vivisection: An Evolutionist to Evolutionists," *Contemporary Review,* 41 (May 1882): 788–811;

"Impersonality and Evolution in Music," *Contemporary Review,* 42 (December 1882): 840–858;

"The Transformations of Chivalric Poetry," *National Review,* 2 (December 1883): 341–353;

"Amour Dure: Passages from the Diary of Spiridion Trepka," *Murray's Magazine,* 1 (1887): 49–65, 188–199; reprinted in *Hauntings: Fantastic Stories* (London: Heinemann, 1890; New York: Lovell, 1890);

"Voix Maudite," *Les Lettres et Les Arts* (August 1887): 125–153; reprinted in *Hauntings: Fantastic Stories;*

"A Worldly Woman (Part 1)," *Contemporary Review,* 58 (October 1890): 520–541;

"A Worldly Woman (Part 2)," *Contemporary Review,* 58 (November 1890): 693–711;

"The Moral Teaching of Zola," *Contemporary Review,* 63 (February 1893): 196–212;

Review of *Florentine Painters of the Renaissance,* by Bernard Berenson, *Mind,* new series 2 (1896): 270–272;

"The Sense of Nationality," *Nation* (London), 12 October 1912, pp. 96–98;

"Lessons of History," *Nation* (London), 18 January 1913, pp. 662–663;

"The Heart of a Neutral," *Atlantic Monthly,* 116 (November 1915): 687;

"En renouant la correspondance avec une amie ex-ennemie," *La Révue de Genève,* no. 14 (August 1921): 174–180;

"Dionysus in the Euganean Hills. W. H. Pater in Memoriam," *Contemporary Review,* 120 (September 1921): 346–353.

Violet Paget, who took the pen name of Vernon Lee, was born to English expatriate parents on 14 October 1856 in Boulogne-sur-Mer, France. She was recognized in her time as one of the most impressive intellectuals of Victorian England, noted for her wide-ranging interests–from art and music to contemporary political and scientific theory–and for her brilliant, although at times extremely acerbic, conversation. Henry James, who avoided her after being satirized in a thinly disguised portrait in her short story "Lady Tal" (published in *Vanitas,* 1892), said she was "as dangerous and uncanny as she is intelligent, which is saying a great deal." She was mentioned, in a favorable comparison with John Ruskin, in Robert Browning's poem "Asolando" (1889), and a list of a few of her correspondents suggests the range of her contacts: Bernard Berenson, Henry and William James, Walter Pater, Sarah Orne Jewett, Aldous Huxley, Mrs. Humphry Ward, and H. G. Wells.

Although she is known primarily for her studies in art history and aesthetic theory and for a large body of travel writings, throughout her long writing career Lee produced a small but significant body of fantastic writing, ranging from traditional fairy tales to ghost stories. That her fiction is not well-known is in part because of shifts in her own focus of activity that lessened general interest in her work.

The daughter of Henry Hippolyte Ferguson Paget and Matilda Lee-Ferguson (née Adams), Violet was an only child; Lee's elder brother, Eugene Lee-Hamilton, was the child of her mother's first marriage. The Paget family traveled widely in France and Germany and finally settled in Italy. Lee's father had neglible influence on her upbringing and education, which was managed by a series of governesses and by her mother, who early recognized her daughter's precocity. Lee said of her childhood that she was educated for a literary career the way other girls were educated for marriage and domesticity. Another strong influence, this time for Vernon Lee the travel writer, was a strong sense of the genius loci, the "spirit of place," inspired by the mother of her childhood friend, the artist John Singer Sargent.

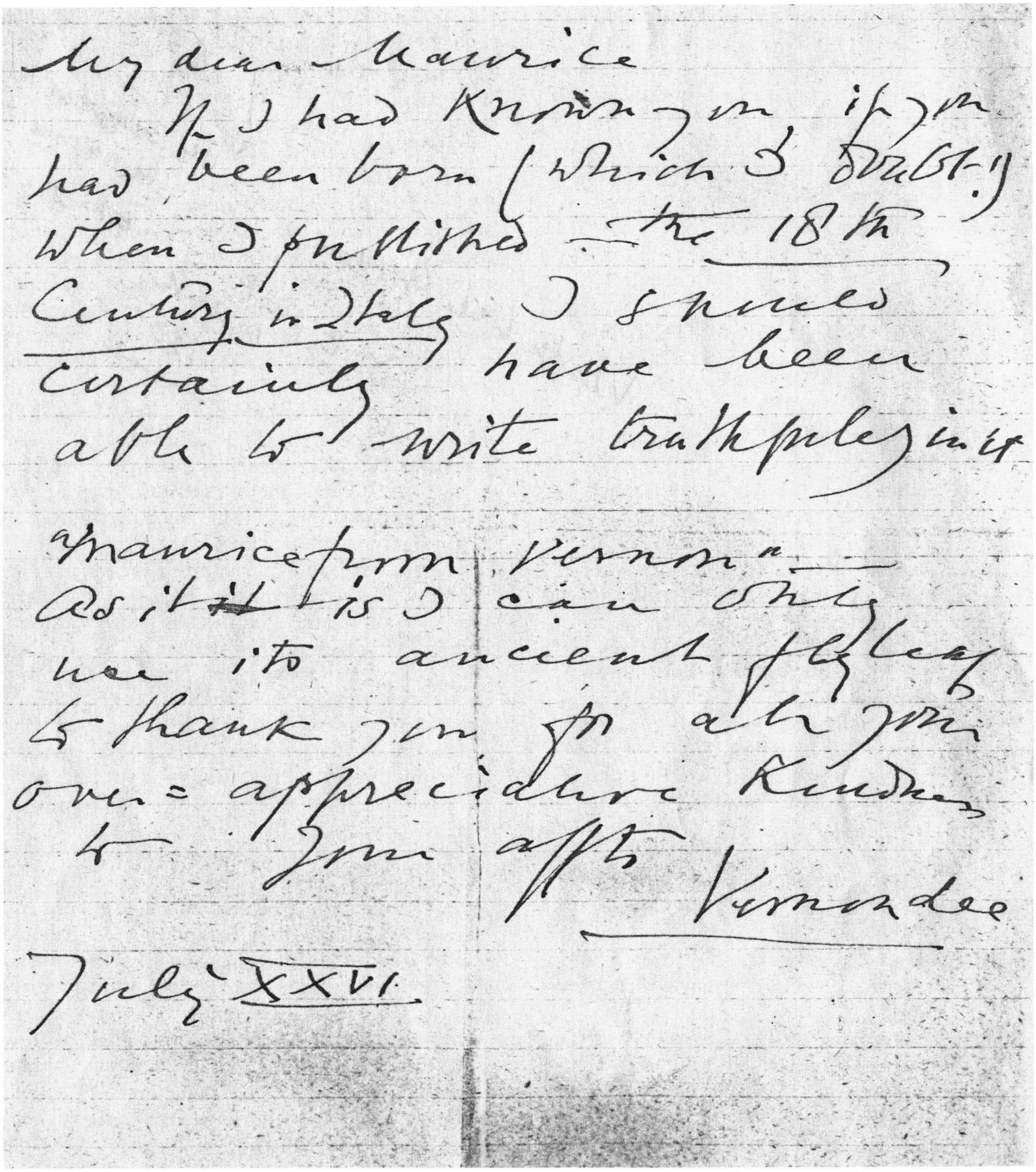

My dear Maurice

If I had known you if you had been born (which I doubt!) when I published the 18th Century in Italy I should certainly have been able to write truthfully in it

"Maurice from Vernon."

As it is I can only use its ancient flyleaf to thank you for all your over = appreciative kindness to

Your affte

Vernon Lee

July XXVI

Lee's inscription to Maurice Baring, inserted into his copy of her first book, Studies of the Eighteenth Century in Italy *(The Colbeck Collection, Special Collections Division, University of British Columbia Library)*

Lee's first published work appeared in a Lausanne periodical when she was thirteen. This short story, "Les aventures d'une pièce de monnaie" (The Adventures of a Coin), traced the history of the coin from Roman times to the present day and showed the attention to historical detail, as well as the narrative fluency, that became hallmarks of Lee's work. She early concentrated her studies on the arts of the eighteenth century, especially sculpture and painting, publishing in 1880, to positive critical response (albeit to "*Mr.* Lee"), her first book, *Studies of the Eighteenth Century in Italy.*

That same year saw her first publication in the literature of fantasy, a collection of fairy tales, *Tuscan Fairy Tales, Taken Down from the Mouths of the People.* She claimed inspiration from the authoritative collections of Domenico Giuseppe Bernoni and Laura Gonzenbach, but it seems likely that the claim she makes in the title can be taken at face value. In a detailed introduction describing her methods, she notes that she has been careful to preserve the unconventional narrative style of her informers. The tales range from purely local folktales, such as "The Beautiful Glutton," to retellings of European classic fairy tales, such as "The Three Cauliflowers," a Bluebeard analog, or "The Glass Coffin," a Snow White story. The collection attests not only to her early interest in the fantastic but also to the impulse to render as fictive narrative what her studies contributed to her historical and theoretical work.

A similar example of this impulse toward fictive embodiment can be found in *The Prince of the Hundred Soups* (1883). Lee had written about Carlo Gozzi and the commedia dell'arte in *Studies of the Eighteenth Century in Italy*, and she uses standard commedia dell'arte characters to tell the story of the triumph of a famed opera singer over plotters who attempt to overthrow the "prince" of the title. The story forms a link between the fairy tales of the 1880 collection and later stories in which opera was a focus.

The stories in *Hauntings* (1890), her first collection of adult fantasy, all involve artists or, in "Amour Dure," a scholar, obsessed with–literally haunted by–their subjects. Critics praised the stories for their Gothic elements and for her intense evocation of place. The works' real power, however, lies in Lee's probing of the psychology of the creative unconscious. One story, "A Wicked Voice," first published as "Voix Maudite" in *Les Lettres et Les Arts* (1887), has been the focus of several recent critical studies; it is especially interesting because its basic plot is a story Lee reworked from her first adult fantasy, "A Culture Ghost," (*Fraser's Magazine,* 1881), and later republished in 1927 as "Winthrop's Adventure" in a collection dedicated to Maurice Baring. In the introduction to that collection she discusses the hold that the story had over her imagination. The protagonist of "A Wicked Voice," Magnus, a self-styled Wagnerian composer, reviles the musical practices of eighteenth-century singers in the presence of a portrait of one of the most famous, the castrato Zaffiro. He soon begins to experience aural hallucinations of the singer and finds himself capable of composing only in the fashion of the eighteenth century.

Lee's literary production tended to follow whatever involved her intellectually at the time; not only had she been intensely involved in research in the art of eighteenth-century Italy, but she had also, from 1872, been fascinated by a portrait of the singer Farinelli, one of the greatest of the eighteenth-century castrati. She and her friend Sargent had studied the protrait together, and she had written the first version of the story at his suggestion. Further, both "A Wicked Voice" and "Dionea," another story in *Hauntings,* reflect another theme in her writing–her conviction of the superiority of the classical and neoclassical to contemporary art; she produced several essays on this subject throughout the 1880s and returned to it several times in later years. In addition to the power of obssession, "A Wicked Voice" and "Dionea" suggest the revenge of the earlier period upon the later in the sufferings of the protagonists. In the same way Spiridion Trepka, the hapless protagonist of "Amour Dure," (the title contains a pun on love that is both lasting as well as harsh) becomes the victim of his passion for the shade of a Renaissance woman famed for her cruelty to her lovers.

The sculptor Waldemar in "Dionea" finds that despite his disdain for the female form, he becomes obsessed with the model reluctantly accepted, at his wife's behest, for a sculpture of Aphrodite. The model, a shipwrecked foundling named Dionea, as was the goddess Aphrodite's mother, has during her upbringing on the island caused several instances of erotic mishap. Waldemar's story ends in disaster, and Dionea is last seen by a village boy, being rowed away in a Greek boat and surrounded by doves as she had always been on the island.

For readers of Lee's nonfiction works, the strong erotic element of her fantastic writing may seem surprising. She was concerned, as were many of her contemporaries, with the moral content of art. Noted for her personal austerity, her position with regard to the freedoms of the "aesthetic" movement was evident in *Miss Brown* (1884), her first novel. She condemned contemporary French fiction for its sexual content, and she considered Richard

Portrait of Lee by John Singer Sargent (Tate Gallery, London)

Wagner's music, especially his *Tristan und Isolde* (1865), morally debilitating. However, the mixed fascination and horror with which the erotic is presented in her stories suggests that sexuality was an area of conflict for Lee, whom Vineta Colby calls a "puritan aesthete." What Lee the moralist disdained was nevertheless articulated in the work of Lee the fantasist.

The story "Oke of Okehurst," also in *Hauntings,* is a reflection of Lee's ambivalence toward "art for art's sake." The story's original title, "The Phantom Lover," reflects the fatal legend that the wife, Mrs. Oke, insists on acting out, with disastrous results. However, the story's true focus is its depiction of the shallow portrait painter who treats the charged domestic situation as pure spectacle.

Lee continued to produce stories based on art and artists in two works of the early 1890s, "Pictor Sacrilegus" and "Ravenna and Her Ghosts." Published in *Renaissance Fancies and Studies: A Sequel to "Euphorion"* (1895), "Pictor Sacrilegus" is the story of a Renaissance painter obsessed with the newly discovered superiority of the classical age's depiction of human form. Frustrated in his attempts to achieve these effects, he reluctantly engages in mock-pagan sacrifices at the behest of his idle humanist patron and suffers the penalties for heresy. "Ravenna and Her Ghosts," published in *Limbo and Other Essays* (1897), includes Lee's retelling of the eighth tale of the fifth day of Giovanni Boccaccio's *Decameron* (1351), in which a young man exploits the dramatic potential of an apparition, staging a pageant to force the hand of his haughty beloved. Both of these stories were incorporated into historical essays, a format that hampers their literary effectiveness.

Although Lee was an agnostic, she had an appreciation for the feeling expressed in religious observances and for depictions of the Madonna, for whom, in her preface to *For Maurice* (1927), she admitted a fondness and admiration. One of the stories she included in the collection *Vanitas,* "Madame Krasinka," suggests that she sympathized with religious sincerity even if she was not a believer herself. The title character, a nun with a mysterious past, is recognized as the former Madame Krasinka, a wealthy young woman rescued from a suicide attempt by the shade of a female vagrant.

Lee did not share the nineteenth century's fascination with its nostalgic reconstruction of the medieval period, but she recognized in medieval art some of the age's religious fervor. Her collection *Pope Jacynth and Other Fantastic Tales* (1904) includes two saint's tales, one of which, the title story, is fantastic only in the presence of miracles; the other, "St. Eudaemon and his Orange-Tree," is an interesting contrast to the Prosper Mérimée novel *The Venus of Ille* (1905). In Lee's story the saint persuades a statue of Venus to give back a ring she holds unrightfully and converts her to benevolence, symbolized by the orange tree that grows from her former shrine. One other story, "The Lady and Death," is inspired by an Albrecht Dürer portrait and is structured after the story of Alcestis and Admetus; although, like many of Lee's stories, it focuses on the power of art, it is less successful than "Dionea" or "St. Eudaemon."

"Prince Alberic and the Snake Lady," a story Lee had originally published in the *Yellow Book* in 1896, is the most complex and intriguing story in *Pope Jacynth.* A version of the Melusine legend, it tells of a young man, Alberic, who is the heir of his negligent, self-indulgent grandfather and the victim of plots by his grandfather's staff. He is consoled by a fairy godmother who first appears as a green garden snake; he resists attempts to force him to wed; and he dies insane after the snake is destroyed. After his death the body of a woman is found in his prison.

Once again there is a an artifact, this time a tapestry depicting the legend of the snake-woman, but it is a minor detail. On the other hand, Lee's blending of godmother, lover, and serpent brings an intriguing ambivalence into the story. The godmother is reminiscent of the erotic, serpentine females in "Der goldene Topf" ("The Golden Pot," 1814), by E. T. A. Hoffmann, but here, unlike most of her other stories, the erotic element, although fatal, is not tinged with vengeance.

Lee's last "miracle story" was *Sister Benvenuta and the Christ Child* (1905), a blend of religious fervor, occultism, and social realism–the disposal of unmarriageable daughters–leavened with irony. Benvenuta, slightly lame and the youngest daughter of a noble family, is sent to a convent, where she becomes extremely devoted to a statue of the Christ Child. Concerned that she is not capable of making appropriately elaborate garments for the statue, she makes a pact with the devil for the necessary skill but is rescued for her devotion. Benvenuta's sincerity is amusingly contrasted with the shallowness of the other noble conventual inmates, and Lee's choice of narrator, a novice near Benvenuta in age, provides an appropriately naive point of view.

In the 1900s Lee's interests began to shift toward aesthetic theory. Her major work in aesthetics was produced in collaboration with Clementina Anstruther-Thomson and is acknowledged today as seminal, despite the inadequacy of its conceptual grounding in the relationships between physiology, psychology, and aesthetic response. Although she continued to produce travel sketches, her general popularity as a writer declined. Lee also, with the onset of World War I, became outspoken in her pacifism, the result of her cosmopolitan upbringing and her keenly analytical intellect, and this alienated friends as well as readers.

Her devotion to pacifism led to *The Ballet of the Nations* (1915), which was expanded as *Satan the Waster* (1920); the latter title refers to Satan as the Waster of All Virtues. For these works Lee chose the marginally fantastic genre of the allegorical drama, casting her polemic as a stage production produced by Satan and his sycophant, Clio, the Muse of history. Although the language is relentlessly satiric, Lee's portraits–of the widow Fear with her twins, Suspicion and Panic, or, even more strongly, of her character Blind Heroism–are sympathetically presented. Despite a predictably negative public response, the work was praised by Bernard Shaw in a lengthy review.

The 1927 collection *For Maurice: Five Unlikely Stories* includes one story, "The Doll," that is not a fantasy, although Lee returns to the theme of obsession–this time of a widower with a doll modeled after his dead wife. To "Winthrop's Adventure" she added three other stories, "The Gods and the Ritter Tannhûser," "Marsyas in Flanders," and the previously published "The Virgin of the Seven Daggers." "Marsyas" is framed as the explanation for an anachronistic effigy on a crucifix in a Flemish church. The original, washed ashore during the twelfth century, had been suppressed, but not without rumors of strange events associated with it. When discovered, it is found to be a pagan figure, the satyr Marsyas, bound in preparation for punishment in a posture similar to that of some crucifixions. In "Marsyas" Lee replays, with originality, the resistance of paganism to the new dispensation.

The same idea is worked into her one purely humorous tale, "The Gods and the Ritter Tannhûser." In her writings Lee never resisted an opportunity to deride Wagner and Wagnerians, to her the epitome of departure from the standards of eighteenth-century musical theory and practice. Here, however, in a story she claims to have composed while at a performance of *Tannhäuser,* she tells the story from the standpoint of the "retired" classical gods. Athena and Apollo agree to accompany Tannhûser (Lee insists on the more "authentic" spelling) to the contest at the Wartburg–Apollo to compete and Athena out of pure intellectual curiosity. Although the denouement follows the plot familiar from Wagner's opera, the humor lies in the vengeance exacted by the gods on the benighted German artists and scholars at the contest, and the staff blossoms as myrtle, the plant sacred to Aphrodite.

"The Virgin of the Seven Daggers" is Vernon Lee's self-conscious contribution to the legend of Don Juan–neither, she asserts, Mozartian nor Byronic. Her Don is saved from damnation by his devotion to the Madonna represented, in a statue in Granada, with her bodice pierced with seven daggers. His oath to remain faithful to her is fulfilled when he is beheaded, in an otherworld adventure, rather than forswear his patron in favor of a Moorish princess; the most effective passage in the story is his slow realization that what he took to be a dream–his execution–is actuality.

Characteristically, and somewhat strangely, Lee justifies her use of such a grim depicition of the Madonna with a reference to her artistic preference for the more benign Northern European or Italian portraits of the Virgin. Despite this preference, the "gloomy" Madonna of her story was the only one by means of whom redemption might come even for Don Juan, as testimony to a vigor of belief lost, in Lee's opinion, since the Counter Reformation. Art and sincerity redeem Don Juan, and it was Lee's implicit belief that therein lay, perhaps, redemption for society as well.

Vernon Lee died in 1935 in San Gervasio, Italy. Her work–dense, idiosyncratic, but always committed and vigorous–is at present being rediscovered and reconsidered. The ambivalence and conflicts of her own psyche, bodied forth in her fantasy, contribute the power and energy that make them engaging fiction even when erudition threatens to outweigh literary art. By the same token, her literary art is a vital index to the ambivalences and conflicts of the best artistic minds of her age.

Letters:

Vernon Lee's Letters, edited by Irene Cooper-Willis (London: Privately printed, 1937).

Bibliographies:

Phyllis F. Mannocchi, "Vernon Lee: A Reintroduction and Primary Bibliography," *English Literature in Transition,* 26, no. 4 (1983): 231–267;

Carl Markgraf, "Vernon Lee: A Commentary and an Annotated Bibliography of Writing about Her," *English Literature in Transition,* 26, no.4 (1983): 268–312.

Biography:

Peter Gunn, *Vernon Lee/Violet Paget, 1856–1937* (London: Oxford University Press, 1964).

References:

Carlo Caballero, " 'A Wicked Voice': On Vernon Lee, Wagner, and the Effects of Music, " *Victorian Studies,* 35 (Summer 1992): 385–408;

Peter G. Christensen, "The Burden of History in Vernon Lee's Ghost Story 'Amour Dure,'" *Studies in the Humanities,* 16, no. 1 (1989): 33–43;

Christensen, "'A Wicked Voice': Vernon Lee's Artist Parable," *Lamar Journal of the Humanities,* 15, no. 2 (1989): 3–15;

Vineta Colby, "The Puritan Esthete: Vernon Lee," in her *The Singular Anomaly: Women Novelists of the Nineteenth Century* (New York: New York University Press / London: University of London, 1970);

Horace Gregory, "The Romantic Inventions of Vernon Lee," introduction to *The Snake Lady, and Other Stories,* by Vernon Lee, edited by Gregory (New York: Grove, 1954).

Papers:

Vernon Lee's papers and unpublished correspondence are held in the Special Collections Department of the Library at Colby College, Waterville, Maine.

Joseph Sheridan Le Fanu

(28 August 1814 – 10 February 1873)

Gary William Crawford

See also the Le Fanu entries in *DLB 21: Victorian Novelists Before 1885; DLB 70: British Mystery Writers, 1860–1919;* and *DLB 159: British Short Fiction Writers, 1800–1880.*

BOOKS: *The Cock and Anchor: Being a Chronicle of Old Dublin City,* anonymous (3 volumes, Dublin: Curry / London: Longmans / Edinburgh: Fraser, 1845; 1 volume, New York: Colyer, 1848); revised as *Morley Court* (London: Chapman & Hall, 1873); republished as *The Cock and the Anchor* (London: Downey, 1895);

The Fortunes of Colonel Torlogh O'Brien: A Tale of the Wars of King James, anonymous (Dublin: McGlashan / London: Orr, 1847; New York: Colyer, 1847);

Ghost Stories and Tales of Mystery, anonymous (Dublin: McGlashan / London: Orr, 1851);

The House by the Churchyard (3 volumes, London: Tinsley, 1863; 1 volume, New York: Carleton, 1866);

Wylder's Hand (3 volumes, London: Bentley, 1864; 1 volume, New York: Carleton, 1865);

Uncle Silas: A Tale of Bartran-Haugh (3 volumes, London: Bentley, 1864; 1 volume, New York: Munro, 1878);

Guy Deverell (3 volumes, London: Bentley, 1865; 1 volume, New York: Harper, 1866);

All in the Dark (2 volumes, London: Bentley, 1866; 1 volume, New York: Harper, 1866);

The Tenants of Malory (3 volumes, London: Tinsley, 1867; 1 volume, New York: Harper, 1867);

A Lost Name, 3 volumes (London: Bentley, 1868);

Haunted Lives, 3 volumes (London: Tinsley, 1868);

The Wyvern Mystery, 3 volumes (London: Tinsley, 1869);

Checkmate (3 volumes, London: Hurst & Blackett, 1871; 1 volume, Philadelphia: Evans, Stoddart, 1871);

The Rose and the Key, 3 volumes (London: Chapman & Hall, 1871);

Chronicles of Golden Friars, 3 volumes (London: Bentley, 1871);

A portrait of Joseph Sheridan Le Fanu by his son Brinsley (Hulton Getty Picture Collection Limited)

In a Glass Darkly, 3 volumes (London: Bentley, 1872);

Willing to Die, 3 volumes (London: Hurst & Blackett, 1873);

The Purcell Papers, 3 volumes (London: Bentley, 1880);

The Poems of Joseph Sheridan Le Fanu, edited by Alfred Perceval Graves (London: Downey, 1896).

Collections: *Madame Crowl's Ghost and Other Tales of Mystery,* edited by M. R. James (London: Bell, 1923);

Best Ghost Stories of J. S. Le Fanu (New York: Dover, 1964);

Ghost Stories and Mysteries (New York: Dover, 1975);

The Collected Works of Joseph Sheridan Le Fanu, 52 volumes (New York: Arno, 1977).

Joseph Sheridan Le Fanu's contribution to supernatural horror fiction is signally important because he was virtually the first writer to produce ghost stories in Britain. Emerging shortly after the heyday of the Gothic novel, Le Fanu's work offered a much more personal, intimate, and realistic use of Gothicism. For the most part avoiding the excesses of the Gothic novel, he transformed the Gothic props, such as castles, dizzying landscapes, and villains, into subtle, haunting, quiet tales of terror utilizing the techniques of realistic fiction. For this reason his prominent position in ghost fiction cannot be underestimated.

The opportunity Le Fanu had to transform the Gothic arose from his position in an Anglo-Irish ascendant class. He was descended from French Huguenots who fled France upon the revocation of the Edict of Nantes, which under Louis XIV exiled all Protestants from the country. On his father's side, he was the grandnephew of the famous dramatist Richard Brinsley Sheridan. His father, Thomas Philip Le Fanu, was a Protestant clergyman in the Church of Ireland who married Emma Lucretia Dobbin. Their first child, Joseph Thomas Sheridan Le Fanu, was born in Dublin. A year later he went with his parents to live at the Royal Hibernian Military School, where his father was chaplain. Here he and his younger brother and sister witnessed the maneuvers of the Irish military and absorbed much of the character of nineteenth-century Dublin. Le Fanu set most of his early fiction in Ireland, and the particular sensibility of the Irish crept into his early ghostly tales.

Irish folktales influenced his writing as well, and Le Fanu heard these stories from the Irish peasants at Abington, County Limerick, where his father was appointed rector in 1826. He was educated by a private tutor and later took honors in classics at Trinity College, Dublin. He went to London to study law at King's Inns and was called to the Irish bar but decided not to practice. Instead, he entered journalism and purchased a leading Irish Protestant newspaper, *The Warder.* He acquired part interest in other Irish newspapers, and his newspaper work dominated most of his life.

However, he turned to the writing of fiction in 1838 with "The Ghost and the Bone-Setter," the first of a series of stories later published in book form as *The Purcell Papers* (1880). These stories purport to be extracts from the papers of Father Francis Purcell, an Irish Catholic priest. The tales, like many linked short stories in the literary magazines of the 1840s, contain a wide variety of themes, but the predominant theme is supernatural horror. "The Ghost and the Bone-Setter" is related by a man whose father, a servant of a prominent squire, experienced his master's return from the grave. Here the squire returns from the dead out of his own portrait. This timeworn Gothic prop is domesticated into a more realistic human, believable level when the subtle ambiguities of the supernatural are explored. Ultimately, the servant's experience of his master's ghost is explained away as a dream. However, such use of the supernatural sets the tone for all of Le Fanu's later tales in which the apparently supernatural events are poised on a dual interpretation–one irrational and one rational.

Le Fanu's lifelong fascination with the spectral stemmed from his reading of the Gothic novel and his frequent use of folklore material. Related to this is the second entry from the papers of Father Purcell, "The Fortunes of Robert Ardagh" (*The Dublin University Magazine,* 1838), in which the story is told in two separate versions, one as a legend, the other as a personal account. The structure of the tale presents two sides of the same event, one fantastic and one more realistic and believable. In this tale Sir Robert Ardagh comes under the power of an evil valet, Jacques, who is described as Satan himself. Eventually Ardagh is killed under mysterious circumstances, which are never fully explained. Here the ambiguous nature of the supernatural–delusion or reality–is explored. Similarly, in "The Drunkard's Dream" (*The Dublin University Magazine,* 1838) Father Purcell is called to the bed of a dying drunkard, who relates to Purcell his vision of hell. The vision may or may not be a dream.

The most celebrated tale from *The Purcell Papers* is "Schalken the Painter" (*The Dublin University Magazine,* 1839), about a young woman, Rose, who marries a ghoulish emissary of Satan by arrangement. Schalken, who loves Rose, sees his love in hell with the corpse of the ghoul and paints a picture based upon his vision. This realistic study of personal terror has been regarded as a new use of the supernatural in fiction, taking the Gothic and giving it a more personal and realistic framework.

In the 1840s Le Fanu turned to the Irish historical novel with the publication of *The Cock and Anchor* (1845) and *The Fortunes of Colonel Torlogh O'Brien* (1847), but bad reviews discouraged him, and he realized that his strength lay in the creation of mystery and supernatural horror. In 1843 Le Fanu met and married Susanna Bennett, the daughter of a promi-

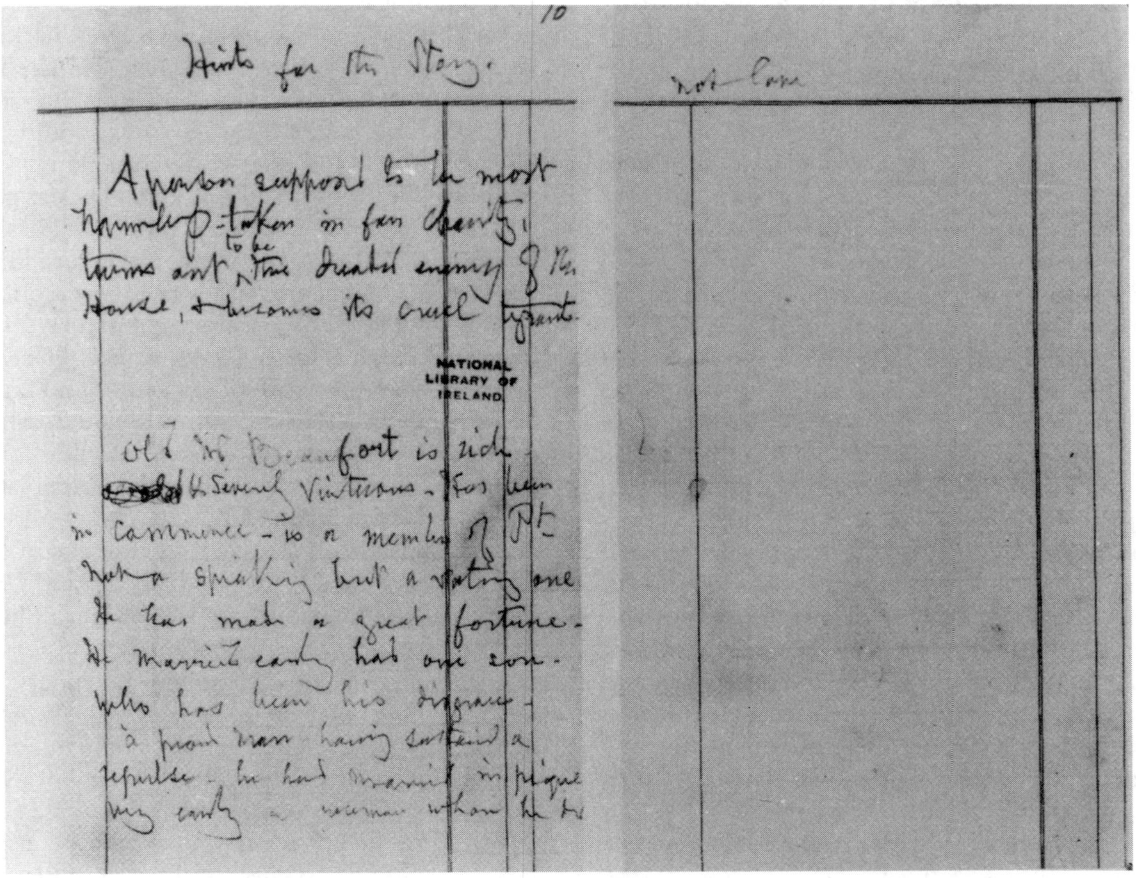
10

Hints for the Story.

not come

A person supposed to be most harmless—taken in from charity, turns out to be the deadliest enemy of the House, & becomes its cruel tyrant.

NATIONAL LIBRARY OF IRELAND

old Mr Beaufort is rich & severely virtuous. Has been in commerce—is a member of Pt—not a speaking but a voting one. He has made a great fortune. He married early had one son—who has been his disgrace—a proud man having suffered a repulse—he had married in pique very early a woman whom he

Page from one of Le Fanu's notebooks (National Library of Ireland)

nent Dublin lawyer, and continued his writing for a time. In 1847 he penned one of the finest ghost stories of the middle of the nineteenth century, "The Watcher" (*The Dublin University Magazine*).

The story concerns Capt. James Barton, a disbeliever in the supernatural and in God, who is haunted by a small man who may or may not be the ghost of Jennings, one of his subordinates whom he ordered punished by death years earlier. The multiple narrators of the story and the distorted quality of what really happens to Jennings is expertly handled by Le Fanu. The final scene suggests that the ghost takes the shape of an owl. The lingering uncertainty about what happens to Jennings invokes a genuine frisson. The tale is remarkably sophisticated for its day, and it presages the elliptical ghost stories of Ramsey Campbell in the twentieth century.

In the three tales in "Ghost Stories of Chapelizod" (*The Dublin University Magazine,* 1851) Le Fanu draws on the oral tradition to produce the short pieces "The Village Bully," "The Sexton's Adventure," and "The Spectre Lovers." Le Fanu was noted for his knowledge of Irish legends and folklore, and many of his stories draw on this wealth of material. "The Spectre Lovers" is the longest of the three and concerns a young Irishman and his vision of a time long past. Here, as in a later writer such as Campbell, the supernatural is given a vague sense of reality by suggestion—a mere hint of terror on the very edges of perception. "The Sexton's Adventure" and "The Village Bully" are both brief sections of "Ghost Stories of Chapelizod" and are written in Le Fanu's typical guilt and retribution mode. In a later tale, "An Account of Some Strange Disturbances in an Old House in Aungier Street," this suggestiveness is present once again. Two students rent an old haunted house and try to rationalize what appears to be supernatural: the ghost of a judge who committed suicide years earlier.

The last-mentioned tale marks the end of the first part of Le Fanu's writing career, because there is no writing by him from 1853 to 1861. During these years he was no doubt heavily involved with his business, but these years were also the most difficult in his personal life. His wife, with whom he had four children, began to exhibit hysterical symptoms, partly exacerbated by the death of her father. She died in 1858, and it is known from a notebook that he wrote shortly after her death that Le Fanu blamed himself for whatever unhappiness she suffered. He describes in great detail the morbidity and hysterical symptoms Susanna displayed, and after her death Le Fanu became more and more reclusive. In 1861 he purchased *The Dublin University Magazine* and set himself up as its editor. He published most of his later fiction in the magazine.

In these years Le Fanu, who was at one time the toast of Dublin society, was nicknamed "The Invisible Prince" because he carried on a secluded existence. Yet this withdrawal from society stimulated his imagination, and the years following 1861 were the most productive for his writing. In the twelve novels produced during this period he attempted to emulate such successful writers of "sensation fiction" as Wilkie Collins and Charles Reade. Most of Le Fanu's novels are mystery and sensation novels, the most famous being *Uncle Silas* (1864), which was influenced by Ann Radcliffe's *The Mysteries of Udolpho* (1794). While not a supernatural novel, there are mystical overtones, for the theological works of the eighteenth century scientist-turned-mystic Emanuel Swedenborg figure in the novel and in some of Le Fanu's other later fiction.

While there is no evidence that Le Fanu was a member of the New Jerusalem Church, as the church founded on Swedenborg's theology was called, he was no doubt impressed by the spectral and ghostly aspects of Swedenborg's works, and perhaps he wanted to believe that Susanna finally reached perfection in the spirit world she entered after death. Susanna's death devastated Le Fanu, and the later works show a preoccupation with female characters who are in some way victims. His male characters as well are tortured by an extraordinary sense of guilt, and suicide often figures in his novels and stories.

In the later works Le Fanu turned his attention to England because his publisher, Richard Bentley, insisted that he would only publish Le Fanu's works if he wrote them for an English audience. This had a profound impact on his later novels, and there is a general weakening and dilation in them. However, the short stories from this period, nearly all supernatural, are marked by a preoccupation with death, and Le Fanu's short fiction generally is better than his novels.

However, *Uncle Silas,* while not a supernatural work, is heavily laden with Swedenborgian mysticism and may be read as a symbolic and metaphorical romance that focuses on death. Maud Ruthyn, who lives with her widower father at Knowl, is left to the care of her Uncle Silas after her father dies. Silas is a profligate and suspected murderer, but it has never been proven that he killed a Mr. Charke at Silas's home some thirty years before. Maud's father, convinced that Silas is innocent, leaves a will that places Maud under the care of Silas should he die. He does, and Maud goes to Bartran-Haugh, where her uncle, her cousin Dudley, and a wicked governess, Madame de la Rougierre, plot to kill Maud for her money. The story is largely a romance in which the settings and characters serve a symbolic function. Knowl is the Swedenborgian natural life, and Bartram-Haugh is the other life of hell. Here Maud journeys after the death of her father and faces the horrors of threatened death. However, Maud is saved because Silas and his son, Dudley, kill Madame de la Rougierre by mistake. The mystery is solved–Silas did indeed kill Mr. Charke–and Maud escapes. The final passage of the novel could have been directly lifted from Swedenborg, and *Uncle Silas* crystallizes Le Fanu's preoccupations in his later fiction, especially the supernatural tales.

One finds a world of guilt in Le Fanu, of gradual darkening and gloom, in which places, objects, people, and even animals serve a symbolic function. This use of symbolism parallels Swedenborg's assertion that the Bible was written in a symbolic language. *Uncle Silas,* while not an overtly supernatural novel, is in its broadest outlines a supernatural drama, and Le Fanu's later ghost stories are much like the novel in their motives and use of setting and character.

In a story such as "Ultor de Lacy" (*The Dublin University Magazine,* 1861) Le Fanu takes a story he heard in his youth and gives it a subtly unnerving dimension when, in a Swedenborgian sense, a ghost returns from "the other life" nearly a century after its death to seek revenge on those who wronged him. Over and over again the primary motive behind Le Fanu's hauntings is guilt, as in "Wicked Captain Walshaw, of Wauling" (*The Dublin University Magazine,* 1864), in which a ghost lives in the wick of a candle used at the burial of his wife. Taking its impetus from Walshaw's own transgressions, his ghost returns to haunt his survivors. Similarly, in "Squire Toby's Will" (*Temple Bar,* 1868) a father dies and leaves his estate to his youngest son, who dies and reappears as a ghost in the form of a bulldog. Here, the animal symbolism common in Swedenborg's works takes on artistic dimensions in Le Fanu's fiction.

"The Haunted Baronet" (*Belgravia,* 1870), a reworking of an earlier tale, "The Fortunes of Sir Robert Ardagh," takes on a larger cast of characters and is influenced to a degree by Swedenborg's mystical-theological writings. A variation of the Faust legend, instead of having its protagonist face the repayment of his final bargain with his death, money is simply exchanged. However, the subtle twists and tangential scenes create a fantastic landscape of evil and sin. The central supernatural incident, the appearance of a ghostly hand from the waters of a lake, takes on the quality of an omen. Indeed, the novella is one of the strangest of Le Fanu's stories, for little of it is actually explained, and it leaves a deep impression of a fantastic realm.

In his later years Le Fanu became more deeply reclusive, frequently complaining of ill health to his friends (he had a heart condition that finally killed him), and a few weeks before he died he refused to see his old friend from his college days, the Irish novelist Charles Lever. But he wrote an extraordinary amount during this time, perhaps because he was freed from business obligations, having divested himself of *The Dublin University Magazine* and his newspaper interests. He thence devoted all of his time to writing.

In 1869 Le Fanu began composing the five tales that comprise his best-known collection of weird stories, *In a Glass Darkly* (1872). These strikingly modern stories are Le Fanu's most complex and sophisticated works. Heavily laden with Swedenborgian mysticism, the stories are unified by the same sources, techniques, and themes. They are linked as case histories from the files of Dr. Martin Hesselius, whose specialty is supernatural phenomena, and the character of Hesselius presages later psychic detectives in British fiction, such as Algernon Henry Blackwood's John Silence and William Hope Hodgson's Carnacki.

In a Glass Darkly is of particular interest because of its principal sources and Le Fanu's use of the structural device of multiple narratives. While critics have approached the most well-known stories of the volume, "Green Tea" and "Carmilla," from a psychological standpoint, the most telling approach to the tales is in terms of an analysis and determination of their structure, which conveys a rather complex irony. *In a Glass Darkly* is about the nature of humanity's relationship to the supernatural and to the unfathomable mysteries of its own existence. Le Fanu's attitude toward the supernatural is one of detached irony toward his occult sources and supernatural incidents. While Le Fanu appears to ascribe an objective reality to the supernatural, he is in fact implying that the supernatural has no rational validity whatsoever.

In "Green Tea," for instance, the Reverend Mr. Jennings is haunted to his death by what appears to be a spirit-demon in the form of a monkey. Yet Le Fanu implies that the monkey has no objective reality by offering several explanations. The appearance of the monkey may be merely a hallucination resulting from the narcotic effects of green tea; it may be, in the scientific-metaphysical jargon of Dr. Hesselius, who diagnoses Jennings's case, an opening of "the interior sense"; it may be one of the demons that Jennings reads about in the occult writings of Swedenborg; and, in psychological terms, it may be the symbolic representation of the Reverend Mr. Jennings's religious doubts. The point of all these conflicting explanations is that no one explanation is satisfactory.

In a Glass Darkly.

BY

J. SHERIDAN LE FANU,

AUTHOR OF "UNCLE SILAS," &C.

IN THREE VOLUMES.

VOL. I.

FIDE ET FIDUCIA
RB

LONDON:
R. BENTLEY & SON, NEW BURLINGTON STREET.
1872.
The right of translation is reserved.

Title page for Le Fanu's fourth book, in which Dr. Martin Hesselius investigates supernatural phenomena

Furthermore, in "Green Tea," as in each story in *In a Glass Darkly,* Le Fanu places these contradictory explanations for what could be termed supernatural phenomena within the narrative viewpoints of several characters. Each of these narrators and their informants contradicts the other and is thereby rendered unreliable. Le Fanu consciously employs multiple narration as the structural device that contributes to his ironic voice. The group of stories that comprise *In a Glass Darkly* is in itself a system of multiple narratives, each presenting different explanations for what could be termed the supernatural.

"Green Tea," written when Le Fanu's heart condition was gradually worsening, explores the tortured character of the Reverend Mr. Jennings. As the first story in the collection, it is a prime example of Le Fanu's ironic distance. The structure of the tale, which is essentially a system of multiple narratives, mirrors the structure of each subsequent tale and, in turn, the structure of the entire book as a system of multiple narratives through a group of linked short stories. The plot of "Green Tea" concerns an apparently pious and gentle unmarried clergyman. While studying ancient religions, he is haunted by a small black monkey until he commits suicide. There are four actual narrators in the story, which places an extreme distance between the reader and the supernatural event. As the reader looks closely into these narratives, he or she discovers that Le Fanu is not saying that the monkey is real, but neither is he saying precisely what drives Jennings to his death. Furthermore, one cannot be sure that one can trust the narrators of the tale, for Le Fanu uses unreliable narrators to present the facts of the story. The character of Hesselius is certainly an unreliable narrator, for he is pompous and arrogant and in fact places the Reverend Mr. Jennings in a dangerous position and opens the door for his death. There is also a psychological loophole as well as a metaphysical explanation–the theological writings of Swedenborg. However, throughout the tale Le Fanu maintains his detachment toward these explanations. Finally, the reader becomes so bewildered by all these possibilities that he or she can only sympathize with Jennings for the terrifying experiences he undergoes.

The second story in the collection, "The Familiar," is a revised version of the earlier tale "The Watcher." However, the story fits remarkably well with the Hesselius tales. The third story in the collection, "Mr. Justice Harbottle," is in its structure more complex and compelling than the first two tales. The reader must contend with a larger group of narrators, all of whom are unreliable, and with what appear to be five separate supernatural manifestations witnessed by five different characters.

The plot as a result is rather complex, but the primary narrative thread of the story centers on an infamous old judge, Elijah Harbottle, a "hanging judge" who at the time of the story is scheduled to try the case of Lewis Pyneweck, who has been accused of forgery. Pyneweck's wife, from whom he is divorced, is Harbottle's mistress and lives in his house under the name of Mrs. Carwell as his housekeeper. Pyneweck, though innocent of forgery, is found guilty by Harbottle and executed. Finally, Harbottle is haunted by a man whom he fears is the ghost of Pyneweck, and ultimately Harbottle himself is tried in a dreamlike court by "a dilated effigy" of himself. Finally, after some complicated incidents, most of which appear to be supernatural, Harbottle hangs himself in his home.

The multiple possible explanations that the events of the tale raise are consistently undercut by the unreliable and distorting narratives through which they are filtered. The effect is the same as in "The Familiar" and "Green Tea": to produce a lingering doubt as to the actual nature of what happens. There are of course the loopholes of psychological explanations. It may be that Harbottle feels guilt for the crimes he perpetrated on the bench, and the seemingly supernatural trial he himself undergoes may be an opening of what Swedenborg calls "internal sight." However, the other characters in the story experience supernatural experiences as well. Dr. Hesselius's explanation is that the nervous energy produced by the other characters radiate outward and infect the others around them. The suggestion of the tale is that the supernatural may be likened to a contagion.

Yet the point of the tale is the same as in "Green Tea." All of the characters are locked in their own minds, victims of their own prejudices and illusions. Each character who experiences so-called supernatural phenomena is also a victim of the prison of the self.

Richard Beckett, the narrator and central character of "The Room in the Dragon Volant" from *In a Glass Darkly,* is certainly a prisoner in his own mind, for he is trapped in an elaborate deception and is the victim of a murder attempt. While not overtly a supernatural tale, this mystery-romance is rich in symbolism. This novella involves the young Englishman Beckett, who travels to Paris shortly after the fall of Napoleon with a great deal of money and plans to make an even greater fortune in gaming. He becomes the dupe of a beautiful woman, the Countess de St. Alyre, who entraps him in an elaborate plot to murder him. At the climax of the tale the countess gives Beckett coffee that, unknown to him, is laced with a drug called *Mortis Imago.* It produces in him the semblance of death in that he can see and hear, but he cannot move. Just before the countess and her accomplices attempt to bury Beckett alive, a French detective rescues him.

Beckett is largely motivated by his imagined love for the countess and her deception that she returns that love. In the context of this symbolic mystery-romance, Beckett is not only deceived by the countess, but by love itself. As the story progresses Beckett begins to believe that love is a genuine supernatural force, a genuine reality, but he discovers at the conclusion that it is not.

Le Fanu uses a system of symbols, much like Swedenborg's symbolic reading of the Bible, that serves to represent in a veiled manner the actual nature of Beckett's adventures. The predominant symbol of the tale is the inn Le Dragon Volant itself. The sign above its entrance bears an image of a flying dragon. The inn itself is reputedly haunted, and at least three wealthy men have disappeared from its haunted chamber. Thus, the inn represents an apparently malignant and terrifying force that is something supernatural like the dragon itself, and Beckett is involved in a quest to conquer it. Ultimately Beckett discovers that love and the inn of the flying dragon are not supernatural, but they remain inexplicable.

The final tale of *In a Glass Darkly* is the vampire novella "Carmilla." Like the preceding tales, it presents a vision of the struggle to come to terms with irrational forces. Similarly, Le Fanu utilizes multiple narrators, all of whom are unreliable, to create a lingering doubt about the true nature of the apparently supernatural experience of Laura, the central character. The suggestion of the tale is, as has been apprehended by several critics, that the vampire Carmilla is not a vampire at all but a mortal who is a lesbian. However, Le Fanu uses verbal subterfuge and the highly subjective and distorted narrative of the young victim, Laura, to create in the reader's mind a lingering doubt as to precisely what kind of being Carmilla is. Thus, Le Fanu raises the question of his narrators' reliability. The reader's perceptions are filtered through theirs, and what one perceives, then, is a very distorted account of Laura's story. However, within the narratives, Le Fanu suggests some possible explanations.

The possibility that Carmilla is a lesbian may be elaborately concealed by Laura's father, who may in fact reinforce this deception by concocting an elaborate explanation of vampirism in accord with the superstitions of Styria and the possibly irrational beliefs of General Spielsdorf and Baron Vordenberg, his associates. In psychological terms, furthermore, Carmilla may be Laura's subconscious projection of her dead mother. Laura reveals that her mother was a Styrian woman who was descended from the Karnstein family, and Carmilla, it is later revealed, is Mircalla, Countess Karnstein.

In the novella Le Fanu tricks the reader into realizing that existence is filled with infinite possibilities, none or all of which may be true. Like all the other tales of *In a Glass Darkly,* one must forever see–as in the allusion to a phrase in the writings of St. Paul from which Le Fanu draws his title–"through a glass, darkly." One can only know reality in part. As with Laura's perceptions, the reader sees the apparent phantasmagoria through a medium–distorted, shifting, and contradictory. Thus, *In a Glass Darkly,* which was Le Fanu's last and most important contribution to supernatural horror fiction, marks the confluence of his lifelong ideas about the nature of the supernatural.

In his final years, from 1870 to 1873, Le Fanu wrote a few novels and a few ghost stories, largely based on Irish folklore. "The Child That Went with the Fairies" (1870) is a story of children carried off by the fairies upon their deaths. Another Irish tale is "The White Cat of Drumgunniol" (1870), about a ghostly white cat who presages the death of anyone who comes upon it. Three short pieces comprise "Stories of Lough Guir" (1870), also drawn from folklore sources.

"The Vision of Tom Chuff" (1870) and "Sir Dominick's Bargain" (1872) are rewritings of earlier tales, "The Drunkard's Dream" and "The Fortunes of Sir Robert Ardagh." All of these last several tales were published in Charles Dickens's magazine *All the Year Round.* In these years Le Fanu wrote while in bed late at night on scraps of notebook paper by the light of two candles. He would go to sleep at two in the morning and wake up later, weaving images from his dreams into his supernatural tales. He would finally rise at noon the next day and attend to his business.

Many writers about Le Fanu have mentioned a legend, which is probably the embellishment of a minor incident. He complained of frequent nightmares about an old house that was about to topple in on him. When Le Fanu died, his doctor looked into the terror-stricken eyes of the dead man and said, "I feared this. That house fell at last." While this story has no basis in fact, it creates an image of Le Fanu as a "ghost-story writer." There is no doubt that Le Fanu retreated to a degree from Dublin society in his last years, perhaps haunted by the death of his wife, Susanna, but he remained devoted to his family.

Le Fanu died on 10 February 1873 and was buried in Mount Jerome Cemetery. The house in which he and Susanna lived still stands, and a commemorative plaque is affixed to the front of the building. While his name is certainly not a household word, he is known and respected among enthusiasts and scholars of Gothic literature, and his quiet, somewhat obscure reputation seems appropriate for a man who cared little for himself and who was not obsessed with his own reputation.

However, Le Fanu has frequently been called "the father of the English ghost story," for no other writer before him maintained such a sensitive fascination with the idea of the supernatural. He was the first writer to realize that there was an aesthetic of supernatural terror in fiction. Taken as a whole, Le Fanu's greatest claim to fame will be the authorship of his supernatural tales, and his influence on later writers, especially M. R. James and Bram Stoker, is considerable.

Bibliography:

Gary William Crawford, *J. Sheridan Le Fanu: A Bio-Bibliography* (Westport, Conn.: Greenwood Press, 1995).

Biography:

W. J. McCormack, *Sheridan Le Fanu and Victorian Ireland* (Oxford: Clarendon Press, 1980).

References:

Michael H. Begnal, *Joseph Sheridan Le Fanu* (Lewisburg, Pa.: Bucknell University Press, 1971);

Nelson Browne, *Sheridan Le Fanu* (London: Barker, 1952);

Patricia Coughlan, "Doubles, Shadows, Sedan-Chairs and the Past: The 'Ghost Stories' of J. S. Le Fanu," in *Critical Approaches to Anglo-Irish Literature,* edited by Michael Allen and Angela Wilcox (Totowa, N.J.: Barnes & Noble, 1989), pp. 17–39;

S. M. Ellis, "Joseph Sheridan Le Fanu," in his *Wilkie Collins, Le Fanu, and Others* (London: Constable, 1931), pp. 140–191;

Ivan Melada, *Sheridan Le Fanu* (Boston: Twayne, 1987);

Harold Orel, "'Rigid Adherence to Facts': Le Fanu's *In a Glass Darkly,*" *Eire-Ireland,* 20 (1985): 65–88;

Peter Penzoldt, *The Supernatural in Fiction* (New York: Humanities Press, 1965);

Jack Sullivan, *Elegant Nightmares: The English Ghost Story from Le Fanu to Blackwood* (Athens: Ohio University Press, 1978);

William Veeder, "'Carmilla': The Arts of Repression," *Texas Studies in Literature and Language,* 22 (1980): 197–223;

Edward Wagenknecht, "J. Sheridan Le Fanu," in his *Seven Masters of Supernatural Fiction* (Westport, Conn.: Greenwood Press, 1991), pp. 1–21.

Papers:

Most of Le Fanu's papers are housed at Cambridge University. His correspondence with his major publisher, Richard Bentley, is housed at the University of Illinois.

Matthew Gregory Lewis

(9 July 1775 – 16 May 1818)

Larry R. Hamm
Iowa State University

Matthew Gregory Lewis (National Portrait Gallery)

See also the Lewis entries in *DLB 39: British Novelists, 1660–1800* and *DLB 158: British Reform Writers, 1789–1832.*

BOOKS: *The Monk: A Romance,* 3 volumes (London: Printed for J. Bell, 1796);

Village Virtues: A Dramatic Satire. In Two Parts (London: Printed for J. Bell, 1796);

The Castle Spectre: A Drama. In Five Acts (London: J. Bell, 1798);

The Love of Gain: A Poem Imitated from the Thirteenth Satire of Juvenal (London: J. Bell, 1799);

The East Indian: A Comedy. In Five Acts (London: Printed by J. Davis for J. Bell, 1800; New York: Reprinted by M. L. & W. A. Davis for H. Caritat, 1800);

Tales of Wonder: Written and Collected by M. G. Lewis, by Lewis, Sir Walter Scott, and Robert Southey, 2 volumes (London: Printed by W. Bulmer, for the author & sold by J. Bell, 1801);

Adelmorn, the Outlaw: A Romantic Drama, in Three Acts (London: Printed for J. Bell, by Wilks & Tayoir, 1801; New York: Longworth, 1805);

Alfonso, King of Castile: A Tragedy, in Five Acts (London: J. Bell, 1801; Philadelphia & Baltimore: Conrad, 1802);

Ruganitino; or, The Bravo of Venice. A Grand Romantic Melo-Drama, in Two Acts (London, 1805);

Adelgitha; or, The Fruits of a Single Error. A Tragedy, in Five Acts (London: Printed by D. N. Shury for J. F. Hughes, 1806; New York: Longworth, 1812);

Romantic Tales (4 volumes, London: Printed by D. N. Shury for Longman, Hurst, Rees & Orme, 1808; 2 volumes, New York: Printed for M. & W. Ward, 1808);

Twelve Ballads, the Words and Music by M. G. Lewis (London, 1808);

Venoni, or, The Novice of St. Mark's: A Drama, in Three Acts (London: Printed by D. N. Shury, for Longman, Hurst, Rees & Orme, 1809; New York: Longworth, 1809);

Monody on the Death of Sir John Moore (London: Printed for Longman, Hurst, Rees & Orme, 1809);

Timour the Tartar: A Grand Romantic Melo-Drama, in Two Acts (London: Lowndes & Hobbs, 1811; New York: Longworth, 1812);

One O'Clock! or, The Knight and Wood Daemon. A Grand Musical Romance, in Three Acts, with music by M. P. King and M. Kelly (London: Lowndes & Hobbs, 1811; New York: Longworth, 1813);

Rich and Poor: A Comic Opera, in Three Acts (London, 1812);

Poems (London: Printed by D. N. Shury & sold by Hatchard, 1812);

The Harper's Daughter; or, Love and Ambition: A Tragedy (Philadelphia: R. & W. Carr, 1813);

The Isle of Devils. A Historical Tale, Founded on an Anecdote in the Annals of Portugal (Kingston, Jamaica, 1827);

Journal of a West India Proprietor Kept during a Residence in the Island of Jamaica (London: John Murray, 1834).

Editions: *The Bravo of Venice: A Romance* (New York: Arno, 1972);

The Monk: A Romance, edited by Howard Anderson (London: Oxford University Press, 1973).

PLAY PRODUCTIONS: *The Castle Spectre: A Drama, in Five Acts,* London, Drury Lane Theatre, 14 December 1797;

The Twins; or, Is It He, or His Brother? A Farce in Two Acts, London, Drury Lane Theatre, 8 April 1799;

The East Indian: A Comedy, in Five Acts, London, Drury Lane Theatre, 22 April 1799;

Adelmorn, the Outlaw: A Romantic Drama, in Three Acts, London, Drury Lane Theatre, 14 May 1801;

Alfonso, King of Castile: A Tragedy, in Five Acts, London, Covent Garden Theatre, 15 January 1802;

The Captive: A Scene in a Private Mad-House, London, Covent Garden Theatre, 22 March 1803;

The Harper's Daughter; or, Love and Ambition. A Tragedy, in Five Acts, London, Covent Garden Theatre, 4 May 1803;

Rugantino; or, The Bravo of Venice. A Grand Romantic Melo-Drama, in Two Acts, London, Covent Garden Theatre, 18 October 1805;

The Wood Daemon; or, The Clock Has Struck. A Grand Romantic Melo-Drama, in Two Acts, London, Drury Lane Theatre, 1 April 1807;

Adelgitha; or, The Fruits of a Single Error. A Tragedy, in Five Acts, London, 30 April 1807;

Venoni; or, The Novice of St. Mark's. A Drama, in Three Acts, London, Drury Lane Theatre, 1 December 1808;

[Temper; or,] The Domestic Tyrant: A Farce, in Two Acts, London, Lyceum Theatre, 1 May 1809;

Timour the Tartar: A Grand Romantic Melo-Drama, in Two Acts, London, Covent Garden Theatre, 29 April 1811;

One O'Clock! or, The Knight and the Wood Daemon. A Grand Musical Romance, in Three Acts, London, Lyceum Theatre, 1 August 1811;

Rich and Poor: A Comic Opera, in Three Acts, London, Lyceum Theatre, 22 July 1812.

TRANSLATIONS: Friedrich Schiller, *The Minister: A Tragedy. In Five Acts. Translated from the German of Schiller* (London: J. Bell, 1797);

August Friedrich Ferdinand von Kotzebue, *Rolla; or, The Peruvian Hero: A Tragedy. In Five Acts. Translated from the German of Kotzebue* (London: Printed for J. Bell, 1799);

Heinrich Zschokke, *The Bravo of Venice. A Romance: Translated from the German* (London: D. N. Shury, 1805);

Christiane Benedicte Eugenie Naubert, *Feudal Tyrants; or, The Counts of Carlsheim and Sargans: A Romance,* 4 volumes (London: J. F. Hughes, 1806).

Matthew Gregory Lewis began writing in an effort to earn money for his mother and stopped his literary efforts because of the burden of managing the wealth inherited from his father. While just nineteen he wrote *The Monk* (1796), his first publication and the book with which he would be forever identified, both morally and creatively. As a result of the immense fame, or notoriety, that he received from *The Monk* he was able to pursue an extremely successful career as a dramatist, although his periodic attempts to escape the Gothic mode resulted in failure. Public taste demanded his outrageous and melodramatic formulas, while critical opinions continually examined his plays for their moral iniquities. Perhaps because he attempted to satisfy both these popular and critical demands, Lewis's plays suffer from a decline in originality, and during the remainder of his life his best work can be found in his poetry, primarily his ballads, and his posthumously published journal. Ironically, Lewis seemed to succeed most when he tried the least, giving in to the powers of his imagination. He darkened the Gothic world of his predecessors and subsequently influenced the creations of George Gordon, Lord Byron, Percy Bysshe Shelley, Sir Walter Scott, and Mary Wollstonecraft Shelley.

Lewis was born in London on 9 July 1775, the first child of four born to his father, Matthew, and his mother, Frances Maria Sewell. Lewis's father was born in Jamaica and owned two large plantations there. At age twenty-two he went into government service with an appointment as chief clerk in the War Office, where he was employed for the next thirty-one years, also holding the concurrent title and responsibilities of the Deputy-Secretary at War beginning in December 1775. Approximately one year after his appointment to the War Office, Matthew married Frances Sewell, a beautiful woman whose family also had a strong connection to Jamaica, where her brother served as attorney general. In addition to being named after his father, Matthew Gregory Lewis was named after his paternal grandmother, Jane Gregory. He disliked both names, preferring to be referred to by his initials until the notoriety of his novel earned him the moniker "Monk."

M. G. Lewis formed a deep attachment to his mother and emulated her love for the theater and even her desire to be a writer. When he was only six his mother left her husband and children, but rather than choose a side in this marital split, he was, according to biographer Margaret Baron-Wilson, "almost an umpire" in maintaining relations with both parents. Although he came close to being disinherited by his father over his father's decision to move his mistress–Mrs. Ricketts, the widow of a family friend–into the household, he continued to live under the financial influence of the elder Lewis. At the same time he maintained a significant correspondence with his mother. Much of Lewis's biography has been constructed from this correspondence.

After his preparatory schooling at Marylebone Seminary, Lewis followed his father's educational course by attending first Westminster School and later Christ Church College, Oxford. Though he did nothing to distinguish himself in his studies, Lewis earned his degree in four years. During this time he also spent summers in Paris and Weimar following his father's wish that he learn languages so that he could become a diplomat at the end of his education. Upon receiving his degree, Lewis, urged on by his father's connections in government and ambitions for him, took a position as attaché to the British embassy in The Hague. In 1796 his father's political influence also obtained a seat for him in Parliament as a member of the House of Commons. This seat had been vacated by William Beckford, the author of the Gothic novel *Vathek* (1786). In a move that prefigured Lewis's own departure from the post, Beckford was forced to resign because of accusations of immoral conduct (he was openly homosexual). Lewis, however, showed little interest in the affairs of government. As his letters from The Hague indicate, he was concerned about earning money to help his mother, eager to begin a literary career, and writing "most furiously" in his efforts to produce a play, a poem, or a novel that would bring him the fame and monetary rewards he sought.

By the time Lewis arrived in The Hague in 1794 he had already written a dramatic farce, "The Epistolary Intrigue" (now lost); composed part of a novel, "The Effusions of Sensibility" (only thirty pages of which remain); and translated *Felix,* a French opera. However, none of these would ever be published or produced. An additional farce, *The East Indian* (1800), and a dramatic satire, *Village Virtues* (1796), were also written prior to 1794, but neither

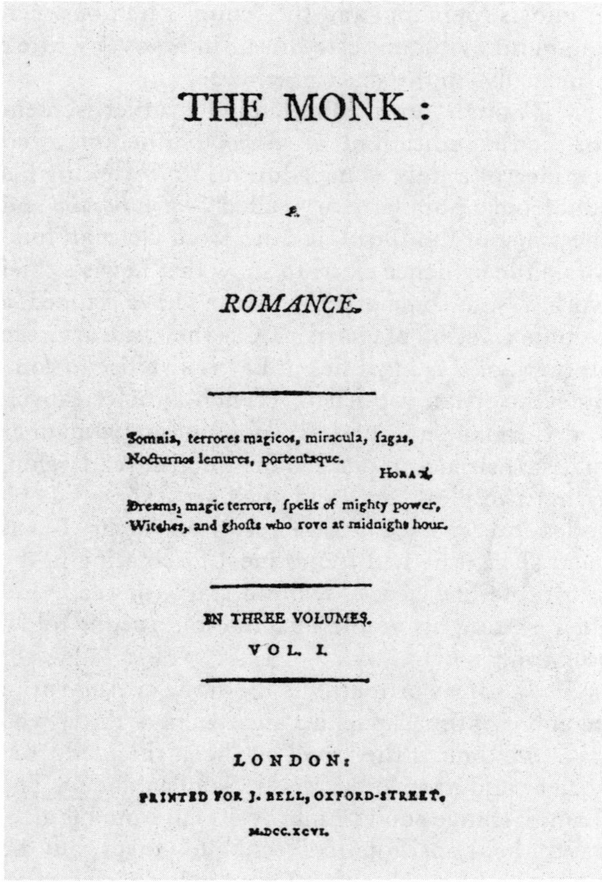
THE MONK:

A

ROMANCE.

Somnia, terrores magicos, miracula, sagas,
Nocturnos lemures, portentaque.
HORAT.

Dreams, magic terrors, spells of mighty power,
Witches, and ghosts who rove at midnight hour.

IN THREE VOLUMES.

VOL. I.

LONDON:

PRINTED FOR J. BELL, OXFORD-STREET.

M.DCC.XCVI.

Title page for Lewis's first book, in which the testing of the vain Ambrosio's shielded innocence results in his committing evil acts, including incest and matricide

appeared until Lewis's popularity was established. While working on these projects aimed at appealing to reason (they were all vehicles of social criticism), Lewis simultaneously indulged himself with the irrational and often unreal world of the Gothic.

As early as March 1792 Lewis had written to his mother of beginning "a romance in the style of the *Castle of Otranto*" (written by Horace Walpole in 1764 and generally considered the first Gothic novel). In his first letter from The Hague he wrote he had "again taken up [the] romance" after having read Ann Radcliffe's *The Mysteries of Udolpho* (1794), which he described as "one of the most interesting books that has ever been published." Six months later Lewis made initial mention of "a romance of between three and four hundred pages octavo . . . called *The Monk.*" Though no clear connection exists between his earlier romance and *The Monk,* Lewis no doubt incorporated parts of the earlier text into either the later one or *The Castle Spectre* (1798), a Gothic drama written at approximately the same time. What is clear, however, is that Lewis's interest in ghosts, demons, and the tenuous line between good and evil manifested itself in a novel written principally "in the space of ten weeks."

Though Lewis admitted in his advertisement for the first edition of *The Monk* that he borrowed the idea of a holy man seduced into forfeiting his soul from a popular story called *Santon Barsisa* and the story of the Bleeding Nun from German folklore, little evidence exists to show that Lewis's chief work was, as some of his detractors have claimed, a complex act of plagiarism. On the contrary, the strength of *The Monk* lies in Lewis's ability to combine European eroticism, French anticlericalism, and German supernaturalism in a unique manner that surprised and pleased his English readership. Using two plots, both of which are a series of episodes interspersed with a series of poems, Lewis used all that he had found most fascinating in the stories he had heard, seen on stage, or read, and then, setting his young imagination free, he told a story of his own.

The Monk primarily is the story of Ambrosio, an abbot of the Capuchins and a man of thirty who until the time of the novel has been shielded from society and its vices by the walls of an abbey. The main narrative shows Ambrosio's fall from his arrogantly held position of esteem and power, but the strength of Lewis's text is found in his presentation of individual scenes. By focusing the story on the villain rather than on his victims, unlike Walpole and Radcliffe, Lewis not only gives a psychologically believable portrait of a vain man's seduction into evil but also shows how attractive temptation can be. Lewis's ability to create scenes of restrained and unleashed desire occurs most effectively in Ambrosio's relationship with Matilda, who initiates him first into carnality and then into necromancy before ultimately encouraging him to sign his soul away to Lucifer. This part of *The Monk* was used as the basis for James Boaden's theatrical adaptation, *Aurelio and Miranda* (1798), but in this drama the demonic character of Matilda (as Miranda) is changed into a chaste virgin; Aurelio is redeemed; and the lovers eventually marry. Audiences apparently disliked the changes, and the play ran only briefly. In *The Monk,* however, Ambrosio sates himself with Matilda's flesh and then pursues the extremely virginal Antonia for the pleasure found in a new form of depravity.

Lewis intuitively understood that his readers possessed the same appetite for new and perverse stimulation. The narrative presents one episode of horror after another. Critics have often sought out sources for Lewis's scenes, and he no doubt borrowed considerably from other authors. However, his willingness to go to extremes to affect his readers is unprecedented. Ambrosio gazes at the naked Antonia in an enchanted mirror; he uses a magic myrtle to gain access to her bedchamber, where he is thwarted by her mother, Elvira, whom he kills; and finally he drugs Antonia so that she appears dead and so that he may later rape her amid the corpses in the vault where she has been interred. Discovered there, he kills her. Having done all of this, Ambrosio learns from Lucifer that Elvira was his mother and Antonia was his sister. As if the crimes themselves are not enough, Lewis gives his readers some ex post facto matricide and incest. Scheduled to be punished for these crimes by the Inquisition, Ambrosio gives his soul to Lucifer, who promptly pierces Ambrosio's skull with his talons, carries him into the sky, and drops him down a rocky precipice to die by a river. Unable to move to satisfy a burning thirst, Ambrosio dies slowly, tormented by insects, over a period of six days.

The second story line involves Raymond and his pursuit of his beloved Agnes. Lewis weaves together scenes utilizing banditti, the Wandering Jew, the Bleeding Nun, a despotic prioress, and a pregnant nun to tell a rather traditional tale of a young woman forced against her will to enter a convent. Lewis consistently embellishes the horror of each scene, and the story ends when Agnes is rescued from her imprisonment in the underground chambers of the convent, where she is found holding the decomposing insect- and reptile-covered corpse of her baby. In a final dramatic scene a mob burns the convent and tears the prioress to pieces.

Though the characters of both narratives intersect at the beginning and end of the novel, they really have little to do with each other. The overall plot of *The Monk* is loosely constructed. One character, Lorenzo, seems significant when he dominates the first chapter, then all but disappears until the end of the text. Matilda appears first as the novice Rosario, then reveals herself as a woman in love with Ambrosio. When this love subsequently turns to lust, Matilda's character moves from sympathetic to predatory. At the end Lucifer tells Ambrosio that Matilda was a demonic agent all along, and Lewis, with this unfortunate addition, thereby undermines his second strongest character in the text. Though the anticlerical theme consistently appears in both narratives, this theme is secondary to the melodrama, grotesqueness, and eroticism in individual scenes. Lewis wanted to be published, for both the money and the fame, so he incorporated all of his reading, his interests in the macabre, his knowledge of human desire, and his ability to set a scene and move an audience into one book of horror and perversion.

He succeeded. *The Monk,* published anonymously in March 1796, was so popular that a second edition was published by 14 September of the same year. Initial critical reactions were mixed, but generally the unknown author was lauded for his imaginative effort. In particular, critics applauded Lewis's poetry, and one ballad, "Alonzo the Brave and Fair Imogene," drew both uncommon critical praise and extremely widespread popular appeal. When the second edition came out Lewis had both his name and his new title, M.P., added to the title page.

In 1797 the two best-known critiques of *The Monk* appeared; both became known not for their discussion of the book's literary value but for their attacks on Lewis for creating an immoral work. The first was published in the *Critical Review* of February and is especially noteworthy because the critic was Samuel Taylor Coleridge. Though he admitted that Lewis's book was the product of "an imagination rich, powerful, and fervid," Coleridge felt a duty to expose the immorality found in the "libidinous minuteness" of Lewis's erotic descriptions. At the same time Coleridge admired Lewis's poetry and specifically praised "The Exile." Lewis's ability to write ballads drew later recognition from Coleridge, but it is significant that one year before the publication of his collaborative effort with William Wordsworth, *The Lyrical Ballads,* Coleridge expressed an admiration for another popular writer of ballads. This admiration nevertheless was tempered by what Coleridge viewed as a book dangerously unfit for young minds.

By far the most damning criticism of Lewis, a satiric poem by the Reverend Thomas James Mathias, was published five months later in the fourth edition of *The Pursuits of Literature.* In this edition of his anonymously published poem Mathias pays special attention to what Coleridge had suggested was a blasphemous passage in *The Monk* where Antonia reads from an expurgated Bible because the original text is described as improper for a young woman. Mathias's comment that Lewis's blasphemy might be indictable began both a movement to censure the book and Lewis, the young member of Parliament, and a countermovement to defend him and his text. While critics argued, the public purchased the book, and by 1800 *The Monk* had been printed in five London and two Dublin editions. Furthermore, unauthorized versions in the form of chapbooks and theatrical productions, as well as parodies and burlesques, emerged using episodes or ballads as bases for their plots.

Though the moral and political discussions surrounding *The Monk* promoted sales, they also affected Lewis as an artist–he never again wrote so freely. His father, who never agreed with his son's choice of a literary career, was especially hurt by the controversy. In Lewis's letter to his father of 23 February 1798 Monk (now his nickname) expresses his "sorrow" for having given him "pain" and discusses a new version of *The Monk* in which he has "expunged" all questionable passages and altered inflammatory words. This version was used for the fourth and fifth London editions of the text. In his letter Lewis blames his youth for his "carelessly chosen" words and expressions but notes that he would be reluctant to trust his own judgment again. Before his next production he would rely on his sister Maria, Lady Lushington, to strike out all passages she deemed offensive.

Perhaps because of these restraints, but more likely because of his sudden demand in social circles, Lewis's originality and popularity never again achieved the levels of his first publication. His subsequent works relied heavily on his earlier attempts to write for the theater and upon his knowledge of French and German literature. His next two publications were plays: *Village Virtues,* written when he was a schoolboy, and *The Minister* (1797), a translation of Friedrich Schiller's tragedy *Kabale und Liebe* (1784). Neither of these plays was ever produced, though Lewis's abridgment of *The Minister, The Harper's Daughter* (1813), was staged 4 May 1803. Contemporary reviews all but ignored *Village Virtues* while responding with relative favor to *The Minister,* a second edition of which was published in 1798.

In order to receive the notice he wished for in the theater Lewis had to return to the Gothic world with which he was already strongly associated. On 14 December 1797 *The Castle Spectre,* a Gothic drama written before or perhaps at the same time as *The Monk,* was first performed at Drury Lane Theatre. In many ways the play was unoriginal. In typical Gothic fashion, the evil earl Osmond preys on his ward, Angela, whose mother, Elvira, and father, Reginald, he believes he has killed. Angela's father (and Osmond's brother), however, has only been imprisoned, and in the end he is freed when Angela kills Osmond. Two innovative alterations by Lewis make the play stand out. The first, an anachronistic use of black servants in a setting that precedes the fifteenth century, shows how little concern Lewis had for actual history. His goal was to achieve an effect, to "give a pleasing variety to the characters and dresses," and when questioned about the anachronism he remarked, "Could I have produced the same effect by making my heroine blue, blue I should have made her." Lewis's second alteration of conventional Gothic drama not only created an

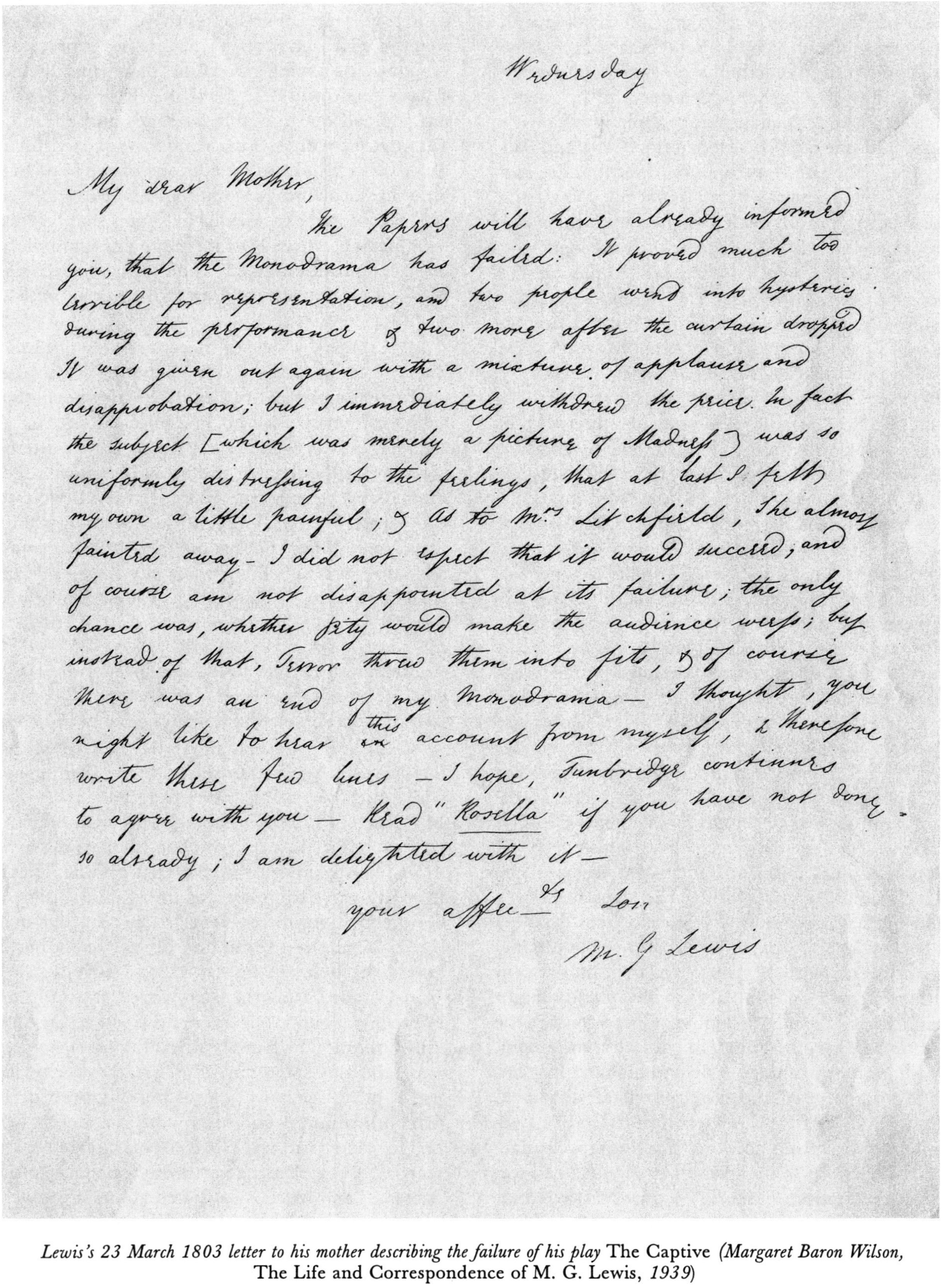

Wednesday

My dear Mother,

The Papers will have already informed you, that the Monodrama has failed: It proved much too terrible for representation, and two people went into hysterics during the performance & two more after the curtain dropped. It was given out again with a mixture of applause and disapprobation; but I immediately withdrew the piece. In fact the subject [which was merely a picture of Madness] was so uniformly distressing to the feelings, that at last I felt it my own a little painful; & as to Mrs Litchfield, she almost fainted away – I did not expect that it would succeed; and of course am not disappointed at its failure; the only chance was, whether pity would make the audience weep; but instead of that, Terror threw them into fits, & of course there was an end of my Monodrama – I thought, you might like to hear this account from myself, & therefore write these few lines – I hope, Tunbridge continues to agree with you – Read "Rosella" if you have not done so already; I am delighted with it –

your affecte Son

M. G. Lewis

Lewis's 23 March 1803 letter to his mother describing the failure of his play The Captive *(Margaret Baron Wilson,* The Life and Correspondence of M. G. Lewis, *1939)*

even greater effect but also established him as the master of that subgenre. Against the strong advice of Richard Brinsley Sheridan, the theater manager, and several others, Lewis insisted on first bringing the ghost of Elvira onto the stage and then using her in the final scene to advance the plot. Critics responded negatively to this use of the supernatural, but audiences loved it. The play was performed forty-seven times in its first season and continued to be a stock piece in subsequent seasons. In the first three months alone *The Castle Spectre* earned £18,000 for Drury Lane. Lewis once again had delivered what the public wanted.

After the failure in 1799 of two farces written in or before 1794, *The East Indian* and *The Twins; or, Is It He, or His Brother?* Lewis returned to his Gothic format and to delivering the outrageous. His next production, *Adelmorn, the Outlaw* (1801), did not succeed in achieving the popularity of *The Castle Spectre,* but it did surpass its predecessor in ghostly appearances and unearthly effects. In his effort to move his audience Lewis continued to employ the character of the guilt-ridden, doomed, and almost sympathetic villain (a predecessor of the Byronic hero) and to expand his use of the supernatural. One visionary scene in *Adelmorn* involves the choral incantation of unseen spirits who chant while demons aid in a ghost's revenge.

In 1807, in what is arguably his second most popular play, *The Wood Daemon; or, The Clock Has Struck* Lewis combined his Gothic machinery with a world of fantasy and fairies. In addition to the wood demon, wood spirits, zephyrs, furies, and a giant appear on stage. The music and ballads at times provide a dreamlike quality, though the settings–a castle and a forest–remain Gothic. By attempting to reach outside of this world for novelty Lewis moved toward an ethereal world similar to that of the later Romantics.

In 1808 Lewis announced that writing for the theater had ceased to amuse him and that he intended to write no more plays. This announcement did not prove entirely true, as he began to experiment more with music and introduced operatic versions of *The Wood Daemon,* retitled *One O'Clock! or, The Knight and the Wood Daemon* (1811) and *The East Indian,* which was retitled *Rich and Poor* (1812). Asked by Thomas Harris, the manager of Covent Garden Theatre, to write an equestrian play (popular at that time), Lewis wrote *Timour the Tartar* (1811), which has a minimum amount of dialogue and a maximum amount of spectacle. Plays such as *Timour the Tartar,* which used trained horses not only to participate in battle scenes and rescues but also to react to human characters, drew strong positive and negative reactions. *Timour the Tartar* was exceptional neither in its appeal nor in the response it generated. It is noteworthy because it is the last of Lewis's eighteen dramatic works. After the elder Matthew Lewis died on 17 May 1812, Monk took over the responsibility of his father's estate, including the two Jamaican plantations. His career as a playwright, the career he had sought as a teenager, was over.

Lewis's reputation today has little to do with his popularity as a dramatist. Known almost solely as the author of *The Monk,* he also receives consideration for his influence on his more famous contemporaries. That Coleridge admired Lewis's ballads has been noted, and some critics have even (perhaps inappropriately) found Coleridgean antecedents in Lewis's work. More directly, Sir Walter Scott credited Lewis with pushing him to improve his own work. Contacted by Lewis in 1798 to contribute to a collaborative collection of ballads–*Tales of Wonder* (1801), which also included the work of Robert Southey and others–Scott maintained that Lewis's sometimes severe judgment directed him as a "young and careless author" toward the discipline necessary for literary success. *Tales of Wonder* was almost unanimously condemned and became the source of several parodies, but its most famous critic, Byron, elevated Lewis to a position amid the best of his contemporaries by satirically attacking him in *English Bards and Scotch Reviewers* (1809). Four years later these two met, and in August 1816, at the Maison Diodati in Geneva, Lewis visited Byron and was introduced to the Shelleys, Claire Clairmont, and John William Polidori. At Geneva Lewis entertained the others with ghost stories and translated Johann Wolfgang von Goethe's *Faust* (1808–1832) to Byron, who was greatly impressed by both Lewis's reading and the work. After Byron's Faustian *Manfred* was published in 1817, the poet credited Lewis for his only contact with Goethe's poem. The influence Lewis had on Percy Bysshe Shelley and the others is difficult to measure. Shelley had already attempted two Gothic novels of his own, *Zastrozzi* (1810) and *St. Irvyne* (1811) and was no doubt familiar with Lewis's work. Because Mary Shelley's *Frankenstein* (1818) and Polidori's *The Vampyre* (1819) were yet to be produced, the "mysteries of his trade," in Shelley's words, that Lewis imparted to this next generation of Gothic writers have been the subject of speculation.

Though in the middle of a Continental tour, Lewis remained involved in his business interests. Prior to the tour he had made his initial trip to Jamaica, and he returned concerned over the conditions under which his slaves lived and worked. In

Geneva, upon the advice of William Wilberforce, a noted philanthropist and advocate of slave emancipation, Lewis added a codicil to his will witnessed by Byron, Percy Bysshe Shelley, and Polidori. The codicil maintained, in the event of Lewis's death, reforms instituted during his first visit. Upon the conclusion of his tour a year later, Lewis almost immediately left again for Jamaica. During this second visit he undertook a dangerous journey to his more remote plantation, where he instituted reforms similar to those at his other plantation. These reforms included a code regulating punishment, but at the time Lewis was attacked for his overprotective and lax treatment of his slaves.

Shortly after leaving Jamaica, on his return voyage Lewis developed complications resulting from yellow fever and died on 16 May 1818. Refusing treatment from the ship's doctor, he gave himself an emetic that apparently increased his suffering and hastened his death. Because of fears of contagion Lewis's body was buried at sea. In a final Gothic irony, the weights attached to the coffin came off, and the coffin rose to the surface and drifted out of sight toward Jamaica.

Lewis kept journals of his Jamaican voyages, and they were published posthumously as *Journal of a West India Proprietor* (1834). Coleridge offered his own unique praise: "It is almost the only unaffected book of travels or touring I have read of late years. You have the man himself . . . a much finer mind than I supposed before from the perusal of his romances." Also published after Lewis's death was a poem found in the journals, *The Isle of Devils* (1827), in which a shipwrecked maiden, held captive and raped by a demon, gives birth to a monster child. She escapes, saved by monks (an interesting twist), and watches from shipboard as the demon and monster child pine for her on shore before plunging to their deaths in the sea. The image evokes the six-year-old Lewis, torn between the father he respected but viewed as cold and autocratic and the mother he adored but whose actions in abandoning her children he could never reconcile. M. G. Lewis, creator of monsters, perhaps felt a monstrous identity of his own.

The Monk provided Lewis with his position in literary history, having changed the Gothic from an episodic portrait of trials and escapes to a psychological display of temptation and terror. In response even Radcliffe altered her formula (and introduced her own villainous monk) in *The Italian* (1797), maybe her best but certainly her darkest novel. Nevertheless, Lewis's chief contribution might have been his own willingness to take chances in order to achieve the effect he desired. His characters could just as easily be ghosts or demons as earls or monks, and one or all of them was almost certain to be capable of necromancy. History was unimportant; his settings were superficial backdrops for a world of his own invention, outside of place and time. He created a world of terror, and readers and audiences responded with delight.

Biographies:

Margaret Baron-Wilson, *The Life and Correspondence of M. G. Lewis,* 2 volumes (London: Henry Colburn, 1839);

Louis F. Peck, *A Life of Matthew G. Lewis* (Cambridge, Mass.: Harvard University Press, 1961).

References:

Syndy M. Conger, *Matthew G. Lewis, Charles Robert Maturin and the Germans: An Interpretative Study of the Influence of German Literature on Two Gothic Novels* (Salzburg: Institut für Englische Sprache und Literatur, Universitat Salzburg, 1977);

Bertrand Evans, *Gothic Drama from Walpole to Shelley* (Berkeley: University of California Press, 1947), pp. 132–161;

Karl S. Guthke, *Englische Vorromantik und deutscher Sturm und Drang: M. G. Lewis' Stellung in der Geschichte der deutsch-englischen Literaturbeziehungen* (Gottingen: Vandenhoeck & Ruprecht, 1958);

Coral Ann Howells, *Love, Mystery, and Misery: Feeling in Gothic Fiction* (London: Athlone Press, 1978), pp. 62–79;

Joseph James Irwin, *M. G. "Monk" Lewis* (Boston: Twayne, 1976);

André Parreaux, *The Publication of "The Monk": A Literary Event 1796–1798* (Paris: Didier, 1960);

William B. Todd, "The Early Editions and Issues of *The Monk,* with a Bibliography," *Studies in Bibliography,* 2 (1949–1950): 3–24.

Papers:

The Victoria and Albert Museum in London holds the largest collection of Lewis manuscripts, including many of his letters; the Henry E. Huntington Library has manuscript copies of several of Lewis's plays; and the British Library possesses the copy of the third edition of *The Monk* in which Lewis made his alterations for the expurgated fourth edition. The original manuscript of *The Monk* is at the Wisbech and Fenland Museum in Wisbech, Cambridgeshire.

George MacDonald

(10 December 1824 - 18 September 1905)

Amelia A. Rutledge
George Mason University

See also the MacDonald entries in *DLB 18: Victorian Novelists After 1885* and *DLB 163: British Children's Writers, 1800-1880.*

BOOKS: *Within and Without: A Poem* (London: Longman, Brown, Green & Longmans, 1855; New York: Scribner, 1872); revised in *Works of Fantasy and Imagination* (London: Strahan, 1871; New York: Routledge, 1871); revised in *Dramatic and Miscellaneous Poems* (New York: Scribner, 1876);

Poems (London: Longman, Brown, Green, Longmans & Roberts, 1857); revised in *A Hidden Life and Other Poems* (London: Longman, 1872; New York: Scribner, Armstrong, 1872);

Phantastes: A Faerie Romance for Men and Women (London: Smith & Elder, 1858; Boston: Loring, 1870);

David Elginbrod, 3 volumes (London: Hurst & Blackett, 1863; New York: Munro, 1879);

Adela Cathcart (3 volumes, London: Hurst & Blackett, 1864; 1 volume, Boston: Loring, 1875?; revised edition, 1 volume, London: Low / New York: Munro, 1887);

The Portent (London: Smith & Elder, 1864; New York: Munro, 1885);

Alec Forbes of Howglen (3 volumes, London: Hurst & Blackett, 1865; 1 volume, New York: Harper, 1867);

Unspoken Sermons, First Series (London: Strahan, 1866; New York: Routledge, 1871); Second Series (London: Longmans, Green, 1886); Third Series (London: Longmans, Green, 1889);

Annals of a Quiet Neighborhood (3 volumes, London: Strahan, 1867; 1 volume, New York: Harper, 1867);

The Disciple and Other Poems (London: Strahan, 1867); revised in *The Poetical Works of George MacDonald* (London: Chatto & Windus, 1893);

Dealings with the Fairies (London: Strahan, 1867; New York: Routledge, 1876);

George MacDonald and his daughter Lilia (photograph by Lewis Carroll)

Robert Falconer (3 volumes, London: Hurst & Blackett, 1867; 1 volume, Boston: Loring, 1870);

The Seaboard Parish (3 volumes, London: Tinsley Brothers, 1868; 1 volume, New York: Routledge, 1868);

Guild Court, (3 volumes, London: Hurst & Blackett, 1868; 1 volume, New York: Harper, 1868);

England's Antiphon (London: Macmillan, 1868; Philadelphia: McKay, n.d.);

The Miracles of Our Lord (London: Strahan, 1870; New York: Randolph, 1870);

At the Back of the North Wind (London: Strahan, 1871; New York: Routledge, 1871);

Ranald Bannerman's Boyhood (London: Strahan, 1871; New York: Routledge, 1871);
The Princess and the Goblin (New York: Routledge, 1871; London: Strahan, 1872);
Works of Fantasy and Imagination, 10 volumes (London: Strahan, 1871; New York: Routledge, 1871);
The Vicar's Daughter (3 volumes, London: Tinsley Brothers, 1872; 1 volume, New York: Roberts Brothers, 1872);
Wilfrid Cumbermede (3 volumes, London: Hurst & Blackett, 1872; 1 volume, New York: Scribner, Armstrong, 1872);
Gutta-Percha Willie: The Working Genius (3 volumes, London: King, 1873; Boston: Hoyt, 1873);
Malcolm (3 volumes, London: King, 1875; 1 volume, Philadelphia: Lippincott, 1875);
The Wise Woman (London: Strahan, 1875); revised as *A Double Story* (New York: Dodd, Mead, 1876; Boston: Lothrop, 1876); revised as *Princess Rosamund: A Double Story* (Boston: Lothrop, 1879); revised as *The Lost Princess; or The Wise Woman* (London: Wells Gardner, Darton, 1895);
Saint George and Saint Michael (3 volumes, London: King, 1876; 1 volume, New York: Ford, 1876);
Thomas Wingfold, Curate (3 volumes, London: Hurst & Blackett, 1876; 1 volume, New York: Routledge, 1876);
The Marquis of Lossie (3 volumes, London: Hurst & Blackett, 1877; 1 volume, Philadelphia: Lippincott, 1877);
Sir Gibbie (3 volumes, London: Hurst & Blackett, 1879; 1 volume, Philadelphia: Lippincott, 1879);
Paul Faber, Surgeon (3 volumes, London: Hurst & Blackett, 1879; 1 volume, Philadelphia: Lippincott, 1879);
A Book of Strife in the Form of the Diary of an Old Soul (London: Unwin, 1880); revised in *Rampolli* (London: Longmans, Green, 1897);
Mary Marston (3 volumes, London: Sampson Low, 1881; 1 volume, Philadelphia: Lippincott, 1881);
Castle Warlock: A Homely Romance (3 volumes, London: Sampson Low, 1882); republished as *Warlock o' Glen Warlock: A Homely Romance* (New York: Munro, 1882);
Weighed and Wanting (3 volumes, London: Sampson Low, 1882; 1 volume, New York: Munro, 1882);
The Gifts of the Child Christ and Other Tales (2 volumes, London: Sampson Low, 1882; 1 volume, New York: Munro, 1882); revised as *Stephen Archer and Other Tales* (London: Sampson Low, 1883; Philadelphia: McKay, 1911);
Orts (London: Sampson Low, Marston, Searle & Rivington, 1882; republished as *The Imagination and Other Essays* (Boston: Lothrop, 1883); revised as *A Dish of Orts, Chiefly Papers on the Imagination and on Shakespere* [sic]*: Enlarged Edition* (London: Sampson Low, Marston, 1893);
The Princess and Curdie (London: Chatto & Windus, 1883; New York: Munro, 1883);
Donal Grant (3 volumes, London: Kegan Paul, Trench, 1883; 1 volume, New York: Harper, 1883);
The Tragedie of Hamlet: With a Study of the Text of the Folio of 1623 (London: Longmans, Green, 1885);
What's Mine's Mine (3 volumes, London: Kegan Paul, Trench, 1886; 1 volume, New York: Munro, 1886);
Home Again: A Tale (London: Kegan Paul, Trench, 1887; New York: Appleton, 1887);
The Elect Lady (London: Kegan Paul, Trench, 1888; New York: Munro, 1888);
A Rough Shaking: A Tale (London: Blackie & Son, 1890; New York: Routledge, 1890);
There and Back (3 volumes, London: Kegan Paul, Trench, 1891; 1 volume, Boston: Lothrop, 1891?);
The Flight of the Shadow (London: Kegan Paul, Trench, 1891; New York: Appleton, 1891);
A Cabinet of Gems, Cut and Polished by Sir Philip Sidney, Now for Their More Radiance Presented Without Their Setting by George MacDonald (London: Elliot Stock, 1892);
The Hope of the Gospel (London: Ward & Low, 1892; New York: Appleton, 1892);
Scotch Songs and Ballads (Aberdeen: John Rae Smith, 1893);
The Poetical Works of George MacDonald, 2 volumes (London: Chatto & Windus, 1893);
Heather and Snow (2 volumes, London: Chatto & Windus, 1893; 1 volume, New York: Harper, 1893);
Lilith (London: Chatto & Windus, 1895; New York: Dodd, Mead, 1895);
Rampolli: Growths from a Long Planted Root, Being Translations New and Old, Chiefly from the German: Along with a "Year's Diary of an Old Soul (London: Longmans, Green, 1897);
Salted with Fire: A Tale (London: Hurst & Blackett, 1897; New York: Dodd, Mead, 1897);
Far Above Rubies (New York: Dodd, Mead, 1899).

SELECTED PERIODICAL PUBLICATIONS–UNCOLLECTED: "My Uncle Peter," *The Queen*, 1 (December 1861): 298–299;

"A Journey Rejourneyed," *Argosy*, 1 (December 1865–January 1866): 127–133;

"Port in a Storm," *Argosy*, 2 (November 1866): 477–486;

"Papa's Story: A Scot's Christmas Story," *Illustrated London News*, 47 (December 1866): 626–628.

OTHER: *Twelve of the Spiritual Songs of Novalis, Done into English by George MacDonald* (Arundel: Mitchell & Son, 1851); revised in *Exotics* (London: Strahan, 1876); revised in *Rampolli* (London: Longmans, Green, 1897);

"Hymns," in *Hymns and Sacred Songs for Sunday Schools and Social Worship*, edited by Rev. G. B. Bubier (Manchester: Fletcher & Tubbs, 1855), pp. 133–135, 176–178, 197–198, 239–240, 246;

Exotics: A Translation of the Spiritual Songs of Novalis, the Hymn Book of Luther, and Other Poems from the German and the Italian (London: Strahan, 1876);

"Preface," in *Letters from Hell*, by L. W. J. S. (London: Richard Bentley & Son, 1884), pp. v–ix;

"Preface," in *For the Right*, by Karl Emil Franzos, translated by Julie Sutter (London: James Clarke, 1887), pp. iii–vi.

Any study that seeks to trace the influences of English and European romanticism in "mythopoeic" fantasy must turn to the works of the Scottish author George MacDonald. He was a contemporary of Lewis Carroll (MacDonald's children read "Alice's Adventures Underground," an early version of the first "Alice" book, and encouraged the author to publish his story) and of Charles Kingsley. MacDonald's fiction has influenced twentieth-century fantasists, most notably C. S. Lewis, and his work has been praised by W. H. Auden. Readers familiar with MacDonald's *The Princess and the Goblin* (1871) will see echoes of that work in J. R. R. Tolkien's *The Hobbit* (1937).

MacDonald, son of George MacDonald and Helen MacKay MacDonald, was born on 10 December 1824 in Huntly, Aberdeenshire, and was one of six children. His mother died when he was eight years old, and his paternal grandmother, Isabella Robertson, provided a strong influence on his development. MacDonald's fantasies abound in wise, benevolent grandmother/fairy-godmother figures, who combine nurturing behavior with strictness. In his "realistic" novels there is more ambivalence about these figures: rigid, pessimistic Sabbatarians, their Scots Presbyterianism nearly overshadows their positive attributes. According to MacDonald's son, Greville, MacDonald's father, revered by his son, inspired MacDonald's personal view of God–a stern but loving father. For MacDonald, Christianity was a religion of hope, a hope grounded in an intense awareness of the contingent nature of human existence (he was tubercular throughout his life). In an autobiographical comment in one of his novels, *Robert Falconer* (1867), he has young Robert declare that he did not wish to be loved by a God who did not love everybody.

This view of God so colored his religious attitudes as to place him very much at odds, in later life, with the religious tenets of his upbringing. He eventually went so far as to advocate the ultimate salvation of all creatures–including animals and Satan himself–a daring step made only in his last fantasy, *Lilith* (1895). In *The Great Divorce* (1946) C. S. Lewis depicts a MacDonald, in the afterlife, recanting this extreme–and to Lewis heretical–view of salvation. Equally important to MacDonald is faith in the power of imagination and a belief in the reality of the unseen. His characters, of which Curdie in *The Princess and the Goblin* is the best-known example, must often proceed in their quest without visible, or even rational, assurance about what is demanded of them, an obvious analogy to Christian doctrines about faith.

MacDonald's education–the Old Town Grammar School in Aberdeen, then Kings College at Aberdeen University starting in 1840, and later Highbury Theological College beginning in 1848–was typical for a Scottish youth of his time. He spent a brief period tutoring and working to catalogue a private book collection before marrying a cousin, Louisa Powell, in 1851. He became minister to Trinity Congregational Church in Arundel that same year; it was his first and last pulpit. He was quickly at odds with the vestry. Disturbed by "German" elements in his preaching, by which they meant the influence of German romantic poets and Swedenborgian mysticism, as well as by MacDonald's doctrinal deviations with regard to predestination and salvation, they first reduced his salary, then released him from employment in 1853.

The MacDonalds, already beginning a large family (eventually numbering eleven children, later joined by two adopted children and a foster child) moved to Manchester, where MacDonald taught in the Ladies College, where he was one of the original lecturers in English literature and in physics. Subsequently the family moved to London, where MacDonald, teaching at Bedford College from 1859 to 1867, began writing to augment his income. His first novel, *Phantastes: A Faerie Romance for Men and Women*, was published in 1858, shortly after the deaths of his father and one of his favorite brothers.

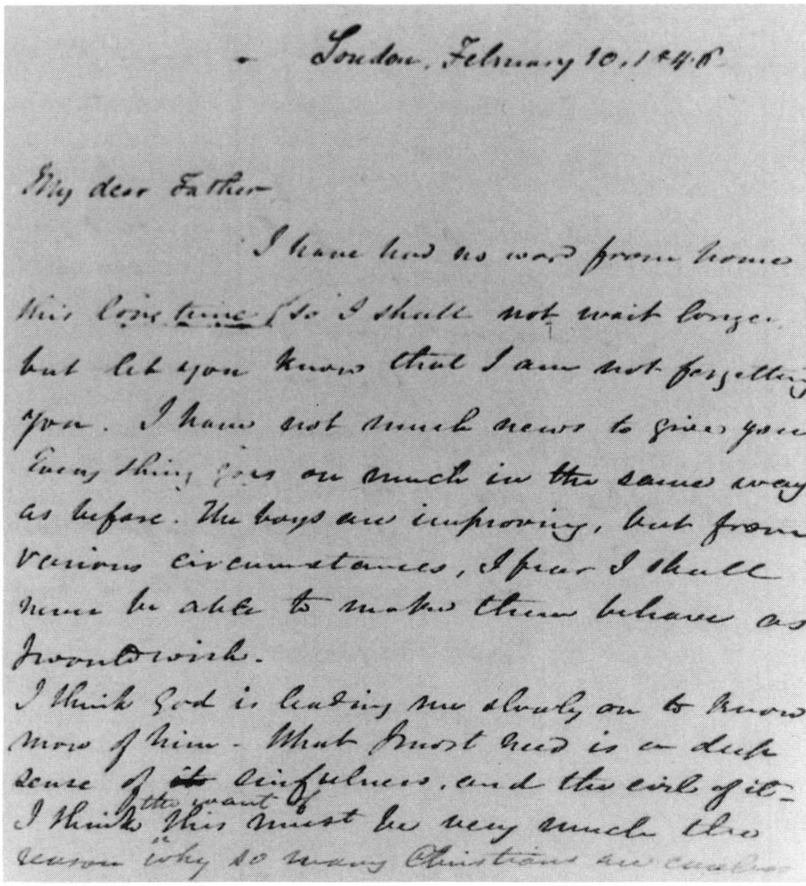

London, February 10, 1846

My dear Father

I have had no word from home this long time & so I shall not wait longer but let you know that I am not forgetting you. I have not much news to give you. Every thing goes on much in the same way as before. The boys are improving, but from various circumstances, I fear I shall never be able to make them behave as I would wish.

I think God is leading me slowly on to know more of him. What I most need is a deep sense of ~~its~~ sinfulness, and the evil of it. I think the want of this must be very much the reason why so many Christians are careless

First page of MacDonald's 10 February 1846 letter to his father (Beinecke Rare Book and Manuscript Library, Yale University)

This novel brought him to the attention of Carroll, Kingsley, his future illustrator Arthur Hughes, Dante Gabriel Rossetti, Dinah Maria Craik, and John Ruskin, for whom he was both a close friend and a counselor during Ruskin's difficult relationship with Rose LaTouche, who lived for a time with the MacDonald family.

Throughout his life fantasy was to be MacDonald's chosen genre and his true métier, but financial exigencies led him to write more than twenty novels in the sentimental, moralistic vein popular at the time. The first of these, *David Elginbrod,* was published with the help of Scottish novelist Margaret Oliphant in 1863, and he published steadily, despite ill health and the demands of a large family, from the 1850s until 1897, when his last major novel, *Salted with Fire,* was published. After 1867 his novels were generally published simultaneously in the United States and in England, and in 1872–1873 he had a very successful tour of the United States, meeting Ralph Waldo Emerson, William Cullen Bryant, and Oliver Wendell Holmes. MacDonald's last fantasy, and to some his greatest, was *Lilith,* published after much effort and soul-searching about its content in 1895. MacDonald died at Ashtead, Surrey, after several years of illness and months of withdrawn silence, in 1905.

MacDonald's influence rests not only upon his fiction but also upon his writings about fantasy and the imagination. In two essays, "The Imagination, Its Function and Culture" (July 1867), published anonymously in *The British Quarterly Review,* and "The Fantastic Imagination," originally the preface to Putnam's edition of *The Light Princess* (1893), MacDonald sets forth his views, which are based on an intensely Christian Neoplatonism with roots in the writings of William Wordsworth and especially of Samuel Taylor Coleridge. The imagination, he writes in the first essay, is "that faculty in man which is likest to the prime operation of the power of God, and has, therefore, been called the creative faculty."

A key element in MacDonald's fantasy is the existence of two realms, the natural and the fantastic/numinous, whose boundaries are subject to rupture and transgression; for some of his characters the fantastic realm is likened a dream. He opens the last chapter of *Phantastes* with this quotation from the German poet Novalis, "Unser Leben is kein Traum, aber es soll und wird vielleicht einer werden" (Our life is no dream, but it ought to become one, and perhaps will). Stephen Prickett, in discussing the import of the "two worlds" that figure so strongly in MacDonald's fantastic fiction, quotes from MacDonald's *The Miracles of Our Lord* (1870) as saying that all of natural creation is "brimful of symbolic and analogical parallels to the goings and comings, the growth and changes in the highest nature of man." This insight also contributes to the power of MacDonald's descriptions of nature and its phenomena in passages that are strong even in his least significant works.

Not only is the natural world a "type" of the real, it is also a stage in the "becoming" that defines human existence. MacDonald's concept of progressive spiritual growth, which has as a logical conclusion these theological views about salvation that caused the minister so much trouble with his parish, becomes a powerful thematic element in his fiction.

Those characters who cannot perceive or who perceive only dimly these vital correspondences between the natural world and the world of the imagination experience difficulties in negotiating the tasks their fantastic experiences set for them. In his realistic novels the protagonists, even when they are not artists, possess strong imaginative faculties and sensitivity to "nature"; this is their version of their counterparts' acceptance of the fantastic realms of their adventures.

MacDonald's nonfantastic novels–conventional in setting but not limited to the strictly mundane–are rarely considered outside of critical and biographical studies; some of them include extended passages in Scots dialect that at least one modern series generally edits severely. In his preface to a MacDonald anthology Lewis judged them, on the whole, to be journeyman work: "If we define Literature as an art whose medium is words, then certainly MacDonald has no place in its first rank–perhaps not even in its second." On the other hand, in the judgment of William Raeper, a contemporary biographer and MacDonald critic, "The world of *Phantastes* bubbles through fissures in the text [of the novels] according to the rules (or lack of them) of a fantasy and not of a realistic novel. The world of the unconscious is opened up and laid bare before the reader." There is in these novels a pervasive tension between everyday details and the urge toward symbolic presentation. Further, some of these novels show a fine grasp of the conventions of Gothic fiction–and, in the case of *David Elginbrod,* written when MacDonald was perhaps still hopeful of continuing to work with fantasy, the line between realism and the fantastic is at times very thin indeed.

Necessity led MacDonald to publish a long series of conventional novels, into which, on occasion, he also interpolates some very good tales, usually ghost stories, notably in *Wilfrid Cumbermede* (1872), *Malcolm* (1875), *Castle Warlock* (1882), and *Donal Grant* (1883). It is not surprising to find, in the novels that follow the publication of *The Princess and the Goblin,* motifs from that novel: for example, wisdom is conveyed to young Cosmo in *Castle Warlock* by an elderly lady at her spinning wheel, and in his eponymous novel Malcolm says that "cellars are the metaphysics, the garrets the poetry of the house."

MacDonald did not reserve fantasy for children only: "I do not write for children, but for the childlike, whether of five, or fifty, or seventy-five," he writes in "The Fantastic Imagination." In fact, three of his novels with chidren as protagonists–*Ranald Bannerman's Boyhood* (1871), *Gutta-Percha Willie* (1873), and *A Rough Shaking* (1890)–are not fantasies at all. All of his stories are bildungsromans, and the process of initiation can be a matter of free participation or of resistance. In his most hopeful characters resistance takes the form of temporary moral stagnation analogous to that of Dante in the prologue of the *Divine Comedy,* a work to which he alludes in *At the Back of the North Wind* (1871). In the fantasies MacDonald is able to be much more explicit about the mystical, numinous aspects of human development, since fantasy accommodates more readily than realism his symbolic presentation of spiritual development. For example, the male characters of his fantastic fiction are "licensed" to be *un*developed, more childlike, and thus less constrained to "worldly activity"–externalized agency–than the male protagonists of his realistic novels. By the same token, his female characters–it is difficult to see them quite as protagonists because of conventional constraints in gender depiction that MacDonald never seriously transgressed–have a relatively greater freedom of agency than all but very few of the females in his realistic fiction. To be sure, the females of MacDonald's fantasies are more icons–princesses or goddess figures–than not, but even so they present a strong contrast to their "realistic" counterparts.

This kind of contrast is evident in the first extended fantasies that MacDonald wrote, *Phantastes,* and the short novel *The Portent* (serialized in 1860 and published with an altered, "happy" ending in 1864). *The Portent* is MacDonald's version, one of many to come, of the Sleeping Beauty story. The passive Lady Alice, a somnambulist treated as men-

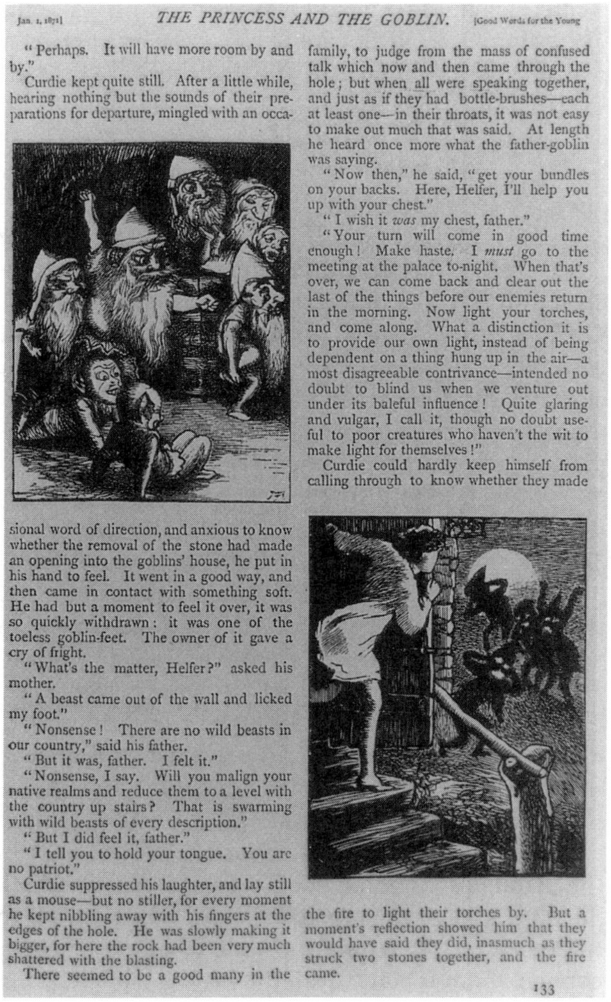

Jan. 1, 1871] *THE PRINCESS AND THE GOBLIN.* [Good Words for the Young

"Perhaps. It will have more room by and by."

Curdie kept quite still. After a little while, hearing nothing but the sounds of their preparations for departure, mingled with an occasional word of direction, and anxious to know whether the removal of the stone had made an opening into the goblins' house, he put in his hand to feel. It went in a good way, and then came in contact with something soft. He had but a moment to feel it over, it was so quickly withdrawn: it was one of the toeless goblin-feet. The owner of it gave a cry of fright.

"What's the matter, Helfer?" asked his mother.

"A beast came out of the wall and licked my foot."

"Nonsense! There are no wild beasts in our country," said his father.

"But it was, father. I felt it."

"Nonsense, I say. Will you malign your native realms and reduce them to a level with the country up stairs? That is swarming with wild beasts of every description."

"But I did feel it, father."

"I tell you to hold your tongue. You are no patriot."

Curdie suppressed his laughter, and lay still as a mouse—but no stiller, for every moment he kept nibbling away with his fingers at the edges of the hole. He was slowly making it bigger, for here the rock had been very much shattered with the blasting.

There seemed to be a good many in the family, to judge from the mass of confused talk which now and then came through the hole; but when all were speaking together, and just as if they had bottle-brushes—each at least one—in their throats, it was not easy to make out much that was said. At length he heard once more what the father-goblin was saying.

"Now then," he said, "get your bundles on your backs. Here, Helfer, I'll help you up with your chest."

"I wish it *was* my chest, father."

"Your turn will come in good time enough! Make haste. I *must* go to the meeting at the palace to-night. When that's over, we can come back and clear out the last of the things before our enemies return in the morning. Now light your torches, and come along. What a distinction it is to provide our own light, instead of being dependent on a thing hung up in the air—a most disagreeable contrivance—intended no doubt to blind us when we venture out under its baleful influence! Quite glaring and vulgar, I call it, though no doubt useful to poor creatures who haven't the wit to make light for themselves!"

Curdie could hardly keep himself from calling through to know whether they made the fire to light their torches by. But a moment's reflection showed him that they would have said they did, inasmuch as they struck two stones together, and the fire came.

133

A page from the serialization of The Princess and the Goblin *in the 1 January issue of* Good Words for the Young, *illustrated by Arthur Hughes*

tally defective by her family, is awakened intellectually and spiritually by the young tutor, Duncan Campbell, who later rescues her from her virtual imprisonment. Only the introduction of the element of "second sight" removes this story from the realm of conventional romance. In contrast to Duncan, however, Anodos, the protagonist of *Phantastes,* is the prototypical MacDonald fantasy hero.

MacDonald took the title *Phantastes* from the name of a character in Phineas Fletcher's *The Purple Island* (1633). The name Anodos (from the Greek) means, variously, "a way back" or "the pathless one"; both meanings are significant in the novel. Anodos, although twenty-one when his adventure begins, has led an aimless existence, and his career, once he enters the fairy realm, is one of seemingly endless blunders founded in naiveté and narcissism. At the end of the twenty-one-day adventure he has only begun to find the path to maturity.

Lewis claimed in the preface cited above that *Phantastes* "baptize[d] his imagination." It was both well received by the public and abused by the critics, notably the *Atheneum* review of 6 November 1858 that called the book "a confusedly furnished second-hand shop of symbols." (The *Atheneum* published a negative review of *Alice in Wonderland* seven years later.) The novel is a map, albeit a blurred and imperfect one, of the MacDonald fantasy; the later fantasies are decidedly more accomplished, with more authorial control of the narratives, but here the fantastic elements, by the compendious nature of their presentation, are readily available to the student of MacDonald's work:

> I saw that a large green marble basin, in which I was wont to wash, and which stood on a low pedestal of the same material in a corner of my room, was overflowing like a spring, and that a stream of clear water was running over the carpet, all the length of the room, finding its outlet I knew not where. And, stranger still, where this carpet, which I had myself designed to imitate a field of grass and daisies, bordered the course of the little stream, the grass-blades and daisies seemed to wave in a tiny breeze that followed the water's flow, while under the rivulet they bent and swayed with every motion of the changeful current. . . .

Anodos's adventures begin with this evocation of nature as his room becomes a forest bower. He learns, slowly, that the natural world is more than a stage or ambience–that "the world is human nature turned inside out," as MacDonald says in "Imagination." He encounters wise women, tree spirits (harmful as well as benevolent), and his own Galatea, whom he, Pygmalion-like, must release from the stone that imprisons her, but only after he learns that love includes the relinquishing of possessive desire. He must also deal with the shadow that is the image of his warped self-love and learn, in his own person, that victory can encompass self-abnegation even to death.

MacDonald experimented with combining the realistic and the fantastic modes in *Adela Cathcart* (1864). Adela, suffering from severe depression, is prescribed a "course" of stories, narrated by a group of friends in the days following Christmas. MacDonald interpolated two nonfantastic stories, "The Broken Swords" (serialized in 1854) and "The Wow o' Rivven" (sometimes titled "The Bell"), but he also included five fantasies: "The Giant's Heart," "The Castle," "The Cruel Painter," "The Shadows," and one of his masterpieces, "The Light Princess." The *Atheneum* (7 May 1864) suggested that MacDonald had ransacked his desk for old stories, and subsequent editions, in which MacDonald substituted other stories, retrospectively offer a validation of this criticism. Indeed, some later editions of *Adela Cathcart* mention but do not print stories that had been collected or published elsewhere. He repeated this practice only once more, interpolating the extended fairy tale "Little Daylight" into *At the Back of the North Wind.*

Despite its formal awkwardnesses *Adela Cathcart* is of interest as a fictive stage for the presentation of MacDonald's ideas about the efficacy of story; Adela's spiritual illness is best addressed by the language of the spirit, that is, symbolically in stories. The novel also stages a polemic of fantasy; Adela's aunt, Mrs. Cathcart, is cast as the perpetual naysayer to the proceedings, and each storyteller succeeds in countering her objections.

Of the stories included in the first edition of *Adela Cathcart,* "The Castle" and "The Giant's Heart" are the least significant. The former is strict allegory, which is surprising since MacDonald was annoyed when readers approached his stories in this way. In "The Castle" a group of siblings yield to the temptations of a life of pleasure in the absence of their eldest brother and are stricken to repentance at his return after their revels become a rout. "The Giant's Heart" presents the worst features of Victorian fairy tales–the protagonists are named Tricksy-Wee and Buffy Bob, and their vengeance against the giant Thunderthump is a sadistic extortion of promises to reform. There may be an element of mockery in the original framing of this story; even though MacDonald did allow it to be reprinted in collections of his stories, the children to whom it is told in *Adela Cathcart* do not hesitate to express their dislike of it.

MacDonald balances moralizing tendencies with fantastic elements to somewhat better effect in "The Shadows," a dream-fantasy in which "shadows" are fairy folk threatened by the technology of artificial light. They make their case to the poet Ralph Rinkelmann by explaining the reforming effects of the images they cast. The premise, the retreat of faerie in the face of modernity, is interesting, but it cannot sustain the very slight narrative.

"The Cruel Painter" shows, in its German setting and its artist protagonist, its debts to E. T. A. Hoffmann, whose fiction MacDonald admired. Karl von Wolkenlicht (the name suggests both clouds and light) apprentices himself, and becomes an unsuspecting experimental subject, to the painter Teufelsburst ("devil's brush"), who is noted for using the face of his beautiful daughter, Lilith, in paintings depicting torture and death. The story is given a happy ending–Teufelsburst is last seen as a doting grandfather to the young couple's children–that is a persistent tendency of MacDonald's even at the expense of narrative credibility.

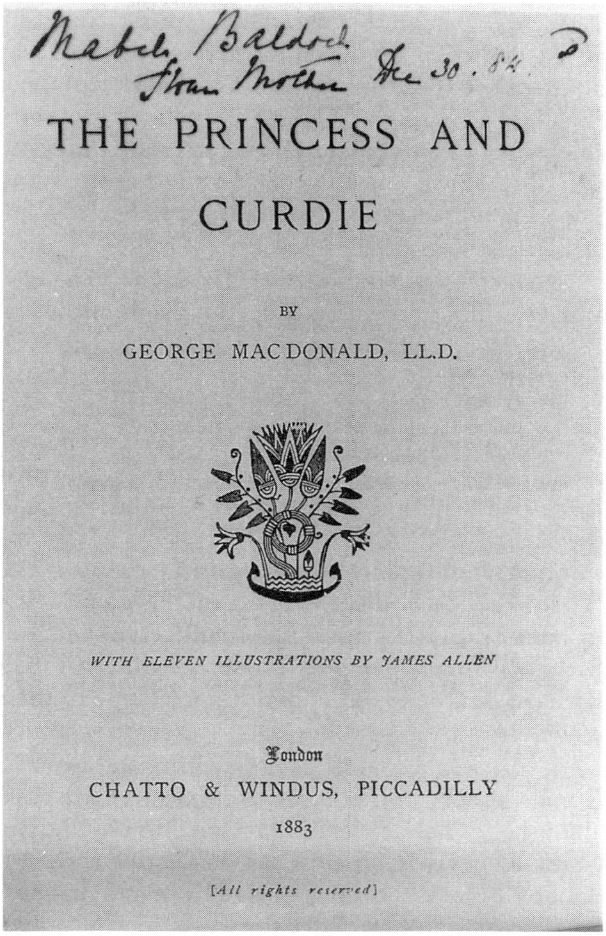

THE PRINCESS AND CURDIE

BY

GEORGE MAC DONALD, LL.D.

WITH ELEVEN ILLUSTRATIONS BY JAMES ALLEN

London

CHATTO & WINDUS, PICCADILLY

1883

[*All rights reserved*]

Title page of MacDonald's second novel about the princess Irene and the miner-boy Curdie

Only one other short fantasy by MacDonald, "The Wise Woman" (variously titled, 1874), has received as much praise as the first interpolated tale, "The Light Princess." In this story MacDonald mingles profundity with play in a retelling of the Sleeping Beauty story that, for all of his returns to the motif, he never surpassed. The princess is cursed at her christening to lose her gravity, both physical and psychological. She regains them to some degree only in water, and when the depletion of the springs in her father's kingdom threatens her life, the prince who loves her all but sacrifices his own. Her first tears, of compassion, negate the curse.

The erotic elements in "The Light Princess" are but thinly disguised, and Ruskin was made somewhat uneasy by their presence. Writing to MacDonald in letters dated 22 July 1863 and 13 April 1864, he claimed that MacDonald had pilloried him in the unsympathetic character of Aunt Cathcart. Ruskin did, in fact, criticize "The Light

Princess" as including material inappropriate for its young audience, but he added that MacDonald was not to delete the erotic elements but to present them "in a less telling way." On the other hand, MacDonald's son and biographer Greville notes, in the context of Ruskin's objection, that his father's reading "The Light Princess" gave him "imperishable happiness."

MacDonald included "The Light Princess" and "The Shadows" in a collection he published in 1867, *Dealings with Fairies,* which also included "The Carasoyn," first serialized in part in *Argosy* (April 1866) as "The Fairy Fleet"; "Cross Purposes"; and "The Golden Key." The first two stories are minor human/fairy encounters. In the former a boy named Colin first helps and then is tricked by the fairies; in the second half of the story he rescues one of his children, abducted by the fairies, in a ten-year quest. "Cross Purposes" deals with the escape from captivity among the fairies of two children, Alice and Richard; Alice, of slightly higher social status than Richard, learns to value him in the course of their adventures.

"The Golden Key" is noteworthy among MacDonald's short stories. It presents, in short compass, all of the formal details of MacDonald's mature fantasies. The wise woman who sets the two children, the boy Mossy and the girl Tangle, on their quest for the land "from whence the shadows fall" anticipates the great-grandmother of the "Princess" novels. Mossy's finding of the golden key at the rainbow's end is the beginning of his adventures, and he and Tangle can only reach their goal by passing through an unnameable valley that transparently symbolizes death. The children's encounters with the wise woman and with the Old Men of the Sea and of Fire are all expressive of MacDonald's concept of life as an endless "becoming," and at the story's end they are ready for the ascent through the rainbow, a "Jacob's Ladder," to their goal.

MacDonald had begun editing the journal *Good Words for the Young* in 1869, and he continued the increasingly uncongenial work, even though the journal began to fail, for the next three years. He was also entering the most productive period of his career, the fifteen years beginning in 1871. In that year he published *At the Back of the North Wind,* the only fantasy in which he addressed contemporary urban life.

In this story Diamond, a coachman's son, is befriended by North Wind, here personified as a beautiful woman, alternately kindly and fierce. In his journeys with North Wind, Diamond is made to consider the vagaries of nature, destructive and helpful by turns. He first journeys to the "back" of the North Wind, a pastoral land surrounded by icy stillness, during a serious illness, and the novel ends: "They thought he was dead. I knew that he had gone to the back of the north wind."

Diamond's innnocent goodness seems at times to be too much insisted upon–he effortlessly charms fretful babies and tames raging drunkards–but MacDonald's fiction includes several examples of "holy strangeness" in those who are afflicted or simply childlike. On the other hand, childhood mortality is handled here with restraint; by the end of the novel Diamond is less the eldest child of a poor family than a confident quester, making the next step in his "becoming."

The social context of *At the Back of the North Wind* is merely a matter of setting for MacDonald; despite his admiration for and friendship with Christian social reformer Frederick Denison Maurice, active engagement in a social agenda was not an aspect of his fiction, and he tended simply to chastise haughty characters from the upper classes while confirming the status quo. The crossing sweepers, Nanny and Jim, function as undercharacterized foils for Diamond and as part of the audience for the interpolated fairy tale "Little Daylight," another of MacDonald's Sleeping Beauty stories, varied this time by a curse that dooms the princess to sleep by day and to wane in strength and beauty with the phases of the moon.

The year 1871 saw not only the publication of *At the Back of the North Wind* but also the publication of "The Gray Wolf" and "Uncle Cornelius: His Story" in *Works of Fancy and the Imagination,* ten volumes that included *Phantastes,* all of his shorter fantasies, and a selection of his poems. "The Gray Wolf," a werewolf story, and "Uncle Cornelius," a ghost story, show MacDonald at work on two standard approaches to the fantastic. More significant was the publication of *The Princess and the Goblin,* which had begun as a serial in November 1870 in *Good Words for the Young.*

In *The Princess and the Goblin* MacDonald achieved a formal perfection of greater complexity than that found in "The Light Princess." The stories of the princess Irene and the miner boy Curdie, her rescuer, rival the best traditional fairy tales, and the great-grandmother, their mentor, is more than a guide encountered and left behind; she is an enigma, ageless and associated with the magic of moonlight, whose elucidation unfolds for the reader as it does for the children in the course of the story. Into the simplicity of the tale–a princess threatened with abduction by troglodytes–MacDonald interweaves various complexities, as Irene and Curdie learn to accept guidance that is not grounded in unalloyed rationality. Irene must trust her grandmother in the face of doubt, and Curdie must learn to prize wisdom even if it is embodied in a child. MacDonald's array of archetypal images–the stairway to the grandmother's garret, the caves in which the goblins plot against the princess and from which both children must escape, and the floods that ultimately cleanse the princess's home of the goblins–works powerfully as a unifying element of the story, effectively countering minor traces of sentimentality in the presentation of Irene and of Curdie's mother.

MacDonald was not so successful with the sequel, *The Princess and Curdie,* serialized in *Good Things* in 1877 and published in 1883. Much of the interest in Irene and Curdie in the earlier book centered on their development; in the sequel Irene is both static and conventional, a princess in need of rescue, this time from her father's treacherous ministers. Curdie, once he advances beyond an initial period of mental stagnation with the help of the great-grandmother, is equally static and at times almost repellent as he becomes the appointed scourge of the traitors and of the idle, self-indulgent staff of the king's household. His assistant in this task is both pathetic and disturbing: Lina, a hideous, doglike creature whose name recalls that of the repentant Mary Magdalene of apocryphal tradition, is working out her penance–Curdie surmises that she was once a woman who was "naughty"–and her eventual reward is dissolution in the cleansing rose-fire of the great-grandmother. MacDonald also inverts his familiar motif of "becoming" to show the devolution of human nature into beast-nature in the story's antagonists. The villains are routed in the end, but long after the deaths of Curdie and Irene the whole town collapses, literally undermined by a ruler greedy for the gold in its foundations. The whole has a vindictive edge that seems an echo of the religion of MacDonald's childhood. His fantasies and his other fiction occasionally tend in this direction when punishment is at issue.

Readers of MacDonald at the time perhaps were prepared for this negative reworking of fairy-tale conventions in *The Princess and Curdie* by a story that had appeared two years earlier, "The Wise Woman" (first serialized in *Good Things* as "A Double Story" in 1875). In the parallel stories impossible children–one a princess, the other the daughter of shepherds, but both the offspring of foolishly indulgent parents–are both taken in hand by a wise woman whose method can best be described as purgatorial in the Dantean sense; each wrong act becomes its own punishment. In the end the self-willed and passionate princess is reformed, and her parents

MacDonald in old age

are punished with blindness that outwardly depicts their true nature. The shepherd's child becomes more like herself, that is, confirmed in self-satisfaction, and becomes her parents' (more precisely, her mother's) punishment. The rigorous logic of punishment, that passion can be directed to love while pride is resistant by nature, does not mitigate the tale's disturbing effect, which is intensified for today's reader by the assignment of the more hopeful role to the princess at the expense of the shepherd's child.

In the December 1879 issue of *The Graphic* MacDonald published "The Day Boy and the Night Girl," known in most modern editions as "Photogen and Nycteris," the names of the hero and the heroine respectively. The powerful female in this story is a witch of an experimental cast of mind: Watho, insatiably curious, sequesters two pregnant women in order to influence their children by controlled environments. One, Vesper, is blind, and Watho surrounds her with soft fabrics and mournful music, never allowing any light either for the mother, who dies in childbirth, or the daughter, Nycteris, "the night-enduring." Aurora, Photogen's mother, received precisely the opposite treatment, as does her son after Watho sends the mother away after falsely reporting the child as dead. The children eventually meet, but each must encounter the fears engendered, in Nycteris by sunlight and in Photogen by darkness. Eventually, as the genre demands, Watho is defeated, killed by Photogen. The events of the story are conventional, but Watho, so tormented by her desire for knowledge that she is described as having "a wolf in her mind" that brings on unexplained "nervous attacks," is unlike any character MacDonald created. She cannot be dismissed as an instance of contempt for female learning, since an intellectually unawakened female in his novels is one spiritually at risk, and women who struggle for learning receive nothing but praise. It is Watho's tyrannical possessiveness, shown first in her treatment of the mothers and then of the children, that merits the censure she receives. Even as early as *David Elginbrod* MacDonald showed his abhorrence for the thirst for personal domination. In 1891 he published *The Flight of the Shadow,* a minor love story most interesting for the hero's domineering mother, Lady Cairnedge, who has broken the bond between the heroine's identical twin uncles by her seductive behavior and who seems a rehearsal for Lilith, the focus of his last fantasy novel. In *Lilith* he presented his most extreme and disturbing version of such a character.

MacDonald wrote to his son Greville (in a letter dated 17 September 1894) that he felt that *Lilith* would be his last novel. As the writing progressed MacDonald worked out multiple drafts, not his usual practice, in a sequence so complex that the relationships have not been fully explained. His wife was disturbed by the dark tale, and he appealed to Greville to adjudicate the disagreement. Greville pronounced the work inspired, and MacDonald completed and published it.

Mr. Vane, the protagonist of *Lilith,* is reminiscent of Anodos, and his name, suggestive of wavering and futility, reinforces the similarity. Through an arrangement of mirrors in his house and guided by Mr. Raven–a former librarian of the house who is sometimes a man, sometimes an actual raven, and ultimately Adam, keeper of the world's cemetery–Vane enters an alternate world of "seven dimensions" that is subject to Lilith. The character Lilith derives from the apocryphal story from the Talmud of Adam's first wife, and MacDonald makes her a shape-changer; further, she is a destroyer because she can acknowledge no will but her own, and she hates children because of a prophecy that a child will be her death. Vane witnesses her eventual submission to redeeming death, a gift he had initially refused, and returns to his world, having confronted the result of his own self-willed behavior, the death of Lilith's daughter at her mother's hands.

In *Lilith* MacDonald explores negative aspects of female sexuality more directly than in any of his works. Lilith exercises every form of domination, not only the political subjection of her land but also the overtly sexual power she uses against Vane, first as a vampire when he discovers her helpless in a forest and later as a capricious object of desire. The maliciously seductive female can be found in MacDonald's realistic fiction–for example, Lady Lufa in *Home Again* (1887)–but nowhere is there a more intense projection of sexual fear in his work than in *Lilith.* Her daughter, Lona, is a foil in her protection of a colony of children to her mother's rapacity, as is Mara, an enigmatic figure who combines attributes of a nature goddess with those of the sorrowing mother, Rachel, who weeps for her children. These two characters thus counterbalance the threatening figure Lilith.

H. G. Wells wrote in a 24 September 1895 letter to MacDonald that MacDonald's use of multiple dimensions was an idea that he was also working on (*The Time Machine* was published in 1895): "I have wanted to get into such kindred worlds for the purposes of romance for several years, but I've been bothered by the way. Your polarization and mirror business struck me as neat in the extreme." Critics remain divided about the work. Some, such as Robert Lee Wolff, see it as senile crankiness; others see it as a rich, densely packed symbolical work; and Auden in a 1954 introduction for the work praised the tightness of its allegorical structure.

MacDonald left behind him a body of work, especially his shorter fantasies, equal to any of the stories produced in that efflorescence of fantasy beginning in the mid 1860s; his literary production was more abundant and of a more consistently sustained quality than that of any of his contemporaries (one need only contrast MacDonald's oeuvre with Lewis Carroll's progression from his "Alice" books to *Sylvie and Bruno*). He also produced, at the beginning and at the end of his career, two "adult" fantasies that, despite critical reservations with regard to formal elements, gain more serious critical attention as time progresses. No study of modern fantasy is complete without reference to George MacDonald, and much contemporary fantasy rests on the bedrock of his best work.

Bibliographies:

J. M. Bullock, "A Centenary Bibliography of George MacDonald," *Aberdeen University Library Bulletin,* 5 (February 1925): 679–747;

George MacDonald: A Bibliographical Catalog and Record (Fairvax, Va.: Privately published for the

Marion E. Wade Collection, Wheaton College, Wheaton, Illinois, 1984).

Biographies:

Greville MacDonald, *George MacDonald and His Wife* (London: Allen & Unwin, 1924);

Kathy Triggs, *The Stars and the Stillness: A Portrait of George MacDonald* (Cambridge, England: Lutterworth Press, 1961);

William Raeper, *George MacDonald* (Tring, Hertfordshire & Batavia, Ill.: Lion, 1987).

References:

Rolland Hein, *The Harmony Within: The Spiritual Vision of George MacDonald* (Grand Rapids, Mich.: Eerdmans, 1982);

C. N. Manlove, "Circularity in Fantasy: George MacDonald," in his *The Impulse of Fantasy Literature* (Kent, Ohio: Kent State University Press, 1983), pp. 70–92;

Manlove, "George MacDonald (1824-1905)," in *Modern Fantasy: Five Studies* (Cambridge: Cambridge University Press, 1975), pp. 55–98;

Roderick F. McGillis, *For the Childlike: George MacDonald's Fantasies for Children* (West Lafayette, Ind.: Children's Literature Association / Metuchen, N.J.: Scarecrow Press, 1992);

McGillis, "George MacDonald and the Lilith Legend in the XIXth Century," *Mythlore,* 10 (Winter 1977);

Karen Michalson, "George MacDonald and *Phantastes*," in her *Victorian Fantasy Literature: Literary Battles with Church and Empire* (Lewiston, N.Y.: Edwin Mellen, 1990), pp. 61–97;

Stephen Prickett, *Romanticism and Religion: The Tradition of Coleridge and Wordsworth in the Victorian Church* (Cambridge: Cambridge University Press, 1976);

William Raeper, *The Gold Thread: Essays on George MacDonald* (Edinburgh: Edinburgh University Press, 1990);

Richard H. Reis, *George MacDonald* (New York: Twayne, 1972);

Robert Lee Wolff, *The Golden Key: A Study of the Fiction of George MacDonald* (New Haven: Yale University Press, 1961).

Papers:

The largest collection of George MacDonald's letters and papers is held by Yale University in the Beinecke Library. There are also a few letters in the National Library of Scotland and at the Houghton Library at Harvard University, with others variously collected at the New York Public Library; the University of California, Los Angeles; and King's College, Aberdeen.

Arthur Llewelyn Jones Machen

(3 March 1863 - 15 December 1947)

Donald M. Hassler
Kent State University

See also the Machen entries in *DLB 36: British Novelists, 1890–1929: Modernists* and *DLB 156: British Short-Fiction Writers, 1880–1914: The Romantic Tradition.*

Arthur Machen (photograph by E. O. Hoppé)

BOOKS: *Eleusinia, by a Former Member of H.C.S.* (Hereford, Wales: Privately printed, 1881);

The Anatomy of Tobacco: or Smoking Methodised, Divided, and Considered after a New Fashion, by Leolinus Siluriensis, Professor of Fumifical Philosophy in the University of Brentford (London: Redway, 1884; New York: Boni & Liveright, 1923); as Arthur Machen, with a new introduction (New York: Knopf, 1926);

The Chronicle of Clemendy (London: Privately printed, 1888; Carbonnek, N.Y.: Privately printed, 1923);

The Great God Pan and the Inmost Light (London: John Lane, 1894; Boston: Roberts, 1894);

The Three Impostors (London: John Lane, 1895; Boston: Roberts, 1895);

Hieroglyphics (London: Richards, 1902; New York: Kennerley, 1913);

The House of the Hidden Light (London: Privately printed, 1904);

The House of Souls (London: Richards, 1906; Boston: Estes, 1906; revised edition, New York: Knopf, 1922);

Dr. Stiggins: His Views and Principles (London: Griffiths, 1906; New York: Knopf, 1925);

The Hill of Dreams (London: Richards, 1907; Boston: Estes, 1907);

The Angels of Mons: The Bowmen and Other Legends of the War (London: Simpkin, Marshall, Hamilton, Kent, 1915; New York: Putnam, 1915);

The Great Return (London: Faith Press, 1915);

The Terror: A Fantasy (London: Duckworth, 1917); republished as *The Terror: A Mystery* (New York: McBride, 1917);

War and the Christian Faith (London: Skeffington, 1918);

The Secret Glory (London: Secker, 1922; New York: Knopf, 1922);

Far Off Things (London: Secker, 1922; New York: Knopf, 1922);

Caerleon Edition of the Works, 9 volumes (London: Secker, 1923);

Things Near and Far (London: Secker, 1923; New York: Knopf, 1923);

The Shining Pyramid (Chicago: Covici-McGee, 1923; revised edition, London: Secker, 1925; New York: Knopf, 1925);

The Collector's Craft (London: Privately printed, 1923);

Strange Roads (London: Classic Press, 1923);

The Grande Trouvaille: A Legend of Pentonville (London: Privately printed, 1923);

Dog and Duck (New York: Knopf, 1924; London: Cape, 1924);

The London Adventure; or, The Art of Wandering (London: Secker, 1924; New York: Knopf, 1924);

The Glorious Mystery, edited by Vincent Starrett (Chicago: Covici-McGee, 1924);

Precious Balms (London: Spurr & Swift, 1924);

Ornaments in Jade (New York: Knopf, 1924);

The Canning Wonder (London: Chatto & Windus, 1925; New York: Knopf, 1926);

Dreads and Drolls (London: Secker, 1926; New York: Knopf, 1927);

Notes and Queries (London: Spurr & Swift, 1926);

A Souvenir of Cadby Hall (London: Lyons, 1927);

Tom O'Bedlam and His Song (Westport, Conn.: Apellicon, 1930);

Beneath the Barley: A Note on the Origins of Eleusinia (London: Privately printed, 1931);

In the 'Eighties (Amersham: Privately printed, 1931);

The Green Round (London: Benn, 1933; Sauk City, Wis.: Arkham House, 1968);

The Cosy Room and Other Stories (London: Rich & Cowan, 1936);

The Children of the Pool and Other Stories (London: Hutchinson, 1936);

Holy Terrors (London: Penguin, 1946);

Tales of Horror and the Supernatural, edited by Philip Van Doren Stern (New York: Knopf, 1948; London: Richards, 1949);

Bridles and Spurs (Cleveland, Ohio: Rowfant Club, 1951);

A Critical Essay (Lakewood, Ohio: Privately printed, 1953);

A Note on Poetry (Wichita, Kans.: Four Ducks Press, 1959);

Eleusinia and Beneath the Barley (West Warwick, R.I.: Necronomicon Press, 1988);

Ritual and Other Stories (Lewes: Tartarus Press, 1992);

The Secret of the Sangraal (Lewes: Tartarus Press, 1995).

OTHER: Marguerite, Queen of Navarre, *The Heptameron,* translated by Machen (London: Dryden Press, 1886; New York: Knopf, 1924);

Beroalde de Verville, *Fantastic Tales or The Way to Attain,* translated by Machen (London: Privately printed, 1890; New York: Boni & Liveright, 1923);

Jacques Casanova, *The Memoirs of Jacques Casanova,* 12 volumes, translated by Machen (London: Privately printed, 1894);

Casanova, *Casanova's Escape from the Leads,* translated by Machen (London: Casanova Society, 1925; New York: Knopf, 1925);

Tobias Smollett, *The Expedition of Humphry Clinker,* edited, with an introduction, by Machen (New York: Modern Library, 1929);

Lady Hester Lucy Stanhope, *Remarks upon Hermodactylus,* translated by Machen (London: Privately printed, 1933);

Charles Dickens, *A Handy Dickens,* edited, with an introduction, by Machen (London: Constable, 1941).

Not only is Arthur Machen remembered for his supernatural war story about Anglo-Saxon archers, "The Bowmen" (published in 1914 in *The Evening News*), and as an important influence on modern fantasy writers such as H. P. Lovecraft and Ray Bradbury, but also much of his writing expresses a wonderful vision of mystical union that is a foundation for the fantastic in literature. As Machen recollects in *Beneath the Barley* (1931), he experienced an early moment of insight, just before beginning the poem *Eleusinia* (1881), in which he sensed that there exists a timeless and spiritual reality underneath what humans perceive in the material world. A major proportion of his large literary output was devoted to the skillful and repeated expression of this vision.

For one who grew up to believe in and to write about this mystical union of all things hidden beneath the surface of matter, Machen's date of birth, 3 March 1863, has a sort of numerological unity, like the stanzas in Dante that are always adding up to the mystical nine. Machen was the only child of a Church of England clergyman in South Wales, and his loneliness and sense of mystical union with the past could not help but be nurtured by his growing knowledge of both the Roman and the Arthurian legends associated with his hometown of Caerleon. Lovecraft thought of himself as linked mysteriously through time back to ancient Rome, a concept he may have derived partly from his favorite writer, Machen.

Born Arthur Llewelyn Jones, Machen began to sign himself in the familiar way after his father assumed the name Jones-Machen in 1874, probably in the hope of receiving some support from his wife's family for young Arthur's education. Machen was given a solid classical training at Hereford Cathedral School and became skilled at the ancient languages, French, and theology. But he had to leave formal schooling after this because of lack of funds; he failed the entrance exams for medical school because of a poor aptitude for arithmetic. Through the 1880s he alternated between living in London, where he did hackwork for booksellers, and wandering the countryside near his family home in South Wales. His mother, an invalid, died in 1885, and his father two years later. Machen received enough of an inheritance

Cover for Machen's 1915 collection titled The Angels of Mons: The Bowmen and Other Ledgends. *The title story is about the British Army receiving supernatural assistance during World War I*

to support himself and his writing for the next fourteen years. He married Amelia Hogg on 31 August 1887.

In the 1890s Machen began to move freely in the London artistic world of decadence and experimentation. He established an acquaintance with Oscar Wilde and became good friends with A. E. Waite, the occultist and mystic who later was an influence on the Christian fantasy writer Charles Williams through the Fellowship of the Rosy Cross. Machen did most of his strongest fantasy writing in this decade, though much of it was not published until later.

The death of Machen's wife at the turn of the century probably caused him to join the Order of the Golden Dawn; though his ideas had always been mystical, they tended to find their best expression in writing rather than in ritual, so his participation in the "orders" was always somewhat skeptical. Also at this time he began a twenty-year period of regular employment, in contrast to the major portion of his long writing life, which was freelance. He acted with the Benson Shakespeare Repertory Company from 1901 until 1909 and then was a London journalist until 1921. His sense of personal presence and social skills acquired from these jobs transformed the shy mystic into the impressive literary lion he remained for the rest of his life.

He married Dorothie Purefoy Hudleston, an actress in the Benson Company, on 25 June 1903; she became a major part of the literary circle around Machen. His writings were rediscovered in the 1920s and were greatly admired in America. He was given a civil list pension in 1932 that was increased in 1938. Machen died on 15 December 1947, preceded only a few months earlier by his second wife.

At the age of seventeen, Machen set out to write a verse pamphlet based on the ancient Greek Eleusinian mysteries. In a story that he simply called "N," which finally appeared in *The Cosy Room and Other Stories* (1936), Machen coins a word derived from Greek that best explains both his technique and his ideas from this early fantasy poem onward. *Perichoresis* refers to the notion of an interpenetration of times and spaces and is one of the ways in which people can touch the "hidden language" underneath the material world. Not only did ancient mystery religions seek to invoke these "lost words," but Machen's own imaginings also reach backward and down to such a fantastic world. But his verse-writing skills never were equal to his mystical insight and fervor of image. Though it is a valuable pamphlet for collectors, no critic praises the text of *Eleusinia.* Some later Machen doggerel that concludes the long story "A Fragment of Life" from *The House of Souls* (1906) reads as better verse on the same theme:

So that, where'er I look, I see
A world of a Great Majesty.
The sullen river rolls all gold,
The desert park's a faery wold,
When on the trees the wind is borne
I hear the sound of Arthur's horn
I see no town in grim grey ways,
But a great city all ablaze
With burning torches, to light up
The pinnacles that shrine the Cup.

Those few lines of octosyllabic couplets contain all the ingredients for Machen's fantasy stories and longer fictions of the 1890s and beyond, in which contemporary life is connected to the mythic past, including the legends of King Arthur and the Holy Grail quest. Machen's thought and writing contains, also, a deeply evil side. But before he fully explored these themes, Machen had to try out what he calls "the grand manner" and to see how funny such fantasy could be.

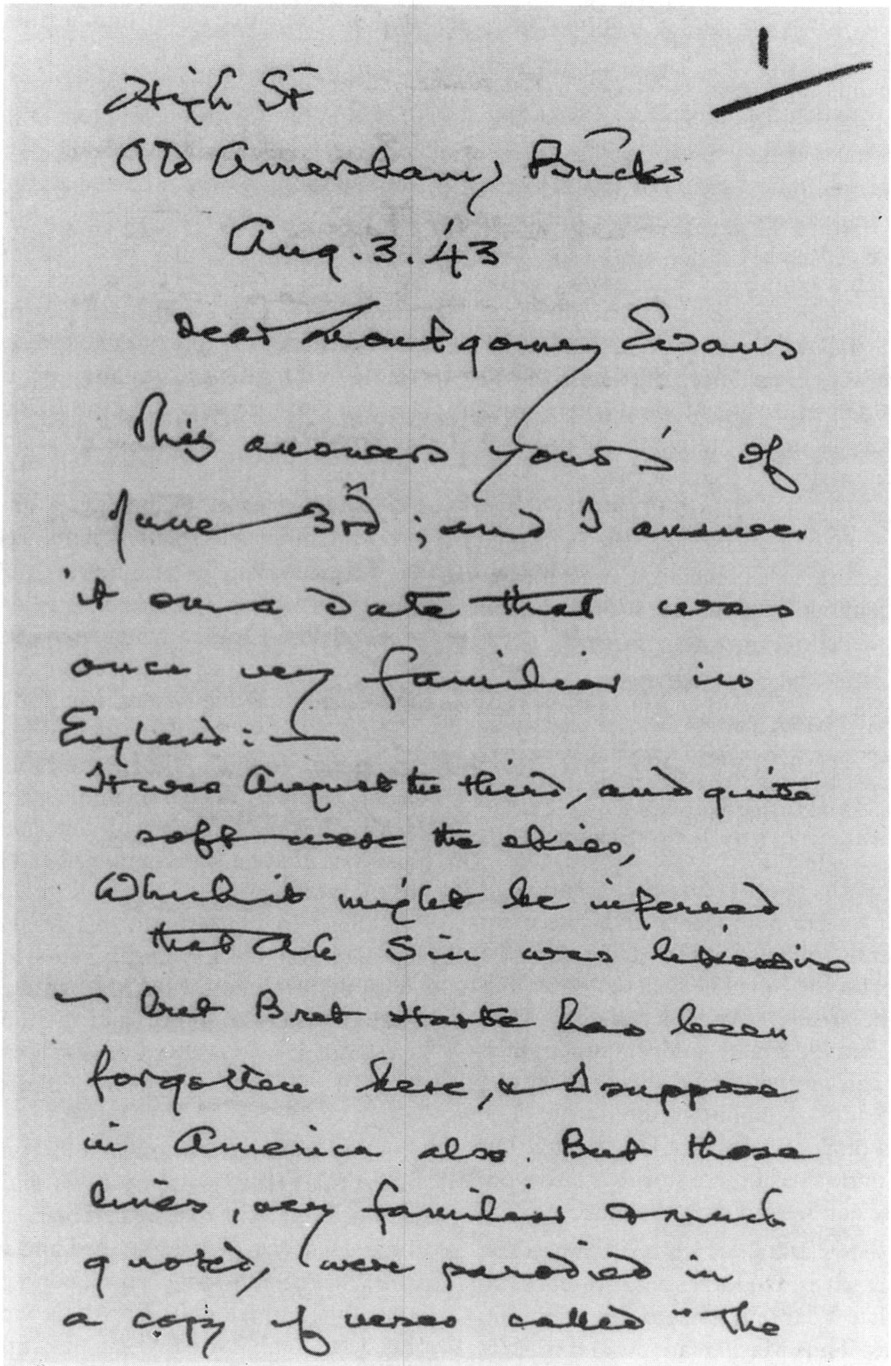

1

High St
Old Amersham, Bucks
Aug. 3. '43
Dear Montgomery Evans
This answers yours of June 3rd; and I answer it on a date that was once very familiar in England:—
It was August the third, and quite soft were the skies,
Which it might be inferred that Ah Sin was likewise
— but Bret Harte has been forgotten here, & I suppose in America also. But these lines, very familiar & much quoted, were parodied in a copy of verses called "The

The first page of Machen's 3 August 1943 letter to Montgomery Evans (Sue Strong Hassler and Donald M. Hassler, Arthur Machen and Montgomery Evans: Letters of a Literary Friendship, 1923–1947, *1994)*

The Chronicle of Clemendy (1888) is a collection of fantastic tales set in medieval Gwent, which is the older name for the region of South Wales where Caerleon is. In these tales Machen tries to sound like Rabelais or Margaret of Navarre in a pastiche of medieval style–"How a Knight of Uske Kept Guard over a Tree" is the title of one of the tales. Machen was beginning to do translating work in London then. In a 1923 introduction to the book, Machen recalls how much he wanted to write "in the grand manner," and that *The Chronicle of Clemendy* was originally to be called "The Glory of Gwent" to indicate the majesty and the lost or hidden dream. Machen also calls himself in this book the "Silurist"; the Silures were the ancient Celtic tribe in

Gwent. The overall effect is a sort of romping facetiousness.

His more serious mysticism, or perichoresis of the hidden, comes in the shorter tales that he published next. Most of these tales of the fantastic and the horrible, from the 1890s, combine a vision of sanctity with evil, and the evil often has sexual overtones. The most well known of these is the title story of *The Great God Pan* (1894), in which a young girl is made to see the hideous reality of Pan, or the devil. Machen is fascinated by, or chose to make a matter of fascination for his readers, the necessity of exploring the depths of evil as one seeks the hidden, inner life. Modern fantasy continues to do just this as an alternative to the uniform grayness of modern urban and technological life.

In the equally well-known story "The White People," which appears with a reprinting of the Pan story in *The House of Souls,* the recluse intellectual, Ambrose, who is almost like a monk, states his quest clearly at the outset:

> "Sorcery and sanctity," said Ambrose, "these are the only realities. Each is an ecstasy, a withdrawal from the common life. . . . I think you are falling into the very general error of confining the spiritual world to the supremely good; but the supremely wicked, necessarily, have their portion in it."

Objects, books often but also gems and important chemicals or minerals, often vaporize at the end of a Machen story from this period. The final sentence from "The Inmost Light" reads: "And then a thin white flame burst forth from the smoke, and shot up into the air and vanished; and on the ground there lay a thing like a cinder, black and crumbling to the touch." The reader is reminded of alchemy and of Dr. Faustus, but the overlay in the narrative of these stories is also wonderfully modern, scientific, and even skeptical.

The Three Impostors (1895), which is structured as a set of fantastic narratives, characterizes rational London gentlemen of late Victorian times as the tellers and listeners of the tales. Thus Machen adopts and masters the technique of H. G. Wells and of Robert Louis Stevenson in telling a fantastic story from a skeptical point of view in order to heighten the wonder. Machen says in *Hieroglyphics* (1902) that he imitated Stevenson's matter-of-fact style, a departure from the "grand manner" of *The Chronicle of Clemendy,* and that as he wrote later in the decade he made a conscious effort to create more of a style of what he calls "ecstasy." Nevertheless, *The Three Impostors* remains Machen's most popular book and has been reprinted more than any other. Some of the mystery in the book comes from his evoking of pre-Celtic "little people" as a horrible continuation of ancient times into modern England. Hence he makes "the glory of Gwent" a fine mix of both sorcery and sanctity.

Machen's notion of "ecstasy," however, is presented in his one sustained book of literary theory and a theory of the fantastic, *Hieroglyphics.* He writes with wit and satiric power as he attacks the "greyness" of realistic writing or of "small" writing. His candidates for creators of ecstasy are writers such as Rabelais or Miguel de Cervantes. This tendency toward traditional literary Romanticism in Machen, which would explain the inner life, the role of evil, and the medievalism, is reinforced by his several attempts to produce a long autobiographical novel. In the 1920s, when collectors and readers wanted to know more about the writer, he did produce two fine, book-length autobiographies. *The Hill of Dreams* (1907), however, on which he began working in the late 1890s, is fantasized autobiography.

Lucian Taylor is much like Machen: he has a Welsh background; he attempts to write in London; he depends on a legacy when the writing does not pay well; and he dreams about ancient Rome, the hills back in Wales, and King Arthur. Machen first titled the text "The Garden of Avallaunius," as he explains in his introduction to the book. He took the word *Avallaunius* from a Roman-British name he had uncovered in research; his version means "the man of Avalon." He also says that he wanted to write a "Robinson Crusoe of the soul," in which a man is lonely in "the midst of millions." In any case, the images and writing are rich in this novel, but it is finally a story of the decay of the spirit that ends in suicide. Machen survived all the spiritual and literary agony, and this may be because he wrote of it so vividly. Some critics have called this novel a decadent book that represents the ending of the 1890s.

Machen's other long autobiographical novel, *The Secret Glory* (1922), attempts two major literary effects and uses material that he had been gathering for some time. Hence it is hardly unified and not as powerful as *The Hill of Dreams.* Machen spends a good bit of time in the novel satirizing the British public schools in which his hero, Ambrose Meyrick, has studied. The fantasy element is that Meyrick sets out on a quest for something much like the Holy Grail, or "graal," as Machen liked to spell it, and becomes a martyr to his quest. The book is filled with evocations of the Welsh countryside and of the magical power that it exercised on Machen's imagination. Clearly there is a lot of repetition in Machen's fantasy writing.

But his work in journalism and World War I resulted in some significant departures from this dream-vision sort of fantastic romance writing, and these new departures helped promote his large degree of fame in the 1920s. Alfred Turner, who was one of Machen's editors at the *Evening News,* assigned him features. Following the desperate retreat at the Battle of Mons early in the war, Machen, lacking any firsthand knowledge, invented a fantastic story of a soldier who has a vision of Saint George leading ghostly archers from England's glorious past against the thousands of Germans. During the months after the appearance of "The Bowman" in the newspaper, Machen created some other legends about fantastic appearances at the front, which were collected the following year in *The Angels of Mons* (1915).

The Terror (1917) also grew from a wartime feature story for Turner and is a well-plotted fantastic mystery about animals suddenly attacking men. Critics have noted that this story is different from Machen's usual mode of dream-vision fantasy. It moves well in the realistic setting of the war and contains some mysterious puzzles that have to be unraveled. Machen even sold the film rights for this story in 1946. The film was never made, but Alfred Hitchcock's film *The Birds,* which was based on Daphne du Maurier's story, is indebted partly to Machen for its central idea. Another journalistic fantasy from this same period is *The Great Return* (1915), in which Machen returns to his favorite Grail story and tells a well-plotted story about visions of some simple people in western Wales.

After the war Machen enjoyed great popularity and in a sense was discovered in America. Much of his work was reprinted at this time, and he was able to publish *The Secret Glory* and write two autobiographies. But he was writing less and enjoying his celebrity. Finally, he did return briefly to the dream-vision mode of fantasy and produced the story "N," collected in *The Cosy Room,* which contains his best expression of the notion of perichoresis, his theoretical foundation for the existence of the hidden life beneath the material. Unlike his firm, journalistic fantasies, this story lacks a conclusive plot or ending, but it is powerful, and critics have compared it to Edgar Allan Poe's "The Narrative of Arthur Godon Pym."

Machen produced another fantasy book, *The Green Round* (1933), the name of which has been picked up as the name of a small press devoted to publishing Machen materials in recent years. The familiar dream-vision hero in this book, who wanders the gray city thinking back to his green western England or Welsh origins much like Lucian Taylor or Ambrose Meyrick in Machen's earlier fictions, has been given the name Hillyer by Machen. An American poet named Robert Hillyer had become one of Machen's most devoted disciples by this time, so it seems evident that the older Machen was broadening his vision to embrace others than the wandering hack writer from Caerleon-on-Usk who had come to London fifty years earlier. *The Green Round* is a beautiful book with some of the most eloquent expressions of the hidden life beyond and beneath London. As a last evidence of perichoresis Machen invents a fictional book entitled *A London Walk,* by Rev. Thomas Hampole, and then has that book fade mysteriously away both in this novel and in the story "N."

It must have seemed rather magical to Machen to watch the work of Lovecraft and other modern fantasists who said they had learned from him flourish as he himself grew older. He really lived a sort of perichoresis himself as if he were an ancient from the past in modern times. He lived on through another great war, where no miraculous events occurred–except that the old master was still alert and writing interesting letters about ancient things. He was surrounded by a wife, two thriving children, and a wide circle of admirers and drinking friends. Surely he was a much happier Welshman, or Silurian, in England than his sad hero in the romance *The Hill of Dreams* had been. So the question concerns what is more fantastic, the happy ending or the sad, sanctity or sorcery? The young classical scholar from Hereford Cathedral School had come a long way when friends gave him his testimonial dinner back in Caerleon on his birthday in 1937, the year Lovecraft died.

Machen was a tough and amusing journalist, a charming autobiographer and letter writer who knew his own real world, and a skillful translator who knew the real worlds of the sixteenth and eighteenth centuries. But what he loved best was the mystical and mysterious world beneath and beyond the surface of the material world that made him an influential creator of fantasy. And like his oldest literary friend, A. E. Waite, he loved old rituals that led him to what he called "ecstasy" in literature. Modern fantasy looks back to him as a stylist and a creator of ecstasy in his best dream-vision texts, often blending in the presence of evil or horror. Beyond this seminal position with respect to later fantasy writing, Machen ought to be more highly valued as a versatile man of letters for whom the writing and the effects from the writing are of the greatest interest.

Letters:

A Few Letters from Arthur Machen (Cleveland, Ohio: Rowfant Club, 1932);

Starrett vs. Machen, A Record of Discovery and Correspondence, edited by Michael Murphy (Saint Louis, Mo.: Autolycus Press, 1977);

Arthur Machen, Selected Letters, edited by Roger Dobson, Godfrey Brangham, and R. A. Gilbert (Wellingborough: Aquarian Press, 1988);

Arthur Machen and Montgomery Evans: Letters of a Literary Friendship, 1923–1947, edited by Sue Strong Hassler and Donald M. Hassler (Kent, Ohio: Kent State University Press, 1994).

Bibliographies:
Henry Danielson, *Arthur Machen: A Bibliography* (London: Henry Danielson, 1923);
Adrian Goldstone and Wesley Sweetser, *A Bibliography of Arthur Machen* (Austin: University of Texas Press, 1965).

Biographies:
Aidan Reynolds and William Charlton, *Arthur Machen: A Short Account of His Life and Work* (London: Richards, 1963);
Purefoy Machen, *Where Memory Slept,* edited by Godfrey Brangham (Caerleon: Green Round Press, 1991).

References:
Arthur St. John Adcock, "Arthur Machen," in his *The Glory That Was Grub Street* (London: Sampson Low, Marston, 1928; New York: Stokes, 1928), pp. 213–244;
James Branch Cabell, *Beyond Life* (New York: McBride, 1919);
Hereward Carrington, *Psychical Phenomena and the War* (New York: Dodd, Mead, 1918);
William Francis Gekle, *Arthur Machen: Weaver of Fantasy* (Millbrook, N.Y.: Round Table Press, 1949);
John Gunther, "The Truth about Arthur Machen," *Bookman,* 61 (July 1925): 571–574;
Donald M. Hassler, "Arthur Machen and Genre: Filial and Fannish Alternatives," *Extrapolation,* 33 (Summer 1992): 115–127;
Robert Hillyer, "Arthur Machen," *Atlantic Monthly,* 179 (May 1947): 138–140;
Hillyer, "Arthur Machen," *Yale Review,* 13 (October 1923): 174–176;
Paul Jordan-Smith, "Black Magic: An Impression of Arthur Machen," in *On Strange Altars* (New York: A. & C. Boni, 1924), pp. 214–235;
S. T. Joshi, "Arthur Machen: The Mystery of the Universe," in his *The Weird Tale* (Austin: University of Texas Press, 1990), pp. 12–41;
Joseph Wood Krutch, "Tales of a Mystic," *Nation* (13 September 1922): 258–259;
H. P. Lovecraft, *The Outsider and Others* (Sauk City, Wis.: Arkham House, 1939);
Lovecraft, *Supernatural Horror in Literature* (New York: Abramson, 1945);
Helen Lynch, "Arthur Machen," Sewanee Review, 47 (July–September 1939): 424–427;
Robert S. Matteson, "Arthur Machen: A Vision of an Enchanted Land," *Personalist,* 46 (Spring 1965): 253–268;
Hamish Miles, "Machen in Retrospect," Dial, 74 (June 1923): 627–630;
R. Ellis Roberts, "Arthur Machen," *Bookman,* 62 (September 1922): 240–242;
Brocard Sewell, ed., *Arthur Machen* (Llandeilo, Wales: St. Albert's Press, 1960);
M. P. Shiel, "On Scholar-Artistry," in his *Science, Life and Literature* (London: Williams & Norgate, 1950), pp. 95–100;
Vincent Starrett, "Arthur Machen: A Novelist of Ecstasy and Sin," in his *Buried Caesars* (Chicago: Covici-McGee, 1923), pp. 1–31;
Wesley D. Sweetser, *Arthur Machen* (New York: Twayne, 1964);
Julian Symons, "Horror in the Nineties," *Books and Bookmen,* 10 (March 1965): 12–37;
Myles Tierney, "The Highway to Avalon," *Bookman* (New York), 60 (October 1924): 224–226;
Robert L. Tyler, "Arthur Machen: The Minor Writer and His Function," *Approach* (Spring 1960): 21–26;
Mark Valentine and Roger Dobson, eds., *Arthur Machen: Apostle of Wonder* (Oxford: Caermaen, 1985);
Valentine and Dobson, eds., *Arthur Machen: Artist and Mystic* (Oxford: Caermaen, 1986);
Carl Van Vechten, "Arthur Machen: Dreamer and Mystic," in his *Excavations* (New York: Knopf, 1926), pp. 162–269;
Edward Wagenknecht, *Seven Masters of Supernatural Fiction* (Westport, Conn.: Greenwood Press, 1991);
A. E. Waite, *Shadows of Life and Thought* (London: Selwyn & Blount, 1938);
Geoffrey H. Wells, "A Welch Border Writer," *Welch Outlook* (January 1924): 15–16.

Papers:
Yale University Library has a large collection of Machen letters under the collection of his friend and Yale alumnus Charles Parsons. The Harry Ransom Humanities Research Center at the University of Texas at Austin, has the Adrian Goldstone collection of Machen papers. Kent State University Library's Special Collections Department has a considerable collection of Machen materials from the David Borowitz collection that includes the notebooks of the Machen correspondence with American Montgomery Evans. The Stanford University Library also contains a considerable collection of Machen materials, gathered by the original Arthur Machen Society, which is now located in Caerleon, Wales, along with the Green Round Press.

Charles Robert Maturin

(25 September 1780 – 30 October 1824)

Darren Harris-Fain
Shawnee State University

BOOKS: *Fatal Revenge; or, The Family of Montorio,* as Dennis Jasper Murphy (3 volumes, London: Longman, Hurst, Rees & Orme, 1807; 2 volumes, New York: Riley, 1808);

The Wild Irish Boy, as "the Author of *Montorio,*" 3 volumes (London: Longman, Hurst, Rees & Orme, 1808); as Dennis Jasper Murphy, 2 volumes (New York: Sargeant & Ward, 1808);

The Milesian Chief: A Romance, as "the Author of *Montorio,*" 4 volumes (London: Colburn, 1812); as Denis [*sic*] Jasper Murphy, 2 volumes (Philadelphia & New York: Bradford & Inkeep, 1812);

Bertram; or, The Castle of St. Aldobrand (London: John Murray, 1816; Boston: Monroe & Francis, 1816);

Manuel: A Tragedy (London: John Murray, 1817; Baltimore: Robinson, 1817);

Women; or, Pour et Contre, as "the Author of *Bertram*" (3 volumes, Edinburgh: Constable / London: Longman, Hurst, Rees, Orme & Brown, 1818; 2 volumes, New York: Wiley, 1818);

Fredolfo: A Tragedy (London: Longman, Hurst, Rees, Orme & Brown / Edinburgh: Constable, 1819; Philadelphia: Carey, 1819);

Sermons (Edinburgh: Constable / London: Longman, Hurst, Rees, Orme & Brown, 1819);

Melmoth the Wanderer: A Tale, as "the Author of *Bertram*" (4 volumes, Edinburgh: Constable / London: Hurst, Robinson, 1820; 2 volumes, Boston: Wells & Lilly, 1821);

The Albigenses: A Romance (4 volumes, London: Hurst, Robinson / Edinburgh: Constable, 1824; 3 volumes, Philadelphia: Bradford & Laval, 1824);

Five Sermons on the Errors of the Roman Catholic Church (Dublin: Folds, 1824; revised edition, Dublin: Curry / London: Hamilton, Adams, 1826).

PLAY PRODUCTIONS: *Bertram,* London, Drury Lane Theatre, 9 May 1816;

Manuel, London, Drury Lane Theatre, 8 March 1817;

Fredolfo, London, Covent Garden, 19 April 1819.

Charles Robert Maturin

OTHER: "Leixlip Castle: An Irish Family Legend," in *The Literary Souvenir; or, Cabinet of Poetry and Romance* (London: Hurst, Robinson, 1825), pp. 211–232;

"The Sybil's Prophesy: A Dramatic Fragment," in *The Literary Souvenir; or, Cabinet of Poetry and Romance,* 10 volumes, edited by A. Alaric Watts (London: Hurst, Robinson / Edinburgh: Constable, 1826), II: 128–136.

SELECTED PERIODICAL PUBLICATIONS–UNCOLLECTED: "The Tragic Drama–*The Apostate:* a Tragedy, in Five Acts, by Richard Sheil," *Quarterly Review,* 17 (April 1817): 248–260;

"*Harrington and Ormond, Tales.* By Maria Edgeworth," *British Review and London Critical Journal,* 9 (February 1818): 37–61;

Excerpts from "Osmyn, the Renegade; or, The Siege of Salerno: A Tragedy," *Edinburgh Library Journal,* 3 (24 April 1830): 242–243.

In a letter of 2 July 1816 to Sir Walter Scott, Charles Robert Maturin writes:

> I think whatever tends to Efface the radical distinctions of intellectual character, and reduce all the wild and wayward shoots of Mind, stubbed, unsightly and grotesque as they may be, to one smooth-shaven Level, by the ponderous operation of the Critical Roller–Whoever does this, does–mischief–The original workings of mind are so rarely seen, we are all so shaped & moulded by Education and society, so "graven by Art and Man's device"–that an original impression however wild or Rude or fantastic as it may be, is worth something, though it should not have the good fortune like Pyrrhus's Ring to bear for its Legend the Muses and Apollo on the Gem–. . . I have no power of affecting, no hope of instructing, no play or other production of mine will ever draw a tear from the eye, or teach a lesson to the Heart, so I wish they would let me do what I am good for, sit down by my magic Cauldron, mix my dark ingredients, see the bubbles work, and the spirits rise, and by the pale and mystic light, I might show them "the best of my delights"–.

A minor writer in his lifetime and an obscure one beyond it, Maturin nonetheless knew what made him distinctive: a fiery imagination, a sense of the dramatic and melodramatic, and an emotional, unrestrained style. Scott, who encouraged and supported the junior writer, recognized this aspect of Maturin's work and noted its attendant flaws. Maturin was aware of these flaws but nonetheless insisted on the quality of his work. As he wrote in the preface to *The Milesian Chief: A Romance* (1812), "If I possess any talent it is that of darkening the gloomy, and of deepening the sad; of painting life in extreme, and representing those struggles of passion when the soul trembles on the verge of the unlawful and the unhallowed."

Like Mary Wollstonecraft Shelley, Maturin was one of the last writers to work in the field of Gothic romance during its height at the end of the eighteenth century and the beginning of the nineteenth; and like her, he produced one of the most significant Gothic novels in his *Melmoth the Wanderer: A Tale* (1820), published two years after her *Frankenstein.* Like her masterpiece, his is a tale of a Faustian quest for power that exceeds natural boundaries and leads to depths of fantastic horror. He produced other novels and plays along similar lines as well, each combining Gothic themes with his religious concerns and written in his rapid, digressive style.

Charles Robert Maturin was born in Dublin on 25 September 1780 into a well-to-do family. The son of William Maturin and Fidelia Watson Maturin, he was raised in a family that included three sisters and two brothers. Maturin later claimed that his French Huguenot ancestor Gabriel Maturin, a pastor who had left France for Ireland during the reign of Louis XIV, was found by a French lady on the Rue des Mathurines in Paris and that he had been imprisoned by the Catholics in the Bastille for twenty-six years–a story that more likely reveals Maturin's imagination and anti-Catholicism than it does historical truth. Maturin's father encouraged his youthful literary efforts and supported his entrance into Trinity College, Dublin, at age fifteen. Maturin graduated in 1800, and three years later he was ordained in the Anglican Church of Ireland. He was assigned the curacy of rural Loughrea, but instead he spent much of his time at Cloghan Castle in western Ireland. On 25 October 1803 he married Henrietta Kingsbury. Two years later Maturin became one of the curates of Saint Peter's parish in Dublin, a position he retained for the remainder of his life. Apart from the few letters that survive, little is known about much of his life, despite the fact that the rumor of his son William destroying many of his papers after his father's death because he was ashamed of his literary efforts is false.

Maturin was poorly paid as a curate, and his efforts to rise in the church were thwarted by his own eccentric personality and ideas. Impulsive and at times paranoid, Maturin enjoyed the good things in life but handled money badly and alienated those who could improve his position. Some in Dublin, in fact, thought him insane–an impression aided by his love of dancing and his tendency to knock on the doors of abandoned houses–although many of his parishioners praised him for how conscientiously he performed his duties. Most of his letters include complaints about his situation. He often asserted that he wrote for money; for instance, Maturin claims in his preface to *Melmoth the Wanderer* that he wrote romances from financial necessity, which failed to spare him from the disapprobation of critics and clerics who found such an activity inappropriate for a man of the cloth. However, it is clear from his letters that he also desired literary fame for its own sake.

Maturin's first novel, *Fatal Revenge; or, The Family of Montorio* (1807), was published at his own

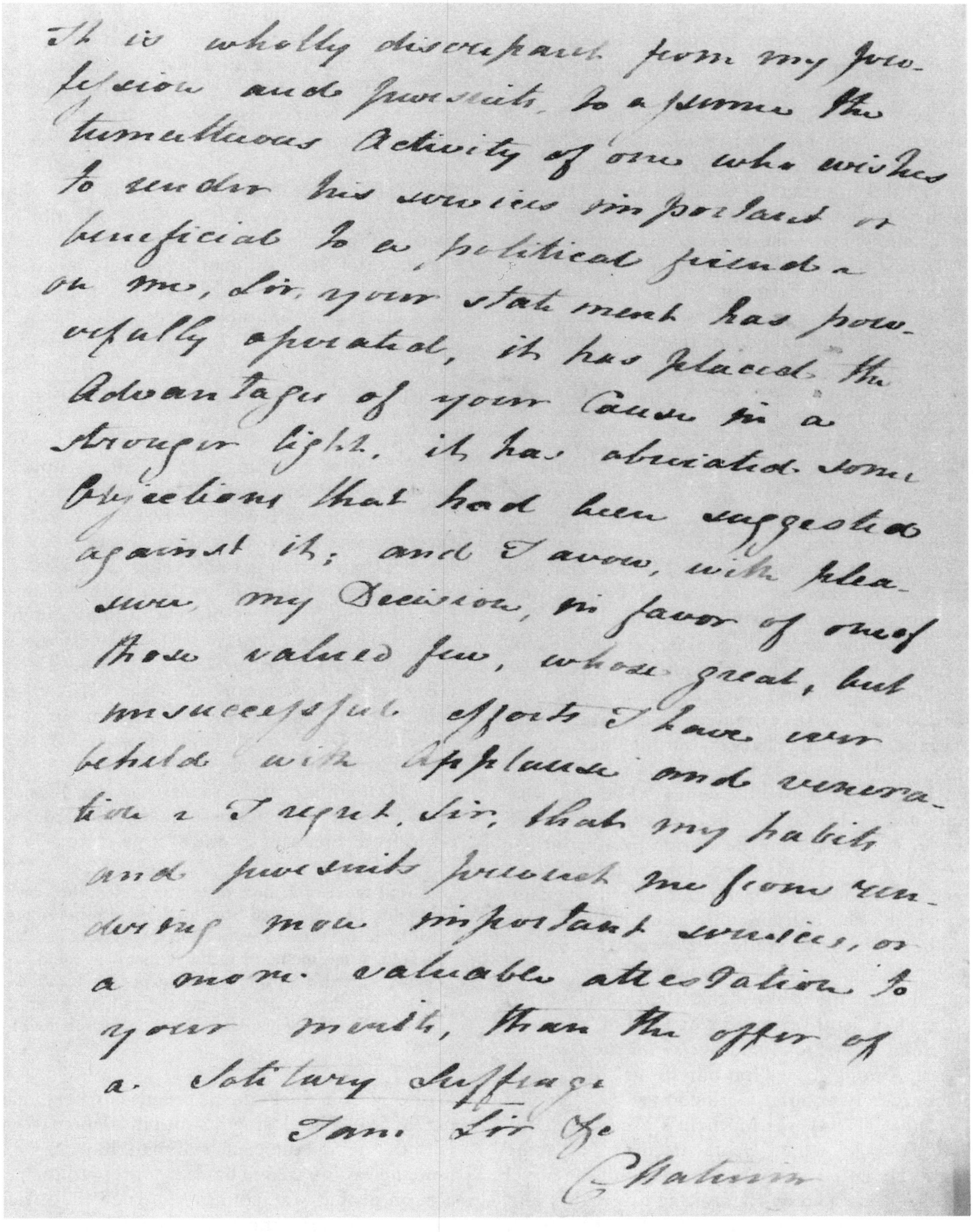

It is wholly discrepant from my pro-
fession and pursuits, to assume the
tumultuous Activity of one who wishes
to render his services important or
beneficial to a political friend —
on me, Sir, your statement has pow-
erfully operated, it has placed the
Advantages of your Cause in a
stronger light, it has obviated some
Objections that had been suggested
against it; and I avow, with plea-
sure, my Decision, in favor of one of
those valued few, whose great, but
unsuccessful efforts I have ever
beheld with Applause and venera-
tion — I regret, Sir, that my habits
and pursuits prevent me from ren-
dering more important services, or
a more valuable attestation to
your merit, than the offer of
a Solitary Suffrage
I am, Sir &c
C Maturin

Last page of Maturin's 15 July 1801 letter to the Irish politician William Conyngham (MA 2640, Pierpont Morgan Library)

expense under the pseudonym Dennis Jasper Murphy because he feared that its publication might damage his position in the church. It was a reasonable concern for the time and place, and in fact his literary ambitions may have hurt his chances at preferment as much as his prickly personality. In addition, Maturin made no efforts to hide his doctrinal differences with his superiors in the established church. It did not help matters that the novel contains all the apparatus of Gothic novelists such as Matthew Gregory Lewis and Ann Radcliffe, including seemingly haunted castles and monastic horrors. As in most of Radcliffe's works, the supernatural elements–including demons and living skeletons, among others–are explained away at the end as the tricks of an ambitious and corrupt man, disguised as a monk. Though the story is difficult to follow, the novel is striking, as Scott would later point out, for its language and atmosphere.

Maturin also published his next novel, *The Wild Irish Boy* (1808), as "the Author of *Montorio,*" at his expense. Like *Fatal Revenge,* it includes such Gothic elements as extravagant plots, atmospheric settings, and villains who bring about incredible suffering for the protagonists. *The Wild Irish Boy* also includes another element, Irish nationalism, found in much of Maturin's work.

In November 1809 disaster struck the Maturin family. His father, one of the clerks of the road, lost his position when he was charged with wrongdoing. An investigative board later found him innocent of the charges, but the damage was done, and the whole family suffered from the loss of his salary–including the younger Maturin, who depended on his father's help to supplement his meager earnings from the church. Maturin then attempted to turn his avocation, writing, into a means of supporting himself and his growing family, which eventually included four children. He also began tutoring students to prepare them for university studies, a task he abhorred.

Maturin first came to the attention of Scott in 1810, when Scott reviewed a handful of "the newest and most fashionable novels" for the *Quarterly Review.* Among these "flat imitations" of Radcliffe's extremely popular Gothic novel *The Mysteries of Udolpho* (1794) was Maturin's pseudonymous *Fatal Revenge,* which Scott discussed at some length. He noted several problems with the novel but also remarked on its striking originality and encouraged the author to seek a literary guide. Maturin did so, writing to Scott himself in December 1812 after learning of the anonymous reviewer's identity, seeking his advice on writing and selling what he wrote. Scott graciously offered what help he could, even to the point of asking influential friends to intercede on Maturin's behalf for a better position within the church. The two writers never met, but they carried on a correspondence that lasted until Maturin's death.

Maturin's next novel–*The Milesian Chief,* another book about Ireland–was his first actual sale. Though Gothic in atmosphere and characterization, it is less Gothic in plot than his other novels. He failed to make much money from his pen, however, until his next work, the 1816 production and publication of *Bertram; or, The Castle of St. Aldobrand,* which Scott diligently promoted while encouraging Maturin to make certain revisions. The play, attacked in Samuel Taylor Coleridge's *Biographia Literaria* (1817) but enthusiastically praised by George Gordon, Lord Byron and George Lamb, was the most critically and financially successful work of Maturin's career, but he spent most of the £1,000 he earned from the play on his family, including helping a bankrupt relative for whom Maturin had pledged security.

Maturin continued to strive to increase his earnings with his writing, often working far into the night after tending to his religious duties. The last eight years of his life are filled with literary efforts ranging from less-successful plays, such as *Manuel* (1817) and *Fredolfo* (1819); novels with religious themes such as *Women; or, Pour et Contre* (1818) and *The Albigenses: A Romance* (1824); and collections of his sermons. His best-known novel, *Melmoth the Wanderer,* also dates from this busy period of Maturin's career.

Maturin's preface to *Melmoth the Wanderer* notes that the "hint of this Romance (or Tale) was taken from a passage in one of my Sermons":

> At this moment is there one of us present, however we may have departed from the Lord, disobeyed his will, and disregarded his word–is there one of us who would, at this moment, accept all that man could bestow, or earth afford, to resign the hope of his salvation?–No, there is not one–not such a fool on earth, were the enemy of mankind to traverse it with the offer!

Working from the fantastic premise of a man gaining the world but losing his soul in a Faustian pact with the devil, "the enemy of mankind, the novel nonetheless is based in part, according to the preface, on people Maturin has known. Still, a friend to whom he read the novel found part of it, according to Maturin, "too much an attempt at the revivification of the horrors of Radcliffe-Romance." While Maturin defends himself against this charge, asserting that his depiction of monastic life in one of the subplots of the novel is more realistic than those found in other romances, his narrator of this tale, in relating his attempted escape to a British listener, says, "Romances have made your country, Sir, familiar with tales of subterranean passages, and supernatural horrors," adding, "All these, painted by the most eloquent pen, must fall short" of his experience. Maturin also alludes early in the novel itself to Lewis's *The Monk* (1796).

Melmoth the Wanderer is a novel of suspense and supernatural horror. It is also a compendium of Maturin's far-reaching but often imperfect knowledge of a wide range of subjects, including classical literature, William Shakespeare, Miguel de Cervantes, and Restoration drama, as well as an ironic attack on Puritanism and the Roman Catholic Church. It also exhibits a wide range of narrative ability, from skillfully paced plotting to seemingly interminable subplots.

The novel deals with a man named Melmoth who in the seventeenth century sold his soul to the devil in exchange for immortality. Granted everlasting life and supernatural powers, he comes to realize that these are a curse and traverses the globe in search of someone whose sufferings are so great that they will take Melmoth's curse upon themselves and free him from immortality and damnation, an escape clause in his bargain with Satan. The lengthy novel, which owes as much to legends of the Wandering Jew as to stories about Faust, contains several first-person stories-within-stories, and sometimes stories within those stories, ranging from modern Ireland to Spain during the Inquisition to India, and all of them reveal the wanderings and quest of Melmoth.

The fascination of the novel stems from two characteristics that contribute to its rank as one of the most notable of the Gothic novels. First, its abundance of romantic horror–set in such locales as madhouses, monasteries, prison cells, exotic islands, isolated towns, and so on, depending on the story within the story–reach heights of beauty mixed with suffering, pleasure combined with pain, that Edmund Burke labeled the sublime in his 1757 work on the subject, *Philosophical Enquiry into the Origin of Our Ideas of the Sublime and the Beautiful.* Second, the figure of Melmoth is a striking one. Though villainous in his earthly damnation and in his efforts to transfer his diabolical fate to another suffering soul, he is also a fascinating figure, especially as Maturin reveals him to the reader often through indirect glimpses, with a complete picture of Melmoth coming only at the end. A classic Gothic hero-villain, Melmoth is both attractive and terrifying.

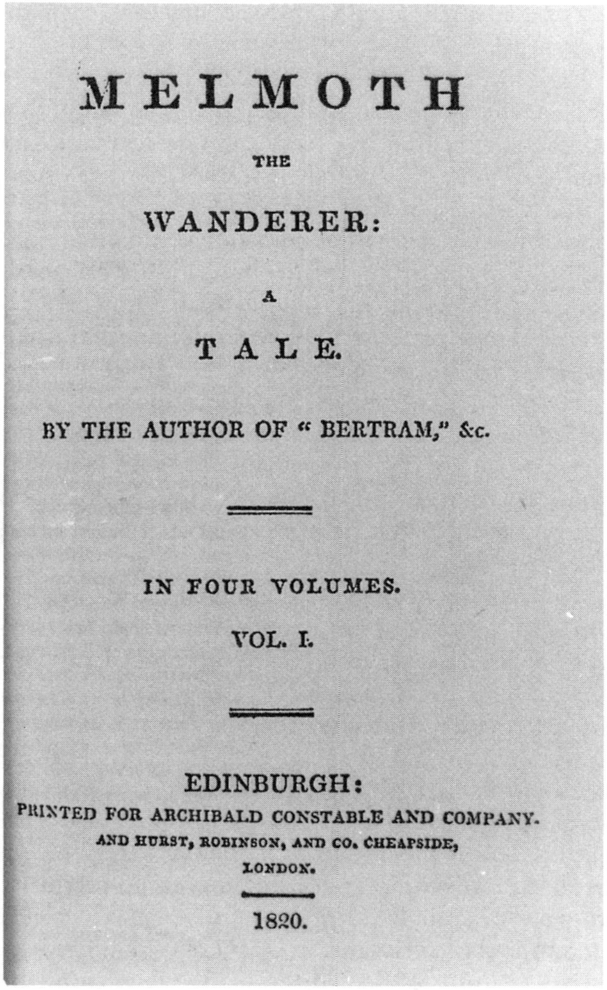
MELMOTH

THE

WANDERER:

A

TALE.

BY THE AUTHOR OF "BERTRAM," &c.

IN FOUR VOLUMES.

VOL. I.

EDINBURGH:

PRINTED FOR ARCHIBALD CONSTABLE AND COMPANY.

AND HURST, ROBINSON, AND CO. CHEAPSIDE,

LONDON.

1820.

Title page for Maturin's ninth book, in which a man who has sold his soul to the devil seeks to escape the curse of immortality

In addition, *Melmoth the Wanderer* is interesting for how it relates its themes of suffering and search with religious issues. As in much of his work, Maturin here offers a scathing critique of much organized religion and of how religious institutions often depart from the gospel of love for the sake of power–a theme to be found in much Gothic fiction and in much Romantic literature in general. Melmoth, through his deal with the devil, has achieved a supernatural knowledge of human nature, and this Faustian knowledge contributes to the despair that impels him to seek someone to relieve him of his powers. The picture of humanity that emerges from the novel is one of general depravity, and it is suggested that most of the human race is as far removed from salvation as Melmoth is. Through suffering, ironically, characters may be purified to a point where they can accept divine grace.

Though Melmoth himself never achieves this, perhaps the most telling reinforcement of this theme is revealed in the subplot concerning Immalee, a young woman Melmoth discovers on an Indian island during his wanderings. Lost and abandoned as a small child, she has grown up in a state of nature, an ironic incarnation of Jean-Jacques Rousseau's "noble savage," and is worshipped as a goddess by the nearby mainland villagers. Hoping he can seduce her to accept his fate, he instead raises her consciousness to a knowledge of God and inadvertently reinforces her natural goodness. So loving is she, in fact, that she falls in love with Melmoth; nonetheless, she cannot accept what he offers despite the fact that she weds him.

Melmoth the Wanderer achieved a wide readership, though it did not equal the commercial success of *Bertram.* Reviews were often negative, but the novel proved compelling enough to take on a life of its own for many readers, and it was routinely mentioned in surveys of British literature throughout the nineteenth century and into the twentieth. Moreover, with the scholarly interest in the Gothic novel and in fantastic fiction in general after World War II came increased attention to what became labeled as Maturin's masterpiece, and modern editions–the University of Nebraska Press edition (1961) edited by W. F. Axton, the Oxford University Press edition (1968) edited by Douglas Grant, and the Penguin Books edition (1977)–have brought the novel to many contemporary readers.

Maturin's last novel, *The Albigenses,* also includes Gothic elements but is less ambitious in scope as a Gothic novel. Instead, it is modeled as much on the historical romances of Scott, whose authorship of the *Waverley* novels was suspected soon after they began appearing anonymously in 1814. Unlike Scott's works, it was unsuccessful. Taking as its setting the destruction of the Albigenses, a heretical sect, early in the thirteenth century, the novel tells of two brothers, one good and one evil, who are reconciled in the end through their marriages. The good brother, Sir Paladour, marries into a family cursed by lycanthropy, but Maturin does little with this supernatural element or with others in the novel.

Maturin's frantic productivity in his battle with poverty proved to be detrimental to his health, and he died at age forty-four on 30 October 1824 in Dublin; he is buried at St. Peter's. There are rumors that his death was a suicide through overdose, but these are unsubstantiated. Scott attempted to publish an edition of Maturin's works after his death to aid Maturin's family but was thwarted by the decline of his publishing interests. Apart from a handful of scholars, most readers who know Maturin are familiar with *Melmoth the Wanderer,* which, although it did not match the contemporary reputation of *Bertram,* has proved to be the work that would secure what little posthumous reputation he has achieved. The work has been praised by such diverse writers as Charles Baudelaire, Lord Byron, Johann Wolfgang von Goethe, Nathaniel Hawthorne, Edgar Allan Poe, Aleksandr Pushkin, Sir Walter Scott, and Oscar Wilde, whose *The Picture of Dorian Gray* (1891) shows Maturin's influence. In addition, the novel was praised by Honoré de Balzac, who hailed Melmoth as a modern mythic character on a level with Goethe's Faust and Byron's Manfred and wrote a sequel, *Melmoth reconcilié* (1835; translated as *Melmoth Reconciled*). *Melmoth the Wanderer* has seldom been out of print since the author's death.

Letters:

Fannie E. Ratchford and William H. McCarthy, eds., *The Correspondence of Sir Walter Scott and Charles Robert Maturin, With a Few Other Allied Letters* (Austin: University of Texas Press, 1937).

Biographies:

"Memoir of Charles Robert Maturin," in Maturin's *Melmoth the Wanderer* (London: Bentley, 1892);

Niilo Idman, *Charles Robert Maturin: His Life and Works* (London: Constable, 1923);

Claude Fierobe, *Charles Robert Maturin (1780–1824): L'Homme et L'Oeuvre* (Lille: Université de Lille III / Paris: Editions Universitaires, 1974).

References:

W. F. Axton, "Charles Robert Maturin," in *Supernatural Fiction Writers: Fantasy and Horror,* volume 1, edited by E. F. Bleiler (New York: Scribners, 1985), pp. 161–168;

Edith Birkhead, *The Tale of Terror: A Study of the Gothic Romance* (London: Constable, 1921; New York: Russell & Russell, 1963);

Syndy M. Conger, *Matthew G. Lewis, Charles Robert Maturin and the Germans: An Interpretative Study of the Influence of German Literature on Two Gothic Novels* (Salzburg: Institut für Englische Sprache und Literatur, Universität Salzburg, 1977);

Leven M. Dawson, "*Melmoth the Wanderer:* Paradox and the Gothic Novel," *Studies in English Literature,* 8 (Autumn 1968): 621–632;

John B. Harris, *Charles Robert Maturin: The Forgotten Imitator* (New York: Arno, 1980);

Mark M. Hennelly Jr., "*Melmoth the Wanderer* and Gothic Existentialism," *Studies in English Literature,* 21 (Fall 1981): 665–679;

Robert Kiely, *The Romantic Novel in England* (Cambridge, Mass.: Harvard University Press, 1972);

Dale Kramer, *Charles Robert Maturin* (New York: Twayne, 1973);

Robert E. Lougy, *Charles Robert Maturin* (Lewisburg, Pa.: Bucknell University Press, 1975);

Jack Null, "Structure and Theme in *Melmoth the Wanderer,*" *Papers in Language and Literature,* 13 (1977): 136–147;

Mario Praz, *The Romantic Agony,* translated by Angus Davidson (Oxford: Oxford University Press, 1933);

Willem Scholten, *Charles Robert Maturin: The Terror-Novelist* (Amsterdam: Paris, 1933);

Shirley Clay Scott, *Myths and Consciousness in the Novels of Charles Robert Maturin* (New York: Arno, 1980);

[Walter Scott,] "Fatal Revenge or the Family of Montorio, a Romance by Dennis Jasper Murphy," *Quarterly Review,* 3 (May 1810): 339–347;

Joseph Arthur Soldati, *Configurations of Faust: Three Studies in the Gothic* (1798–1820) (New York: Arno, 1980);

Devendra P. Varma, *The Gothic Flame* (London: Barker, 1957).

Papers:

Collections of Maturin's papers are located at the National Library of Scotland in Edinburgh; the library at Abbotsford, Scotland; the British Museum, London; and the Bodleian Library, Oxford University.

William Morris

(24 March 1834 - 3 October 1896)

Corbin Scott Carnell
University of Florida

See also the Morris entries in *DLB 18: Victorian Novelists After 1885, DLB 35: Victorian Poets After 1850, DLB 57: Victorian Prose Writers After 1867,* and *DLB 156: British Short-Fiction Writers, 1880–1914: The Romantic Tradition.*

William Morris

BOOKS: *The Defence of Guenevere and Other Poems* (London: Bell & Daldy, 1858; Boston: Roberts, 1875);

The Life and Death of Jason: A Poem (London: Bell & Daldy, 1867; Boston: Roberts Brothers, 1867);

The Earthly Paradise: A Poem, 3 volumes (London: Ellis, 1868–1870; Boston: Roberts Brothers, 1868–1870);

The Lovers of Gudrun: A Poem (Boston: Roberts Brothers, 1870);

Love Is Enough; or, The Freeing of Pharamond: A Morality (London: Ellis & White, 1873; Boston: Roberts Brothers, 1873);

The Story of Sigurd the Volsung and the Fall of the Niblungs (London: Ellis & White, 1876; Boston: Roberts Brothers, 1876);

Hopes and Fears for Art: Five Lectures Delivered in Birmingham, London, and Nottingham 1878–81 (London: Ellis & White, 1882; Boston: Roberts Brothers, 1882);

A Summary of the Principles of Socialism Written for the Democratic Federation, by Morris and H. M. Hyndman (London: Modern Press, 1884);

Textile Fabrics: A Lecture (London: Clowes, 1884);

Art and Socialism: A Lecture; and Watchman, What of the Night? The Aims and Ideals of the English Socialists of Today (London: Reeves, 1884);

Chants for Socialists: No. 1. The Day Is Coming (London: Reeves, 1884);

The Voice of Toil, All for the Cause: Two Chants for Socialists (London: Justice Office, 1884);

The God of the Poor (London: Justice Office, 1884);

Chants for Socialists (London: Socialist League Office, 1885; New York: New Horizon Press, 1935);

The Manifesto of the Socialist League (London: Socialist League Office, 1885);

The Socialist League: Constitution and Rules Adopted at the General Conference (London: Socialist League Office, 1885);

Address to Trades' Unions (The Socialist Platform–No. 1) (London: Socialist League Office, 1885);

Useful Work v. Useless Toil (The Socialist Platform–No. 2) (London: Socialist League Office, 1885);

For Whom Shall We Vote? Addressed to the Working-Men Electors of Great Britain (London: Commonweal Office, 1885);

What Socialists Want (London: Hammersmith Branch of the Socialist League, 1885);

The Labour Question from the Socialist Standpoint (Claims of Labour Lectures–No. 5) (Edinburgh: Cooperative Printing, 1886);

A Short Account of the Commune of Paris (The Socialist Platform–No. 4) (London: Socialist League Office, 1886);

The Pilgrims of Hope: A Poem in Thirteen Books (London: Buxton Forman, 1886; Portland, Maine: Mosher, 1901);

The Aims of Art (London: Commonweal Office, 1887);

The Tables Turned; or, Nupkins Awakened: A Socialist Interlude (London: Commonweal Office, 1887);

True and False Society (London: Socialist League Office, 1888);

Signs of Change: Seven Lectures Delivered on Various Occasions (London: Reeves & Turner, 1888; New York: Longmans, Green, 1896);

A Dream of John Ball and A King's Lesson (London: Reeves & Turner, 1888; East Aurora, N.Y.: Roycroft, 1898);

A Tale of the House of the Wolfings and All the Kindreds of the Mark (London: Reeves & Turner, 1889; Boston: Roberts Brothers, 1890);

The Roots of the Mountains wherein Is Told Somewhat of the Lives of the Men of Burgdale, Their Friends, Their Neighbours, Their Foemen, and Their Fellows in Arms (London: Reeves & Turner, 1890; New York: Longmans, Green, 1896);

Monopoly; or, How Labour Is Robbed (The Socialist Platform–No. 7) (London: Commonweal Office, 1890);

News from Nowhere; or, An Epoch of Rest: Being Some Chapters from a Utopian Romance (Boston: Roberts Brothers, 1890; London: Reeves & Turner, 1891);

Statement of Principles of the Hammersmith Socialist Society, anonymous (Hammersmith: Hammersmith Socialist Society, 1890);

The Story of the Glittering Plain Which Has Been Also Called the Land of Living Men or the Acre of the Undying (Hammersmith: Kelmscott Press, 1891; London: Reeves & Turner, 1891; Boston: Roberts Brothers, 1891);

Poems by the Way (Hammersmith: Kelmscott Press, 1891; London: Reeves & Turner, 1891; Boston: Roberts Brothers, 1892);

Address on the Collection of Paintings of the English Pre-Raphaelite School (Birmingham: Osborne, 1891);

Under an Elm-Tree; or, Thoughts in the Country-side (Aberdeen: Leatham, 1891; Portland, Maine: Mosher, 1912);

Manifesto of English Socialists, anonymous, by Morris, Hyndman, and G. B. Shaw (London: Twentieth Century Press, 1893);

The Reward of Labour: A Dialogue (London: Hayman, Christy & Lilly, 1893);

Concerning Westminster Abbey, anonymous (London: Women's Printing Society, 1893);

Socialism: Its Growth and Outcome, by Morris and E. B. Bax (London: Sonnenschein, 1893; New York: Scribners, 1893);

Help for the Miners: The Deeper Meaning of the Struggle (London: Baines & Searsrook, 1893);

Gothic Architecture: A Lecture for the Arts and Crafts Exhibition Society (Hammersmith: Kelmscott Press, 1893);

The Wood beyond the World (Hammersmith: Kelmscott Press, 1894; London: Lawrence & Bullen, 1895; Boston: Roberts House, 1895);

The Why I Ams: Why I Am a Communist, with L. S. Bevington's Why I Am an Expropriationist (London: Liberty Press, 1894);

Child Christopher and Goldilind the Fair (2 volumes, Hammersmith: Kelmscott Press, 1895; 1 volume, Portland, Maine: Mosher, 1900);

Gossip about an Old House on the Upper Thames (Birmingham: Birmingham Guild of Handicraft, 1895; Flushing, N.Y.: Hill, 1901);

The Well at the World's End: A Tale (1 volume, Hammersmith: Kelmscott Press, 1896; 2 volumes, London: Longmans, Green, 1896);

Of the External Coverings of Roofs, anonymous (London: Society for the Protection of Ancient Buildings, 1896);

How I Became a Socialist (London: Twentieth Century Press, 1896);

Some German Woodcuts of the Fifteenth Century, edited by S. C. Cockerell (Hammersmith: Kelmscott Press, 1897);

The Water of the Wondrous Isles (Hammersmith: Kelmscott Press, 1897; London: Longmans, Green, 1897);

The Sundering Flood (Hammersmith: Kelmscott Press, 1897; London: Longmans, Green, 1898);

A Note by William Morris on His Aims in Founding the Kelmscott Press, Together with a Short Description of the Press by S. C. Cockerell and an Annotated List of the Books Printed Thereat (Hammersmith: Kelmscott Press, 1898);

Address Delivered at the Distribution of Prizes to Students of the Birmingham Municipal School of Art on 21 February 1894 (London: Longmans, Green, 1898);

Art and the Beauty of the Earth (London: Longmans, Green, 1899);

Some Hints on Pattern-Designing (London: Longmans, Green, 1899);

Architecture and History, and Westminster Abbey (London: Longmans, Green, 1900);

Art and Its Producers, and the Arts and Crafts of Today (London: Longmans, Green, 1901);

Architecture, Industry, and Wealth: Collected Papers (London: Longmans, Green, 1902);
Communism (Fabian Tract No. 113) (London: Fabian Society, 1903);
The Hollow Land and Other Contributions to the Oxford and Cambridge Magazine (London: Longmans, Green, 1903);
The Unpublished Lectures of William Morris, edited by Eugene D. LeMire (Detroit: Wayne State University Press, 1969);
Icelandic Journals of William Morris (Fontwell: Centaur Press, 1969; New York: Praeger, 1970);
A Book of Verse: A Facsimile of the Manuscript Written in 1870 (London: Scolar Press, 1980);
Socialist Diary, edited by Florence Boos (Iowa City: Windhover Press, 1981);
The Novel on Blue Paper, edited by Penelope Fitzgerald (London: Journeyman Press, 1982);
The Ideal Book: Essays and Lectures on the Arts of the Book, edited by William S. Peterson (Berkeley & London: University of California Press, 1982);
The Juvenilia of William Morris, edited by Boos (New York: William Morris Society, 1983);
The Widow's House by the Great Water, edited by Helen A. Timo (New York: William Morris Society, 1990).

Collections: *The Early Romances of William Morris* (London: Dent, 1907);
The Collected Works of William Morris, edited by May Morris, 24 volumes (London & New York: Longmans, Green, 1910–1915; New York: Russell & Russell, 1966);
William Morris: Artist, Writer, Socialist, edited by May Morris, 2 volumes (Oxford: Blackwell, 1936; New York: Russell & Russell, 1966).

OTHER: "Mural Decoration," by Morris and J. H. Middleton, in *Encyclopaedia Britannica,* ninth edition, volume 17 (Edinburgh: Black, 1884; New York: Allen, 1888);
John Ruskin, *The Nature of Gothic: A Chapter of The Stones of Venice,* preface by Morris (London: George Allen, 1892);
Thomas More, *Utopia,* foreword by Morris (Hammersmith: Kelmscott Press, 1893);
Robert Steele, ed., *Medieval Lore,* preface by Morris (London: Stock, 1893; Boston: Luce, 1907);
Arts and Crafts Exhibition Society, *Arts and Crafts Essays* (London: Rivington, Percival, 1893; New York: Scribners, 1893)–includes a preface and three articles by Morris.

TRANSLATIONS: *The Story of Grettir the Strong,* translated by Morris and Eiríkr Magnússon (London: Ellis, 1869; New York: Longmans, Green, 1901);
The Story of the Volsungs and the Niblungs, translated by Morris and Magnússon (London: Ellis, 1870; New York: Longmans, Green, 1901);
Three Northern Love Stories, and Other Tales, translated by Morris and Magnússon (London: Ellis & White, 1875; New York: Longmans, Green, 1901);
The Aeneid of Virgil Done into English Verse (Boston: Roberts Brothers, 1875; London: Ellis & White, 1876);
The Odyssey of Homer Done into English Verse, 2 volumes (London: Reeves & Turner, 1887);
The Saga Library, translated by Morris and Magnússon, 6 volumes (London: Quaritch, 1891–1905);
The Tale of King Florus and the Fair Jehane (Hammersmith: Kelmscott Press, 1893);
Of the Friendship of Amis and Amile (Hammersmith: Kelmscott Press, 1894);
The Tale of the Emperor Coustans and of Over Sea (Hammersmith: Kelmscott Press, 1894);
The Tale of Beowulf, translated by Morris and A. J. Wyatt (Hammersmith: Kelmscott Press, 1895; London & New York: Longmans, Green, 1898);
Old French Romances (London: Allen, 1896; New York: Scribners, 1896);
The Story of Kormak, the Son of Ogmund, translated by Morris and Magnússon, edited by Grace Calder (London: William Morris Society, 1970).

One of the most versatile and gifted of all the Victorians, William Morris has had an important influence on literature, printing, interior decoration, and thought on the relation of government and the arts. He wrote poems and stories and translated the *Aeneid* (1875) and the *Odyssey* (1887). He modeled in clay, carved wood and stone, illuminated manuscripts, and did embroidery. He designed wallpapers, textiles, stained glass, and furniture for his flat in Red Lion Square in Oxford, and he loved printing and bookbinding. Yet with all his talent he left Oxford without a degree. He also contributed significantly to modern fantasy and science fiction with his medieval romances and his utopian novel *News from Nowhere* (1890).

Morris was born on 24 March 1834 in Walthamstow, and he grew up next to Epping Forest, a unique tract of land unchanged since the Middle Ages. Early in his life he began to acquire a detailed knowledge of trees, flowers, and birds. He had two older sisters to pamper him, and he seems to have had his own way most of the time, a fact that would help to explain his frustration and rage in later years when he was crossed by anyone. He learned to read at an early age, being able to read Sir Walter Scott's Waverley novels at age four. This dose of literary medievalism colored his later tastes and writing.

His father died when Morris was thirteen, but the family was well provided for because of good investments. He entered Oxford in 1853 and soon became a close friend of Edward Burne-Jones; this friendship was lifelong and did much to stimulate Morris's interests in the arts, especially painting.

According to *The Dictionary of National Biography* Morris was brought up in a "stagnant Evangelicalism." He came under the influence of Anglo-Catholic thought and practice at Oxford and considered becoming a Roman Catholic but turned instead to the artistic and social enthusiasms of Thomas Carlyle, John Ruskin, and Charles Kingsley. He considered seeking holy orders and did not give up that possibility until after leaving Oxford, when he decided to become an architect. Burne-Jones and Morris had a long talk on an August night when they decided not to be priests but rather artists–Burne-Jones as a painter, Morris as an architect. Morris was articled to a Gothic revivalist architect in Oxford, though shortly after, in 1857, he took up painting at age twenty-three.

Morris helped decorate the new Union Society building at Oxford. Some of the pictures quickly faded, as the artists did not prepare the surface beforehand. Morris finished his picture first and then painted flowers and foliage on the beams of the roof. This turned out to be the most successful part of the whole project.

As early as 1853 Morris was thinking of founding a monastic community in which he and his friends might combine an ascetic life with the production of religious art. He continually returned to this idea, and in his friendships with artists such as Ford Madox Brown, Dante Gabriel Rossetti, and Burne-Jones he sought to develop a sense of community with them.

At first, poetry was Morris's relaxation, but in his twenties he began writing it seriously. His efforts are wholly in his own voice and style, though there may be some slight influence of Alfred Tennyson. He also contributed five poems, two essays, a review, and eight prose tales to the *Oxford and Cambridge Magazine,* which he helped to found, in January 1856. Though it lasted only twelve issues, the magazine was a remarkable expression of the youthful idealism of Morris and his friends, who at this time called themselves the Brotherhood.

All but one of these eight tales, Morris's earliest published work, are romantic fantasies set either

Morris's painting Queen Guenevere, *for which his future wife, Jane Burden, served as model (Tate Gallery)*

in medieval Europe or in otherworldly places; and though they are influenced by Gothic fiction, they offer an early example of Morris's romance subjects and technique as dreamlike tales of death, guilt, initiation, and sexuality with mysterious or even dead narrators. In "The Story of the Unknown Church" (January 1856), for instance, the work of a thirteenth-century French stonemason mysteriously vanishes, only to be replaced by the beauty of nature. The story is told by the stonemason, who claims he has been dead for six hundred years. "The Dream" (March 1856) is a story about a young woman and a knight who meet briefly at intervals over the centuries.

Morris's next three stories for the *Oxford and Cambridge Magazine* reveal his building interest in the traditional sagas of northern Europe. The two-part "Gertha's Lovers" (July–August 1856) describes a love triangle in which a young king and his friend vie for the hand of the beautiful young Gertha, who becomes queen upon their deaths and who joins the king in death. For later readers the story is interesting for its foreshadowing of the relationship between Morris, his wife, and Rossetti. "Svend and His Brethren" (August 1856) tells of the escape from a mob to a western land of a prince, his followers, and his brothers; centuries later, knights from these lands return and find the bloody mob frozen in place. In addition, the story tells of how the central character moves from unfulfilled love to arts and crafts and then to revolutionary utopianism, a path Morris himself would travel. A nineteenth-century man is transformed into a thirteenth-century priest in "Lindenborg Pool" (September 1856) in a bizarre punishment for his crimes; like many of these tales, this shows the influence of Morris's reading in Gothic fiction, as well as the direct influence here of Benjamin Thorpe's *Northern Mythology* (1851). Somewhat similar is "Golden Wings" (December 1856), in which a young knight narrates his violent adventures.

Because of its strongly dreamlike quality and its anticipations of Morris's later romances, the most significant of these early fantasies is "The Hollow Land" (September–October 1856). In this long tale reminiscent of Robert Browning's "Childe Roland to the Dark Tower Came" (1855), a knight, Florian of the Lilies, has participated in the death of an evil queen named Swanhilda. After fighting with her son, Red Harald, he enters the Hollow Land and meets Margaret, with whom he falls in love. He again battles Harald, and the two mysteriously become "painters" of "God's judgments" on the walls of Florian's castle. The story concludes with Florian's return to the Hollow Land and his winning of Margaret.

The various contributions to the *Oxford and Cambridge Magazine* were all anonymous, and in addition to the seven romances and the less-successful contemporary story "Frank's Sealed Letter" (April 1856), all of which are known to have been written by Morris, two others have been convincingly attributed to him: "The Two Partings" (February 1856) and "A Night in the Cathedral" (May 1856). Like "Frank's Sealed Letter," they are largely realistic stories with contemporary settings.

In 1859 Morris, at age twenty-five, married the beautiful Jane Burden, the uneducated daughter of an Oxford stableman and one of Rossetti's models. None of Morris's relatives attended the wedding. Jane was tall and large-boned, with a long graceful neck, pale ivory skin, and a wealth of crinkly black hair. Some thought that she had Gypsy blood. Three years into the marriage the couple had two daughters, Jane (Jenny) and Mary (May).

Morris's wife was the subject of his painting *Queen Guenevere,* and she was also painted by Rossetti and Madox Brown. She sometimes painted with her husband. Jane was as responsive to fun as a child and was liked for her unaffected ways. But after ten years the marriage was in serious trouble. She had spent many years as a semi-invalid. After Rossetti's wife, Elizabeth Siddal, died, he repeatedly drew and painted Jane in art that has a half-mystical sexuality. Rossetti and Jane spent many hours together, sometimes going away together for many days at a time. Morris did not seem to object.

In 1859 Morris became interested in designing, with Phillip Webb, Red House in Upton, Kent. Red House was so called because it was of red brick at a time when stucco was in fashion. Located ten miles from London, it was both a romantic and a practical house. It had high ceilings, exposed beams, brick arches, a brick fireplace, and a medieval garden. It had no wallpaper, as Morris was not yet working with wallpapers. It was the scene of many parties during the Morrises' six years there.

Morris was surprised at how difficult it was to find well-designed, reasonably priced furniture, and he thought again of how good it would be to have a brotherhood making beautiful things. The furniture he designed was not successful. The textiles and wallpapers were another matter, for once Morris embarked on designing them he was remarkably successful.

By the early 1860s Morris was in his prime, by far the best all-round artist in a group that included Rossetti, Burne-Jones, and Madox Brown. He was widely praised for his magnanimity and generosity. He could make fun of himself, but he was strong willed and perhaps a bit masochistic in regard to his wife. He had little use for holidays, much preferring to work. William Butler Yeats wrote that he had never known any man as well loved as Morris, though Yeats admired only isolated lines in Morris's poetry. Morris was also handsome in an unusual way. His friends thought he looked like a medieval king. He was below average height and deep chested, with a powerful physique.

Morris could become irritable and moody, particularly because he could not be both an architect and a painter. He was accommodating to those interested in his ideas, teaching women to embroider his designs. He could be loud, boisterous, and quick tempered. He once flung his dinner out the window because it was poorly prepared. If a door did not open at once, he might wrench off its handle or kick in one of its panels. He may have had a tendency toward epilepsy. (This malady afflicted his older daughter, Jenny, and kept her close to home.) Morris might start work at five in the morning or paint all night. Most of his writing was done at night or in the small hours of the morning.

In art Morris admired Hans Memling and Jan van Eyck. He loved the cathedral in Rouen. He recalled it forty years later with great poignancy. He preferred Oxford to Cambridge architecturally. When he entered Amiens Cathedral he wanted to shout with pleasure. He loved Thomas Malory's *Morte D'Arthur* (1485), regarding it as next to the Bible. He admired Rossetti's sonnets, though he sometimes found them obscure.

Morris's early poetry, such as is found in *The Defence of Guenevere and Other Poems* (1858), came easily to him. Some of it, mostly unpublished, he realized was not good, although the long narrative title poem had a deep influence on a limited audience. Though Rossetti admired Morris's poetry, he discouraged his friend from pursuing a literary career. Nonetheless, Morris dedicated *The Defence of Guenevere* to Rossetti. In turn, A. C. Swinburne was an admirer of Morris and dedicated his *Astrophel* (1894) to him. As Frederick Kirchhoff points out, the poems for which Morris is best known today met with the least immediate favor. *The Defence of Guenevere* had a few favorable reviews, but often the work was condemned for its Pre-Raphaelite atmosphere. Certainly Morris's depiction of sexuality and death took his work past the limits of Victorian taste. Also, Morris's Guenevere is confusing because he was not certain what attitude to adopt toward her. There are three groupings of the poems in *The Defence of Guenevere:* those based on Arthurian materials, a rather realistic group based on Foissart's *Chronicles,* and a third group, which Kirchhoff describes as fantasies. Many use the dramatic monologue form, which Morris knew well through having read Browning.

Almost a decade passed between the publication of *The Defence of Guenevere* and the publication of *The Life and Death of Jason* (1867) and the first two parts of *The Earthly Paradise* (1868–1870). Morris had originally planned for "The Deeds of Jason" to head the twenty-four stories in *The Earthly Paradise.* All of the stories in both works present the classical past through the lens of the Middle Ages. The tales in *The Earthly Paradise* are often remarkably similar, though some critics have argued that the recurring themes and situations can be seen as a rhythmic device to adumbrate Morris's ideas. Some of the most memorable stories are "Atalanta's Race," "The Wanderers," "Story of Cupid and Psyche," "The Love of Alcestis," "The Land East of the Sun and West of the Moon," and "The Hill of Venus."

The last, a retelling of the Tannhäuser legend, is especially noteworthy. A man named Walter enters a cavern that leads to Venus's underground

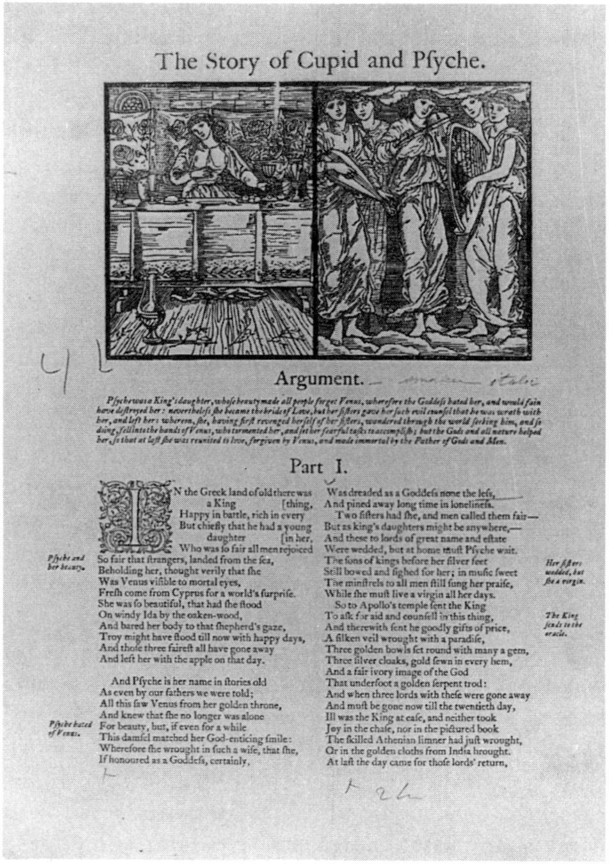

The Story of Cupid and Psyche.

Argument.

Psyche was a King's daughter, whose beauty made all people forget Venus, wherefore the Goddess hated her, and would fain have destroyed her: nevertheless she became the bride of Love, but her sisters gave her such evil counsel that he was wroth with her, and left her: whereon, she, having first revenged herself of her sisters, wandered through the world seeking him, and so doing, fell into the hands of Venus, who tormented her, and set her fearful tasks to accomplish; but the Gods and all nature helped her, so that at last she was reunited to love, forgiven by Venus, and made immortal by the Father of Gods and Men.

Part I.

IN the Greek land of old there was a King [thing,
Happy in battle, rich in every
But chiefly that he had a young daughter [in her,
Who was so fair all men rejoiced
So fair that strangers, landed from the sea, *Psyche and her beauty.*
Beholding her, thought verily that she
Was Venus visible to mortal eyes,
Fresh come from Cyprus for a world's surprise.
She was so beautiful, that had she stood
On windy Ida by the oaken-wood,
And bared her body to that shepherd's gaze,
Troy might have stood till now with happy days,
And those three fairest all have gone away
And left her with the apple on that day.

And Psyche is her name in stories old
As even by our fathers we were told;
All this saw Venus from her golden throne,
And knew that she no longer was alone
For beauty, but, if even for a while *Psyche hated of Venus.*
This damsel matched her God-enticing smile:
Wherefore she wrought in such a wife, that she,
If honoured as a Goddess, certainly,
Was dreaded as a Goddess none the less,
And pined away long time in loneliness.
Two sisters had she, and men called them fair—
But as king's daughters might be anywhere,—
And these to lords of great name and estate
Were wedded, but at home must Psyche wait.
The sons of kings before her silver feet *Her sisters wedded, but she a virgin.*
Still bowed and sighed for her; in music sweet
The minstrels to all men still sung her praise,
While she must live a virgin all her days.
So to Apollo's temple sent the King
To ask for aid and counsell in this thing, *The King sends to the oracle.*
And therewith sent he goodly gifts of price,
A silken veil wrought with a paradise,
Three golden bowls set round with many a gem,
Three silver cloaks, gold sewn in every hem,
And a fair ivory image of the God
That underfoot a golden serpent trod:
And when three lords with these were gone away
And must be gone now till the twentieth day,
Ill was the King at ease, and neither took
Joy in the chase, nor in the pictured book
The skilled Athenian limner had just wrought,
Or in the golden cloths from India brought.
At last the day came for those lords' return,

One of Morris's trial pages, illustrated by Edward Burne-Jones, for a proposed edition of The Earthly Paradise *that was never published (William S. Peterson,* The Kelmscott Press: A History of William Morris's Typographical Adventures, *1991)*

bower. He becomes her lover, but Venus will not answer his questions, and he leaves her. He finds himself an alien in the everyday world and joins a pilgrimage to Rome, where he talks with the Pope. The Pope tells Walter he will not find happiness with Venus any more than his staff will grow flowers. Later, flowers do sprout from the staff, and the Pope dies, rejoicing in the miracle. Walter never learns of the miracle, however. Troubled by moral ambiguities, he turns toward Venus, realizing her realm is now but a dream.

The poetic narratives in *The Earthly Paradise* include some of the most significant and now least fully appreciated works of the nineteenth century. Here and in *The Life and Death of Jason* there is a feel of primal myth, the sort of thing that J. R. R. Tolkien and C. S. Lewis came to admire in Morris and that influenced their fiction.

As a poet Morris sometimes overdoes quaint archaisms and prettiness in *The Life and Death of Jason* and *The Earthly Paradise.* Some of his verse narratives are extremely long, and *The Life and Death of Jason* is one of the longest poems in English. *The Life and Death of Jason* and *The Earthly Paradise,* with its illustrations by Burne-Jones, sold well and gave him wide popularity, though his literary fame in his lifetime was obscured by the variety of his interests and accomplishments. In 1861, at age twenty-seven, Morris had started the firm that eventually was called Morris and Company, a firm that employed artists to design things for everyday use. It was not well managed, which was unfortunate since Morris came to depend on it more and more financially as his inherited investments declined in value. His designs for the firm had the freshness of nature about them, doubtless because of his earlier keen observations of nature. Because of the Anglo-Catholic movement there was a demand for ecclesiastical art, which Morris and Company endeavored to meet. They created furniture, stained glass, tapestries, paintings, embroidery, and tiles for churches.

An important influence during these years was Ruskin. He had come often to Morris's rooms at Red Lion Square and was fond of Morris and Burne-Jones. Ruskin influenced Morris to work with his hands before he pronounced on art. Interest in the arts at this time was also connected with Morris's political interests. He later joined the Democratic Federation, a league of London working men's clubs with socialist leanings. The Federation became increasingly socialist, and Morris went right along. He later adopted a more passive socialism because he feared violent revolution.

In 1871 Morris visited Iceland, an experience that stimulated his imagination greatly. He loved the north and its culture and had two years before published translations of Icelandic poetic narratives. Surprisingly, he was not fond of Richard Wagner, who, he believed, degraded his subject; Morris thought opera was a rococo and decadent form of art. Morris's Iceland journal contains straightforward, vigorous English prose, especially compared to the fancy prose of the saga translations. Morris sees Iceland in a clear-eyed, unsentimental way, yet it takes on a mysterious, almost numinous aura through his eyes.

Morris's marriage continued to decline. It is difficult to assess the triangular relationship of Morris, Jane, and Rossetti. On the one hand Morris seemed glad of Rossetti's interest in Jane, and Rossetti's involvements with women were often platonic in spite of their appearance otherwise. Rossetti indulged Jane's apparent hypochondria and continued to use her as a model for his paintings. Morris was such a frank pagan in his views on love and sex that he might well have countenanced a sexual relationship between Jane and Rossetti. The real situation, however, may not have been so sanguine for Morris, as he seems in his writing to align himself with the betrayed or hopeless lover.

In 1871, at age thirty-seven, Morris joined Rossetti in leasing Kelmscott Manor on the Thames in Oxfordshire. The setting evoked for Morris the medieval past he loved so well. Four years later he pressured Rossetti to give up his share of the lease, and Morris and Company was reorganized under Morris's sole control. In 1881, six years later, he moved his company to Merton Abbey, Surrey, just south of London. Here, on the banks of the River Wandle, Morris created his ideal workplace, his employees working harmoniously in pleasant surroundings. Kelmscott Manor was Morris's refuge from crowded, noisy London. The 1880s were the busiest decade of his life and probably the most productive.

Morris paid his workers well and took care that the repetitive jobs were passed around so as not to become onerous to any individual. He knew how to do each job that he assigned to another. In setting standards of beauty and quality in household goods, Morris and Company was a great success. In making them more available to the general public, it was a failure. Only the well-off could afford Morris's products.

In 1873 Morris published *Love Is Enough; or, The Freeing of Pharamond: A Morality*, the first of his imaginative works to reflect his interest in Iceland. Pharamond, a mighty ruler, dreams of a maiden in a northern valley; this leads him to go in search of her. After years of searching, Pharamond finds the land he had dreamed of and the fair Azalais. He returns to his old kingdom and finds he has been replaced. The story ends with Pharamond planning to return to Azalais. The valley where the hero finds his love is surrounded by mountains, and the descriptions sound like Iceland. Here Pharamond finds fulfillment, and he blithely leaves his kingdom behind.

Morris's next work was *The Story of Sigurd the Volsung and the Fall of the Niblungs* (1876). He had translated the Völsunga Saga six years before, attracted to its stylistic restraint. The 1876 performance of Wagner's *Der Ring des Nibelungen* may have motivated Morris to finish his story, and Morris apparently saw Wagner as a rival. Morris also felt that the English language had been degraded by casual misuse, and he tried to invent a new English style in *The Story of Sigurd the Volsung and the Fall of the Niblungs*. He drew not only on the *Völsunga* but also on other Icelandic sources, such as the *Elder Edda* and the *Younger* or *Prose Edda,* as well as the

Germanic *Nibelungenlied,* on which the closing parts of Morris's story are based.

Signy, daughter of Volsung, is to marry Siggeir, king of the Goths. When Siggeir comes to take his bride, Odin plants a sword in oak wood, and only Sigmund is able to pull it out. Later Sigmund is the only one of the Volsungs to escape Siggeir's slaughter. Signy comes to Sigmund in his cave, and the child of their union is Sinfiotli, who is raised by Sigmund. With Signy's help Sigmund and their son kill Siggeir and his children. Sigmund returns to his own land and marries Borghild. Sinfiotli kills Borghild's brother in a dispute over the spoils of a war, and the queen then poisons Sigmund's son. Sigmund banishes Borghild and marries a younger woman, Hiordis. Later Odin breaks Sigmund's sword, and he is killed soon afterward. Hiordis bears Sigmund's child and finds refuge in another kingdom.

Wagner streamlines the story; Morris tells it in a more diffuse way, yet he captures a northern atmosphere of swift action and revenge. Morris was also interested in the idea of Ragnarok, the destruction of the world and the return of a golden age. He alludes to such events in *The Story of Sigurd the Volsung and the Fall of the Niblungs.* The role of Baldur in this apocalyptic material links this work with Joseph Campbell's monomyth, in which the hero suffers for his people and is glorified.

In the late 1870s Morris became increasingly interested in Ruskinian socialism. He had begun to be an activist when he joined those who opposed Turkish aggression. Morris also had a letter published in a newspaper expressing his support of a law protecting free speech. There had been police attacks on demonstrators for the unemployed.

In 1879 Morris became treasurer of the National Liberal League, which joined with the Labour support of William Gladstone, the prime minister. In 1882 Morris began reading socialist literature, and he attended a series of public lectures on socialism. In 1883 he joined the Democratic Federation and began reading Karl Marx's *Das Kapital* (1867) in a French translation. In late 1883 he announced his conversion to socialism.

Morris gave almost two hundred lectures in the last twenty years of his life. Some are simply responses to current events. Others were written for publication and help to make up the two essay collections published during Morris's lifetime: *Hopes and Fears for Art: Five Lectures Delivered in Birmingham, London, and Nottingham, 1878–81* (1882) and *Signs of Change: Seven Lectures Delivered on Various Occasions* (1888). Morris's approach in these essays is heavily influenced by Ruskin, who believed that one cannot be concerned with art without being concerned with the social order. Morris argues against elitism in art in favor of an equality between the maker and the user of art. He believed that if art became too expensive for the masses, it would lead to an alienation between artist and user–which is ironic since the products of Morris and Company were too costly for the common household. This was the case because of Morris's insistence that the artists receive fair wages; he was caught between a sense of duty to the artist and duty to the user.

Morris also believed in the utilitarian approach to art, saying that art must be useful, and he stressed simplicity in art. This view seems overstated, though, because he also believed in "high art," though he argued that the high arts could be reborn only after the "lesser arts" have flourished.

Morris's views led him to teach that working conditions should be as pleasant as possible and that life should be so arranged that the common people can enjoy and possess art. He feared that technology might impair the quality of life, but he was no machine smasher. Though he said he was opposed to machines, as an artisan he knew the good potential of machines. He also feared the Victorian tendency to treat human beings as if they existed for commerce rather than as if commerce existed for the sake of human convenience and betterment.

In his essays Morris is openly socialist, stressing the value of cooperation over individual accomplishment. He revered Gothic architecture as the greatest example of cooperation. Artists who are too individualistic to cooperate harm the social fabric and art itself, he believed.

This view is at odds, however, with Morris's interest in the role of special men in his late romances, to which he turned in his last years. Among the first of these was *A Dream of John Ball,* published with *A King's Lesson* in 1888. In a dream vision the narrator of *A Dream of John Ball* finds himself in fourteenth-century Kent. John Ball was a real priest who was one of the leaders of a peasant rebellion. He was imprisoned, released when the peasants briefly gained power, and then recaptured and executed. The title of Morris's work has a threefold meaning, referring to Morris's dream of the past, to John Ball's dream of revolution, and to the utopian dream of a socialist future. Ball is against villenage, in which freemen were elevated above their fellow serfs in a feudal society, and hopes that abolishing it will bring fellowship among men. Morris has him say, "Fellowship is heaven, and lack of fellowship is hell." Ball believes that "heaven and earth are not two but one," and his hope is something like the dream of the classless society in communism.

Frontispiece depicting Kelmscott Manor for the 1892 Kelmscott Press edition of Morris's News from Nowhere

Because of dissension in the socialist Democratic Federation, a group withdrew and founded the Socialist League, and Morris went along. This group's great weakness was their belief that as middle-class theorists they could impose socialism on the lower classes. A riot in 1887, which began as a protest against the police ban on demonstrations, confirmed Morris's fear that revolution would involve bloodshed. He soon withdrew from active participation in the Socialist League. He continued to write socialist essays and to preside at meetings of the Hammersmith Socialist Society. Morris's second daughter, May, also took an active part in the socialist cause.

Morris's next romance, *A Tale of the House of the Wolfings and All the Kindreds of the Mark* (1889), is set in the ancient world and tells of the struggle of a people who are threatened by the legions of Rome. Their leader is Thiodolf, who loves a Valkyrie named Wood-Sun. She gives him a hauberk to wear in battle, but he does not know there is a curse on the hauberk. The wearer will not perish in battle but will lose his honor. Thiodolf refuses to wear the hauberk and is killed by the Romans, who are nonetheless beaten. Thiodolf dies believing that he will live on in the hearts of his people.

The Roots of the Mountains wherein Is Told Somewhat of the Lives of the Men of Burgdale, Their Friends, Their Neighbours, Their Foemen, and Their Fellows in Arms (1890) is another prose romance. Set in the fifth century A.D., it is similar to *The House of the Wolfings* in that the hero must struggle against the outside world to protect himself and his tribe. The protagonist, Face-of-God, while wandering the forest, meets a beautiful woman and her brother, who encourage him to join them in an attack on the Huns. They are victorious, and the two tribes discover that they are related. Face-of-God and the woman are then joined in marriage. In style *The Roots of the Mountains* is full of medievalisms and uses verse for important speeches. A sanguine work, it illustrates Morris's belief that the clan must decline and be reborn in a socialist ideal world.

News from Nowhere is regarded by many scholars as the central work of Morris's career. It portrays daily life in a communist utopia and is similar to works such as Sir Thomas More's *Utopia* (1516), Edward Bellamy's *Looking Backward 2000–1887* (1888), and B. F. Skinner's *Walden Two* (1948) in describing ideal living and working conditions. *News from Nowhere* appeared serially in *Commonweal* and has the inconsistencies and loose form of serial fiction. Morris wrote here not only against capitalism but also against the Socialist League, with which he had recently broken. Also, the early chapters were written to counter *Looking Backward,* for Morris was troubled by Bellamy's version of socialism. He disagreed with Bellamy's mechanized approach to work and with his liking for cities.

In the beginning of *News from Nowhere* the narrator, a thinly fictionalized version of Morris, wakes to find himself in the Hammersmith Guest House, but in 1962. He has a communal breakfast and spends most of the day discussing politics with a man who may be Morris's grandson. The later chapters of the work take place the next day, when the narrator and two friends journey up the Thames to take part in harvesting the hay in Oxfordshire. For many chapters the narrator and his hosts make their way up the river. At the end he wakes in his house in Hammersmith unsure whether all these events have been a dream or whether he indeed has traveled in the future.

In this utopian world there is no private property, no class distinctions, no divorce, no parliament, no law courts or police, little violence and crime, and no politics. There is true equality, and women are emancipated, though many freely choose to be housewives and bear children. The people are mostly good-looking, and they dress lightly but often with colorful adornment. The narrator learns of a murderer who is punished only by his own great remorse. The people believe that heaven and hell are here on earth and can be created by human beings.

In *News from Nowhere* life is decentralized: the old big cities have been broken up, and most people live in villages. Wild places have been preserved, as have some old buildings. Work is highly valued, and the people fear running short of it. Machines are used for boring kinds of work, but the machine is not revered. It is understood that if one concentrates on making work pleasant, the result will be high quality in the product.

News from Nowhere is wordy and diffuse by contemporary standards, but like many utopian and dystopian novels it is full of ideas and is stimulating to read. In a sense Morris's vision of the future is a further development of the beliefs of Morris and Company. He emphasized good craftsmanship, which is the foundation for happiness in his views. Using his hands gave such pleasure to Morris that he assumed that everyone would enjoy what he did.

In this work Morris tells of class war in the twentieth century, which results in victory for the workers and a decline in the power of the state as it had existed under capitalism. He naively assumes that the reasonableness of the workers will make a legal system unnecessary and that government in general will be minimal. Morris also reflects a Wordsworthian belief in the restoring powers of nature. He seems to argue for a childlike openness fused with the strength of maturity.

In 1891 Morris established the Kelmscott Press in Hammersmith, a suburb of London. He had seen an exhibition three years earlier in which he admired the Gothic fonts of early German printing, which were really an extension of manuscript writing. In addition to the Kelmscott Press, the other great achievement of Morris's last years were his remaining prose romances, including *The Story of the Glittering Plain Which Has Been Also Called the Land of Living Men or the Acre of the Undying* (1891), *The Wood Beyond the World* (1894), and *The Well at the World's End: A Tale* (1896). During this later period he also participated in the Society for the Protection of Ancient Buildings.

Between 1891 and 1897 the Kelmscott Press published fifty-three works in sixty-six volumes, including Morris's works and editions of poets he admired, including Samuel Taylor Coleridge, John Keats, Percy Bysshe Shelley, Tennyson, Rossetti, and Swinburne. Morris designed the typefaces himself. Perhaps the most important work published was the Kelmscott edition of the works of Geoffrey Chaucer.

The five romances Morris wrote in his last years show a return to a form he loved and a departure from a preoccupation with the social order so evident in *News from Nowhere.* While he had been working on this utopian novel he had begun *The Story of the Glittering Plain,* which was one of the first of his Kelmscott editions. *The Story of the Glittering Plain* also has utopian features: an economy with no buying or selling and inhabitants who experience a complete harmony with nature and great freedom. They live like children, yet with complete sexual freedom. There is no love or fellowship, however. Hallblithe, the hero, rejects this society. Yet there seems to be no escape, and he turns to artistic craftsmanship for his salvation.

The Wood beyond the World is another quest romance. The hero is Walter, who is more complicated than Hallblithe. He goes in search of a woman he has seen in a vision. He and his new wife start a new life in a new land. Morris's treatment of physical love here is frank by Victorian standards.

The Well at the World's End is the longest of the prose romances, and Morris regarded his work on it as interminable. The hero, Ralph, searches for a magical well. He leaves his family, encountering two women in the wild who give him the ability to return to his home and save it from a dark fate.

The Water of the Wondrous Isles (1897) has as its protagonist a heroine, Birdalone, who was taken from her mother as an infant and raised by a witch. When she comes of age she begins to assert herself against the witch, who is not at home with nature, whereas Birdalone gains power from nature. After a series of adventures, mostly escapes from various captors, Birdalone is united with a man named Arthur. This romance was published after Morris's death.

The posthumous *The Sundering Flood* (1897) was not quite finished at the time of Morris's death. Osberne and Elfhild, though they grew up on opposite sides of a large river called the Sundering Flood, fall in love and communicate across what seems an impassable barrier. They go through separate journeys and tests yet are ultimately united, and they return to Osberne's family's place. Like Morris, Osberne is a poet and a craftsman–according to Kirchhoff, the most satisfactory male in the late romances.

From 1894 to 1896 Morris suffered from illness and feebleness. In the summer of 1896 he made his last voyage to the north, traveling along the coast of Norway. A few weeks after his return he was stricken with lung congestion and died on 3 Oc-

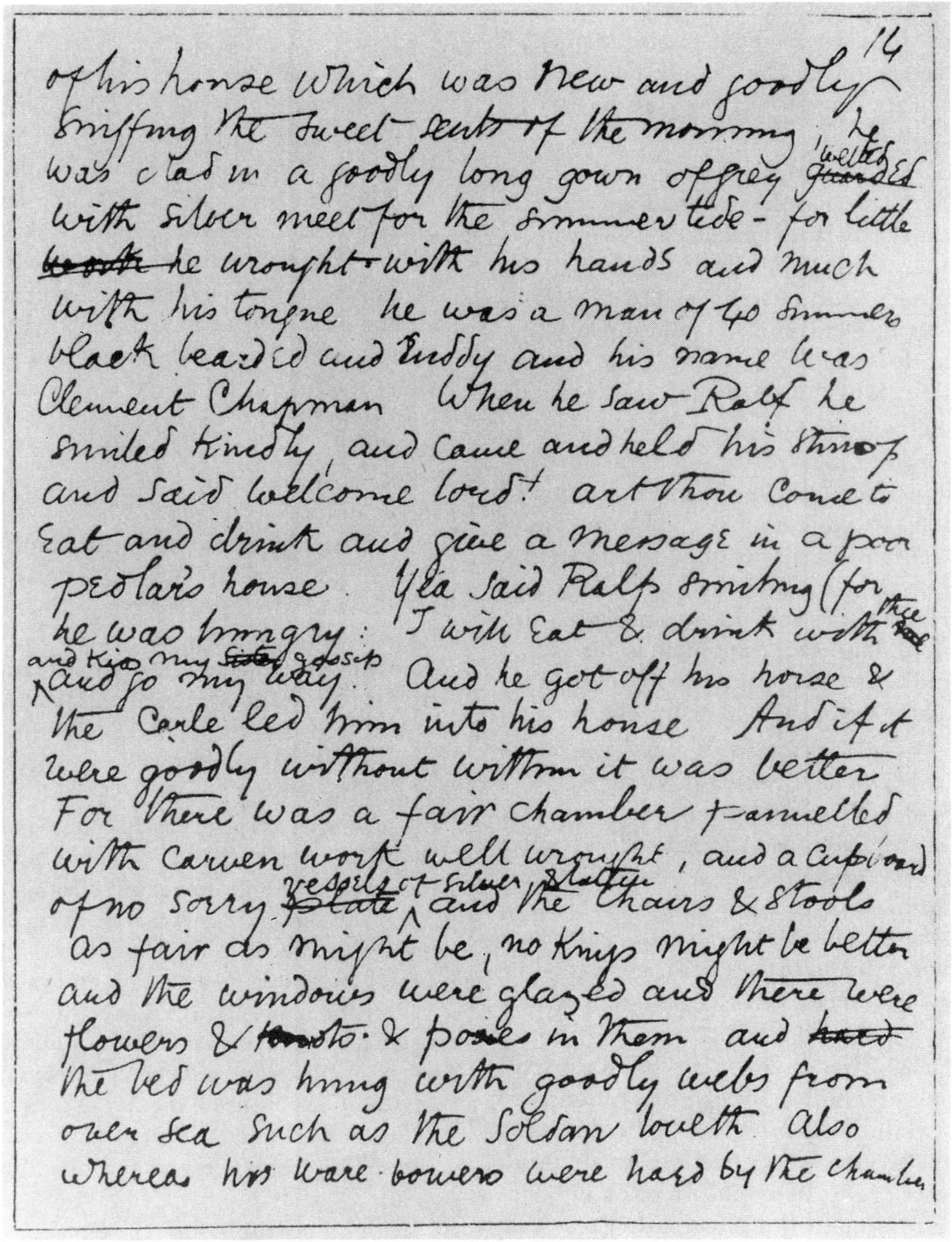
14

of his house which was new and goodly
sniffing the sweet scents of the morning; he
was clad in a goodly long gown of grey welted
with silver meet for the summer tide - for little
he wrought with his hands and much
with his tongue he was a man of 40 summers
black bearded and ruddy and his name was
Clement Chapman When he saw Ralf he
smiled kindly, and came and held his stirrup
and said welcome lord! art thou come to
eat and drink and give a message in a poor
pedlar's house. Yea said Ralf smiling (for
he was hungry: I will eat & drink with thee
and kiss my gossip and go my way. And he got off his horse &
the carle led him into his house And if it
were goodly without within it was better
For there was a fair chamber panelled
with carven work well wrought, and a cupboard
of no sorry vessels of silver & latten and the chairs & stools
as fair as might be, no kings might be better
and the windows were glazed and there were
flowers & knots & posies in them and
the bed was hung with goodly webs from
over sea such as the Soldan loveth. Also
whereas his ware-bowers were hard by the chamber

A page from the first draft of Morris's The Well at the World's End *(H. Buxton Forman,* The Books of William Morris Described, with Some Account of His Doings in Literature and in the Allied Crafts, *1897)*

tober 1896 at age sixty-two. His wife survived him by seventeen years, dying in 1914.

Morris is in many ways a contradictory and baffling character. He is like a Janus facing two ways: toward the Middle Ages, which he loved, and toward the socialist future, toward which he felt a duty to press. His literary legacy for fantasy and science fiction is equally double-sided: in his romantic fantasies he offered medieval or otherworldly settings and stories, while in *News from Nowhere* he presents a pastoral yet futuristic socialist utopia. For both aspects of his writings he will long be remembered and read.

Letters:

The Letters of William Morris to His Family and Friends, edited by Philip Henderson (London & New York: Longmans, Green, 1950);

The Collected Letters of William Morris, edited by Norman Kelvin, volume 1, 1848–1880 (Princeton: Princeton University Press, 1984); volume 2, part A, 1881–1884; volume 2, part B, 1885–1888 (Princeton: Princeton University Press, 1987).

Bibliographies:

H. Buxton Forman, *The Books of William Morris Described, with Some Accounts of His Doings in Literature and in the Allied Crafts* (London: Hollings, 1897; Chicago: Way & Williams, 1897);

Temple Scott (J. H. Isaacs), *A Bibliography of the Works of William Morris* (London: Bell, 1897);

Theodore G. Ehrsam, *Bibliographies of Twelve Victorian Authors* (New York: Wilson, 1936), pp. 162–187;

William E. Fredeman, "William Morris and His Circle: A Selective Bibliography of Publications," *Journals of the William Morris Society,* 1 (Summer 1964): 23–33; 2 (Spring 1966): 13–26;

Fredeman, *Pre-Raphaelitism: A Bibliocritical Study* (Cambridge, Mass.: Harvard University Press, 1965; London: Oxford University Press, 1965);

John J. Walsdorf, *William Morris in Private Press and Limited Editions: A Descriptive Bibliography of Books by and about William Morris, 1891–1981* (Phoenix: Oryx Press, 1983);

K. L. Goodwin, *A Preliminary Handlist of Manuscripts and Documents of William Morris* (London: William Morris Society, 1984);

Gary L. Aho, *William Morris: A Reference Guide* (Boston: G. K. Hall, 1985);

David Latham and Sheila Latham, *An Annotated Critical Bibliography of William Morris* (London: Harvester Wheatsheaf, 1991; New York: St. Martin's Press, 1991).

Biographies:

John W. Mackail, *The Life of William Morris,* 2 volumes (London & New York: Longmans, Green, 1899);

Margaret R. Grennan, *William Morris: Medievalist and Revolutionary* (New York: King's Crown Press, 1945);

Philip Henderson, *William Morris: His Life, Work and Friends*, second edition (Harmondsworth: Penguin, 1973);

Jack Lindsay, *William Morris: His Life and Work* (London: Constable, 1975);

E. P. Thompson, *William Morris: Romantic to Revolutionary,* revised edition (London: Merlin Press, 1977; New York: Pantheon, 1977);

Fiona MacCarthy, *William Morris: A Life for Our Time* (New York: Knopf, 1995).

References:

Patrick Brantlinger, "'News from Nowhere': Morris's Socialist Anti-Novel," *Victorian Studies,* 19 (September 1975): 35–49;

Blue Calhoun, *The Pastoral Vision of William Morris* (Athens: University of Georgia Press, 1975);

Lin Carter, *Imaginary Worlds* (New York: Ballantine, 1973);

Peter Faulkner, ed., *William Morris: The Critical Heritage* (London & Boston: Routledge & Kegan Paul, 1973);

Frederick Kirchhoff, *William Morris* (Boston: Twayne, 1979; London: Prior, 1979);

C. S. Lewis, "William Morris," in his *Rehabilitations and Other Essays* (London: Oxford University Press, 1939; New York: Folcroft, 1973);

Roderick Marshall, *William Morris and His Earthly Paradises* (Tisbury: Compton Russell, 1979);

Richard Mathews, *Worlds beyond the World: The Fantastic Vision of William Morris* (San Bernardino, Cal.: Borgo Press, 1978);

Carole Silver, *The Romance of William Morris* (Athens: Ohio University Press, 1982);

Silver and Joseph R. Dunlap, eds., *Studies in the Late Romances of William Morris* (New York: William Morris Society, 1976);

Lionel Trilling, "Aggression and Utopia: A Note on William Morris's 'News from Nowhere,'" *Psychoanalytic Quarterly*, 42 (1973): 214–225.

Papers:

The British Library in London holds the majority of Morris's manuscripts, letters, and other papers. Other collections include the Beinecke Rare Book Room and Manuscript Library, Yale University; the Sanford and Helen Berger Collection in Carmel, California; the Birmingham City Museum and Art Gallery in England; the Bodleian Library, Oxford University; the Fitzwilliam Museum, Cambridge University; the Henry E. Huntington Library, San Marino, California; the International Institute of Social History, Amsterdam; the Pierpont Morgan Library, New York City; the William R. Perkins Library, Duke University; the Harry Ransom Humanities Research Center at the University of Texas, Austin; the Victoria and Albert Museum Library, London; and the William Morris Gallery, Walthamstow.

E. Nesbit

(15 August 1858 – 4 May 1924)

Amelia A. Rutledge
George Mason University

See also the Nesbit entries in *DLB 141: British Children's Writers, 1880–1914* and *DLB 153: Late-Victorian and Edwardian British Novelists, First Series.*

E. Nesbit

BOOKS: *Fading Light: Verses by E. Nesbit* (London & New York: Hagelberg, n.d.);

Apple Pie (n.p., n.d.);

Miss Mischief (London: Nister, n.d.; New York: Dutton, n.d.);

Fairies (London: Tuck, n.d.);

The Prophet's Mantle, by Nesbit and Hubert Bland, as Fabian Bland (London: Henry J. Drane, 1885; Chicago & New York: Bedford Clarke, 1889);

Lays and Legends (London: Longmans, Green, 1886); republished as *Lays and Legends, First Series* (London: Longmans, Green, 1892);

The Lily and the Cross (London: Griffiths, Farran, Okeden & Welsh, 1887; New York: Dutton, 1887);

The Star of Bethlehem (London: Nister, 1887; New York: Dutton, 1887);

The Better Part, and Other Poems (London: Henry J. Drane, 1888);

Easter-Tide: Poems by E. Nesbit and Caris Brooke (London: Henry J. Drane; New York: Dutton, 1888);

Landscape and Song (London: Henry J. Drane; New York: Dutton, 1888);

The Message of the Dove (London: Henry J. Drane; New York: Dutton, 1888);

All Round the Year, by Nesbit and Caris Brooke (London: Von Portheim, 1888);

Leaves of Life (London & New York: Longmans, Green, 1888);

The Lilies Round the Cross, by Nesbit and Helen Wood (London: Nister, 1889; New York: Dutton, 1889);

Corals and Sea Songs (London: Nister; New York: Dutton, 1889);

Songs of Two Seasons (London: Tuck, 1890);

The Voyage of Columbus: Discovery of America (London & New York: Tuck, 1892);

Sweet Lavender (London: Nister, 1892; New York: Dutton, 1892);

Lays and Legends, second series (London & New York: Longmans, Green, 1892);

Grim Tales (London: A.D. Innes, 1893);

Something Wrong (London: A.D. Innes, 1893);

The Girls' Birthday Book (London: Henry J. Drane, 1894);

The Butler in Bohemia, by Nesbit and Oswald Barron (London: Henry J. Drane, 1894);

The Marden Mystery (Chicago: Bedford Clarke, 1894);

Pussy Tales (London: Marcus Ward, 1895); republished with *Doggy Tales* as *Pussy and Doggy Tales* (London: Dutton, 1899);

Doggy Tales (London: Marcus Ward, 1895), republished with *Pussy Tales* as *Pussy and Doggy Tales* (London: Dutton, 1899);

Rose Leaves (London: Nister, 1895);

Holly and Mistletoe, A Book of Christman Verse, by Nesbit, Norman Gale, and Roger le Gallienne (London: Marcus Ward, 1895);

A Pomander of Verse (London: John Lane, 1895; Chicago: McLury, 1895);
In Homespun (London: John Lane, 1896; Boston: Robert Brothers, 1896);
As Happy as a King (London: Marcus Ward, 1896);
The Children's Shakespeare, edited by Edric Vredenberg (London: Tuck, 1897; Philadelphia: Altemus, 1900);
Romeo and Juliet, and Other Stories (London: Tuck, 1897);
Royal Children in English History (London: Tuck, 1897; New York: Mershon, 1903);
Dog Tales, and Other Tales, by Nesbit, A. Guest, and Emily R. Watson (London Tuck, 1898; New York: Tuck, 1900);
A Book of Dogs, Being a Discourse on Them, with Many Tales and Wonders Gathered by E. N. (London: Dent, 1898; New York: Dutton, 1898);
Songs of Love and Empire (London: Constable, 1898);
The Secret of Kyriels (London: Hurst & Blackett, 1898; Philadelphia: Lippincott, 1899);
The Book of Dragons (London: Harper, 1899; New York: Harper, 1900);
The Story of the Treasure-Seekers, Being the Adventures of the Bastable Children in Search of a Fortune (London: Unwin, 1899; New York: Frederick Stokes, 1899);
Nine Unlikely Tales for Children (London: Unwin, 1901; New York: Dutton, 1904);
The Wouldbegoods, Being the Further Adventures of the Treasure-Seekers (London: Unwin, 1901; New York: Harper, 1901);
To Wish You Every Joy (London: Tuck, 1901);
Thirteen Ways Home (London: Anthony Treherne, 1901);
Five Children and It (London: Nister, 1902; New York: Dodd, Mead, 1905);
The Revolt of the Toys and *What Comes of Quarrelling* (London: Nister, 1902; New York: Dutton, 1904);
The Red House (London: Methuen, 1902; New York: Harper, 1902);
The Literary Sense (London: Methuen, 1903; New York: Macmillan, 1903);
The Rainbow Queen and Other Stories (London: Tuck, 1903);
Playtime Stories (London: Tuck, 1903);
The New Treasure-Seekers (London: Unwin, 1904; New York: Frederick Stokes, 1904);
The Phoenix and the Carpet (London: George Newnes, 1904; New York: Macmillan, 1904);
Cat Tales, by Nesbit and Rosamund Bland (London: Nister, 1904; New York: Dutton, 1904);
The Story of the Five Rebellious Dolls (London: Nister, 1904; New York: Dutton, 1904);
Oswald Bastable and Others (London: Wells, Gardner, Darton, 1905; New York: Coward-McCann, 1960);
Pug Peter (Leeds & London: Alfred Cooke, 1905);
The Rainbow and the Rose (London & New York: Longmans, Green, 1905);
Man and Maid (London: Unwin, 1906);
The Story of the Amulet (London: Unwin, 1906; New York: Dutton, 1907);
The Railway Children (London: Wells, Garner, Darton, 1906; New York: Macmillan, 1907);
The Incomplete Amorist (London: Constable, 1906; New York: Doubleday, Page, 1906);
The Enchanted Castle (London: Unwin, 1907; New York: Harper, 1908);
Twenty Beautiful Stories from Shakespeare (Chicago: Jenkins, 1907);
The Old Nursery-Stories (London: Henry Frowde & Hodder & Stoughton, 1908; New York: Doran, 1912);
The House of Arden (London: Benn, 1908; New York: Coward-McCann, 1908);
Jesus in London (London: Fifield, 1908);
Ballads and Lyrics of Socialism (London: Fifield, 1908);
Harding's Luck (London: Benn, 1909; New York: Frederick Stokes, 1910);
These Little Ones (London: George Allen, 1909);
Salome and the Head (London: Doubleday, Page, 1909); published as *The House with No Address* (New York: Doubleday, Page, 1909);
Daphne in Fitzroy Street (London: George Allen, 1909; New York: Doubleday, Page, 1909);
Garden Poems (London: Collins, 1909);
Cinderella, a Play with Twelve Songs to Popular Airs (London: Sidgwick & Jackson, 1909);
Children's Stories from English History, by Nesbit and Doris Ashley (London: Tuck, 1910);
Fear (London: Stanley Paul, 1910);
The Magic City (London: Benn, 1910; New York: Macmillan, 1910);
Dormant (London: Methuen, 1911); published as *Rose Royal* (New York: Dodd, Mead, 1912);
Children's Stories from Shakespeare, with "When Shakespeare was a Boy," by Dr. F. J. Furnivall (London: Tuck, 1911; Philadelphia: McKay, 1912); republished as *Shakespeare Stories for Children* (London: Tuck, 1925);
Ballads and Verses of the Spiritual Life (London: Elkin Mathews, 1911);
The Wonderful Garden, or the Three Cs (London: Macmillan, 1911; New York: Coward-McCann, 1935);
The Magic World (London & New York: Macmillan, 1912);

Wet Magic (London: T. Werner Laurie, 1913; New York: Coward-McCann, 1937);

Wings and the Child, or the Building of Magic Cities (London: Hodder & Stoughton, 1913; New York: Doran, 1913);

The Incredible Honeymoon (New York: Harper, 1916; London: Hutchinson, 1921);

The Lark (London: Hutchinson, 1922);

Many Voices (London: Hutchinson,1922);

To the Adventurous (London: Hutchinson,1923);

Five of Us, and Madeline, edited by Rosamund (Bland) Sharp (London: Unwin, 1925; New York: Adelphi, 1926).

PLAY PRODUCTIONS: *Cinderella*, Deptford Board School, December 1892;

A Family Novelette, by Nesbit and Oswald Barron, New Cross, February 1894;

Sleeping Beauty, Deptford Board School, December 1895;

Aladdin, Deptford Board School, December 1896;

The King's Highway, by Nesbit and Dorothea Deakin, Woolwich, Freemasons' Hall, 13 May 1905;

The Philandrist, or the London Fortune Teller, by Nesbit and Dorothea Deakin, Woolwich, Freemasons' Hall, 13 May 1905;

The Magician's Heart, London, St. George's Hall, 14 January 1907;

Unexceptionable References, London, Royalty Theatre, Autumn 1912.

OTHER: *Spring Songs and Sketches, Summer Songs and Sketches, Autumn Songs and Sketches*, and *Winter Songs and Sketches; Morning Songs and Sketches, Noon Songs and Sketches, Evening Songs and Sketches*, and *Night Songs and Sketches*, 8 volumes, selected and arranged by Nesbit and Robert Ellice Mack (London: Griffith, Farran, Okeden & Welsh, 1886–1887; New York: Dutton, 1886–1887);

River Sketches, edited by Nesbit (London: Von Portheim, 1887; New York: Dutton, 1887);

The Time of Roses, edited by Nesbit and Caris Brooke (London: Henry J. Drane, 1888);

By Land and Sea, edited by Nesbit (London: Henry J. Drane, 1888);

Autumn Leaves, edited by Nesbit (London: Henry J. Drane, 1888; New York: Dutton, 1888);

Winter Snow, edited by Nesbit (London: Henry J. Drane, 1888; New York: Dutton, 1888);

Lilies and Heartsease, edited by Nesbit (London: Griffith, Farran, Okeden & Welsh, 1888; New York: Dutton, 1888);

Falling Leaves, edited by Nesbit (London: Griffith, Farran, Okeden & Welsh, 1888);

Daisy Days, edited by Nesbit (London: Griffith, Farran, Okeden & Welsh, 1888);

Evergreen from the Poet's Corner, edited by Nesbit (London: Nister, 1889; New York: Dutton, 1889);

Songs of Scotland, edited by Nesbit (London: Nister, 1890; New York: Dutton, 1890);

Life's Sunny Side (London: Nister, 1890; New York: Dutton, 1890)–includes poems by Nesbit;

Told by the Fireside (London: Griffith, 1890)–includes "The Excursion";

Twice Four (London: Griffith, Farran, Browne, 1891)–includes "Finding a Sister";

The Poets and Poetry of the Century, edited by Alfred H. Miles (London: Hutchinson, 1891)–includes eleven poems by Nesbit;

Story Upon Story and Every One True (London: Tuck, 1892)–includes "Allie's House-Keeping";

Flowers I Bring and Songs I Sing (London: Tuck, 1893)–includes seven poems by Nesbit, as E. Bland;

Our Friends and All About Them (London: Tuck, 1893)–includes poems by Nesbit and "The Self Respecting Pussies"; "Down at Grannie's"; "Mabel's Pussy";

Listen Long and Listen Well (London: Tuck, 1893)–includes "Midsummer Day" and "The Oak Panel";

Sunny Tales for Snowy Days (London: Tuck, 1893)–includes "Ella's Adventure";

Told by the Sunbeams and Me (London: Tuck, 1893)–includes "Dorothy's Birthday" and "Being Bandits";

What Really Happened (London: Tuck, 1893)–includes "The Babe in the Wood, or What Happened at Kitty's Party";

We've Tales to Tell (London: Tuck, 1893)–includes "How Jack Came to Tea" and "A Crooked Tail";

Hours in Many Lands (London: Tuck, 1894)–includes "The Little Heroine" and "Effie's Birthday";

Tales That Are True for Brown Eyes and Blue (London: Tuck, 1894)–includes one story by Nesbit, "Lonely Mabel";

Tales to Delight from Morning till Night (London: Tuck, 1894)–includes "Mother's Present";

Fur and Feathers, Tales for All Weathers (London: Tuck, 1894)–includes "More Haste, Less Speed";

All But One, Told by the Flowers (London: Tuck, 1894)–includes "Hot Pies";

Lads and Lasses (London: Tuck, 1894)–includes stories by Nesbit;

The Girls' Own Birthday Book, edited by Nesbit (London: Henry J. Drane, 1894);

Tick Tock, Tales of the Clock (London: Tuck, 1895)–includes "The Glordy John";

Stories in a Shell (London: Tuck, 1895)–includes "The Rainbow Queen";

Treasures from Storyland (London: Tuck, 1895)–includes "Linda and the Prince";

Friends in Fable, a Book of Animal Stories (London: Tuck, 1895)–includes "A House of Her Own";

Dulcie's Lantern, and Other Stories (London: Tuck, 1895)–includes "Dulcie's Lantern";

Once Upon a Time, The Favourite Nursery Tales (London: Nister, 1897; New York: Dutton, 1897);

Dinna Forget (London: Nister, 1897; New York: Dutton, 1897)–includes two poems by Nesbit;

Father Tuck's Annual (London: Tuck, 1900)–includes one story, "The Fairy Godmother," and "Fairies";

My Sea-side Story Book, by Nesbit and G. Manville Fenn (London: Nister, 1911; New York: Dutton, 1911)–includes "Prince Feather-Head and the Mer-Princess";

Our New Story Book (London: Nister, 1913; New York: Dutton, 1913)–includes "Our Black Cat" and "The Likeness";

Battle Songs, edited by Nesbit (London: M. Goschen, 1914);

Essays by Hubert Bland, edited by Nesbit (London: M. Goschen, 1914);

The New World Literary Series, Book Two, edited by Henry Cecil Wylde (London & Glasgow: Collins, 1921)–includes ten short stories and ten poems for children by Nesbit.

SELECTED PERIODICAL PUBLICATIONS–UNCOLLECTED: "Something Wrong," by Nesbit and Hubert Bland, as Fabian Bland, *Weekly Dispatch,* 6 January–23 March 1884;

"In the Queen's Garden" (or "Absolom"), *Neolith,* no. 4 (August 1908);

"The Haunted House," *Strand Magazine* (December, 1913).

Edith Nesbit, one of the most prolific writers of fantasy both for children and adults, is best known for two series of children's stories, the Bastable books and her "magic" series, which were praised in her own time by Rudyard Kipling and H. G. Wells. She was a pioneer in the use of time travel in children's fantasies, and her work influenced the writings of C. S. Lewis, Charles Williams, and Edward Eager. Her stories distinguish themselves from many of the children's fantasies produced in the nineteenth century in their focus on children as members of families, in contrast to the solitary adventures of Lewis Carroll's Alice or the various heroines and heroes of George MacDonald's stories. Her stories are the literary heirs of the tradition of believable, that is, mischievous, child protagonists begun by Catherine Sinclair in *Holiday House* (1839) and continued in Nesbit's own time by the domestic and historical tales of Charlotte Mary Yonge.

Nesbit was the youngest of the six children of John Collis Nesbit and Sarah Green (née Alderton) Nesbit. Her father, who single-handedly administered an agricultural college–the first of its kind, founded by his father–died when Nesbit was three years old. Although she could not have had many memories of her father, the return of the absent father becomes a poignant moment in many of her fantasies. Her mother–indulgent toward all her children, and especially toward the youngest, known as "Daisy"–took over her husband's work for a time. Failing finances and the onset of tuberculosis in her oldest child, Mary, occasioned a series of moves, both in England and on the Continent; Nesbit was ever after to be concerned with stability of place, and her nostalgia for the scenes of childhood play and relative calm was to remain intense throughout her life.

Nesbit published "My School-Days" in a series of articles for *The Girl's Own Paper* during 1896–1897; many of these memories–adventures with her much-loved elder brothers, Henry and Alfred–were to be transformed into the escapades of her fictional children. Her school experiences, which were much less positive, informed several short stories that focused on the child's desire for freedom from the petty tyrannies of schoolmates and schoolmasters.

In 1880 Nesbit married Hubert Bland. Shortly after their marriage Bland contracted smallpox, and during his illness his business partner abandoned him, taking their joint capital. Nesbit, with their first child as well as her husband to support, began writing verses and painting pictures for greeting cards. She began writing short stories as well, the first of which was accepted by Alice Hoatson, a manuscript reader for a minor publication, *Sylvia's Home Journal.* Hoatson later gave up her job and lived with the Blands, giving Nesbit much-needed assistance in her writing and with household tasks.

Nesbit's first published novel, *The Prophet's Mantle* (1885), was written in collaboration with her husband under the pseudonym Fabian Bland; it was not well received. Although Bland, after his recovery, began a successful journalistic career, in which Nesbit also collaborated, it was her steady and in-

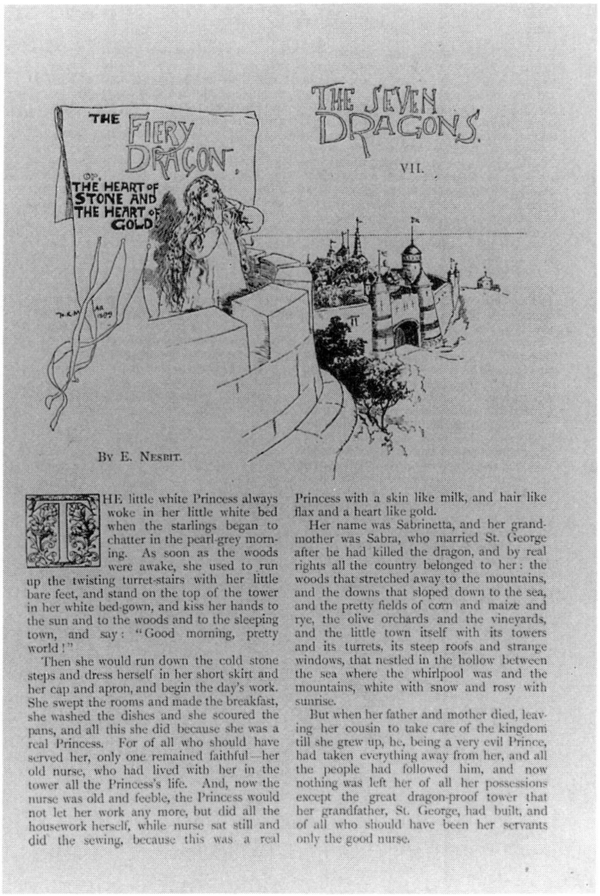

THE SEVEN DRAGONS.

VII.

THE FIERY DRAGON, OR, THE HEART OF STONE AND THE HEART OF GOLD

BY E. NESBIT.

THE little white Princess always woke in her little white bed when the starlings began to chatter in the pearl-grey morning. As soon as the woods were awake, she used to run up the twisting turret-stairs with her little bare feet, and stand on the top of the tower in her white bed-gown, and kiss her hands to the sun and to the woods and to the sleeping town, and say: "Good morning, pretty world!"

Then she would run down the cold stone steps and dress herself in her short skirt and her cap and apron, and begin the day's work. She swept the rooms and made the breakfast, she washed the dishes and she scoured the pans, and all this she did because she was a real Princess. For of all who should have served her, only one remained faithful—her old nurse, who had lived with her in the tower all the Princess's life. And, now the nurse was old and feeble, the Princess would not let her work any more, but did all the housework herself, while nurse sat still and did the sewing, because this was a real Princess with a skin like milk, and hair like flax and a heart like gold.

Her name was Sabrinetta, and her grandmother was Sabra, who married St. George after he had killed the dragon, and by real rights all the country belonged to her: the woods that stretched away to the mountains, and the downs that sloped down to the sea, and the pretty fields of corn and maize and rye, the olive orchards and the vineyards, and the little town itself with its towers and its turrets, its steep roofs and strange windows, that nestled in the hollow between the sea where the whirlpool was and the mountains, white with snow and rosy with sunrise.

But when her father and mother died, leaving her cousin to take care of the kingdom till she grew up, he, being a very evil Prince, had taken everything away from her, and all the people had followed him, and now nothing was left her of all her possessions except the great dragon-proof tower that her grandfather, St. George, had built, and of all who should have been her servants only the good nurse.

THE SEVEN DRAGONS. 113

great powers are on our side: the power of Love and the power of Arithmetic. Those two are stronger than anything else in the world."

So when the tide began to go down Nigel and the Princess ran out on to the sands, and there, full in sight of the terrace where the dragon kept watch, Nigel took his Princess in his arms and kissed her. The griffin was busy sweeping the stairs of the Lone Tower, but the dragon saw, and he gave a cry of rage—and it was like twenty engines all letting off steam at the top of their voices inside Cannon Street Station.

And the two lovers stood looking up at the dragon. He was dreadful to look at. His head was white with age—and his beard had grown so long that he caught his claws in it as he walked. His wings were white with the salt that had settled on them from the spray of the sea. His tail was long and thick and jointed and white, and had little legs to it, any number of them—far too many—so that it looked like a very large fat silkworm; and his claws were as long as lessons, and as sharp as bayonets.

"Good-bye, love!" cried Nigel, and ran out across the yellow sand towards the sea. He had one end of a cord tied to his arm.

The dragon was clambering down the face of the cliff, and next moment he was crawling and writhing and sprawling and wriggling across the beach after Nigel, making great holes in the sand with his heavy feet—and the very end of his tail, where there were no legs, made as it dragged a mark in the sand such as you make when you launch a boat; and he breathed fire till the wet sand hissed again, and the water of

"HE BREATHED FIRE TILL THE WET SAND HISSED AGAIN."

Vol. xviii.—15

"COCK-A-TRICE-A-DOODLE-DOO."

Illustrations by H. R. Millar for the dragon stories published in The Strand Magazine *during 1899*

creasing production of verse and short narratives for edited collections of children's ephemera that supported their growing family, eventually including the two children of Bland and Alice Hoatson. This was not Bland's first infidelity; he had maintained a mistress during his courtship of Nesbit and continued to have affairs throughout his life. Nesbit's reaction to the revelation of the paternity of Hoatson's children was complex, yet she acquiesced in Bland's insistence that Hoatson remain with them. It is possible that Nesbit realized, although not consciously, that Hoatson, by taking on the household management and a great deal of the child rearing upon herself, was helping to facilitate Nesbit's increasingly demanding career as the latter slowly established herself with a series of publishers, first for her verse and later for her fiction.

Nesbit and Bland were active members in the Byron and the Shelley societies, and they became influential in the newly founded socialist group the Fabian Society. These activities brought them into contact with many of the leading intellectuals and literati of their time, notably Wells and Bernard Shaw. Nesbit and Bland were respected for their commitment to the ideals of the Fabian Society, but they were also recognized for their great personal charm; Nesbit's biographers cite several letters and reminiscences–from Shaw and Wells, for example–that comment on her own vitality and beauty. The couple considered themselves advanced in terms of social mores, despite their repudiation of universal suffrage, and Nesbit engaged in discreet affairs of her own. The young women of her mainstream novels reflect, more often than not, idealized self-portraits, as well as wish fulfillment, in their successful love experiences.

Nesbit was almost forty before she began to publish fiction outside of serial collections edited by others. Throughout her career her work manifested varied, almost conflicting, impulses as she produced fiction in every available mode. Her own first ventures were two collections, *Grim Tales* and *Something Wrong,* both published in 1893; these books included stories from various serial publications. Both collections received cautiously positive reviews and are the earliest evidence of Nesbit's work as a writer of the fantastic. Her adult fantasies tended to be most successful when she mined her personal fears of death and of premature burial, but most of the tales lack intensity and stop short of the truly horrific. For example, "John Charrington's Wedding" in *Grim Tales* is a conventional ghost story, and "The Ebony Frame" and "Uncle Abraham's Romance," from the same collection, are stories of amorous female revenants. On the other hand, modern critics consider "Man-Size in Marble," also from this collection, to be her most accomplished piece in the Gothic mode. The story describes the fate of a young wife, left alone by her husband who has gone to investigate a village superstition that two marble sculptures in the church become animated on one night of the year; it is both admirably paced and a powerful evocation of the horror of physical violation.

Nesbit produced no large collections of fantastic tales for the adult audience until *Fear* (1910); after 1893 she published them intermittently–only a few, the minor "Barring the Way" from *In Homespun* (1896) and two in *Man and Maid* (1906), "The Power of Darkness" and "The House of Silence," are particularly noteworthy: "The Power of Darkness" is an ironic tale of "the biter bit" since the attempt to use the victim's fear of darkness recoils on the attacker. "The House of Silence" is reminiscent of Edgar Allan Poe's use of claustrophobic panic.

The Secret of Kyriels (1898), is a Gothic novel in the tradition of Charlotte Brontë's *Jane Eyre* (1847), with which it shares some plot elements. It is located less definitely in the fantastic mode than the later novel, *Dormant* (1911), which engages both fantasy and science fiction. A mother/daughter romance, *The Secret of Kyriels* is the story of Edith, a woman who flees with a former lover in distress at false information of her husband's infidelity. Imprisoned and driven mad by the lover, she is only freed much later by her resourceful daughter, Esther, who also recovers her real father–her mother's husband–who was in fact guiltless; both mother and daughter are thinly disguised portraits of Nesbit. In the fashion of Ann Radcliffe, Nesbit tends to find a rational solution for the enigmas she presents, and, like Brontë, much use is made of incarceration and enigmatic dwellings. *The Secret of Kyriels,* characteristic of most of Nesbit's fantastic tales for adults, is less thoroughgoing fantasy than sensational fiction of the kind produced by Wilkie Collins.

Between 1894 and 1899 Nesbit published more verse, including a few poems for the *Yellow Book*, and continued to produce minor children's books such as *Pussy Tales* and *Doggy Tales* (both 1895), which resembled Beatrix Potter's more-famous animal stories, although Nesbit's characteristic, astringent tone was already present in parent/child exchanges; these were almost the last such books she produced. She also wrote children's versions of William Shakespeare's plays and a series of historical narratives, *Royal Children in English History* (1897), although her own historical novels would not appear for several years.

Illustration by C. Sheppardson for "The Blue Mountain," one of the stories in Nesbit's Nine Unlikely Tales for Children

Nesbit's versatility as a writer was something of a hindrance to her development as an artist, especially in the early years when she was writing to survive. She very much wanted a career as a poet, but critics saw in her work more facility of expression than profound poetic inspiration. One, W. E. Henley, warned her in a letter dated 17 April 1899 to resist the impulse simply to imitate her quick successes. Her strongest gifts–the powerful evocation of place and a wry, ironic sense of play–came together best in her children's fiction, but she was always aware of children's literature as a marginal genre. She made a serious attempt, in her collection *The Literary Sense* (1903), to imitate the style of Henry James, whose fiction she greatly admired. However, her depictions of preciosity seem, if not superficial, at least unconvincing.

The deep fund of memory tapped first by her Bastable stories, beginning with *The Story of the Treasure-Seekers* and continued in *The Wouldbegoods* (1901), could not be ignored nor denied; the success of the stories was instantaneous and lasting. The highest royalties Nesbit ever received were the £1,100 she earned for *The Wouldbegoods* in its first year. It was not until she was commissioned by the editors of *The Strand Magazine* to write a series of stories, at £30 per episode (as opposed to £50 for a single book), that she began the series of fantastic tales upon which much of her fame as a writer would rest.

These seven stories, collected in *The Book of Dragons* (1899), are Nesbit's playful variations on dragon stories, and they contain almost all of the elements, excepting only time travel, that were to become the hallmarks of Nesbit's fantasies for children. "Kind Little Edmund," for example, includes a helpful cockatrice whose behavior toward the child is that of an authoritarian, rather crotchety adult, a characteristic that would be more fully developed in the Psammead of *Five Children and It* (1902). Nesbit also favored, on occasion, a mathematical or logical solution to the narrative dilemma, and an early case in point is "The Island of the Nine Whirlpools," in which the dragon can be defeated only when all of the whirlpools are stilled. The hero discovers the equation that determines the crucial moment and is able to claim that he has won the princess by "love and mathematics."

Among the most interesting stories in this collection are those in which gender differences are submerged by cooperation. Nesbit's literary approach to matters of gender is complex and ambivalent; when she uses a character-narrator the persona is male, the most famous being Oswald Bastable. Further, in her novels for children most of her female characters conform to contemporary stereotypes; they are the repositories of empathy and the voices of moral principle. In her short stories, however, gender roles are less rigid; the girls are as enterprising and forthright as the boys. Sabrinetta, the granddaughter of St. George in "The Fiery Dragon," not only saves the life of her usurping cousin, Tiresome, when he fails to capture a dragon but also works with the pig keeper, Elfin, to save the kingdom. The Princess of Cornwall in "The Last of the Dragons," originally published in a later issue of *The Strand Magazine* in 1899 than the other seven dragon stories but posthumously included with them, is stronger and a better fencer than any of the eligible princes. She conquers the dragon with affection, but only after she has convinced the bookish and ineffectual prince that she is the expedition's leader. Despite the princess's evident superiority, Nesbit's conservatism is strongly evident in her allowing a "feminine" quality in her sword-wielding princess to be the deciding factor.

Nesbit called her next collection of stories *Nine Unlikely Tales for Children* (1901), and the title is appropriate since Nesbit exploits the fairy tale for structure while interpolating her own, distinctively improbable, content. These stories may have been, in part, a reaction against the conventionality of the tales she had contributed to other collections. Critics also mention their similarity to William Makepeace Thackeray's *The Rose and the Ring* (1855). Her own approach tends toward the ironic and the satiric, as two of the unlikely tales, "The Cockatoucan" and "Whereyouwanttogoto," illustrate. In the former, the heroine finds herself in a land whose inner logic is scrambled by the intermittent laughter of the "Cockatoucan." She discovers that the princess, excessively thin–resembling a "paper tape"–from refusing meals, has spitefully provoked the bird to laughter in vengeance against those who have refused her her favorite food. The amusing reversals that need righting coexist uneasily with the alliance between the malicious creature and a self-destructively disaffected child. Nesbit departs almost completely from the fairy tale into the fable in "Whereyouwanttogoto, Or, the Bouncible Ball," the story of two children who spoil a perfect vacation by bickering and, in a fit of pique, cutting open the magical ball that has provided their adventures.

The *Nine Unlikely Tales* have few themes in common; instead, Nesbit provides a series of jeux d'esprit. Some of these express her irritation with urban society, such as "Fortunatus Rex," a satire on the developers who were ruining her cherished rural landscape; "The Blue Mountain" with agressively busy antlike creatures whom the protagonist must put in their places; or "The Plush Usurper," whose crime seems to be the excessive use of bright color and plush fabrics. When Nesbit does rework conventional fairy tales, she heightens the absurdities of selected fairy-tale plots, as in two variations on the "christening curse" motif: the prince in "The Prince, Two Mice, and Some Kitchen Maids" is rescued from his curse by one of Nesbit's resourceful females. "Melisande," called by one critic her "anti-Rapunzel," is also a mathematical puzzle, since the prince must calculate the moment in which Melisande's alternation between constantly growing hair or constantly increasing height can be stopped. Despite the conventional resolution to the story, it is Melisande who provides all the interim solutions to her own problems.

Nesbit's biographer Julia Briggs calls the 1900s "the crowded years." Not only was Nesbit completing the Bastable books with *The Wouldbegoods* and *The New Treasure-Seekers* (1904) and writing several minor stories such as *The Revolt of the Toys, and What Comes of Quarrelling* (1902), *The Rainbow Queen and Other Stories* (1903), and *The Story of the Five Rebellious Dolls* (1904), she also embarked on a new children's series, her magical series, beginning with *Five Children and It.* This book, again a collection of tales first published in serials, was the beginning of Nesbit's most influential contribution to fantasy literature.

The series' roots in her earlier stories are obvious–the large family of children and the plots and escapades that never have quite the desired results–but her narrative strategies are different. Instead of the character-narrator of the Bastable books, she chooses a first-person external narrator, although this allows an element of adult condescension to enter the stories. Nesbit's plots are now motivated by the desire not merely for amusement but for marvels. She introduces her fantastic creatures into the contemporary reality of her characters, whose adventures are inspired by their reading–books about Atlantis or Babylon, beseiged castles, or the novels of James Fenimore Cooper. Finally, she adds the element of time travel; her fantastic voyages are inspired by works of F. Anstey, such as *Tourmalin's Time Cheques* (1891) and *The Brass Bottle* (1900) as much as by Wells.

Humphrey Carpenter argues that Nesbit's Psammead, or sand fairy–the *it* of the title–is much more memorable than the five children, but this is Nesbit's narrative economy. The commonplace children contrast effectively with their uncommon adventures; thus, the author avoids one oddity too many. Her children are more types than individuals: Anthea, the eldest, upon whom the role of "mother" devolves; Robert, the stubborn, forthright elder brother; Cyril, a younger brother, inventive, but less self-confident; and a younger sister, Jane, whose chief characteristic is timidity. The fifth child, an infant called "the Lamb," has only a negligible role after the first novel. Nesbit's magic works with great economy; her creatures grant wishes, and the everyday impulses of her children take on a "logic" of absurdity. A wish to be as "beautiful as the day" leads to frustrating nonrecognition by their nurse, and the wish for wings leaves them stranded on a church tower when the wish lapses, according to rule, at sunset. The reader, with a sense of superiority, waits to see how the irascible Psammead's predictions of disaster will come out each time and how the children will be extricated from their current predicament.

One testimony to children's response to *Five Children and It* is found in a letter from Rudyard Kipling to Nesbit, dated 11 March 1903. Speaking of his children, aged five and seven, he writes: "Their virgin minds never knew one magazine from another till it dawned upon Elsie that 'a thing called the *Strand*' 'with a blue cover and a cab' was where the Psammead tales lived . . . I have been sent for *Strands* in the middle of the month, I have had to ex-

Cover for Nesbit's 1902 book featuring Psammead, a sand fairy who grants wishes that last only a day

plain their non-arrival; and I have had to read them when they came. They were a dear delight to the nursery."

Unlike the Bastable stories, which suffer from a lack of real variety in adventures, the magic series shows Nesbit at her most consistently inventive. Especially effective is the distinctive character of each magical creature. The phoenix of *The Phoenix and the Carpet* (1904) is just as touchy as the Psammead, but its dominant trait is vanity, which makes careful flattery the children's most effective way of getting wishes granted but which also leads to their final adventure, one which makes them resolve to rid themselves of their ally. The phoenix, convinced that a theater to which the children have taken it is one of its own temples, starts a fire that brings the adventures to the brink of disaster. Nesbit's daring in placing children at risk succeeds in a way that simply exhausting the Psammead's patience did not. The children are forced to become objectively critical, and the renunciation of magic becomes as inevitable as their invoking it.

The Psammead was too popular for Nesbit to leave behind when she began the third book of the series, *The Story of the Amulet* (1906), although the children had been forced to promise never to ask it to grant another wish. Its gratitude for their rescuing it from a pet shop leads them to find half of an Egyptian amulet, and with the help of a fellow boarder, an Egyptologist who decodes the amulet's inscription, the series of adventures begins. Nesbit was inspired in part by her fondness for historical fiction but also by information provided by Wallis Budge, a close friend and a curator at the British Museum. The time-travel motif dominates here because the full power of the amulet is dependent on their finding it at a moment when it is unbroken. Their adventures take them from Atlantis to early Britain, and suspense is heightened as they become aware of being followed by Rehk-mara, a former possessor of the amulet. The resolution of the chase through time, although logical–the Egyptologist, a solitary scholar, permits the time-fugitive Rekh-mara to share his body and thus live in a time when his quest for knowledge can be fulfilled–has resonances beyond those expected in a book directed to a child audience: "But the eyes of Rekh-mara and the learned gentleman met, and were kind to each other, and promised each other many things, secret and sacred and very beautiful."

Part of Nesbit's success in these fantasies is her willingness to depict harshness as well as to provide purely escapist charm–for example, the Babylonian queen's massacre on the stock exchange, the destruction of Atlantis, or the self-sacrifice of the Tyrian sailors in *The Story of the Amulet*. This novel is the culmination of a progression in the element of danger from the inconvenience in *Five Children and It* and the near misses of *The Phoenix and the Carpet*. The criticism that Nesbit's children always have a safety net may be valid, but if one views the stories as narratives of suspense and of the close call, the predictable outcomes are less a weakness. It is less easy to condone her persistent condescension toward or sentimentality about lower-class characters, as with the cook whom the children, as a gesture of esteem, transport to a Pacific island where she will reign, as a cook, over the natives, or the London orphan relocated to a surrogate mother in early Britain.

Nesbit was at the height of her powers as a writer of extended fantasies when she wrote the last of the magic series, and the stories that appear in the collection *Oswald Bastable and Others* (1905) suggest that most of her creative energy focused in the novels. This collection features the last Bastable adventures, and the others form an assortment of short tales similar to those in the *Nine Unlikely Tales*. The results are even more mixed here than in the earlier collection, although "Billy the King," whose hero remains in the fantasy world because of the degradations of slum life, and "The Charmed Life: or, the Princess and the Lift-Man," with its portrait of an impoverished king who turns inventor and whose elevator allows his son to win a princess, are equal in quality to the best of the earlier short tales.

Oswald Bastable and Others shows contradictory tendencies in Nesbit's fiction; certain tales seem to look beyond affirming the status quo, while elsewhere in the collection the reader encounters stories that are conventional, if not reactionary. For example, the story "The Ring and the Lamp" presents a twist in the usual "genie" story in that the servants of the two magical objects reject their assigned roles–"No one really *likes* being in service," form a company, and employ the fathers of the two girls who originally summoned them. On the other hand, "Muscadel" shows social stereotyping made even more distasteful by an added element of ridicule. What starts as a foolish birthday wish made by a princess turns into a series of amorous misadventures as the hero, a fly (hence his name, from *musca*, Latin for "fly") transformed into an unprepossessing archer, wins the hand of the shallow princess who complicates matters by first wishing she could be the kind of person who could love Muscadel. Her transformation, into a slatternly milkmaid, is unrelieved caricature. Even the resolution, one based as much on a transformation in Muscadel's appearance as a development of sympathy in the princess, fails both as fairy tale and as satire. The same canceling effect occurs between "The Twopenny Spell," an interesting reversal of gender behavior in which the boy learns consideration and the girl learns self-assertion, and "The Princess and the Cat," in which the princess is characterized by her softheartedness and the prince by his iron resolve (literal in his case, since each failure has burdened him with a piece of scrap metal).

Some of Nesbit's later fantasies also tend to rely more heavily on her established formulas, but her historical novels–*The House of Arden* (1908) and *Harding's Luck* (1909)–and *The Enchanted Castle* (1907) show disciplined evolution in her artistry. *The Enchanted Castle* brings into the foreground an interrogation of fantastic experiences that had been obscured by the episodic abundance of adventures in the earlier series fiction.

The Enchanted Castle uses a magical object, a ring whose properties depend on the claims made by its holder. The magic in this novel also proceeds according to rules, a departure from the randomness of the adventures of the magic series; the duration of the spell is always some multiple of seven. The story is an exploration of how the magic works as well as a demonstration of what happens when it is invoked, and a subplot of the novel is the way the children, three siblings and an orphan, slowly work out the rules.

There is another departure from Nesbit's past practice in the novel. She had always exploited the potential for mishaps concomitant with the granting of wishes; in *The Enchanted Castle,* however, she adresses the matter directly. Near the novel's end the French governess, reunited after a long separation with her lover through the efforts, both fantastic and mundane, of the children, explains to him the magic at the heart of the castle and their adventures: "Except from children . . . the ring exacts a payment. You paid for me, when I came by your wish, by this terror of madness that you have since known. Only one wish is free." The only free wish for an adult is the end of the magic.

Nesbit had shown adult wishes being granted through a child's–or a creature's–intercession, but she had not before intruded such an adult perspective so explicitly. She had also not brought horror, as opposed to terror, into her fantasies, as she does here. Several critics have cited the uniqueness of the "Ugly-Wuglies" episode, about scarecrow creatures constructed by the children as an audience for their theatricals and animated, as usual, by a disastrous wish. The Ugly-Wuglies are monstrous and uncanny, their speech distorted because they have no roofs to their mouths, and capable of savage hostility when frustrated. This is the horror of "Man-Size in Marble" inserted into a children's fantasy–there is also something of that element in the moonlight-animated statuary of the castle ground–yet with the same effect of deepening the fantastic presence rather than disrupting it for sensational effect. What would provide merely a genteel frisson in her adult fantasies is here in its proper setting and function. Despite Nesbit's own ambivalence about her work for children the successful interjection of an element of seriousness into *The Enchanted Castle* attests again to the true locus of her best gifts, accepted or not.

When Nesbit tried her hand at extended historical fiction, she wisely divided the action between two of her novels, but the interconnections between the adventures of Elfrida and Edred her brother in *The House of Arden* with those of Dickie Harding in *Harding's Luck* are so complex that reading the former is enhanced by having read the latter. Nesbit is not fully in control of the complexities of this new narrative technique, which combines time travel

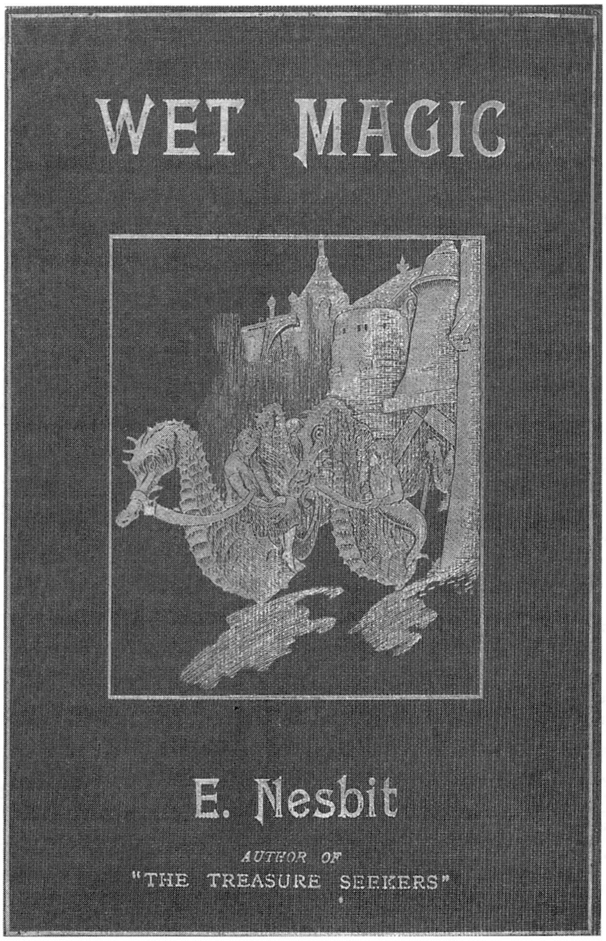

Cover for Nesbit's final novel for children, in which a mermaid is the focus of the adventure (courtesy of the Lilly Library, Indiana University)

with episodic adventures and a double plot involving the children's own adventures and the restoration of the House of Arden. However, she was proudest of *Harding's Luck* among all of her other fiction.

The adventures in *The House of Arden* are not a major departure from what Nesbit's readers had encountered before. Two children, whose father has disappeared while exploring and who seek to restore the fortunes of their family, have time-travel adventures that set them on the road to maturity. Their arrival during the English civil war tests their fortitude and good faith, and they encounter their counterpart in Richard, the young lord of Arden. Nesbit insists throughout the adventures that Edred is Elfrida's inferior in imagination and in moral development, and to the end he is never as sympathetic a character as his sister or Richard, who adamantly refuses to travel with them to their own time, hinting that he knows too much about it.

Richard's peculiar knowledge of the twentieth century is explained in *Harding's Luck*. Dickie, a slum child lame from infancy, discovers a charm and a spell that send him from 1906 to 1606 into a world in which he is not lame and is the heir to the House of Arden. Dickie's desire to stay in the past is complicated by moral obligations to a tramp in the twentieth century, but ultimately his empathy with Edred and Elfrida shows him that their mutual happiness depends on his permanent exile from the present, since his claim on the fortunes of Arden takes precedence over their father's.

The wish fulfillment presented here is complex, a compendium of Nesbit's dominant motifs: the quest for the absent father, the quest for a stable place and the security of privileged class status, and a nostalgia for rural beauty vanishing in the face of development (the Arden lands are in the hands of a soap magnate). The presentation of these matters overwhelms the magic symbolized in the second novel by the heraldic symbol of the Ardens, "The Mouldiwarp" (north-country dialect for "mole"). This creature provides one of Nesbit's humorous touches, since it can only be summoned by poetry, a gift the children lack. The overall effect of the interlinked novels, however, is an impression of potential rather than thoroughgoing accomplishment.

Nesbit's output was large and varied in 1909, although its quality was mixed. The collection *These Little Ones,* published that year and dedicated to Alice Hoatson, finds Nesbit once again writing conventional sentimental fiction, largely about impoverished children and social injustice. The collection includes her only ghost story involving children, "Lucy," presented as the memories of an elderly lawyer of a strange playmate whom he eventually learned had died three days before their games began. The other fantasies are insignificant: "The Little Chap" is a typical Christmas fantasy, and "The Left-Handed Sword" is a frank imitation of the treasure-hiding episode in *The House of Arden.*

The course of Nesbit's fiction from 1909 until 1923 recapitulates the sequence of her earlier career. She skirted the fantastic in *Salome and the Head* (1909), which is a sensational story with Gothic elements, and published the collection *Fear,* reprinting five stories from *Grim Tales.* Of the newly collected stories one, "The Shadow," rivals the best of the earlier short fantasies, in the doubts the reader experiences about the innocence of the withdrawn, reticent housekeeper who narrates the story that triggers a fatal heart attack in the daughter of her long-dead rival. A new type of character, the experimental scientist, appears in two of the short stories, "The Five Senses," and "The Three Drugs." These are Nesbit's variations on the "Jekyll and Hyde" story, although the scientist in "The Five Senses" renounces his efforts and remains happily married. Both of these stories suggest that Nesbit was working with ideas that would reappear in *Dormant* her first novel-length foray into Gothic fantasy for adults since *The Secret of Kyriels.*

Prior to the publication of *Dormant* Nesbit wrote *The Magic City* (1910), a story based on her fascination with model cities made with household objects. The story introduces little that is new, except that Nesbit is more insistent on female self-reliance and has one character claim, when her boy protagonist protests that his role is being usurped, "here . . . the girls are expected to be brave and the boys kind." She does not qualify this assertion as she tended to in the past. Nesbit's ambivalence about women's political involvment never changed, but she intermittently engaged in her own questioning of established roles in her fiction as well as in her actions, although that questioning never emerged as any definite commitment. On this issue *Dormant* is a case in point.

Rose Royal, the protagonist of *Dormant,* is the most explicit of Nesbit's self-portraits–the "advanced," artistic, socially conscious young woman who has less control over her choices in love than she might like. Her fiancé, Anthony Drelincourt, becomes heir to a fortune and an estate and, upon discovering the laboratory of an ancestor, also finds a young woman who is inexplicably catatonic. He succeeds in reviving her, falling in love in the process, but fails in his partly scientific, partly magical attempt to join her in her now-immortal state. Rose, unlike the "sleeping beauty" who has ensnared her lover, will be content to marry her second choice. *Dormant* manages to intermingle elements of fairy tale with hints of Mary Wollstonecraft Shelley's *Frankenstein* (1818); indeed, Anthony differs little from Victor Frankenstein in his self-absorption.

With the exception of *To the Adventurous* (1923), Nesbit's final works of fantasy were destined for children. *The Wonderful Garden, or the Three Cs* (1911) is uninspired and repetitive of her tested plots and features group adventures with magic, this time spells based on a book. Of greater interest than the fantasy is her portrait of the child, Rupert, who slanders his guardian in order to be taken in by the children Charlotte, Charles, and Caroline, the "three Cs."

Although Nesbit had written stories involving alienated or troublesome children earlier, fully one-half of the stories in her last major collection, *The Magic World* (1912), involve children who are well intentioned but misunderstood or in Coventry. Typically, these children seek solace in a fantastic place and return either reformed, if they have truly been at fault, or to a world that has recovered its goodwill toward them. One of the most successful, "The Cathood of Maurice," involves an exchange of circumstances between a cat, Lord Hugh Cecil, and his tormentor, Maurice; the cat's excruciated sensibilities, now Maurice's, are both intensely and sympathetically depicted. Nesbit also includes "christening curse" stories, such as "Belinda and Bellamant," "The Princess and the Hedgehog," and "The Magician's Heart." The latter features a social-climbing magician who specializes in ruining christenings, and "Belinda and Bellamant" unites the disastrous christening with a conundrum only a scientist can solve.

Nesbit's last extended fantasy, *Wet Magic* (1913), inverts the classic mermaid story of Hans Christian Andersen by making the mermaid the object of unrequited love. The novel is an oddly assorted work that never achieves unity of focus; it is full of motifs and situations from past novels. Its frame story, the quest of the circus boy Reuben for his true parents, is only occasionally integrated into the story as a whole. The mermaid, rescued by four children with Nesbit's usual assortment of character traits, is as difficult at first as the Psammead or the Phoenix, but later, once the undersea portion of the adventures begins, she too is simply one of a larger cast of characters than Nesbit usually marshals. The central action of the story is the defeat, with the children's help, of an invasion of the water realm by "under-people," one of whom is the mermaid's rejected lover. Among the invaders are also characters from books the children have read, but their connections with events are less clear than similar instances in earlier works.

Works such as *The Magic City*, *The Wonderful Garden*, and *Wet Magic,* although less inventive than their predecessors, are nevertheless worthy of attention. They depart, as does all of Nesbit's best work, from the conventional ephemerae she produced almost as a trade while finding herself as an artist. Nesbit never forgot how to think, feel, and speak in writing with the children of her audience, and although these late novels would not have made the reputation she gained as a writer of fantasy, they do not detract from it.

Nesbit died on 4 May 1924 of heart disease and possibly lung cancer; she was buried in Jesson St. Mary's, Kent. Her daughter Rosamund published a posthumous collection *The Five of Us–and Madeline* in 1925. In it were one of the dragon stories from 1899, "The Last of the Dragons"; a fairy tale, "Princess Zuleika of Rosyposia"; Nesbit's last "christening curse" story; and a story, "Tammy

Nesbit's grave

Lee's Jack," co-authored by T. T. Tucker, her second husband, whom she had married in 1917.

Contemporary critics such as Eleanor Cameron and Roger Lancelyn Green attest not only to Nesbit's influence on other writers but also to the number and diversity of testimonials her work has received–for example, it has been highly praised both by Noel Coward and by Gore Vidal. Edward Eager's frank imitations, such as *Half Magic* (1954) or *The Time Garden* (1958) are perhaps the most notable borrowings, but C. S. Lewis used a slightly disguised version of Nesbit's Babylonian queen from *the Story of the Amulet* as the witch in *The Magician's Nephew* (1955). Valuable for its own sake, Nesbit's fantastic fiction has much to offer to those who study the craft of fiction either as readers or as makers, as well as to those who seek insight into the literary culture of the late nineteenth century.

Bibliography:

Selwyn Goodacre, comp., "Bibliography," in *A Woman of Passion: the Life of E. Nesbit, 1858–1924,* by Julia Briggs (London: Hutchinson/New York: New Amsterdam Books, 1987), pp. 453–465.

Biographies:

Anthea Bell, *E. Nesbit* (London: Bodley Head/New York: Walck, 1964);

Doris L. Moore, *E. Nesbit; A Biography*, revised edition (New York: Chilton, 1966);

Julia Briggs, *A Woman of Passion: The Life of E. Nesbit, 1858–1924* (London: Hutchinson/New York: New Amsterdam Books, 1987).

References:

Julia Briggs, "Women Writers and Writing for Children: From Sarah Fielding to E. Nesbit," in *Children and Their Books: A Celebration of the Work of Iona and Peter Opie,* edited by Gillian Avery and Briggs (Oxford: Clarendon Press, 1989), pp. 221–250;

Eleanor Cameron, "The Green and Burning Tree: A Study of Time Fantasy," in her *The Green and Burning Tree: On the Writing and Enjoyment of Children's Books* (Boston: Little, Brown, 1962), pp. 71–134;

Humphrey Carpenter, *Secret Gardens: The Golden Age of Children's Literature* (Boston: Houghton Mifflin, 1985);

Marcus Crouch, *Treasure Seekers and Borrowers: Children's Books in Britain 1900–1960* (London: Library Association, 1962);

Roger Lancelyn Green, *Tellers of Tales: Children's Books and Authors from 1800–1968* (London: Kaye & Ward, 1969);

Junior Bookshelf, special Centenary issue on Nesbit, 22 (October 1958);

U. C. Knoepflmacher, "Of Babylands and Babylons: E. Nesbit's Reclamation of the Fairy Tale," *Tulsa Studies in Women's Literature,* 6 (2) (Fall 1987): 299–325;

C. N. Manlove, "The Union of Opposites in Fantasy: E. Nesbit," in his *The Impulse of Fantasy Literature* (Kent, Ohio: Kent State University Press, 1983), pp. 45–69;

Stephen Prickett, "Worlds Within Worlds: Kipling and Nesbit," in his *Victorian Fantasy* (Brighton: Harvester Press, 1979), pp. 190–235;

W. W. Robson, "E. Nesbit and *The Book of Dragons,*" in *Children and Their Books: A Celebration of the Work of Iona and Peter Opie,* edited by Avery and Briggs (Oxford: Clarendon Press, 1989), pp. 251–270;

Elmar Schenkel, "Domesticating the Supernatural: Magic in E. Nesbit's Children's Books," in *The Victorian Fantasists: Essays on Culture, Society, and Belief in the Mythopoeic Ficton of the Victorian Age,* edited by Kath Filmer (New York: St. Martin's Press, 1991), pp. 205–216;

Noel Streatfeild, *Magic and the Magician: E. Nesbit and Her Children's Books* (London: Benn, New York: Abelard-Schuman, 1958).

Papers:

Most of Nesbit's correspondence is in private hands, although a great deal of it has been incorporated into the biographies by Moore and Briggs. There are collections of her professional correspondence in the Berg Collection of the New York Public Library and in the British Museum, as well as in the John Lane Papers in the Harry Ransom Humanities Research Center at the University of Texas at Austin. The H. G. Wells collection in the Wells Papers at the University of Illinois at Urbana-Champaign holds some of Wells's correspondence with Nesbit. One of the few existing collections of manuscript materials can be found in England in the Greenwich Local History Library.

Ann Radcliffe

(9 July 1764 - 7 February 1823)

Deborah D. Rogers
University of Maine

See also the Radcliffe entry in *DLB 39: British Novelists, 1660–1800.*

BOOKS: *The Castles of Athlin and Dunbayne* (London: Hookham, 1789; Philadelphia: Thomas Bradford, 1796);

A Sicilian Romance, anonymous (London: Hookham, 1790; Baltimore & Philadelphia: Rice, 1795);

The Romance of the Forest (London: Hookham, 1791; Boston: Samuel Etheridge, 1795);

The Mysteries of Udolpho (London: Robinson, 1794; Boston: Printed by Samuel Etheridge for White, Spotswood, Thomas & Andrews, West, Larkin, Blake, Folsom, etc., 1795);

A Journey Made in the Summer of 1794 through Holland and the Western Frontier of Germany with a Return down the Rhine: To Which Are Added Observations during a Tour to the Lakes of Lancashire, Westmoreland and Cumberland (London: Robinson, 1795);

The Italian, or The Confessional of the Black Penitents (London: Cadell & Davies, 1797; Philadelphia: Campbell & Rice, 1797);

Gaston de Blondeville, or The Court of Henry III Keeping Festival in Ardenne. A Romance; St. Albans Abbey. A Metrical Tale (London: Colburn, 1826; Philadelphia: Carey & Lea, 1826).

One of the most popular novelists of her era, Ann Ward Radcliffe created a female Gothic that transformed the emotional extravagances of the classic male Gothic novel, pioneered by Horace Walpole's *The Castle of Otranto* (1764). Usually set in haunted castles, graveyards, ruins, or wild landscapes, the traditional male form inspires fear by emphasizing the supernatural and the macabre. Although Radcliffe's novels are replete with fantastic details, she uses such elements to create female fantasies that deal with issues such as female isolation and dependence in a patriarchal society.

Radcliffe's use of the unconscious, her development of fear and suspense, and her vivid poetic descriptions of architecture, landscape, weather, and lighting created a craze for the picturesque, sentimental, and sublime Gothic novel. Her widely translated work, which left its mark permanently on detective and horror literature, inspired countless adaptations and imitations and influenced such authors as Charles Robert Maturin; Jane Austen; Matthew Gregory "Monk" Lewis; William Hazlitt; Sir Walter Scott; George Gordon, Lord Byron; Percy Bysshe and Mary Wollstonecraft Shelley; William Wordsworth; John Keats; Samuel Taylor Coleridge; William Makepeace Thackeray; Charles Dickens; the Brontës; Nathaniel Hawthorne; and Edgar Allan Poe.

The novelist who created such a sensation was born in London on 9 July 1764, the only child of William Ward and Ann Oates Ward. Although William was a tradesman, the owner of a haberdashery, the family was well connected: William Ward was related to the internationally famous William Cheselden, surgeon to King George II; and Ann Oates was descended from the respected De Witt family of Holland. As a girl Radcliffe made frequent, extended visits to her uncle, Thomas Bentley, of Wedgwood and Bentley. When she was seven or eight, her family moved to Bath, where William Ward would manage a branch of the pottery firm. At Bath, Ann met the Lee sisters and was greatly impressed by Sophia Lee's *The Recess* (1783–1785).

Ann was brought up in and remained faithful to the Established Church. Though not schooled in the classics, she received a solid education. On 15 January 1787 Ann Ward married William Radcliffe, an Oxford graduate trained to be a lawyer. Apparently disenchanted with the law, he became editor and proprietor of a newspaper, the *English Chronicle.* The young couple moved to London. Although the Radcliffes had no children, she was quite domestic. She spent her leisure hours reading, singing, and attending operas, oratorios, and plays.

By all accounts Radcliffe, who was formal and reserved, withdrew from society to lead an uneventful, retired life. Radcliffe's seclusion and her insistence on privacy relate to her excessive shyness. She was especially reticent about female authorship. Her delicacy, however, failed to prevent contemporaries from speculating that she was insane.

Six of Radcliffe's works appeared in her lifetime in the eight years from 1789 to 1797. (A seventh was published posthumously.) Starting with *The Castles of Athlin and Dunbayne* (1789), which appeared when she was only twenty-five years old, Radcliffe published her first three novels in as many years.

The Castles of Athlin and Dunbayne is a short, mediocre work set in the Highlands. This clunky, melodramatic attempt to join the modern and the medieval develops neither atmosphere (rugged and picturesque though the landscape may be) nor characters. It received little contemporary notice. Unenthusiastic early reviewers greeted Radcliffe's debut with criticism of historical inaccuracies (in this case the inauthentic rendering of the Highlands) that would continue to plague her. Modern critics seldom pay attention to Radcliffe's unremarkable first experiment.

Only one year after *The Castles of Athlin and Dunbayne* was published, Radcliffe's second novel, *A Sicilian Romance* (1790), appeared anonymously. Here, as in her works to follow, Radcliffe develops the Gothic atmosphere, building suspense by increasing supernatural effects that are, in the end, shown to have ordinary causes. Most Gothic novelists explain little in an attempt to make readers believe that their mysteries are real. Radcliffe, however, in all of her novels except *Gaston de Blondeville* (1826), employs the "explained supernatural," providing the supernatural with banal causality that explains it away.

In *A Sicilian Romance* Radcliffe develops themes that will figure in her later novels, such as mother-daughter reconciliation and parental tyranny, anticipating both father-villain and mother-villainess characters. The first of her novels set in Italy, this attempt, like *The Castles of Athlin and Dunbayne*, received relatively little attention. Contemporaries praised this novel for its decorum, morality, originality, and elegant language, despite its improbabilities. Modern critics seldom study this work except for primarily historical purposes.

If *The Castles of Athlin and Dunbayne* and *A Sicilian Romance* are generally considered experimental, Radcliffe's reputation rests primarily on her next three novels. Published the year after *A Sicilian Romance, The Romance of the Forest* (1791) is Radcliffe's first major work. *The Romance of the Forest* not only displays for the first time the full power of Radcliffe's landscape descriptions, but it also develops her first important central character. This novel established Radcliffe's reputation.

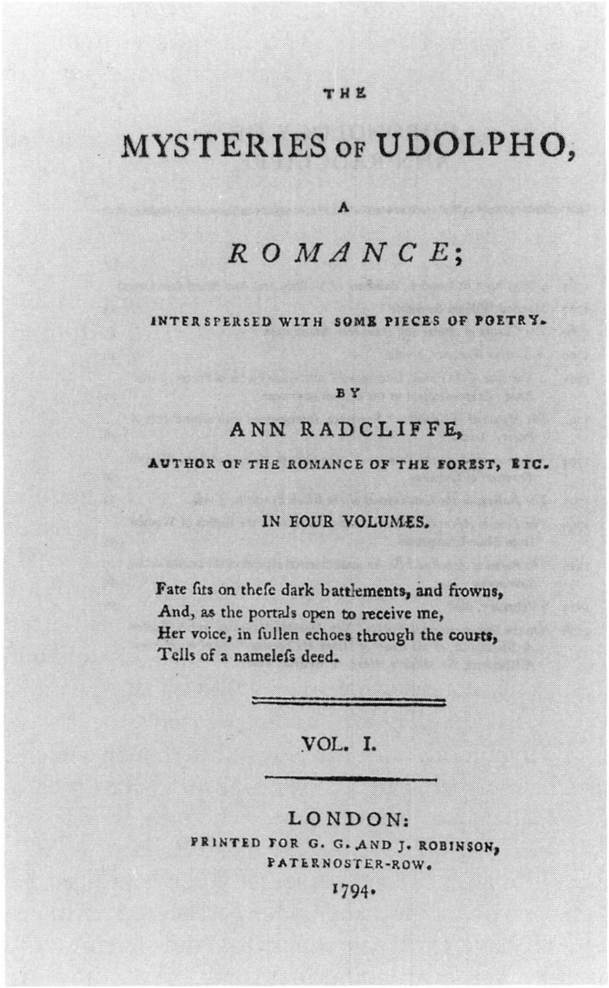
THE

MYSTERIES OF UDOLPHO,

A

ROMANCE;

INTERSPERSED WITH SOME PIECES OF POETRY.

BY

ANN RADCLIFFE,

AUTHOR OF THE ROMANCE OF THE FOREST, ETC.

IN FOUR VOLUMES.

Fate fits on thefe dark battlements, and frowns,
And, as the portals open to receive me,
Her voice, in fullen echoes through the courts,
Tells of a namelefs deed.

VOL. I.

LONDON:
PRINTED FOR G. G. AND J. ROBINSON,
PATERNOSTER-ROW.
1794.

Title page for Ann Radcliffe's fourth book, in which apparently supernatural horrors are shown to have a human source

Approaching poetry, Radcliffe's prose descriptions of scenery were influenced by Edmund Burke's concept of the sublime, by travel literature, and by painters such as Nicolas Poussin, Claude Lorrain, and Salvator Rosa. Her landscapes reflect spirituality, even as they define character.

A female-centered novel, *The Romance of the Forest* concerns Adeline's search for her true identity, as she discovers two false fathers. Real horrors parallel supernatural terrors, which turn out to have natural causes.

With the publication of this Gothic, which met with rave reviews, the anonymous author became one of the most popular novelists of the day. The success of this novel may have encouraged the reticent Radcliffe, who had not previously acknowledged any of her work, to acknowledge the second edition of *The Romance of the Forest* when it appeared in 1792. If Radcliffe's first two novels are mediocre, she comes of age with *The Romance of the Forest.* Even so, today *The Romance of the Forest* is virtually ignored–despite its contemporary status and popularity.

The Romance of the Forest may have established Radcliffe's reputation, but it was *The Mysteries of Udolpho* (1794) that catapulted her to fame. Another female-centered novel, *The Mysteries of Udolpho* centers on the testing of Emily as she moves from the pastoral, sentimental, sunlit atmosphere of her home to the dark, terrifying recesses of Udolpho, the prototype of the Gothic castle. Here Emily experiences supernatural horrors, which are later explained to have natural causes. For Radcliffe human beings are the source of all evil.

Although *The Mysteries of Udolpho* was popular with the public, the immediate response in journals was decidedly mixed. They praised Radcliffe's hallmark and pioneering poetical descriptions of landscapes and her creation of suspense but criticized the interspersed verses, anachronisms, improbabilities, excessive descriptions, explained supernatural events, and inadequate characterization.

In general, with the exception of such violent, overbearing villains as Montoni in *The Mysteries of Udolpho* and her masterpiece, Schedoni, in *The Italian, or The Confessional of the Black Penitents* (1797), Radcliffe's characters–especially her feminized heroes–are undistinquished. Her persecuted heroines, who all have excessive sensibility and sensitivity to nature, are also one-dimensional. Yet they are strong, resilient figures who learn to control imagination with reason. Best of all, they are not rescued by heroes and maintain their independence in romantic relationships.

Recently *The Mysteries of Udolpho* and Radcliffe's other novels have been approached with new seriousness, as Radcliffe's heroines have been analyzed in their role as women and in their relationships with their mothers. Feminist critics have recently observed that although the traditional, male gothic novel reflects patriarchal violence, Radcliffe's female Gothic emphasizes the mother-daughter bond in terms of the daughter's issues of separation and female identity. Radcliffe delineates the daughter's conflict with maternal figures who symbolize the daughter's own fate if she, too, becomes a mother.

After the publication of *The Mysteries of Udolpho,* in the summer of 1794 the Radcliffes traveled to Holland (the home of the De Witts, her maternal ancestors), western Germany, and the Rhine. This trip represented the only time she ever went abroad, although her vivid descriptions of regions such as southern France and Italy, which she never visited, would seem to argue otherwise. Returning from the Continent, the Radcliffes toured the English Lake District. The following year Radcliffe published an account of this excursion, *A Journey Made in the Summer of 1794* (1795). If early reviewers found the book less exciting than Radcliffe's novels, it was generally praised for its reflections and descriptions. Even so, it is discussed little today except in terms of Radcliffe's other works.

After her first and only excursion abroad, Radcliffe traveled at least once a year within England with her husband. In the fall of 1797 the Radcliffes visited the coast of Kent. The same year saw the publication of Radcliffe's next novel.

Radcliffe appears to have taken the criticism of her novels, especially *The Mysteries of Udolpho,* seriously. Radcliffe's darkest and most powerful work, *The Italian, or The Confessional of the Black Penitents,* takes place in the present, so it contains no anachronisms. Gone also are the interspersed verses. The explained supernatural is less prominent, and the landscape descriptions are more succinct. Characterization also improved with the creation of Radcliffe's most fully developed character, the nearly superhuman, manipulative, revengeful Schedoni. For these reasons, *The Italian* has been regarded as an improvement on Radcliffe's previous romances, even if some personally prefer earlier works.

From the beginning *The Italian* invited comparisons with other Gothic works. As might be expected since *The Mysteries of Udolpho* inspired Lewis's *The Monk* (1795), which in turn inspired a reaction in the form of *The Italian,* Lewis, with his unexplained supernatural, has often been contrasted with Radcliffe, with her explained mysteries.

Ellena, Radcliffe's heroine in *The Italian*, receives relatively little discussion, even though she is confined and harassed in a Gothic environment. Like Emily in *The Mysteries of Udolpho,* Ellena is not rescued by her hero.

Critics observe that *The Mysteries of Udolpho* is Radcliffe's most female-centered novel, while *The Italian* is Radcliffe's least female-centered novel insofar as its focus is on the developing consciousness of Schedoni, the villain, rather than the heroine. Nevertheless, much as they did with *The Mysteries of Udolpho,* modern critics are beginning to focus on the female relationships in *The Italian.* By reuniting with her mother Ellena discovers her identity, which ultimately allows her to marry Vivaldi.

Shortly after the publication of *The Italian,* on 24 July 1798, Radcliffe's father died, leaving her interests in the rents of property in Houghton-on-the-Hill, near Leicester. Her mother inherited their house in Bath and the right to raise money on their

Frontispieces for The Mysteries of Udolpho *from the four-volume London edition (1799)*

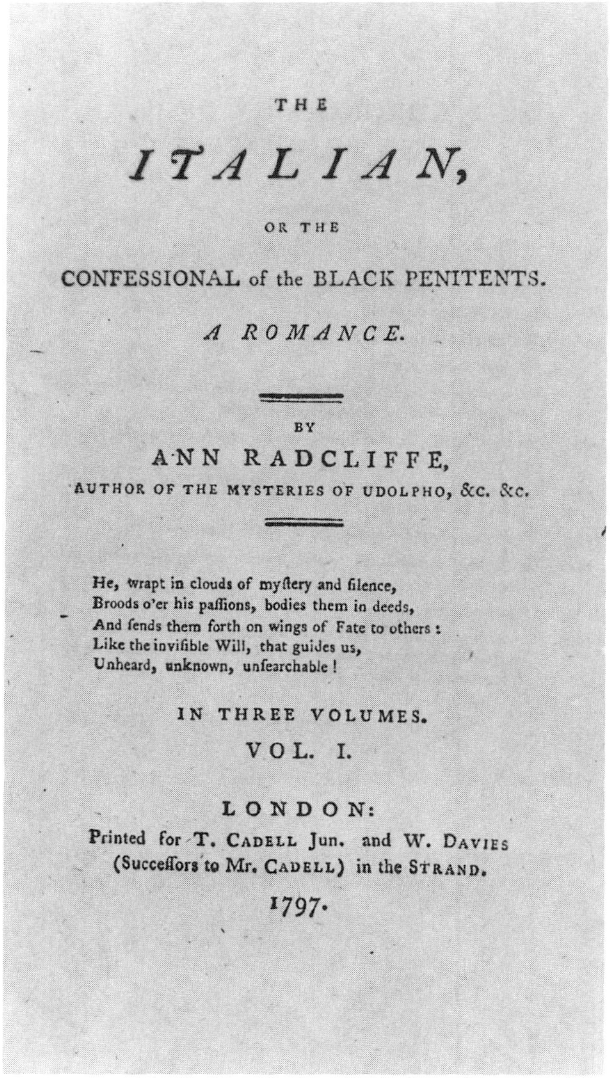

THE

ITALIAN,

OR THE

CONFESSIONAL of the BLACK PENITENTS.

A ROMANCE.

BY

ANN RADCLIFFE,

AUTHOR OF THE MYSTERIES OF UDOLPHO, &c. &c.

He, wrapt in clouds of myſtery and ſilence,
Broods o'er his paſſions, bodies them in deeds,
And ſends them forth on wings of Fate to others:
Like the inviſible Will, that guides us,
Unheard, unknown, unſearchable!

IN THREE VOLUMES.

VOL. I.

LONDON:

Printed for T. CADELL Jun. and W. DAVIES
(Succeſſors to Mr. CADELL) in the STRAND.

1797.

Title page for Radcliffe's sixth book, which focuses on the mind of Schedoni, the villain

house in Leicester. William Ward also provided for his son-in-law in the event that William Radcliffe should outlive Ann. Perhaps to assuage the grief caused by her father's death, two months later, in September 1798, the Radcliffes visited Portsmouth, the Isle of Wight, and Winchester.

These years were full of sickness and death. In the spring of 1799 William Radcliffe was apparently ill. The following year Ann Oates Ward died, leaving most of her property to her daughter, interestingly cutting William off with the provision that nothing be left to him and that none of the bequest be used to pay off his debts.

Radcliffe's inheritance and the considerable profits from her novels left her in comfortable circumstances. At least by this time she did not need to publish to generate income. Indeed, she published no more novels during her lifetime. For several years after the death of her parents, Radcliffe wrote nothing of major significance, spending most of her time secluded at home or traveling around England.

In the summer of 1800, perhaps again to escape feelings of grief, the Radcliffes toured the southern coast of England, taking along Chance, Radcliffe's favorite dog. For an only child to have lost both father and mother in such a relatively short span must have been devastating. According to Thomas Noon Talfourd's memoir, in her private journal of the trip Radcliffe poignantly expressed her sense of loss and abandonment: "In this month, on the 24th of July, my dear father died two years since: on the 14th of last March, my poor mother followed him: I am the last leaf on the tree!"

That fall the Radcliffes spent two weeks at Little Hampton. In autumn of the following year, 1801, they traveled to Southampton and Lymington and returned to the Isle of Wight. The next fall the Radcliffes ventured to Leicester, Warwick, Woodstock, and Oxford to see such attractions as Warwick Castle, Blenheim, and Kenilworth Castle. Kenilworth Castle inspired *Gaston de Blondeville,* her most historically accurate novel, which was written in the winter of 1802 and published posthumously. Although this was Radcliffe's last novel, she continued to write some poetry.

Perhaps in reaction to criticism of her explained mysteries, in *Gaston de Blondeville* Radcliffe for the first time offers no explanations for the supernatural. Reviewers, however, found Radcliffe's specter troubling, weak, and unconvincing. Today Radcliffe's last work is almost totally ignored.

Several years after she wrote *Gaston de Blondeville,* in June 1805 Radcliffe visited Lord Eardley's mansion, Belvedere House, and in the fall of 1807 the Radcliffes visited Knole House. Apparently four years elapsed before she journeyed out again, returning to Portsmouth and the Isle of Wight and visiting Penshurst, the seat of the Sydneys, and Malvern.

During this time Radcliffe stopped writing for publication. Although *The Italian* had been published to great acclaim in 1797, Radcliffe declined to publish *Gaston de Blondeville* during her lifetime. In the twenty-six years between the publication of *The Italian* and Radcliffe's death in 1823, mysteriously, she published no new romances. In fact, she virtually stopped publishing at the age of thirty-four, after achieving her greatest success. Many assumed that Radcliffe had died or gone insane. So reticent and self-effacing was she that she never corrected such rumors. More likely reasons that Radcliffe stopped publishing are her sensitivity to criticism, her declining health, her financial security, and the parodies as well as the many inferior imitations of her work.

During her last twelve years Radcliffe apparently suffered respiratory and intestinal problems, which may at times have left her too debilitated to write. Her husband seems to have devotedly cared for her during this period.

In the beginning of October 1822 Radcliffe was well enough to travel to Ramsgate. Although fog exacerbated her respiratory condition and her dietary concerns persisted, she began to feel stronger. After her return from Ramsgate, however, in early January 1823, she relapsed, finally dying on 7 February 1823 at age fifty-eight. She was interred in a vault at Bayswater in the Chapel of Ease, which belongs to the parish of St. George's, Hanover Square.

So died Ann Radcliffe, the mother of the female gothic. She may have been criticized both in her own day and in the present for her characterizations, anachronisms, and explained supernatural events. Yet her pioneering poetic landscape descriptions and her powerful development of the psychology of terror left a permanent mark on the development of horror fiction, of romantic and Victorian literature, and, perhaps most important, of female psychology and fantasy.

Bibliographies:

Montague Summers, *A Gothic Bibliography* (New York: Russell & Russell, 1964);

Dan McNutt, *The Eighteenth-Century Gothic Novel* (New York & London: Garland, 1975);

Frederick Frank, *Guide to the Gothic* (Metuchen, N.J. & London: Scarecrow Press, 1984);

Robert Donald Spector, *The English Gothic: A Bibliographic Guide to Writers from Horace Walpole to Mary Shelley* (Westport, Conn. & London: Greenwood Press, 1984);

Frank, *The First Gothics: A Critical Guide to the English Gothic Novel* (New York & London: Garland, 1987);

Benjamin Franklin Fisher IV, *The Gothic's Gothic: Study Aids to the Tradition of the Tale of Terror* (New York & London: Garland, 1988);

Frank, *Gothic Fiction: A Master List of Twentieth-Century Criticism and Research* (Westport, Conn.: Meckler, 1988);

Deborah D. Rogers, "Ann Radcliffe in the 1980s: An Annotated Bibliography of Criticism," *Extrapolation,* 32 (1991): 343–349;

Rogers, *Ann Radcliffe: A Bio-Bibliography* (London & Westport, Conn.: Greenwood Press, 1996).

Biographies:

Aline Grant, *Ann Radcliffe: A Biography* (Denver: Swallow Press, 1951);

Pierre Arnaud, *Ann Radcliffe et le fantastique: Essai de psychobiographie* (Paris: Aubier Montaigne, 1976);

Deborah D. Rogers, *Ann Radcliffe: A Bio-Bibliography* (London & Westport, Conn.: Greenwood Press, 1996).

References:

Barbara M. Benedict, "Pictures of Conformity: Sentiment and Structure in Ann Radcliffe's Style," *Philological Quarterly,* 68 (1989): 363–377;

Marilyn Butler, "The Woman at the Window: Ann Radcliffe in the Novels of Mary Wollstonecraft and Jane Austen," *Women and Literature,* 1 (1980): 128–148;

Margaret L. Carter, *Specter or Delusion? The Supernatural in Gothic Fiction* (Ann Arbor, Mich.: UMI Research Press, 1987);

Syndy M. Conger, "Sensibility Restored: Radcliffe's Answer to Lewis's *The Monk,*" in *Gothic Fictions: Prohibition/Transgression,* edited by Kenneth W. Graham (New York: AMS, 1989), pp. 113–149;

Daniel Cottom, *The Civilized Imagination: A Study of Ann Radcliffe, Jane Austen, and Sir Walter Scott* (New York: Cambridge University Press, 1985);

William Patrick Day, *In the Circles of Fear and Desire: A Study of the Nineteenth-Century Gothic* (Chicago & London: University of Chicago Press, 1985);

Eugenia C. DeLamotte, *Perils of the Night: A Feminist Study of the Nineteenth-Century Gothic* (New York: Oxford University Press, 1990);

David Sedgwick Durant Jr., *Ann Radcliffe's Novels: Experiments in Setting* (New York: Arno, 1980);

Kate Ferguson Ellis, *The Contested Castle: Gothic Novels and the Subversion of Domestic Ideology* (Urbana & Chicago: University of Illinois Press, 1989);

Mary Laughlin Fawcett, "*Udolpho*'s Primal Mystery," *Studies in English Literature,* 23 (1983): 481–494;

Rhoda L. Flaxman, "Radcliffe's Dual Modes of Vision," in *Fetter'd or Free? British Women Novelists, 1670–1815,* edited by Mary Anne Schofield and Cecilia Macheski (Athens: Ohio University Press, 1986), pp. 124–133;

Frederick Garber, "Meaning and Mode in Gothic Fiction," *Studies in Eighteenth Century Culture,* 3 (1973): 155–169;

John Garrett, *Gothic Strains and Bourgeois Sentiments in the Novels of Mrs. Ann Radcliffe and Her Imitators* (New York: Arno, 1980);

Susan C. Greenfield, "Veiled Desire: Mother-Daughter Love and Sexual Imagery in Ann Radcliffe's *The Italian*," in *The Critical Response to Ann Radcliffe,* edited by Deborah D. Rogers (London & Westport, Conn.: Greenwood Press, 1994), pp. 57–70;

Jean H. Hagstrum, "Pictures to the Heart: The Psychological Picturesque in Ann Radcliffe's *The Mysteries of Udolpho*," in *Greene Centennial Studies: Essays Presented to Donald Greene in the Centennial Year of the University of Southern California,* edited by Paul J. Korshin and Robert R. Allen (Charlottesville: University Press of Virginia, 1984), pp. 434–441;

Lynne Epstein Heller, *Ann Radcliffe's Gothic Landscape of Fiction and the Various Influences upon It* (New York: Arno, 1980);

Tamar Heller, *Dead Secrets: Wilkie Collins and the Female Gothic* (New Haven: Yale University Press, 1992);

Norman N. Holland and Leona F. Sherman, "Gothic Possibilities," *New Literary History,* 8 (1977): 279–294;

Coral Ann Howells, *Love, Mystery and Misery: Feeling in Gothic Fiction* (London: Athlone, 1978);

Howells, "The Pleasure of the Woman's Text: Ann Radcliffe's Subtle Transgressions in *The Mysteries of Udolpho* and *The Italian*," in *Gothic Fictions: Prohibition/Transgression,* edited by Graham, pp. 151–162;

Claire Kahane, "Gothic Mirrors and Feminine Identity," *Centennial Review,* 24 (1980): 43–64;

Frederick R. Karl, *The Adversary Literature* (New York: Farrar, Straus & Giroux, 1974);

Robert Kiely, *The Romantic Novel in England* (Cambridge, Mass.: Harvard University Press, 1972);

Elizabeth MacAndrew, *The Gothic Tradition in Fiction* (New York: Columbia University Press, 1979);

D. L. Macdonald, "Bathos and Repetition: The Uncanny in Radcliffe," *Journal of Narrative Technique,* 19 (1989): 197–204;

Mark S. Madoff, "Inside, Outside, and the Gothic Locked-Room Mystery," in *Gothic Fictions: Prohibition/Transgression,* edited by Graham, pp. 49–62;

Elizabeth Wheeler Manwaring, *Italian Landscape in Eighteenth Century England: A Study Chiefly of the Influence of Claude Lorrain and Salvator Rosa on English Taste 1700–1800* (New York: Oxford University Press, 1925);

Roderick Marshall, *Italy in English Literature 1755–1815* (New York: Columbia University Press, 1934);

Clara Frances McIntyre, *Ann Radcliffe in Relation to Her Time* (New Haven: Yale University Press, 1920);

Alan D. McKillop, "Mrs. Radcliffe on the Supernatural in Poetry," *Journal of English and Germanic Philology,* 31 (1932): 352–359;

Anne McWhir, "The Gothic Transgression of Disbelief: Walpole, Radcliffe and Lewis," in *Gothic Fictions: Prohibition/Transgression,* edited by Graham, pp. 29–47;

Robert Miles, *Gothic Writing 1750–1820* (London & New York: Routledge, 1993);

Raymond Mise, *The Gothic Heroine and the Nature of the Gothic Novel* (New York: Arno, 1980);

Ellen Moers, *Literary Women* (New York: Oxford University Press, 1985);

David Morse, *Romanticism: A Structural Analysis* (Totowa, N.J.: Barnes & Noble, 1982);

E. B. Murray, *Ann Radcliffe* (New York: Twayne, 1972);

Elizabeth R. Napier, *The Failure of Gothic: Problems of Disjunction in an Eighteenth-Century Literary Form* (Oxford: Clarendon Press, 1987);

Nina da Vinci Nichols, "Place and Eros in Radcliffe, Lewis and Brontë," in *The Female Gothic,* edited by Juliann E. Fleenor (Montreal & London: Eden Press, 1983), pp. 187–206;

Mary Poovey, "Ideology and *The Mysteries of Udolpho*," *Criticism,* 21 (1979): 307–330;

David Punter, *The Literature of Terror: A History of Gothic Fictions from 1765 to the Present Day* (New York: Longman, 1980);

Robert Princeton Reno, *The Gothic Visions of Ann Radcliffe and Matthew G. Lewis* (New York: Arno, 1980);

Bette B. Roberts, "The Horrid Novels: *The Mysteries of Udolpho* and *Northanger Abbey*," in *Gothic Fictions: Prohibition/Transgression,* edited by Graham, pp. 89–111;

Deborah D. Rogers, ed., *The Critical Response to Ann Radcliffe* (London & Westport, Conn.: Greenwood Press, 1994);

Ann Ronald, "Terror-Gothic: Nightmare and Dream in Ann Radcliffe and Charlotte Brontë," in *The Female Gothic,* edited by Fleenor, pp. 176–186;

Victor Sage, *Horror Fiction in the Protestant Tradition* (New York: St. Martin's Press, 1988);

Dorothy Scarborough, *The Supernatural in Modern English Fiction* (New York: Putnam, 1917);

Eve Kosofsky Sedgwick, *The Coherence of Gothic Conventions* (New York: Methuen, 1986);

Leona F. Sherman, *Ann Radcliffe and the Gothic Romance: A Psychoanalytic Approach* (New York: Arno, 1980);

Nelson Charles Smith, *The Art of Gothic: Ann Radcliffe's Major Novels* (New York: Arno, 1980);

Patricia Meyer Spacks, *Desire and Truth: Functions of Plot in Eighteenth-Century English Novels* (Chicago: University of Chicago Press, 1990);

Jane Spencer, *The Rise of the Woman Novelist* (New York & Oxford: Blackwell, 1986);

Dale Spender, *Mothers of the Novel* (London & New York: Routledge & Kegan Paul, 1986);

John Andrew Stoler, *Ann Radcliffe: The Novel of Suspense and Terror* (New York: Arno, 1980);

Montague Summers, *The Gothic Quest: A History of the Gothic Novel* (New York: Russell & Russell, 1938);

Ford Harris Swigart Jr., *A Study of the Imagery in the Gothic Romances of Ann Radcliffe* (New York: Arno, 1980);

Janet Todd, "Posture and Imposture: The Gothic Manservant in Ann Radcliffe's *The Italian*," in *Men by Women,* edited by Todd (New York & London: Holmes & Meier, 1981), pp. 25-38;

J. M. S. Tompkins, *Ann Radcliffe and Her Influence on Later Writers* (New York: Arno, 1980);

Tompkins, *The Popular Novel in England 1770–1800* (Lincoln: University of Nebraska Press, 1961);

Devendra P. Varma, *The Gothic Flame* (London: Barker, 1957);

Alida Alberdina Sibbellina Wieten, *Mrs. Radcliffe–Her Relation towards Romanticism* (Amsterdam: Paris, 1926);

Judith Wilt, *Ghosts of the Gothic* (Princeton: Princeton University Press, 1980);

Cynthia Griffin Wolff, "The Radcliffean Gothic Model: A Form for Feminine Sexuality," *Modern Language Studies*, 9 (1979): 98–113;

Susan Wolstenholme, *Gothic (R)evisions: Writing Women as Readers* (Albany: State University of New York Press, 1993).

Papers:

Radcliffe's journals, letters, manuscripts, and other documents are for the most part nonextant. The one substantial extant Radcliffe manuscript, a forty-two-leaf commonplace book, is in the Chamberlain Collection of the Boston Public Library (MS Ch.K.1.10). Other notable exceptions include a copy of the contract for *The Mysteries of Udolpho* in the Sadleir-Black Collection at the University of Virginia Library, and a note to a Miss Williamson in the Robert H. Taylor Collection of the Princeton University Library. The only known surviving portrait of Radcliffe is in the Bettman Archives.

Mary Wollstonecraft Shelley

(30 August 1797 – 1 February 1851)

Darren Harris-Fain
Shawnee State University

See also the Shelley entries in *DLB 110: British Romantic Prose Writers, 1789–1832, Second Series; DLB 116: British Romantic Novelists, 1789–1832;* and *DLB 159: British Short-Fiction Writers, 1800–1880.*

BOOKS: *Mounseer Nongtongpaw; or, The Discoveries of John Bull in a Trip to Paris* (London: Proprietors of the Juvenile Library, 1808);

History of a Six Weeks' Tour through a Part of France, Switzerland, Germany, and Holland: With Letters Descriptive of a Sail Round the Lake of Geneva, and the Glaciers of Chamouni, by Shelley and Percy Bysshe Shelley, anonymous (London: Hookham & Ollier, 1817);

Frankenstein; or, The Modern Prometheus (3 volumes, London: Lackington, Hughes, Harding, Mayor & Jones, 1818; revised edition, 1 volume, London: Colburn & Bentley / Edinburgh: Bell & Bradfute / Dublin: Cumming, 1831);

Valperga; or, The Life and Adventures of Castruccio, Prince of Lucca (3 volumes, London: Whittaker, 1823; 1 volume, New York: Woodstock, 1995);

The Last Man (3 volumes, London: Colburn, 1826; 1 volume, Philadelphia: Carey, Lea & Blanchard, 1840);

The Fortunes of Perkin Warbeck, 3 volumes (London: Colburn & Bentley, 1830);

Lodore (3 volumes, London: Bentley, 1835; 1 volume, New York: Wallis & Newell, 1835);

Lives of the Most Important Literary and Scientific Men of Italy, Spain, and Portugal, three volumes, by Shelley, James Montgomery, and Sir David Brewster, volumes 86–88 of *The Cabinet Cyclopædia,* edited by Dionysus Lardner (London: Longman, Rees, Orme, Brown, Green, Longman & Taylor, 1835–1837); abridged as *Lives of Eminent Literary and Scientific Men of Italy,* 2 volumes (Philadelphia: Lea & Blanchard, 1841);

Mary Wollstonecraft Shelley, as painted by Richard Rothwell in 1841 (National Portrait Gallery, London)

Falkner: A Novel (3 volumes, London: Saunders & Otley, 1837; 1 volume, New York: Harper, 1837);

Lives of the Most Eminent Literary and Scientific Men of France, 2 volumes, volumes 102 and 103 of *The Cabinet Cyclopædia,* edited by Lardner (London: Longman, Orme, Brown, Green, Longmans & Taylor, 1838–1839); republished as *Lives of the Most Eminent French Writers,* 2 volumes (Philadelphia: Lea & Blanchard, 1840);

Rambles in Germany and Italy, in 1840, 1842, and 1843, 2 volumes (London: Moxon, 1844; Folcroft, Pa.: Folcroft Library Editions, 1975, 1977);

Proserpine and Midas: Two Unpublished Mythological Dramas, edited by A. Koszul (London: Milford, 1922);

Mary Shelley's Journal, edited by Frederick L. Jones (Norman: University of Oklahoma Press, 1947);

Mathilda, edited by Elizabeth Nitchie (Chapel Hill: University of North Carolina Press, 1959);

The Journals of Mary Shelley, 1814–44, 2 volumes, edited by Paula R. Feldman and Diana Scott-Kilvert (Oxford: Clarendon Press, 1987; Baltimore: Johns Hopkins University Press, 1995).

Editions and Collections: *The Last Man,* edited by Hugh J. Luke Jr. (Lincoln: University of Nebraska Press, 1965);

Collected Tales and Stories, edited by Charles E. Robinson (Baltimore & London: Johns Hopkins University Press, 1976);

The Annotated Frankenstein, edited by Leonard Wolf (New York: Potter, 1977);

Frankenstein; or, The Modern Prometheus: The 1818 Text, edited by James Rieger (Chicago: University of Chicago Press, 1982);

The Last Man, edited by Brian W. Aldiss (London: Hogarth, 1986);

The Mary Shelley Reader, edited by Betty T. Bennett and Robinson (New York & Oxford: Oxford University Press, 1990);

Frankenstein: The 1818 Text, Contexts, Nineteenth-Century Responses, Modern Criticism, edited by J. Paul Hunter (New York & London: Norton, 1996);

The Frankenstein Notebooks, edited by Robinson (New York & London: Garland, 1996).

OTHER: *Posthumous Poems of Percy Bysshe Shelley,* edited by Shelley (London: Hunt, 1824);

"Lacy de Vere," anonymous, in *Forget-Me-Not* (London: Ackermann, 1826), pp. 275–294;

"The Ritter von Reichenstein," as M.S., in *The Bijou* (London: Pickering, 1827), pp. 114–138;

"The Sisters of Albano," as "the Author of *Frankenstein,*" in *The Keepsake for MDCCCXXIX* (London: Hurst, Chance, 1828), pp. 80–100;

"Ferdinando Eboli: A Tale," as "the Author of *Frankenstein,*" in *The Keepsake for MDCCCXXIX* (London: Hurst, Chance, 1828), pp. 195–218;

"The Mourner," as "the Author of *Frankenstein,*" in *The Keepsake for MDCCCXXX* (London: Hurst, Chance, 1829), pp. 71–97;

"The Evil Eye," as "the Author of *Frankenstein,*" in *The Keepsake for MDCCCXXX,* pp. 150–175;

"The False Rhyme," as "the Author of *Frankenstein,*" in *The Keepsake for MDCCCXXX,* pp. 265–268;

"Transformation," as "the Author of *Frankenstein,*" in *The Keepsake for MDCCCXXXI* (London: Hurst, Chance, 1830), pp. 18–39;

"The Swiss Peasant," as "the Author of *Frankenstein,*" in *The Keepsake for MDCCCXXXI,* pp. 121–146;

"The Dream," as "the Author of *Frankenstein,*" in *The Keepsake for MDCCCXXXII* (London: Longman, Rees, Orme, Brown & Green, 1831), pp. 22–38;

Edward John Trelawny, *Adventures of a Younger Son,* edited by Shelley (London: Colburn & Bentley, 1831);

"The Brother and Sister, An Italian Story," as "the Author of *Frankenstein,*" in *The Keepsake for MDCCCXXXIII* (London: Longman, Rees, Orme, Brown, Green & Longman, 1832), pp. 105–141;

"The Invisible Girl," as "the Author of *Frankenstein,*" in *The Keepsake for MDCCCXXXIII,* pp. 210–227;

"The Smuggler and His Family," in *Original Compositions in Prose and Verse* (London: Lloyd, 1833), pp. 27–53;

"The Mortal Immortal: A Tale," as "the Author of *Frankenstein,*" in *The Keepsake for MDCCCXXXIV* (London: Longman, Rees, Orme, Brown, Green & Longman, 1833), pp. 71–87;

"The Trial of Love," as "the Author of *Frankenstein,*" in *The Keepsake for MDCCCXXXV* (London: Longman, Rees, Orme, Brown, Green & Longman, 1834), pp. 70–86;

"The Elder Son," in *Heath's Book of Beauty, 1835* (London: Longman, Rees, Orme, Green & Longman, 1834), pp. 83–123;

William Godwin Jr., *Transfusion; or, The Orphans of Unwalden,* edited by Shelley (London: Macrone, 1835);

"The Parvenue," in *The Keepsake for MDCCCXXXVII* (London: Longman, Rees, Orme, Green & Longman, 1836), pp. 209–211;

"Euphrasia: A Tale of Greece," in *The Keepsake for MDCCCXXXIX* (London: Longman, Rees, Orme, Green & Longman, 1838), pp. 135–152;

The Poetical Works of Percy Bysshe Shelley, 4 volumes, edited by Shelley (London: Moxon, 1839);

The Poetical Works of Percy Bysshe Shelley, 1 volume, edited by Shelley (London: Moxon, 1840);

Percy Bysshe Shelley, *Essays, Letters from Abroad, Translations and Fragments,* 2 volumes, edited by Mary Shelley (London: Moxon, 1840);

Thomas Hutchinson, ed., *The Complete Poetical Works of Percy Bysshe Shelley,* with prefaces and notes by Shelley (London: Oxford University Press, 1965).

SELECTED PERIODICAL PUBLICATIONS–UNCOLLECTED: "The Clouds: A Dream," anonymous, *London Magazine,* 8 (November 1823): 482–484;

"On Ghosts," *London Magazine,* 9 (March 1824): 253–256.

When they hear the name "Shelley," most students and scholars of literature usually think of the great Romantic poet Percy Bysshe Shelley. Students and scholars of those particular types of literature labeled in modern times as fantasy and science fiction, however, when they hear the name "Shelley," think not of the great Romantic poet but of his great wife, Mary Wollstonecraft Shelley. For it was she who gave readers the unforgettable characters of Victor Frankenstein and his monstrous creation in one of the greatest of the Gothic novels and in what many science-fiction scholars, following the lead of Brian W. Aldiss in *Billion Year Spree: The True History of Science Fiction* (1973), claim to be the first authentic science-fiction novel in history. Few literary works have had such a profound impact on the development of science fiction and horror and on Western thought in general.

That Shelley was able to emerge as an author in her own right is a testimony to her talent. Besides her marriage to one of the best-known poets in the English language, she was also the daughter of well-known writers. She was born on 30 August 1797 to William Godwin, a political philosopher and author of the novel *Things as They Are; or, The Adventures of Caleb Williams* (1794), which employed the Gothic mode for ideological purposes. Her mother was Mary Wollstonecraft, whose *A Vindication of the Rights of Women* (1792) is an early classic of feminism. From the start Mary's origins were unconventional: her parents married when Wollstonecraft was five months pregnant with Mary, and her mother died eleven days after Mary's birth.

Although her father was not particularly affectionate or attentive, she did not grow up alone. Her mother had had another child, Fanny Imlay, from an earlier affair, and after Godwin remarried in 1801 she gained more siblings. Her stepmother, Mary Jane Clairmont, favored her own two children and the son she and Godwin shared over Wollstonecraft's daughters, and Mary's childhood was not happy. At one point she was sent to live with family friends in Scotland for two years, probably because of conflict between her and her stepmother. Although she received no formal education, she found consolation in intellectual pursuits, especially books–including her mother's.

Her father hosted many of the prominent intellectuals and writers of the day, including poets such as Samuel Taylor Coleridge and Percy Bysshe Shelley. She first met Percy when she was fifteen, but their relationship did not fully develop until two years later. Despite the fact that Percy was six years her senior and was married with one child and another on the way, he and Mary left for the Continent in July 1814; soon thereafter Mary became pregnant. Although Percy's father, Sir Timothy Shelley, stopped his son's allowance, they traveled in Switzerland, Germany, and Italy, with a few brief visits to England. In November 1816, a month after the suicide of Fanny, Mary's half-sister, Percy's wife Harriet also killed herself. The couple married on 30 December 1816, but Percy was unable to obtain custody of his two children with his first wife. The Shelleys tried to have children of their own, but their first three died as infants, which plunged her into depression. In 1819 their only surviving child, Percy Florence, was born.

Their marriage was troubled and short-lived. In addition to the deaths of their first three children, the marriage was hurt by Percy's infidelity. Finally, the marriage came to a sudden end when Percy drowned while sailing in a storm in 1822. Mary remained in Italy for a year, then returned to England with her son to make her living as a writer. She was helped somewhat by her father-in-law, but his insistence on keeping the family name out of print resulted in the anonymous publication of her work. Her financial stresses were gradually relieved as her son became older and received an ever-increasing allowance from her father-in-law, which allowed them to travel to Germany and Italy. She died, at age fifty-three, on 1 February 1851.

Although it seems incredible that Shelley could write such a well-crafted novel as *Frankenstein* in her late teens and early twenties, she had been writing since childhood and had already published a book of juvenilia through her father's Juvenile Library and written a travel book with her husband before her first novel appeared. The story behind *Frankenstein; or, The Modern Prometheus* (1818) is well known, and Shelley gives her version in her preface to the 1831 edition. Gathered together in 1816 at Lake Leman in Switzerland, the Shelleys, Lord Byron, and John Polidori challenged each other to compose a work of Gothic fiction in a sort of contest following a visit from Matthew G. Lewis, author of the quintessential Gothic novel *Ambrosio; or, The Monk* (1796). While little came of the poets' efforts, Polidori completed what would prove to be an influential work of horror fiction with *The Vampyre* (1819), while Mary Shelley wrote an even more important book.

Her influences were many and have been well documented by literary scholars. They include the myth of Prometheus, of course, as well as the Bible, John Milton's *Paradise Lost* (1667), the works of Jo-

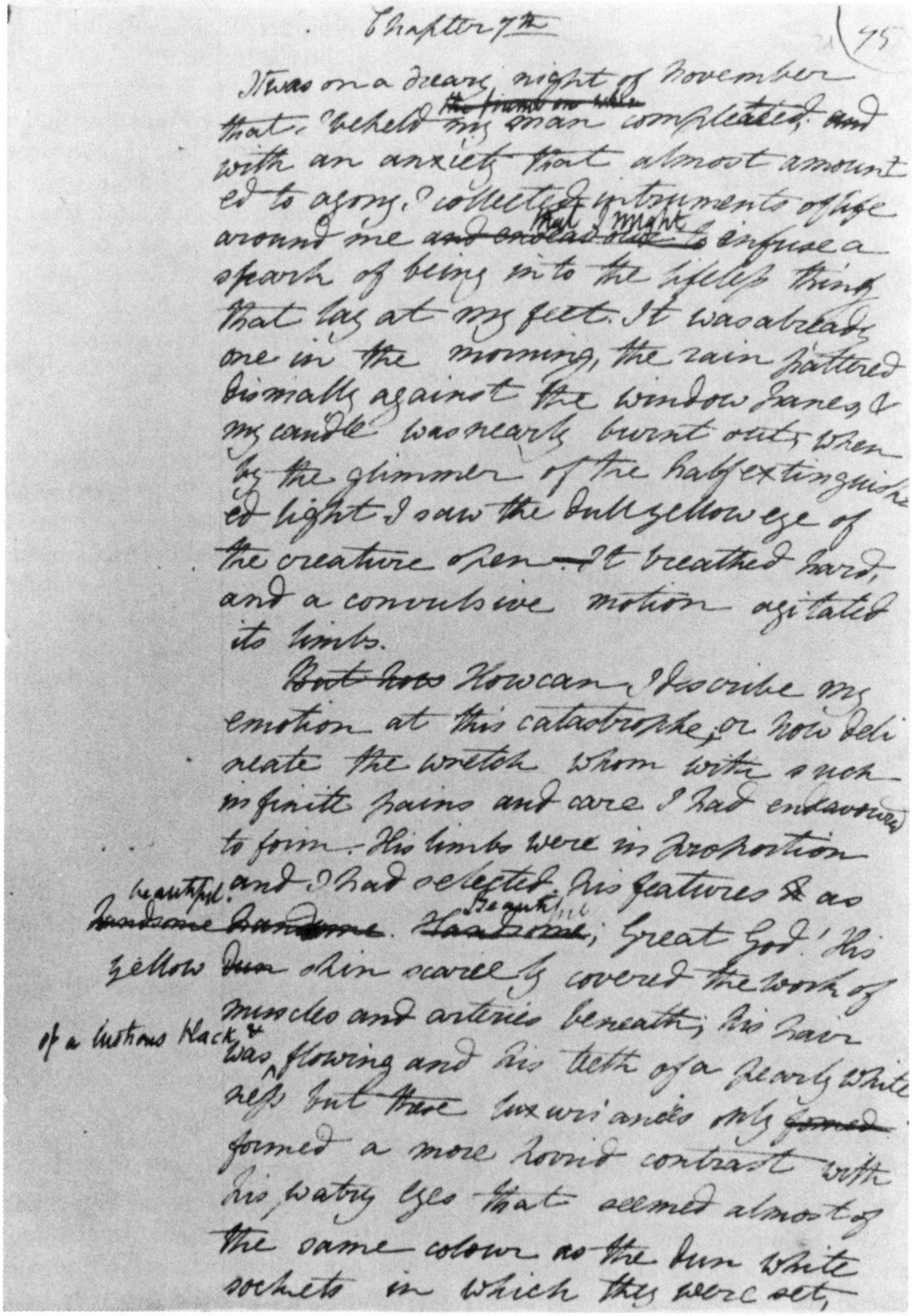

Chapter 7th (95)

It was on a dreary night of November that I beheld ~~the frame on whic~~ my man compleated; ~~and~~ with an anxiety ~~that~~ almost amounted to agony I collected instruments of life around me ~~and endeavoured to~~ that I might infuse a spark of being into the lifeless thing that lay at my feet. It was already one in the morning, the rain pattered dismally against the window panes & my candle was nearly burnt out, when by the glimmer of the half extinguished light I saw the dull yellow eye of the creature open – It breathed hard, and a convulsive motion agitated its limbs.

~~But how~~ How can I describe my emotion at this catastrophe, or how delineate the wretch whom with such infinite pains and care I had endeavoured to form. His limbs were in proportion and I had selected his features ~~h~~ as ~~handsome handsome~~ beautiful. ~~Handsome~~ Beautiful; Great God! His yellow ~~dun~~ skin scarcely covered the work of muscles and arteries beneath; his hair was of a lustrous black & flowing and his teeth of a pearly whiteness but ~~these~~ luxuriances only ~~formed~~ formed a more horrid contrast with his watry eyes that seemed almost of the same colour as the dun white sockets in which they were set,

A page from the manuscript of Frankenstein; or, The Modern Prometheus *(courtesy of the Bodleian Library)*

hann Wolfgang von Goethe and Erasmus Darwin, and then-modern ideas about electricity, chemistry, and anatomy, among others. The Gothic tradition of the late eighteenth and early nineteenth centuries, to which *Frankenstein* is a late contribution, is another. Yet another major influence was the Jewish legend of the golem, a stone figure animated by supernatural means. Shelley's depiction of Frankenstein's efforts to bring his creature to life, however, differ from the golem legend in that, though he is a former student of alchemy, Frankenstein instead relies on natural or scientific means to animate his composite man. Of course readers must nonetheless suspend their disbelief of this account, but Shelley's effort to describe Frankenstein's endeavors through scientific, rather than supernatural, agencies is the

Frontispiece for the 1831 revised edition of Frankenstein; or, The Modern Prometheus *(British Library, London)*

primary reason Aldiss and others have labeled *Frankenstein* a seminal work of science fiction.

However, in many ways *Frankenstein* is unlike much science fiction published since. In style and structure it is much closer to its eighteenth-century forebears. An epistolary novel told in increasingly tightening circles or frames and interspersed with poetry (including that of Shelley's husband), *Frankenstein* often surprises modern readers whose knowledge of the story is limited to generally unfaithful movie adaptations and vague appropriations of Frankenstein's monster in popular culture. It also differs from much science fiction in its use of Gothic conventions. While Shelley departed from many of the characteristics of the mode, with its haunted castles and threatened maidens, she nonetheless successfully conveys a Gothic atmosphere in *Frankenstein,* which stands in sharp contrast both to Enlightenment rationality and the scientific objectivity of modern science fiction in its sense of the strange and the irrational.

The novel, as is widely known, concerns a driven doctor, Victor Frankenstein, and his desire to bypass God and create human life in the laboratory. This he accomplishes by piecing together parts from corpses. Shelley does not shirk from the horrific elements of her story, but neither are these the sole focus of the work, as they tend to be in adaptations of *Frankenstein* into other media. Instead, she is more concerned with the creation's desire for a sense of human identity and belonging–Shelley was as much a Romantic as her husband–and with its creator's abdication of his responsibility and its tragic results. While to simplify the novel as a philosophical or theological allegory would be to miss its richness as a work of fiction, *Frankenstein* is not just a story about a mad scientist and a rampaging monster.

Some controversy exists among Shelley scholars about which version of *Frankenstein,* the 1818 edition or the 1831 revision, is more authoritative. Both have been used as the copy text for critical editions of the novel, and scholars continue to debate the merits of arguments for preferring one or the other.

The first reviewers of the novel generally missed the novel's depth, branding it as simply another Gothic novel and thus ascribing the generally low critical status of Gothic fiction to the book. Many early Victorian critics had the same reaction, but later Victorian readers began to look beyond the horror of *Frankenstein* to see other aspects of Shelley's creation. Since then the novel's reputation has grown generation by generation, with each new critical school finding ample material in the book and in the life of its author to explore.

Although *Frankenstein* is thus much more complex than its early critics and those unfamiliar with the novel have tended to believe, Shelley did contribute to a stereotype of science fiction in almost single-handedly creating the trope of the scientist who, in probing the secrets of the universe, goes too far and who, like the hubristic figures of ancient Greek tragedy, is punished for his arrogance by the forces he has unleashed upon the world. In her other fantastic novel, *The Last Man* (1826), she also was one of the first writers to deal with what would become another common science-fiction motif, the destruction of the world.

The Last Man, as its title indicates, is about a man who believes himself the sole survivor of the utter destruction of civilization. While this novel was doubtless influenced by the works of other authors–most especially Jean-Baptiste Cousin de Grainville's 1805 novel *Le Dernier homme,* translated in 1806 as *The Last Man*–again Shelley incorporates influences into a work uniquely her own. While the Romantic poets often wrote of isolated souls in a fallen world, Shelley in this novel made the subject literal. Set in the twenty-first century, the novel depicts a plague that devastates Europe and the efforts of "the last man," Lionel Verney, to reach Rome in a search for other survivors. The novel also includes another speculative element early in the story: the last king of England, who abdicates before the plague hits the country.

The Last Man is Shelley's best-known work after *Frankenstein*. Contemporary readers found her other novels–*Valperga; or, The Life and Adventures of Castruccio, Prince of Lucca* (1823); *The Fortunes of Perkin Warbeck* (1830); *Lodore* (1835); and *Falkner* (1837)–more conventional and less well written, and they fared poorly then and have fared poorly ever since.

Shelley also wrote many shorter works for periodicals. Much of this she dismissed as hackwork, but many are worth reading, especially those in which she employs her talent for the fantastic. For instance "Transformation" (1830) is a story about the exchange of personalities, a common idea in nineteenth-century fantasy but one to which Shelley gives her characteristic bent. Another common theme, the search for immortality, is addressed in "The Mortal Immortal" (1833), in which a man who thinks an elixir of youth to be a blessing finds it a curse instead. Many of Shelley's shorter works are collected in *Collected Tales and Stories* (1976), edited by Charles E. Robinson, and in *The Mary Shelley Reader* (1990), edited by Robinson and Betty T. Bennett. Shelley's "hideous progeny," as she once labeled *Frankenstein,* truly has taken on a life of its own. It has inspired plays, an opera, several movie adaptations of varying degrees of faithfulness to the book, sequels, and scores of imitators, direct or indirect. In science fiction *Frankenstein* is an archetypal work, its subject and themes frequent touchstones for writers concerned about the use and abuse of scientific knowledge. *The Last Man* is also significant in the history of science fiction for its depiction of massive catastrophe. It is likely that both works will continue to be read and studied in the future.

Letters:

Mrs. Julian Marshall, *The Life and Letters of Mary Wollstonecraft Shelley*, 2 volumes (London: Bentley, 1889);

The Letters of Mary W. Shelley, 2 volumes, edited by Frederick L. Jones (Norman: University of Oklahoma Press, 1946);

New Shelley Letters, edited by W. S. Scott (London: Bodley Head, 1948);

The Letters of Mary Wollstonecraft Shelley, 3 volumes, edited by Betty T. Bennett (Baltimore: Johns Hopkins University Press, 1980–1983); abridged as *Selected Letters of Mary Wollstonecraft Shelley* (Baltimore: Johns Hopkins University Press, 1994).

Bibliography:

W. H. Lyles, *Mary Shelley: An Annotated Bibliography* (New York & London: Garland, 1975).

Biographies:

Mrs. Julian Marshall, *The Life and Letters of Mary Wollstonecraft Shelley*, 2 volumes (London: Bentley, 1889);

R. Glynn Grylls, *Mary Shelley: A Biography* (London: Oxford University Press, 1938; New York: Haskell House, 1969);

Elizabeth Nitchie, *Mary Shelley: Author of Frankenstein* (New Brunswick, N.J.: Rutgers University Press, 1953);

Margaret Leighton, *Shelley's Mary: A Life of Mary Godwin Shelley* (New York: Farrar, Straus & Giroux, 1973);

Jane Dunn, *Moon in Eclipse: A Life of Mary Shelley* (New York: St. Martin's Press, 1978);

Bonnie Rayford Neumann, *The Lonely Muse: A Critical Biography of Mary Wollstonecraft Shelley* (Salzburg: Institut für Angelstik und Amerikanstik, 1979);

Muriel Spark, *Mary Shelley* (New York: Dutton, 1987);

Anne K. Mellor, *Mary Shelley: Her Life, Her Fiction, Her Monsters* (New York & London: Routledge, 1988);

William St. Clair, *The Godwins and the Shelleys: The Biography of a Family* (London: Faber & Faber, 1989);

Emily W. Sunstein, *Mary Shelley: Romance and Reality* (Boston: Little, Brown, 1989).

References:

Brian W. Aldiss and David Wingrove, "On the Origin of Species: Mary Shelley," in their *Trillion Year Spree: The History of Science Fiction* (New York: Atheneum, 1986), pp. 25–52;

Harold Bloom, Afterword to Shelley's *Frankenstein; or, The Modern Prometheus* (New York: Signet, 1965), pp. 212–223;

Bloom, ed., *Mary Shelley's* Frankenstein (New York: Chelsea House, 1987);

David Ketterer, *Frankenstein's Creation: The Book, the*

Monster, and Human Reality (Victoria, B.C.: University of Victoria Press, 1979);

George Levine and U. C. Knoepflmacher, eds., *The Endurance of Frankenstein: Essays on Mary Shelley's Novel* (Berkeley & Los Angeles: University of California Press, 1982);

H. P. Lovecraft, "The Aftermath of Gothic Fiction," in his *Supernatural Horror in Literature* (New York: Abramson, 1945), pp. 36–44;

Allene Stuart Phy, *Mary Shelley* (Mercer Island, Wash.: Starmont, 1988);

Burton R. Pollin, "Philosophical and Literary Sources of *Frankenstein*," *Comparative Literature*, 17 (1965): 97–108;

James Rieger, "Dr. Polidori and the Genesis of *Frankenstein*," *Studies in English Literature*, 3 (1963): 461–472;

Sir Walter Scott, "Remarks on 'Frankenstein; or, The Modern Prometheus: A Novel,' " anonymous, *Blackwood's Edinburgh Magazine*, 2 (March 1818): 613–620;

George Slusser, "The *Frankenstein* Barrier," in *Fiction 2000: Cyberpunk and the Future of Narrative*, edited by Slusser (Athens: University of Georgia Press, 1992), pp. 46–71;

Christopher Small, *Ariel like a Harpy: Shelley, Mary and Frankenstein* (London: Gollancz, 1972);

Martin Tropp, *Mary Shelley's Monster: The Story of "Frankenstein"* (Boston: Houghton Mifflin, 1976);

William A. Walling, *Mary Shelley* (New York: Twayne, 1972).

Papers:

The Pforzheimer Collection at the New York Public Library includes an extensive collection of materials related to Mary Wollstonecraft Shelley.

Bram Stoker

(8 November 1847 – 20 April 1912)

Stephanie Moss
University of South Florida

See also the Stoker entries in *DLB 36: British Novelists, 1890–1929: Modernists* and *DLB 70: British Mystery Writers, 1860–1919.*

BOOKS: *Duties of Clerks of Petty Sessions in Ireland* (Dublin: Privately printed by J. Falconer, 1879);

Under the Sunset (London: Low, Marston, Searle & Rivington, 1882);

A Glimpse of America: A Lecture Given at the London Institution, 28th December, 1885 (London: Low, Marston, Searle & Rivington, 1886);

The Snake's Pass (New York: Harper, 1890; London: Low, Marston, Searle & Rivington, 1891);

Crooken Sands (New York: De Vinne, 1894);

The Man From Shorrox's (New York: De Vinne, 1894);

The Watter's Mou' (New York: De Vinne, 1894; London: Constable, 1895);

The Shoulder of Shasta (London: Constable, 1895);

Dracula: The Undead (London: Constable, 1897; Garden City: Grosset & Dunlap, 1897);

Miss Betty (London: Pearson, 1898);

Sir Henry Irving and Miss Ellen Terry In Robespierre, Merchant of Venice, The Bells, Nance Oldfield, The Amber Heart, Waterloo, etc., Drawn by Pamela C. Smith (New York: Doubleday & McClure, 1899);

The Mystery of the Sea: A Novel (New York: Doubleday, Page, 1902; London: Rider, 1913);

The Jewel of Seven Stars (London: Heinemann, 1903; New York: Harper, 1904);

The Man (London: Heinemann, 1905);

Personal Reminiscences of Henry Irving, 2 volumes (London: Heinemann, 1906; New York: Macmillan, 1906);

Lady Athlyne (London: Macmillan, 1908; New York: Reynolds, 1908);

The Gates of Life (New York: Cupples & Leon, 1908);

Snowbound: The Record of a Theatrical Touring Party (London: Collier, 1908);

The Lady of the Shroud (London: Heinemann, 1909);

Famous Imposters (London: Sidgwick & Jackson, 1910; New York: Sturgis & Walton, 1910);

The Lair of the White Worm (London: Rider, 1911);

Dracula's Guest, and Other Weird Stories (London: Routledge, 1914; New York: Hillman-Curl, 1937);

The Bram Stoker Bedside Companion: Ten Stories by the Author of Dracula, edited by Charles Osborne

(London: Gollancz, 1973; New York: Taplinger, 1973).

Edition: *The Essential Dracula: The Definitive Annotated Edition of Bram Stoker's Classic Novel,* edited by Leonard Wolf (New York: Plume, 1993).

SELECTED PERIODICAL PUBLICATIONS–UNCOLLECTED: "Actor-Manager's, *Nineteenth Century,* 27 (June 1890): 1040–1051;

"The Red Stockade," *Cosmopolitan,* 17 (1894): 619+;

"The Art of Ellen Terry," *Cosmopolitan,* 31 (1901): 241–250;

"Question of a National Theatre," *Nineteenth Century,* 63 (May 1908): 734–742;

"De Morgan's Habits of Work," *World's Work,* 16 (June 1908): 10337–10342;

"The Censorship of Fiction," *Nineteenth Century,* 64 (September 1908): 479–487;

"Americans as Actors," *Fortnightly Review,* 91 (February 1909): 243–252;

"Dead-Heads," *Fortnightly Review,* 92 (October 1909): 646–658;

"American Tramp Question and the Old English Vagrancy Laws," *North American,* 190 (November 1909): 605–614;

"The Censorship of Stage Plays," *Nineteenth Century,* 66 (December 1909): 974–989;

"Real Mlle. De Maupin," *Bookman,* 32 (January 1911): 456–457;

"Irving and Stage Lighting," *Nineteenth Century,* 69 (May 1911): 903–912.

Like a creature from his own imagination, Abraham (Bram) Stoker is frequently described by biographers as a red-haired giant, a man of prodigious strength and energy. During his lifetime Stoker's name was most often linked to that of his famous partner, the actor Sir Henry Irving. Although Stoker was a well-known participant in the world of late-Victorian theater, he is primarily acknowledged today for a single creation: *Dracula* (1897), a terrifying tale of vampires and Victorians. One hundred years old in 1997, this classic novel, like the count himself, has lived several lives: in print, film, television, and in thousands of popular permutations. It has spawned fan clubs and societies all over the world. Stoker wrote nineteen other books, several short stories, and periodical articles, but modern critical appraisal has remained contained and confined to *Dracula;* Stoker's other writings remain largely unexplored. *Dracula* was and continues to be a best-seller, and Stoker was a familiar figure in the most rarefied Victorian social circles, but there is no mention of his name in any nineteenth-century biographical entry.

Bram Stoker was born on 8 November 1847 in Dublin, the third of seven children of Charlotte Thornley Stoker and her civil-servant husband, Abraham Stoker; Stoker remained bedridden with an undiagnosed illness for the first eight years of his life. During that period, according to biographers Harry Ludlam and Daniel Farson, Charlotte told stories of the cholera epidemic that swept Ireland in 1832. Both Ludlam and Farson, Stoker's grandnephew, believe those stories influenced Stoker's adult fascination with terror.

Stoker's childhood illness, which had hysteria-like symptoms, may have led him to imagine the predicament he would later create for his vampire victims: a helpless malaise and paralysis similar to the author's early infirmity. As Stoker himself states in *Personal Reminiscences of Henry Irving* (1906), "I was naturally thoughtful and the leisure of long illness gave opportunity for many thoughts which were fruitful according to their kind in later years." The novel's fascination with medical phenomena may also have been nourished by Stoker's relationship to his two physician brothers, George and William.

Stoker entered Trinity College in Dublin at age seventeen in 1864. There, completely recovered from his childhood debilitation, he became known for his athletic and debating skills and overall keen intellect. While at school he made his first acquaintance with the controversial poetry of Walt Whitman; he would later befriend the poet. His passion for Whitman's work presaged, as David Skal notes, his later passion for the work of Irving. Indeed, Stoker's first glimpse of Irving occurred at nearly the same time as his early attachment to Whitman's writing.

On 28 August 1867 Stoker saw Irving perform as Anthony Absolute in Richard Brinsley Sheridan's *The Rivals* at the Theatre Royal. Because he had long been a theater fan, Stoker was well acquainted with the technicalities of acting and recognized that Irving's talent, as a stock player in a provincial touring company, was well above the norm. Although Irving was limited by strict necessity to re-create the details formalized and adhered to by previous actors in the role, Stoker noted in *Personal Reminiscences* that Irving's Absolute was "different from any performance of the same part which [he] had seen."

Stoker graduated from Trinity with honors in mathematics in 1868, obtained his M.A., and assumed a job as civil servant, which had been procured by his father. In May 1871 Irving again played in Dublin at the Theatre Royal, in James Albery's comedy *Two Roses.* Stoker records in *Personal Reminiscences* his continued delight in Irving's talent: "Irving impersonated in a manner so complete that to this day I have been unable to get it out of my mind as a reality." Stoker's admiration, however, was not echoed by the local reviewers. To remedy this situation he took a job at the Dublin newspaper the *Mail* in November 1871 as an unpaid reviewer.

By 1876, when Irving returned to Dublin in *Hamlet,* the actor had accumulated a string of London successes. Stoker notes that Irving's new method of "thoughtful acting," as opposed to the predominant histrionic approach, was winning approval. Stoker's reputation at the time rested upon his contributions to theater; therefore, it is important to note that what he was attracted to in Irving's theatrical style was its break from the commonplace, its revolutionary naturalism, which was an early exploration of acting realism. Stoker wrote in *Personal Reminiscences,* "Irving always helped his audience to think."

Irving and Stoker finally met after Irving read Stoker's review of *Hamlet;* Stoker was thirty years old and writing what he termed his "dry-as-dust" text: *The Duties of Clerks of Petty Sessions* (1879). The mutual attraction was cemented when, after dining together, Irving recited Thomas Hood's poem "The Dream of Eugene Aram" and Stoker was so moved that he "burst into something like hysterics," surprising both himself and his friends.

In June 1877 Irving revisited Dublin for a promised theatrical reading. The future relationship of the two men was presaged when Stoker helped Irving through the dangerous throng of mobbing fans to safety. Thirteen days later Stoker went to London on holiday, and Irving encouraged him to relocate and try his fortune as a writer. Through Irving, Stoker made a connection with John Knowles, the editor of the influential Victorian periodical *Nineteenth Century.* Although Stoker returned to Dublin after meeting with Knowles and did not contribute to the periodical until 1890, an important literary friendship was initiated.

In 1878, when Irving again visited Dublin, the actor finally made the offer Stoker had been waiting for; he asked Stoker to give up his post in public service and "share his fortunes." Irving had just taken over management of the Lyceum theater and wanted Stoker to take charge of business as acting manager. On 9 December 1878 Stoker met Irving in Birmingham to assume responsibility for the Lyceum's business. Stoker notes in *Personal Reminiscences,* "I left Glasgow on November 25. . . . Having in the meantime altered my whole business life . . . and last, not least, having got married."

The thirty-one-year-old Stoker married nineteen-year-old Florence Anne Lemon Balcombe on 4 December 1878. His new bride, a neighbor and the daughter of a lieutenant colonel, had been previously courted by Oscar Wilde. According to Skal, Florence made the transition from a penniless Irish girl to one who inflamed the imaginations of at least three nineteenth-century literati: Stoker, Wilde, and George Du Maurier, who considered her "one of the three most beautiful women he had ever seen." Farson characterizes Florence as "an elegant, aloof woman, more interested in her position in society" than in her son, Noel Thornley Stoker, who was born in 1879. Many years later Noel's daughter reported to Farson that "she doubted if [Florence] . . . was really capable of love." Stoker's granddaughter continued, "She was cursed by her great beauty and the need to maintain it. In my knowledge now, she was very anti-sex. After having my father in her early twenties, I think she was quite put off." However, Skal notes that the essayist and contemporary observer of the late-Victorian social scene, Horace Wyndam, remembers Florence as "a charming woman and full of Irish wit and impulsiveness."

Stoker's partnership with Irving lasted twenty-seven years, during which time he was the workhorse who kept the theater running smoothly while also managing to maintain his fairly prodigious literary output. All this was accomplished while attending each of Irving's daily performances and rehearsals, keeping the business of the theater in running order, and often staying until dawn with Irving and friends at the Beefsteak Room.

In 1882 Stoker published his first book of stories, *Under the Sunset,* a collection for children dedicated to his son. Described by some modern critics as bizarre or macabre, Phyllis A. Roth reports that Douglas Street finds them true to the genre in their exploration of family relationships and the morality of good and evil. Ludlam quotes Victorian reviews that called them "delicate and forcible allegories" characterized by "purity and grace." Today they are primarily examined as predecessors to Stoker's later horror stories. Joseph S. Bierman sees a strong correlation between some of these tales, the vampire novel, and the period of Stoker's childhood immobility. As in *Dracula,* the forces of good and evil line up to do battle. According to Roth, Stoker's primary themes of father/son relationships, ambivalence toward women, male rivalry, and the blurred line between life and death may be glimpsed. Like most Victorian children's stories, these contain mythical folk images and moralizing.

During Stoker's tenure as Irving's manager he not only saw to all the Lyceum's business–which totaled £2 million during Irving's occupancy–and hovered over the actors to make sure nothing distracted them from the concentration necessary to perform, but also arranged and participated in

Cover for Stoker's 1897 novel, which contrasts Victorian culture to the folk culture of Transylvania

seven American tours. From 1883 to 1900 Irving's troupe played the former colonies; the knowledge and affection Stoker acquired for America became a recurrent theme in his stories and novels. One of his early publications was a pamphlet about the country called *A Glimpse of America* (1886). This short book reflects Stoker's infatuation with America–its lack of class consciousness, the easy and inexpensive accessibility to its land, and its chivalrous treatment of women. The pamphlet was successful and elicited admiration from, among others, the explorer Henry M. Stanley.

Also written during Stoker's tenure as Irving's manager was his first full-length novel, *The Snake's Pass,* published in 1890 but written in the late 1880s. A portion, titled "The Gombeen Man," is often excerpted and included in anthologies. The novel, which was unanimously praised by the Victorian press, is one of several Stoker narratives written in the romance genre. Roth notes an elitism that many perceive in Stoker's inclination to social climb. She also remarks that *The Snake's Pass* contains unfortunate references to "niggers." This racial bigotry can be seen in other works, most notably in the character of Oolanga in *The Lair of the White Worm* (1911). In marked contrast to the usual concern Stoker evinced toward others, particularly the poor, his reactionary thinking again surfaced in his later advocacy of branding vagrants. Furthermore, Ludlam reports that among the books lost when the Stokers made their final move to St. George's Square was a volume devoted to the history of the Ku Klux Klan.

Like *Dracula, The Snake's Pass* employs legend, myth, suspense, mystery, quest, and the use of the double. A love story, the novel is also preoccupied, like *Dracula,* with science. It uses a most traditional plot theme: boy meets girl; boy gets girl and treasure. While traveling in Ireland the narrator, Arthur Severn, meets Norah, the daughter of a local landowner, Phelim Joyce. The Gombeen Man, or usurer, Black Murdock, hopes to get Joyce's land, but Arthur employs an engineer, Dick Sutherland, and locates the mythical treasure believed to be buried in the area. The descriptive passages, which correlate to actual locations visited by Stoker, ring with a beauty and authenticity that is consistently praised in Stoker's writing despite critical condemnation of his tendency toward cliché and sentimentalism.

The Watter's Mou', a novel published in 1894, was generally favorably received. Set in Stoker's vacation spot, Cruden Bay, Scotland, this tale of smuggling centers around Maggie Macwhirter. Again, the author's descriptive style was praised, this time by Sir Arthur Conan Doyle. One popular periodical of the era, the *Athenaeum,* extolled the descriptive passages but attacked the novel's excessive melodrama and stagy speeches.

In 1895 *The Shoulder of Shasta* was published and, according to both Farson and Ludlam, the novel was dismissed by contemporary critics as being written in haste and carelessly constructed. Set in California, the story concerns a young San Francisco girl's attraction to Grizzly Dick, a mountain man. This novel's lack of success contrasts markedly to the next book Stoker published. In 1897, two years after the appearance of *The Shoulder of Shasta,* Stoker's masterpiece of horror and psychological terror appeared.

The working notes for *Dracula,* which now reside in the Rosenbach Museum and Library in Philadelphia, show that, unlike the careless construction that marks *The Shoulder of Shasta,* Stoker labored over the text of his horror novel for seven years. A record of this labor appears in the book's typed manuscript, which surfaced in 1984. Once believed lost, this shows Stoker's painstaking and careful editing and revisions.

Dracula continues the prose exploration of the mythical death demon begun in *The Vampyre* (1819) by John William Polidori, George Gordon, Lord Byron's physician. James Malcolm Rymer's potboiler *Varney the Vampyre* (1847) followed Polidori's story, and Joseph Sheridan Le Fanu perfected the genre in his classic"Carmilla" (1871), a novella that heavily influenced Stoker. The folk mythology in *Dracula,* based on Transylvanian legend, was discovered by Stoker during his research in the British Museum and in the library at Whitby. Much of this material has been traced to an excerpted 1895 article in *The Nineteenth Century* of E. Gerard's 1888 travel text *Land beyond the Forest.* Stoker was also influenced by Armenius Vambery, a professor from the University of Budapest who visited the Lyceum in 1890 and is mentioned both in *Personal Reminiscences* and in *Dracula.* Much of the character of Van Helsing may be patterned after him. The count's relationship to a historical eastern European figure, the Wallacian prince Vlad Tepes, or Vlad Dracula, son of Vlad Dracula, was explored in depth in 1972, when Raymond T. McNally and Radu Florescu published *In Search of Dracula.*

After the publication of *In Search of Dracula,* worldwide economic repercussions from Stoker's novel, begun with the 1922 pirated German film *Nosferatu,* spread like the disease of vampirism itself to the tourist office in Transylvania. After the release of Francis Ford Coppola's 1992 film, *Bram Stoker's Dracula,* a small financial war broke out between the seaside communities of Whitby, England, and Cruden Bay, Scotland, both vacation homes of Stoker's. Each town claims to be the site of original inspiration, although the discovery of the novel's working notes (the first published reference to them was in 1977) verified the ancient seaport of Whitby as the place of genesis. The notes pinpoint the initiation of the novel to the year 1890, concurrent with Stoker's summer vacation in the English seaport whose history dates back to the Anglo-Saxon era.

The novel begins with the journey of a young solicitor, Jonathan Harker, to Transylvania, where he has been employed by Dracula to purchase real estate in London. Upon landing in Whitby, Dracula preys upon Lucy Westenra and later Lucy's best friend, Mina, who is engaged to Jonathan. Van Helsing, a scholar steeped in supernatural tradition, is brought in to help slay the vampire. Also involved are Lucy's three suitors: the wealthy Arthur Holmwood, the alienist John Seward, and American Quincy Morris. The final participant is Seward's mental patient Renfield.

Stoker locates his story first in Transylvania before moving his vampire into contemporary Victorian England. These contrasting locales establish the polarity in late-nineteenth-century culture of technological science and the superstition of an earlier age. Gerard portrays Transylvania as a land where the "curious crooked plant of delusion flourished so insistently and in such bewildering variety" that superstition survived the development of technology. Here the demons, pixies, witches, and hobgoblins driven from the rest of Europe by the "wand of science" found a "secure lurking place whence they might defy their persecutors yet awhile."

When first entering the Carpathians, Jonathan clings to his Western artifacts. Using a dictionary he attempts to translate the peasants' warnings *ordog, pokol, stregoica, vrolok,* and *vlkoslak.* Although his Enlightenment dictionary translates words, it keeps Jonathan from grasping the supernatural reality of his circumstances. The peasants, steeped in folkloric vampirism, are better equipped than the Victorian middle-class male to combat a creature of the superstitious past. In the end the ancient icon rejected by England, the crucifix, wields more power than a dictionary full of definitions.

By opening the novel in the exotic world of Transylvania, Stoker provides distance from modern Victorian England. This allows the integration of the soulless incubus and the barbarous rites of penetration and beheading into the technological mindset of the nineteenth-century reader. The bloody and religious metaphors that run throughout *Dracula* gain further credence because they emanate not from the soul of civilized society but from the bowels of a strange and unfamiliar region of the earth.

Victorian critical response was generally favorable, often glowing. Ludlam quotes the *Daily Mail:* "In seeking a parallel to his weird, powerful and horrible story, our minds revert to such tales as *The Mysteries of Udolpho, Frankenstein, Wuthering Heights,* [and] *The Fall of the House of Usher.*" A fully annotated version of the text, compiled by Leonard Wolf, was published in 1973 and was followed, in 1993, by a revised and substantially different version.

Wolf was among the first who responded to the 1970s critical fascination with *Dracula.* He noticed the polymorphous sexuality that lies beneath the surface of what appears to be a straightforward horror tale. Most critics find this sexual theme to be unconscious on Stoker's part even though *Dracula* was published during the same period as Sigmund Freud's texts on hysteria. To date no proof has surfaced that Stoker was exposed to Freud's theories. However, the author was insatiably curious; although Freud's earliest texts did not appear in English until 1908, some material on hysterical paralysis was published in French in 1893, and Stoker might have encountered it. Indeed, the similarities between the symptoms of the vampire attack and a hysterical episode, to which so many Victorian women were susceptible, is quite striking.

Another sexual reading of the text is Robert Tracy's interpretation of the female vampire as prostitute. The wide acceptance of this reading is in part because of Farson's conclusion that Stoker died of the ravages of syphilis. Undoubtedly, the disease of vampirism is spread much like the venereal diseases that were closely associated with the prostitutes who preyed and were preyed upon by Victorian males.

The critical revival of *Dracula* in the early 1970s turned a trickle of literary criticism into a deluge. Most of these exegeses view the novel as a repository of dark Victorian psychology rather than as a text of literary merit. The most frequently mentioned psychological aspect is the madonna/whore schism within Victorian perceptions of women, seen most clearly in Lucy's transformation from aristocratic female to vampire. The frank nature of the liberated women is also tentatively explored through the character of Mina. Most critics, however, elide the artistic importance of Stoker's accurate scenic and scientific portrayal, which enables the "willing suspension of disbelief " necessary for reader acceptance of supernatural occurrences.

In 1898 Stoker published *Miss Betty,* a "short and lightweight novel suffering from lack of consideration and revision," according to Roth. This romance of regeneration lacks the complexity of *Dracula.* Set in the aristocratic milieu of the early 1700s, *Miss Betty* concerns the relationship between the pure Betty Pole and her miscreant fiancé, Rafe Otwell. Otwell's kinsman, the historic statesman Robert Walpole, hopes to marry Otwell to the wealthy but old and ugly Lady Mary. Betty's cousin Fenton remains true to Betty while Otwell strays. As in *Dracula's* use of Vlad Tepes, Stoker weaves historic figures into his fiction.

The Mystery of the Sea (1902) concerns the narrator, Archie Hunter, and his adventures with an American heiress. Employing Gothic apparatus–including a witch figure, second sight, and Gaelic runes–*The Mystery of the Sea* details Stoker's love for Cruden Bay, his seafaring experiences, and historical data on the Spanish Armada. According to Ludlum, Sir Arthur Conan Doyle praised Stoker's use of cryptograms, which Conan Doyle also used in his Sherlock Holmes stories.

The Jewel of Seven Stars (1903) evolved out of Stoker's interest in archaeological exploration and tells the story of the resurrection of a queen who ruled Egypt twenty-five centuries before Christ. The mummy of Queen Tara is brought back to London by the Egyptologist Abel Trelawny. His daughter, Margaret, slowly merges personalities with the dead queen. The story is narrated by barrister Malcolm Ross, who is in love with Margaret.

As in *Dracula,* Stoker uses his modern technological world as a foil to the ancient and mysterious world that lies beyond scientific understanding. The characters prepare to raise the mummified queen and, as a modern talisman to guard against any "disaster that human could foresee," the area has been extensively fortified with electric lights; the ceremony is approached scientifically. However, the group soon learns that "the bounds of human thought [are] but narrow. Beyond it [lies] the vast of Divine wisdom, and Divine power!"

Farson feels this text is next to *Dracula* in readability. Roth discerns Stoker's talent for narrative and description, adding that "Stoker's power lies with the mysterious and uncanny" and "his novels succeed to the degree that the romance components are subordinated to those of the fantastic."

The Man was published in 1905 and is generically another romance. Roth characterizes it as Stoker's most sophisticated examination of character and motivation. The female hero, Stephen Norman, is named to satisfy her father's desire for a son. While Stephen rejects stereotypically female roles by proposing to Leonard Everard, the "Man" of the title is not Stephen but Harold An Wolf. Harold helps the heroine assume her female role within strictly gendered Victorian society. Although *The Man,* like *Dracula,* reasserts Victorian norms in the end, the gender twisting noted by critics in *Dracula* is even more overt. The same fascination with the liberated female seen in the character of Mina Harker, the same rejection of sexual forwardness exemplified by Lucy Westenra is repeated, while self-sacrifice becomes a virile virtue rather than a feminine one.

It is a testament to Stoker's stamina that, as he was writing these novels, he made his own contribution to theatrical art by providing freedom from extraneous worries for Irving. Stoker describes the necessity for this freedom in *Personal Reminiscences:* "an actor has to keep himself, in certain ways at least, for his work. When in addition he has the cares and worries and responsibilities and labours and distractions of management to encounter daily and hourly, it is vitally necessary that he has trustworthy and, to him, sufficing assistance." Management of a theater and its often temperamental actors is a job that even today requires vast resources of time and energy.

Stoker reports in *Personal Reminiscences* that a humorous incident occurred when one reporter, offended because Stoker refused to allow him to accompany the troupe on tour, summed up Stoker's managerial responsibilities: "[he] seems to occupy some anomalous position between secretary and valet; whose manifest duties are to see that there is mustard in the sandwiches and to take the dogs out for a run." This is a flippant although accurate statement of the petty chores to which a theatrical manager must often attend. Ludlam describes Stoker's responsibilities this way: "he was a kind of welfare officer, public relations man and a half dozen other of today's familiar titles in one." Indeed, Ellen Terry, Irving's leading lady for much of his career, affectionately referred to Stoker as "mama," an expression of Stoker's centrality to Irving's theatrical family.

Stoker in return was delighted with his exposure to late-nineteenth-century British society. Irving threw enormous after-theater dinners, and the list of guests included many legendary figures. Jan Paderewski, J. M. Barrie, P. T. Barnum, Ethel Barrymore, James Whistler, Thomas Hardy, Arthur Wing Pinero, the explorer Sir Richard Francis Burton, Mark Twain, William Gladstone, and Alfred Tennyson among others attended these lavish dinners. Many acquaintances Stoker made during these parties developed into important friendships.

In 1898 Irving fell ill and sold his interest in the Lyceum to a syndicated company. Foreseeing financial disaster, Stoker protested but was shut out of all proceedings. The company began to lose money while the syndication profited. Finally, the new owners chose to convert the Lyceum into a music hall, and Irving became an actor without a theater. His health began to fail; he became frail; and in 1904 he opted for retirement, embarking upon a farewell tour that lasted two years. Irving died on 13 October 1905 in his hotel following a performance.

Stoker, bereft, suffered a stroke. However, he continued to write and in the years following Irving's death completed five novels, his two-volume remembrances of Irving, and many periodical articles. *Personal Reminiscences of Henry Irving* was written in a passion of loss after Irving's death. Unlike the biographical material written to cash in on Irving's popularity, Stoker's text is a personal memoir drawn from his journal notations. Extremely valuable as his most autobiographical material, the narrative is also fascinating in its detailing of the world of late-Victorian theater, Irving's theatrical pyrotechnics, and his approach to acting. A description of Irving's performance of *Macbeth* shows that in emphasizing Macbeth's inner psychology rather than the witches' prophecy, the actor came to a critical understanding of the tragedy that presaged later scholarly interpretation.

A portrait of Vlad the Impaler, the inspiration for Dracula, who is believed to have put thousands to death

All Stoker's literary and theatrical ventures after Irving's death failed to make money, and the Stokers moved down the economic and social ladder to smaller and smaller London houses. He attempted a lecture tour and a return to theatrical management, both of which were failures. From Stoker's vast coterie of friends the only one who came to his financial aid in these final years was Sir Hall Caine–a formidable figure, a best-selling author, and the "Hommy Beg" to whom *Dracula* is dedicated. There were reports, according to Farson's mother, that Stoker became "dotty." He suffered from the kidney ailment, Bright's disease (also known as nephritis), from failing eyesight, and from gout.

In 1908 Stoker published *Lady Athlyne,* another example of his fascination with character doubling, here based on false identity. Joy Ogilvie, the daughter of a Kentucky colonel, assumes the name of a British lord. The novel has been all but forgotten by modern critics.

The Lady of the Shroud (1909) is more a mystery story than a tale of the supernatural. The protagonist, Rupert, is placed in opposition to the greedy Ernest Roger Halbard Melton, nephew of the late

Roger Melton. Written, like most of *Dracula,* in journal form, the novel concerns Rupert's voyage to Transylvania, where he falls in love with and marries the Lady of the Shroud, the Voivode Tueta, last of a royal line. Janet Mackelpie, a character with second sight, functions as the novel's only representation of the supernatural in a text that mixes the uncanny with the mysterious. One of the novel's most notable features, however, is its prescient use of airplanes to wage war.

Roth praises the plot, narration, and description but rates the text below *The Jewel of Seven Stars* and *Dracula.* Carol A. Senf finds that Stoker's use of the same supernatural elements employed in *Dracula,* which were so troubling on a psychological level, are peacefully resolved in this text. She attributes this resolution partially to the deaths of both Irving and the earlier death of Charlotte Stoker in Dublin in 1900 but also suggests that Stoker may have begun to shy away from sexually charged material. In making this comment Senf elides the overt sexual symbolism in Stoker's final novel, *The Lair of the White Worm.* However, "The Censorship of Fiction," an article Stoker wrote for the *Nineteenth Century* in 1908 that has caught the attention of numerous critics, advocates purging literature of material dealing with sexual impulses. Farson uses this article to help justify his conclusion that Stoker's final illness was tertiary syphilis. Indeed the article carries an oblique reference to Henrik Ibsen, whose problems plays, including *Ghosts* (1882), had pushed lighter fare from the British stage. Stoker expressed an early distaste for Ibsen's plays, attacking, however, inconsistency of character rather than the use of sensational material.

Famous Imposters (1910) is Stoker's third nonfiction book. The text examines hoaxes, deceptions, and fraudulent identities of both historical and mythical characters. Roth claims the book is often "witty and vivid" and fits into Stoker's canon thematically through its demonstrations of dualism and the "mixture of the practical and mundane."

The Lair of the White Worm has garnered critical fascination because of the overt sexual symbolism of the foul female worm, which supports the misogynistic subtext many see in *Dracula.* The protagonist, Adam Salton, visits his granduncle Richard. Sir Nathaniel de Salis, Richard's oldest friend, becomes the Van Helsing figure, and Edgar Caswell, the villainous descendant of a ruling house, possesses mesmerizing talents. Also involved are Lilla, the daughter of a tenant farmer, and her cousin Mimi. Lady Arabella, a human manifestation of the white worm of the title, is a wealthy aristocrat who becomes the focus of the text. As this was Stoker's final literary work, written one year before his death, Farson speculates that the novel was written under the influence of drugs taken to alleviate pain, adding, "[W]ithout a vestige of humour, [the novel] is extremely funny." Indeed, it evinces a loss of literary power in its disjointed narration and clumsy wording. Roth politely calls it "the weakest of all [Stoker's] novel-length fictions."

Bram Stoker died on 20 April 1912 at age sixty-four, leaving £4,723 sterling as his estate. The cause of death is reported as "exhaustion" on the death certificate. However, this document also lists "Locomotor Ataxy 6 month Granular Contracted Kidneys," which Farson contends is a medical euphemism for tabes dorsalis and general peresis, or general paralysis of the insane–the terminal symptoms of tertiary syphilis.

Primarily because of subliminal sexual readings of *Dracula,* much speculation has been made about Stoker's psychological makeup. Because of the lack of concrete biographical data, however, very little of the more lurid elements of his speculation can be verified. Farson's conclusion that the real nature of Stoker's last illnesses and death was tertiary venereal disease remains loosely based upon an amalgam of reinterpreted facts.

Steven Marcus's examination of the Victorian period describes sexual ambivalence as a prominent psychological feature. Prostitution and its shadow, syphilis, were rife. The image of woman as the "angel of the house" vied with the emerging New Woman. Hysterical symptoms left middle-class women prone upon their couches. How this zeitgeist affected Stoker remains mere conjecture; everything known of the man is recorded in his autobiographical observances of his contemporaries. What remains true of Stoker, after all speculation has been completed, is that he was a man who managed a demanding career while continuing to create a large, but largely ignored, volume of literary material.

Ultimately, the lasting legacy of *Dracula* lies in the brilliant manipulation of the fantastic. Although in his article on censorship Stoker acknowledged a peril in the use of imagination, what remains true of his only critically acknowledged work is that this tale of the twilight world of shadows allows a momentary catharsis that liberates nightmares from the unconscious. The deeply repressed chaos of death, apparently comfortable and confined in the coffin of literature, sleeps undisturbed until one opens the pages and Dracula escapes his textual prison.

Biographies:

Harry Ludlam, *A Biography of Dracula: The Life Story of Bram Stoker* (London: Foulsham, 1962);

Daniel Farson, *The Man Who Wrote Dracula: A Biography of Bram Stoker* (New York: St. Martin's Press, 1976);

Barbara Belford, Bram Stoker: *A Biography of the Author of* Dracula (New York: Knopf, 1996).

References:

Joseph S. Bierman, "*Dracula:* Prolonged: Childhood Illness, and the Oral Triad," *American Imago,* 29 (Summer 1972): 186–198;

Christopher Craft, "'Kiss Me with Those Red Lips': Gender and Inversion in Bram Stoker's Dracula," in *Dracula: The Vampire and the Critics,* edited by Margaret L. Carter (Ann Arbor, Mich.: UMI Research, 1988), pp. 167–194;

Christopher Frayling, *Vampyres: Lord Byron to Count Dracula* (London: Faber & Faber, 1991);

E. Gerard, "Transylvanian Superstitions," *Nineteenth Century,* 18 (July–December 1895): 130–150;

Clive Leatherdale, *Dracula the Novel and the Legend: A Study of Bram Stoker's Gothic Masterpiece* (Wellingborough: Aquarian Press, 1985);

Steven Marcus, *The Other Victorians: A Study of Sexuality and Pornography in Mid-Nineteenth Century England* (New York: Basic Books, 1964);

Raymond T. McNally and Radu Florescu, *In Search of Dracula: A True History of Dracula and Vampire Legends* (Greenwich, Conn.: New York Graphic Society, 1972);

J. Gordon Melton, *The Vampire Book: The Encyclopedia of the Undead* (Detroit: Gale Research, 1994);

Stephanie Moss, "Performing *Dracula:* Role Playing and the Interpolating Tropology of Hysteria and Hypnosis," thesis, University of South Florida, 1993;

Phyllis A. Roth, *Bram Stoker* (Boston: Twayne, 1982);

Carol A. Senf, *"The Lady of the Shroud:* Stoker's Successor to *Dracula," Essays in Arts and Sciences,* 19 (May 1990): 82–96;

David Skal, *Hollywood Gothic: The Tangled Web of Dracula from Novel to Stage to Screen* (New York: Norton, 1990);

Douglas Street, "Bram Stoker's *Under the Sunset:* An Edition with Introductory Biographical and Critical Material," dissertation, University of Nebraska, 1977;

Robert Tracy, "Loving You All the Way: Vamp, Vampires, Necrophile and Necrofille in Nineteenth-Century Fiction," in *Sex and Death in Victorian Literature,* edited by Regina Barecca (Bloomington: Indiana University Press, 1990), pp. 32–59.

Papers:

Seventy-eight pages of Bram Stoker's diagrams, notes including sources, and outlines for *Dracula* are held by the Rosenbach Museum and Library, Philadelphia.

H. G. Wells

(21 September 1866 – 13 August 1946)

Darren Harris-Fain
Shawnee State University

See also the Wells entries in *DLB 34: British Novelists, 1890–1929: Traditionalists; DLB 70: British Mystery Writers, 1860–1919;* and *DLB 156: British Short-Fiction Writers, 1880–1914: The Romantic Tradition.*

H. G. Wells

BOOKS: *Text-Book of Biology,* 2 volumes (London: Clive, 1893; volume 1 revised, 1894);

Honours Physiography, by Wells and R. A. Gregory (London: Hughes, 1893);

Select Conversations with an Uncle, Now Extinct, and Two Other Reminiscences (London: John Lane, 1895; New York: Merriam, 1895);

The Time Machine: An Invention (London: Heinemann, 1895; New York: Holt, 1895);

The Wonderful Visit (London: Dent, 1895; New York: Macmillan, 1895);

The Stolen Bacillus, and Other Incidents (London: Methuen, 1895);

The Island of Doctor Moreau (London: Heinemann, 1896; New York: Stone & Kimball, 1896);

The Wheels of Chance: A Holiday Adventure (London: Dent, 1896); republished as *The Wheels of Chance: A Bicycling Idyll* (New York: Macmillan, 1896);

The Plattner Story, and Others (London: Methuen, 1897);

The Invisible Man: A Grotesque Romance (London: Pearson, 1897; enlarged edition, New York: Arnold, 1897; London: Pearson, 1900);

Thirty Strange Stories (New York: Arnold, 1897);

Certain Personal Matters: A Collection of Material, Mainly Autobiographical (London: Lawrence & Bullen, 1897);

The War of the Worlds (London: Heinemann, 1898; New York & London: Harper, 1898);

When the Sleeper Wakes: A Story of the Years to Come (London & New York: Harper, 1899); revised as *The Sleeper Awakes* (London: Nelson, 1910);

Tales of Space and Time (London & New York: Harper, 1899; New York: Doubleday, McClure, 1899);

A Cure for Love (New York: Scott, 1899);

Love and Mr. Lewisham (London & New York: Harper, 1900);

The First Men in the Moon (London: Newnes, 1901; Indianapolis: Bowen-Merrill, 1901);

Anticipations of the Reaction of Mechanical and Scientific Progress upon Human Life and Thought (London: Chapman & Hall, 1901; New York & London: Harper, 1902);

The Discovery of the Future: A Discourse Delivered to the Royal Institution on January 24, 1902 (London: Unwin, 1902; New York: Huebsch, 1913);

The Sea Lady: A Tissue of Moonshine (London: Methuen, 1902); republished as *The Sea Lady* (New York: Appleton, 1902);

Mankind in the Making (London: Chapman & Hall, 1903; New York: Scribners, 1904);

Twelve Stories and a Dream (London: Macmillan, 1903; New York: Scribners, 1905);

The Food of the Gods, and How It Came to Earth (London: Macmillan, 1904; New York: Scribners, 1904);

A Modern Utopia (London: Chapman & Hall, 1905; New York: Scribners, 1905);

Kipps: The Story of a Simple Soul (London: Macmillan, 1905; New York: Scribners, 1905);

In the Days of the Comet (London: Macmillan, 1906; New York: Century, 1906);

The Future in America: A Search after Realities (London: Chapman & Hall, 1906; New York & London: Harper, 1906);

Faults of the Fabian (London: Fabian Society, 1906);

Reconstruction of the Fabian Society (London: Privately printed, 1906);

Socialism and the Family (London: Fifield, 1906; Boston: Ball, 1906);

The So-Called Science of Sociology (London: Macmillan, 1907);

This Misery of Boots (London: Fabian Society, 1907; Boston: Ball, 1908);

Will Socialism Destroy the Home? (London: Independent Labour Party, 1907);

New Worlds for Old (London: Constable, 1908; New York: Macmillan, 1908);

The War in the Air, and Particularly How Mr. Bert Smallways Fared While It Lasted (London: Bell, 1908; New York: Macmillan, 1908);

First and Last Things: A Confession of Faith and Rule of Life (London: Constable, 1908; New York & London: Putnam, 1908; revised and enlarged edition, London, New York, Melbourne & Toronto: Cassell, 1917);

Tono-Bungay (New York: Duffield, 1908; London: Macmillan, 1909);

Ann Veronica: A Modern Love Story (London: Unwin, 1909; New York & London: Harper, 1909);

The History of Mr. Polly (London, Edinburgh, Dublin, New York, Leipzig & Paris: Nelson, 1910; New York: Duffield, 1910);

The New Machiavelli (New York: Duffield, 1910; London: John Lane, 1911);

Floor Games (London: Cecil Palmer, 1911; Boston: Small, Maynard, 1912);

The Country of the Blind, and Other Stories (London, Edinburgh, Dublin, Leeds, New York, Leipzig & Paris: Nelson, 1911); revised as *The Country of the Blind* (London, 1939);

The Door in the Wall and Other Stories (New York & London: Kennerley, 1911; London: Richards, 1915);

The Great State: Essays in Construction, by Wells and others (London: Harper, 1912); republished as *Socialism and the Great State* (New York: Harper, 1914);

The Labour Unrest (London: Daily Mail, 1912);

Marriage (London: Macmillan, 1912; New York: Duffield, 1912);

War and Common Sense (London: Daily Mail, 1913);

Liberalism and Its Party: What Are the Liberals to Do? (London: Good, 1913);

Little Wars: A Game for Boys from Twelve Years of Age to One Hundred and Fifty and for That More Intelligent Sort of Girls Who Like Boys' Games (London: Cecil Palmer, 1913; Boston: Small, Maynard, 1913; revised edition, London & Toronto: Dent, 1931);

The Passionate Friends (London: Macmillan, 1913; New York & London: Harper, 1913);

An Englishman Looks at the World: Being a Series of Unrestrained Remarks upon Contemporary Matters (London, New York, Toronto & Melbourne: Cassell, 1914); republished as *Social Forces in England and America* (New York & London: Harper, 1914);

The World Set Free: A Story of Mankind (London: Macmillan, 1914; New York: Dutton, 1914);

The Wife of Sir Isaac Harman (London & New York: Macmillan, 1914);

The War That Will End War (London: Cecil Palmer, 1914; New York: Duffield, 1914);

The Peace of the World (London: Daily Chronicle, 1915);

Boon, the Mind of the Race, the Wild Asses of the Devil, and the Last Trump, Being a First Selection from the Literary Remains of George Boon, Appropriate to the Times, as Reginald Bliss (London: Unwin, 1915; New York: Doran, 1915); as H. G. Wells (New York: Doran, 1920);

Bealby: A Holiday (London: Methuen, 1915; New York: Macmillan, 1915);

The Research Magnificent (London & New York: Macmillan, 1915);

What Is Coming? A Forecast of Things after the War (London, New York, Toronto & Melbourne: Cassell, 1916); republished as *What Is Coming? A European Forecast* (New York: Macmillan, 1916);

Mr. Britling Sees It Through (London, New York, Toronto & Melbourne: Cassell, 1916; New York: Macmillan, 1916);

The Elements of Reconstruction: A Series of Articles Contributed in July and August 1916 to the Times (London: Nisbet, 1916);

War and the Future: Italy, France, and Britain at War (London, New York, Toronto & Melbourne: Cassell, 1917); republished as *Italy, France, and Britain at War* (New York: Macmillan, 1917);

God the Invisible King (London, New York, Toronto & Melbourne: Cassell, 1917; New York: Macmillan, 1917);

A Reasonable Man's Peace (London: Daily News, 1917);

The Soul of a Bishop: A Novel (with Just a Little Love in It) about Conscience and Religion and the Real Troubles of Life (London, New York, Toronto & Melbourne: Cassell, 1917; New York: Macmillan, 1917);

In the Fourth Year: Anticipations of a World Peace (London: Chatto & Windus, 1918; New York: Macmillan, 1918);

Joan and Peter: The Story of an Education (London, New York, Toronto & Melbourne: Cassell, 1918; New York: Macmillan, 1918);

British Nationalism and the League of Nations (London: League of Nations Union, 1918);

The Undying Fire: A Contemporary Novel (London, New York, Toronto & Melbourne: Cassell, 1919; New York: Macmillan, 1919);

The Idea of a League of Nations, by Wells and others (London: Oxford University Press, 1919; Boston: Atlantic Monthly Press, 1919);

The Way to a League of Nations, by Wells and others (London: Oxford University Press, 1919);

History Is One (New York: Ginn, 1919);

The Outline of History, Being a Plain History of Life and Mankind (24 parts, London: Newnes, 1919–1920; 1 volume, London: Newnes, 1920; 2 volumes, New York: Macmillan, 1921);

Frank Swinnerton, Personal Sketches: Together with Notes and Comments on the Novels of Frank Swinnerton, by Wells, Arnold Bennett, and G. M. Overton (New York: Doran, 1920);

Russia in the Shadows (London: Hodder & Stoughton, 1920; New York: Doran, 1921);

The Salvaging of Civilisation (London, New York, Toronto & Melbourne: Cassell, 1921; New York: Macmillan, 1921);

The New Teaching of History: With a Reply to Some Recent Criticisms of The Outline of History (London, New York, Toronto & Melbourne: Cassell, 1921);

Washington and the Hope of Peace (London, Glasgow, Melbourne & Auckland: Collins, 1922); republished as *Washington and the Riddle of Peace* (New York: Macmillan, 1922);

What H. G. Wells Thinks about "The Mind in the Making" (New York & London: Harper, 1922);

The Secret Places of the Heart (London, New York, Toronto & Melbourne: Cassell, 1922; New York: Macmillan, 1922);

The World, Its Debts, and the Rich Men: A Speech (London: Finer, 1922);

University of London Election: An Electoral Letter (London: Finer, 1922);

A Short History of the World (London, New York, Toronto & Melbourne: Cassell, 1922; New York: Macmillan, 1922);

Men Like Gods (London, New York, Toronto & Melbourne: Cassell, 1923; New York: Macmillan, 1923);

Socialism and the Scientific Motive (London: Cooperative Printing Society, 1923);

To the Electors of London University General Election, 1923, from H. G. Wells, B.Sc., London (London: Craig, 1923);

The Labour Ideal of Education (London: Craig, 1923);

The Story of a Great Schoolmaster: Being a Plain Account of the Life and Ideas of Sanderson of Oundle (London: Chatto & Windus, 1924; New York: Macmillan, 1924);

The Dream: A Novel (London: Cape, 1924; New York: Macmillan, 1924);

A Year of Prophesying (London: Unwin, 1924; New York: Macmillan, 1925);

Christina Alberta's Father (London: Cape, 1925; New York: Macmillan, 1925);

A Forecast of the World's Affairs (New York & London: Encyclopaedia Britannica, 1925);

The World of William Clissold: A Novel at a New Angle (3 volumes, London: Benn, 1926; 2 volumes, New York: Doran, 1926);

Mr. Belloc Objects to The Outline of History (London: Watts, 1926; New York: Doran, 1926);

Democracy under Revision: A Lecture Delivered at the Sorbonne, March 15th, 1927 (London: Leonard & Virginia Woolf, 1927; New York: Doran, 1927);

Meanwhile: The Picture of a Lady (London: Benn, 1927; New York: Doran, 1927);

Experiments on Animals: Views For and Against, by Wells and George Bernard Shaw (London: British Union for the Abolition of Vivisection, 1927);

Playing at Peace (London: National Council for the Prevention of War, 1927);

The Short Stories of H. G. Wells (London: Benn, 1927; Garden City, N.Y.: Doubleday, Doran, 1929); republished as *The Complete Short Stories* (London: Benn, 1966);

Wells's Social Anticipations, edited by H. W. Laidler (New York, 1927);

The Way the World Is Going: Guesses and Forecasts of the Years Ahead (London: Benn, 1928; Garden City, N.Y.: Doubleday, Doran, 1929);

The Open Conspiracy: Blue Prints for a World Revolution (London: Gollancz, 1928; Garden City, N.Y.: Doubleday, Doran, 1928; revised edition, London: Leonard & Virginia Woolf, 1930); revised again as *What Are We to Do with Our Lives?* (London: Heinemann, 1931; Garden City, N.Y.: Doubleday, Doran, 1931);

Mr. Blettsworthy on Rampole Island (London: Benn, 1928; Garden City, N.Y.: Doubleday, Doran, 1928);

The King Who Was a King: The Book of a Film (London: Benn, 1929); republished as *The King Who Was a King: An Unconventional Novel* (Garden City, N.Y.: Doubleday, Doran, 1929);

The Common Sense of World Peace: An Address Delivered to the Reichstag at Berlin, on Monday, April 15th, 1929 (London: Leonard & Virginia Woolf, 1929);

The Adventures of Tommy (London: Harrap, 1929; New York: Stokes, 1929);

Imperialism and the Open Conspiracy (London: Faber & Faber, 1929);

The Science of Life: A Summary of Contemporary Knowledge about Life and Its Possibilities, by Wells, Julian Huxley, and G. P. Wells (31 parts, London: Amalgamated Press, 1929–1930; 3 volumes, London: Amalgamated Press, 1930; 4 volumes, Garden City, N.Y.: Doubleday, Doran, 1931);

The Autocracy of Mr. Parham: His Remarkable Adventures in This Changing World (London: Heinemann, 1930; Garden City, N.Y.: Doubleday, Doran, 1930);

Points of View: A Series of Broadcast Addresses, by Wells and others (London: Allen & Unwin, 1930);

The Way to World Peace (London: Benn, 1930);

The Problem of the Troublesome Collaborator: An Account of Certain Difficulties in an Attempt to Produce a Work in Collaboration and of the Intervention of the Society of Authors Therein (Woking: Privately printed, 1930);

Settlement of the Trouble between Mr. Thring and Mr. Wells: A Footnote to The Problem of the Troublesome Collaborator (Woking: Privately printed, 1930);

Divorce as I See It, by Wells and others (London: Douglas, 1930);

The Work, Wealth and Happiness of Mankind (2 volumes, Garden City, N.Y.: Doubleday, Doran, 1931; 1 volume, London: Heinemann, 1932);

The New Russia: Eight Talks Broadcast, by Wells and others (London: Faber & Faber, 1931);

After Democracy: Addresses and Papers on the Present World Situation (London: Watts, 1932);

The Bulpington of Blup: Adventures, Poses, Stresses, Conflicts, and Disasters in a Contemporary Brain (London: Hutchinson, 1932; New York: Macmillan, 1933);

What Should Be Done–Now: A Memorandum on the World Situation (New York: Day, 1932);

The Scientific Romances (London: Gollancz, 1933);

Stories of Men and Women (London: Hutchinson, 1933);

The Shape of Things to Come: The Ultimate Revolution (London: Hutchinson, 1933; New York: Macmillan, 1933);

Experiment in Autobiography: Discoveries and Conclusions of a Very Ordinary Brain (Since 1866), 2 volumes (London: Gollancz & Cresset, 1934; New York: Macmillan, 1934);

Stalin-Wells Talk: The Verbatim Record and a Discussion by G. Bernard Shaw, H. G. Wells, J. M. Keynes, Ernst Toller and Others (London: New Statesman & Nation, 1934);

The New America: The New World (London: Cresset, 1935; New York: Macmillan, 1935);

Things to Come: A Film Story Based on the Material Contained in His History of the Future, "The Shape of Things to Come" (London: Cresset, 1935; New York: Macmillan, 1935);

The Anatomy of Frustration: A Modern Synthesis (London: Cresset, 1936; New York: Macmillan, 1936);

The Croquet Player: A Story (London: Chatto & Windus, 1936; New York: Viking, 1937);

The Idea of a World Encyclopedia: A Lecture Delivered at the Royal Institution, November 20th, 1936 (London: Leonard & Virginia Woolf, 1936);

The Man Who Could Work Miracles: A Film Story Based on the Material Contained in His Short Story (London: Cresset, 1936; New York: Macmillan, 1936);

Star Begotten: A Biological Fantasia (London: Chatto & Windus, 1937; New York: Viking, 1937);

The Informative Content of Education (London: British Association, 1937);

Brynhild (London: Methuen, 1937); republished as *Brynhild; or, The Show of Things* (New York: Scribners, 1937);

The Camford Visitation (London: Methuen, 1937);

The Brothers: A Story (London: Chatto & Windus, 1938; New York: Viking, 1938);

World Brain (London: Methuen, 1938; Garden City, N.Y.: Doubleday, Doran, 1938);

Apropos of Dolores (London: Cape, 1938; New York: Scribners, 1938);

The Holy Terror (London: M. Joseph, 1939; New York: Simon & Schuster, 1939);

Travels of a Republican in Search of Hot Water (Harmondsworth: Penguin, 1939);

The Fate of Homo Sapiens: An Unemotional Statement of the Things That Are Happening to Him Now and of the Immediate Possibilities Confronting Him (London: Secker & Warburg, 1939); republished as *The Fate of Man: An Unemotional Statement of the Things That Are Happening to Him Now and of the Immediate Possibilities Confronting Him* (New York: Alliance/Longmans, Green, 1939);

The New World Order, Whether It Is Attainable, How It Can Be Attained, and What Sort of World a World at Peace Will Have To Be (London: Secker & Warburg, 1940; New York: Knopf, 1940);

The Rights of Man; or, What Are We Fighting For? (Harmondsworth & New York: Penguin, 1940);

Babes in the Darkling Wood (London: Secker & Warburg, 1940; New York: Alliance, 1940);

The Common Sense of War and Peace: World Revolution or War Unending? (Harmondsworth & New York: Penguin, 1940);

All Aboard for Ararat (London: Secker & Warburg, 1940; New York: Alliance, 1941);

Two Hemispheres or One World? (N.p., 1940);

Guide to the New World: A Handbook of Constructive World Revolution (London: Gollancz, 1941);

You Can't Be Too Careful: A Sample of Life 1901–1951 (London: Secker & Warburg, 1941; New York: Putnam, 1942);

The Outlook for Homo Sapiens: An Amalgamation and Modernisation of Two Books, The Fate of Homo Sapiens and The New World Order, Published Severally in 1939 and 1940 (London: Secker & Warburg, 1942);

Modern Russian and English Revolutionaries: A Frank Exchange of Ideas between Commander Leu Uspensky, Soviet Writer, and H. G. Wells (London: Privately printed, 1942);

Science and the World-Mind (London: New Europe, 1942);

Phoenix: A Summary of the Inescapable Conditions of World Reorganisation (London: Secker & Warburg, 1942; Girard, Kans.: Haldeman-Julius, 1942);

A Thesis on the Quality of Illusion in the Continuity of Individual Life of the Higher Metazoa, with Particular Reference to the Species Homo Sapiens (London: Privately printed, 1942);

The Conquest of Time, by H. G. Wells: Written to Replace His First and Last Things (London: Watts, 1942);

The New Rights of Man (Girard, Kans.: Haldeman-Julius, 1942);

Crux Ansata: An Indictment of the Roman Catholic Church (Harmondsworth & New York: Penguin, 1943; New York: Agora, 1944);

The Mosley Outrage (London: Daily Worker, 1943);

'42 to '44: A Contemporary Memoir upon Human Behaviour during the Crisis of the World Revolution (London: Secker & Warburg, 1944);

Reshaping Man's Heritage: Biology in the Service of Man, by Wells, J. S. Huxley, and J. B. S. Haldane (London: Allen & Unwin, 1944);

The Happy Turning: A Dream of Life (London & Toronto: Heinemann, 1945);

Mind at the End of Its Tether (London: Heinemann, 1945);

Mind at the End of Its Tether and The Happy Turning: A Dream of Life (New York: Didier, 1946);

The Desert Daisy, edited by Gordon N. Ray (Urbana, Ill.: Beta Phi Mu, 1957);

Hoopdriver's Holiday, edited by Michael Timko (West Lafayette, Ind.: Purdue University Press, 1964);

The Wealth of Mr. Waddy: A Novel, edited by Harris Wilson (Carbondale: Southern Illinois University Press, 1969; London: Feffer & Simmons, 1969);

H. G. Wells in Love: Postscript to An Experiment in Autobiography, edited by G. P. Wells (London & Boston: Faber & Faber, 1984).

Editions: *The Works of H. G. Wells,* Atlantic Edition, 28 volumes, revised by Wells (London: Unwin, 1924; New York: Macmillan, 1924);

The Works of H. G. Wells, Essex Edition, 24 volumes (London: Benn, 1926–1927);

A Modern Utopia, edited by Mark R. Hillegas (Lincoln: University of Nebraska Press, 1967);

The Definitive 'Time Machine': A Critical Edition of H. G. Wells's Scientific Romance, edited by Harry M. Geduld (Bloomington & Indianapolis: Indiana University Press, 1987);

The Island of Doctor Moreau: A Variorum Text, edited by Robert M. Philmus (Athens & London: University of Georgia Press, 1993);

A Critical Edition of 'The War of the Worlds': H. G. Wells's Scientific Romance, edited by David Y. Hughes and Geduld (Bloomington & Indianapolis: Indiana University Press, 1993);

The War of the Worlds, edited by Hughes (New York: Oxford University Press, 1995);

The First Men in the Moon, edited by David Lake (New York: Oxford University Press, 1995).

Collections: *Journalism and Prophecy, 1893–1946,* edited by W. Warren Wagar (Boston: Houghton Mifflin, 1964);

Early Writings in Science and Science Fiction, edited by Robert M. Philmus and David Y. Hughes (Berkeley, Los Angeles & London: University of California Press, 1975);

H. G. Wells's Literary Criticism, edited by Patrick Parrinder and Philmus (Brighton: Harvester, 1980; Totowa, N.J.: Barnes & Noble, 1980);

The Man with a Nose and Other Uncollected Short Stories, edited by J. R. Hammond (London: Athlone Press, 1984).

OTHER: Amy Catherine Wells, *The Book of Catherine Wells,* edited, with an introduction, by H. G. Wells (London: Chatto & Windus, 1928; Garden City, N.Y.: Doubleday, Doran, 1928).

MOTION PICTURES: *Things to Come,* screenplay by

Wells and Lajos Biro, London Films, 1935; *The Man Who Could Work Miracles,* screenplay by Wells, 1936.

SELECTED PERIODICAL PUBLICATION–UNCOLLECTED: "Utopias," *Science-Fiction Studies,* 27 (July 1982): 117–121.

Fantastic literature has existed as long as literature itself; Renaissance thinkers wrote of utopias and imaginary voyages into space; Jonathan Swift employed the fantastic in *Gulliver's Travels* (1726) to satirize contemporary society; Brian W. Aldiss has claimed that Mary Wollstonecraft Shelley wrote the first science-fiction novel in 1818 with her *Frankenstein; or, The Modern Prometheus;* and through the Victorian era tales based on the fantastic or the scientific or a combination of both found a wide audience, as did the translated novels of Jules Verne. It was H. G. Wells, however, who more than any other author consolidated and transformed these types of writing into the nascent genre of the "scientific romance," in the process changing the nature of science fiction and modern fantasy in his wake. While the term "father of science fiction" has been variously ascribed to different figures by individual historians, Wells undoubtedly deserves to be one of the recipients of that title.

A popular author and a widely respected intellectual of his day, Wells remains a major canonical figure in modern science fiction. Though less influential in fantasy, he nonetheless contributed several well-wrought stories and a few novels to the subgenre. Images, stories, and themes in the well-known scientific romances of his early career–*The Time Machine* (1895), *The Island of Doctor Moreau* (1896), *The Invisible Man* (1897), and *The War of the Worlds* (1898)–have become part of the Western consciousness through movies and other media as well as the books themselves. Some of these works have not gone out of print since their first publication, and in *The Western Canon* (1995) Harold Bloom lists Wells's scientific romances as part of the Western literary heritage.

Such prominence did not come easily to Wells. Born in Bromley, Kent, on 21 September 1866, Wells was the third son of Joseph Wells, a gardener and shopkeeper, and Sarah Wells, a housekeeper. The family's lower-middle-class status was not helped by the fact that Wells's father preferred playing cricket to working as a shopkeeper; when he was injured in Wells's childhood, Wells's mother became the primary breadwinner, working as a housekeeper. While the young Wells inherited his mother's capacity for hard work, he did not share her religious nature. Wells later commented that he found religion of little use during a period of painful convalescence after

Cover for Wells's fourth book, in which the Time Traveller observes the future and the tragic results of human evolution

breaking his leg in 1874. What he did find useful, he said, was the opportunity to read voraciously at this time. In particular, his father was able to procure for him books on science, and Wells later identified his reading as a turning point in his life.

While the world began to open up for the young Wells, however, his parents had other plans. As a child he received a few years of formal education, but this was cut short by his family's need to live on a limited income. Instead, from 1879 to 1882 Wells was apprenticed in different trades, none of which interested him; in particular, he served two apprenticeships in drapers' shops, which he loathed. His parents, unable to provide much in the way of formal education to their son, believed they could best help him by having him learn a trade. Wells, however, wanted more.

Wells struggled to gain an education and finally succeeded in doing so, first starting at age sixteen at Midhurst Grammar School in Sussex, then two years later as a scholarship student from 1884 to 1887 at the Normal School of Science in South Kensington, where he studied the natural sciences, most significantly under the well-known proponent of evolution T. H. Huxley. Wells also became associated with the *Science Schools Journal* as a writer and editor. Wells published several stories in the magazine, including "A Tale of the Twentieth Century" (May 1887), about a high-speed train, and the essay-story "The Chronic Argonauts" (April–June 1888), the first version of the work that would eventually become *The Time Machine.*

In 1887 Wells and his cousin Isabel Mary Wells fell in love while he was living with her family as a student. He left the Normal School before completing his degree so that he could make money and the two could marry. However, his first job, teaching at Holt Academy in North Wales, turned out badly, and his poor health required him to resign. He recuperated from tuberculosis and kidney problems at his mother's home, then in 1888 began teaching at Henley House School in Kilburn, near London. Two years later, after much studying on his own for his examinations, he received his bachelor of science degree and began teaching for the University Correspondence College. In 1891 he and Isabel were married. Their relationship, though at first affectionate, was unsatisfying for Wells, and he seduced one of her friends only a few weeks after their wedding. He left Isabel at the end of 1893 and moved in with Amy Catherine Robbins, one of his students from the University Correspondence College. Wells divorced Isabel in 1895 and later that year married Jane, as he called Robbins.

In the early 1890s Wells also began to publish more widely, from his essay "Rediscovery of the Unique" (1891) in the *Fortnightly Magazine* to journalism and fiction for magazines such as *The Pall Mall Gazette* to science textbooks. He left teaching in 1891, never to return. Instead, early in his career as a writer Wells occupied what his friend George Gissing would label New Grub Street, a world of journalism, essays, and literary effort that enabled writers to make a living at writing provided they were steady producers. While Wells's constant writing again took a toll on his health at this time, he nonetheless became able to earn his living from his pen.

Not content to continue writing for periodicals, Wells turned his attention to books, and a good indication of how prolific he was at this time can be seen in the fact that in 1895 he published four books: *Select Conversations with an Uncle, Now Extinct, and Two Other Reminiscences,* a series of sketches; *The Time Machine: An Invention,* his fictional development of his earlier story about time travel; the contemporary fantasy novel *The Wonderful Visit;* and *The Stolen Bacillus, and Other Incidents,* a collection of stories.

Wells's first book-length scientific romance, *The Time Machine,* is one of his best-known novels. As is typical with Wells's fantasy and science fiction, he eases the reader into the fantastic or the marvelous by first realistically establishing a sense of a contemporary setting and of ordinary characters. The narrator who relates the story of the Time Traveller accomplishes this task in *The Time Machine.* Regardless of the incredible nature of the tale he passes on, he does so as if he were describing any after-dinner conversation.

Having created this narrative frame, Wells then quickly proceeds into the story of the Time Traveller, who has created a device that he believes can pass through the fourth dimension, time. This element of the story is part of the book's significance in the history of science fiction. There had been many time-travel narratives before *The Time Machine;* what Wells did differently was that he explained how time travel might be possible through pseudoscientific discussions, in contrast to such stories in previous fantasy. For example, the protagonist in Mark Twain's *A Connecticut Yankee in King Arthur's Court* (1889) travels to the Middle Ages from the nineteenth century as a result of a blow on the head. Definitions of fantasy and science fiction abound, and often distinctions between the two subgenres are difficult to delineate, but certainly the difference between the fantastic account of time travel in Twain and the scientific account in Wells indicates one distinction between fantasy and science fiction.

Following the introduction of the Time Traveller and his claims, Wells then relates the results of the character's experiments with traveling into the future. With his first effort he finds himself, according to the instruments on his time machine, in the year A.D. 802,701. Here he finds an apparently idyllic human race, whom he calls the Eloi. (Since he at first believes these people to be the lords of their world, the Traveller apparently chooses the word for *Lord* used by Christ on the cross.) Though the remnants of industrial civilization are to be found only in a deserted museum, the life of the Eloi is seemingly paradisiacal, free of disease and the need to work. However, he soon learns otherwise when he discovers another human race, the brutal underground Morlocks (as he calls them), who feed upon the effete Eloi. The Morlocks are afraid of fire, but the Eloi are so degenerate that they do not know how to make fire. The Time Traveller hypothesizes that, after thousands of years of social and genetic evolution, the human race has become two separate species. Moreover, he speculates that the division between the graceful but stupid Eloi and the violent but resourceful Morlocks has its origins in the division between owners and laborers of his own time. Ironically, it is the Morlocks, not the Eloi, who are the true masters of this time.

Reflecting his scientific education, especially the influence of Huxley, Wells repeatedly employs evolution as a motif in his scientific romances. Just as Charles Darwin and other scientists had demonstrated that, in biological terms, there was nothing particularly special about human beings and that species could and had become extinct, so in his early science fiction Wells does not treat the human race as the end of the evolutionary development of life on this planet. This is apparent in the Time Traveller's voyage into the future from 802,701. In this effort he goes even further into the future, only to find no traces of human life and only the most basic of life forms. Apparently in this far future humanity has become extinct. He pushes even further, thirty million years into the future, and finds no sign of life and the sun cooling. Despite these depressing scenarios, after his return again to his own time the Traveller resolves to go even further into the future. After his final departure, according to the narrator, the Time Traveller has not returned in three years.

A lesser-known work published in the same year as *The Time Machine*, the fantasy novel *The Wonderful Visit* shows a different side of Wells, whose few fantasy novels, like the fantasies of contemporary writers such as F. Anstey, tend to be satiric in tone. The plot concerns an angel who, after being mistakenly shot down, finds himself in modern-day England. (The story was probably inspired by John Ruskin's remark that if an angel were to visit the earth, he would likely be shot by a hunter.) The purity, if not naiveté, of the visitor from the Land of Dreams serves as a comic foil to the narrow-mindedness, selfishness, and venality of most of the mortals he encounters; his position in the novel as an outsider also allows him, and indirectly allows Wells, to question social mores and prejudices that strike him as illogical and unkind. Like many writers before and after him, Wells here uses the fantastic as a means of social critique. Later in his career he would largely dispense with such means, instead offering his social and political ideas directly.

Wells employs both fantasy and science fiction in his first collection, *The Stolen Bacillus.* The title story contains little speculation, but others are as imaginative as his book-length scientific romances and fantasy novels. For instance, in "The Remarkable Case of Davidson's Eyes," first published in the *Pall Mall Budget* (21 March 1895), a scientist named Davidson finds himself with the ability to see faraway places as a result of a laboratory accident. Davidson and his friends believe that what he sees are hallucinations, and he gradually recovers his normal sight. Years later, however, he learns that what he had seen was real, made possible through a brief experience of telescopic and X-ray vision.

Another story in *The Stolen Bacillus,* "The Flowering of the Strange Orchid" (first published in the 2 August 1894 *Pall Mall Budget*), concerns a botanist who procures a strange orchid that a collector found in a jungle. Unconcerned that its previous owner was found dead, with all the blood drained from his body, the botanist tends to his new acquisition. When it blooms, however, its odor renders him unconscious, and when his cousin searches for him after he fails to appear for tea, she finds him unconscious, the orchid's tendrils covering his body and slowly draining his blood. She frees him from the deadly plant and he survives. Other stories dealing with interactions between people and strange forms of life include "In the Avu Observatory" and "Aepyornis Island."

Wells juxtaposes a modern symbol of science and power with the traditional in another story in *The Stolen Bacillus,* "The Lord of the Dynamos." Here Azuma-zi, a visitor from "the mysterious East," sees in England a dynamo, which he proceeds to worship. His worship includes human sacrifice to the great machine before he is stopped. Like much of Wells's early science fiction, this story presents modern science and technology in an ambiguous light, implicitly asking whether these are really humanity's servants or its masters.

One of Wells's most interesting fantasy stories can also be found in *The Stolen Bacillus.* In "The Temptation of Harringay," which first appeared in the 9 February 1895 issue of the *St. James Gazette*, an artist's work begins to take on a life of its own thanks to the intercession of a diabolic figure inhabiting the painting who tempts the artist with the possibility of becoming a great artist. The artist rejects this pact, however, and struggles with the painting for control of the canvas, the artist frantically painting over it and the painting fantastically trying to assert its authority.

Largely on the basis of *The Time Machine,* which was a popular sensation during its 1895 serialization in William Ernest Henley's *New Review* and even more popular when published in book form, Wells became an overnight sensation and was compared to a host of other writers. As he notes in his autobiography, he was variously called the next Swift, the next Verne, the next Robert Louis Stevenson, the next Rudyard Kipling, the next J. M. Barrie, and so on. All he wanted, though, was to be recognized for his own sake, and his next few scientific romances would carve out a reputation for him that was distinctly Wellsian.

Wells's next book-length scientific romance, *The Island of Doctor Moreau,* also takes science and evolution as the basis for the story. Again readers are introduced to the story by an outside narrator, Prendick, who is shipwrecked and finds himself on an isolated island populated by the mysterious Doctor Moreau; his assistant, Montgomery; and a host of strange-looking

" I could not distinguish what he said."

The frontispiece for Wells's The Island of Doctor Moreau

and strange-acting men and women. To his horror Prendick eventually discovers that these strange people are not really people at all; instead, they are animals who have been transformed by Moreau's grisly experiments. Just as Victor Frankenstein sought to circumvent God by creating a human being from parts of corpses, so Moreau seeks to accelerate the aeons-long course of evolution by turning animals into grotesque approximations of human beings. (The 1996 film version of the novel, starring Marlon Brando and Val Kilmer, substitutes genetic engineering for vivisection as Moreau's means, consequently lessening the painful horror of the original despite all the extravagance of the movie.)

If Moreau is a twisted God figure in *The Island of Doctor Moreau*, the Beast Folk, as Prendick calls them, offer a savage satire of humanity and civilization similar to that found in *Gulliver's Travels* by Swift, one of Wells's favorite and most influential authors. The assorted Beast Folk live apart from Moreau's compound and ritualistically follow a warped version of biblical law that keeps them from harming their human masters and from reverting to their bestial natures. However, the bestial is so at odds with their more "civilized" qualities that reversion is inevitable, leading to the carnage from which Prendick barely escapes. When he returns to Europe, much like Gulliver after his experience with the Houyhnhnms, Prendick cannot help seeing his fellow human beings as essentially animals. Despite all our seeming civilization, *The Island of Doctor Moreau* tells readers, because of our evolutionary heritage we are more like the Beast Folk than we would care to admit.

While this novel was less well received than *The Time Machine,* Wells nonetheless was on his way up the literary ladder even as he was composing tales of devolution. In 1896 he and Jane moved again, to Worchester Park, the same year that the renowned literary agent J. B. Pinker began to represent him. Wells also began to make the acquaintance of other writers, among them Gissing, Arnold Bennett, Joseph Conrad, Henry James, and Ford Madox Hueffer (later Ford Madox Ford), with all of whom he corresponded a great deal.

In 1897 Wells published another collection, *The Plattner Story, and Others*, which was more widely and positively reviewed than *The Stolen Bacillus.* Again not all of its fifteen stories are fantastic, but the book contains several noteworthy shorter works of fantasy and science fiction. These include "Pollock and the Porroh Man," first published in the *New Budget* (23 May 1895); "Under the Knife," from the January 1896 *New Review;* "The Red Room," originally published in the *Idler* (March 1896); "The Plattner Story," first published in the April 1896 *New Review;* and "The Sea-Raiders," from the *Weekly Sun Literary Supplement* (6 December 1896).

"Pollock and the Porroh Man" involves the results of a conflict between Pollock, an Englishman, and an African witch doctor. Their clash is not only between colonizer and colonized but also between the modern scientific worldview and the witchcraft of the Porroh Man. Here magic is triumphant, as Pollock returns to England plagued by the effects of the native's curse.

The horror is more psychological in "The Red Room," one of Wells's rare ghost stories. The protagonist, dismissing stories about the haunted red room of Lorraine Castle, decides to prove that it is not. Nor is the room haunted by spirits, but as the man's light is extinguished and as he dimly sees what is really his own reflection in the room's massive mirrors, his bravado is replaced by fear, and in his panic he injures himself trying to escape. He concludes that the red room is haunted not by ghosts but by fear. Fear is also a major element in "The Sea-Raiders," a fairly conventional horror story about monstrous ocean creatures.

More typically Wellsian are "Under the Knife" and "The Plattner Story." In the first a man undergoing surgery believes he is not anesthetized but dead, and while unconscious he dreams that his spirit leaves his body and flies over London and then into space. "The Plattner Story" similarly describes astral projection, this time into another dimension as a result of a laboratory accident.

Ambiguity about the consequences of scientific exploration is also present in Wells's next book-length scientific romance, *The Invisible Man: A Grotesque Romance.* In this novel of another scientist who goes too far and whose quest for knowledge becomes warped into a thirst for power, Wells presents a highly imaginative approach toward an old concept in fantasy. As with *The Time Machine,* the novelty lies not in the originality of the idea but in the way in which the idea is explained–in quasi-scientific terms.

The book relates the story of Dr. Griffin, a scientist who cares more about his work than other people and whose misanthropic tendencies prove dangerous when he discovers how to turn himself invisible. The novel uses the concept of an invisible man to great effect in an exciting story. Also, the novel effectively uses invisibility as a metaphor for anarchy, as the Invisible Man's actions exemplify. Conrad was a great admirer of *The Invisible Man,* and his dedication to Wells of his novel *The Secret Agent* (1907), about the activities of an anarchist who moves invisibly through the streets of London, is surely related.

Wells's next scientific romance, *The War of the Worlds,* is also one of his best known. Again, evolution and extrapolation play central roles. The premise is that not only does intelligent life exist on Mars, but that the Martians are superior in intelligence and weaponry to the inhabitants of Earth. Their advanced evolution, once the Martians are finally revealed (Wells was a master of suspense and plotting in his scientific romances), is demonstrated by their enlarged brains and atrophied limbs–hence their reliance on machines. (The Martians' anatomy is based on Wells's speculations about how human beings might evolve in his 1893 essay "The Man of the Year Million.") The plot basically hinges on whether human beings can survive an attack from Martians who seem superior in every way.

The fundamental concept of *The War of the Worlds* again was hardly new. By the time Wells published the book dozens of future-war stories, from *The Battle of Dorking* to the works of George Griffith and William Le Queux, had been read by audiences at first as cautionary tales and later for their vivid scenes of mass destruction. *The War of the Worlds* is also a future-war novel with many scenes of mass destruction; Wells's innovation here consisted of the fact that this was one of the first, if not the first, such works to describe an invasion by beings from another planet. Like *The Time Machine* with its suggestion that the extinction of the human race is possible if not in fact likely, the result is a questioning of humanity's confidence in its supremacy. Wells reinforces this theme with the conclusion of the novel: while some people have fought valiantly against the Martian onslaught, it is not human ingenuity or power that defeats the aliens, but rather human microbes.

Wells was ill once more in 1898, with exhaustion from overwork a large part of the problem. He recuperated on the southern English coast, where he visited James and Conrad, and later in the year he traveled to Italy with Gissing–his first trip outside of England. Still, Wells's illness and his travels did not prevent him from publishing two more books of fiction in 1899: *When the Sleeper Wakes,* a scientific romance, and the collection *Tales of Space and Time.* He also completed one of his first realistic novels, *Love and Mr. Lewisham* (1900).

Like many writers of the nineteenth century and the early twentieth century, Wells saw "romances" and "novels" as two distinct types of fiction. Though both could be book-length works that modern readers would simply identify as novels, for Wells and many of his contemporaries a novel was a realistic work with a contemporary setting, whereas a romance included a setting different from one's own time and place and could involve unrealistic elements. Hence Nathaniel Hawthorne, a writer Wells had admired from his childhood, called *The Scarlet Letter* (1850) a romance, and hence Wells labeled books such as *The Time Machine* and *The War of the Worlds* romances, reserving the term *novel* for more-conventional efforts. Accompanying these ideas was a widespread belief that, though the romance was not inherently disreputable, it was not as important as the novel. Wells tended to accept this valuation, especially while under the influence of Henry James.

Still, Wells continued to explore the particular possibilities of scientific romance. In *When the Sleeper Wakes: A Story of the Years to Come* he reflects on a frequent theme in his nonfiction: the need for humanity to educate itself to adapt to social and technological change. The future the novel depicts is ruled by wealthy capitalists who exploit the workers. Clearly writing about a growing chasm between haves and have-nots, Wells critiques such an unequal society using a common fantastic device, the character who "sleeps" in suspended animation through a great amount of time and awakes in a different world an outsider. He finds that he now owns half the planet, but disaster results when he upsets the status quo in the future world he enters by instigating a socialist revolution.

As the title indicates, the stories in *Tales of Space and Time* are all science fiction or fantasy. For instance, "The Crystal Egg" is about an antique dealer who finds in his shop a marvelous egg that allows him to see Mars and, in return, may allow inhabitants of Mars to see his shop. The most powerful story in the collection is "The Star." Its initial publication in the Christmas 1897 issue of *Graphic* was appropriate, for the story specifically alludes to the star of Bethlehem that plays such an important role in the Christmas story. The story, ironically beginning on New Year's Day, starts with the collision of a vast moving mass with Neptune. The mass takes the planet with it on its path to the sun, and astronomers compute that Earth in within the new star's trajectory. Although Earth is finally spared, the satellite causes all kinds of devastation in its wake, including tidal waves, floods, earthquakes, and volcanoes. Also, as the narrator says, the Sun has become larger with the addition of the great mass, days are warmer than before, and the Moon has been reduced in size. A final ironic twist is added by the fact that the Martians, who were not in the missile's path, watched the possible destruction of Earth as objectively as humans had watched the possible destruction of Neptune. As in *The Time Machine* and much naturalistic fiction of the period, Wells suggests that human beings, in contrast to the message of the biblical story of Christmas, have no special status in an indifferent universe of physical law.

Wells deals explicitly with the human evolutionary heritage in "A Story of the Stone Age." However, the story is more convincing in its archaeological details than in its plot or characterization. In 1955 William Golding reacted against this story and Wells's best-selling *The Outline of History* (1919–1920) in *The Inheritors*, which takes a more sympathetic view of those outstripped by Homo sapiens than Wells's. A companion piece to "A Story of the Stone Age" in *Tales of Space and Time* is "A Story of the Days to Come." Set in a twenty-second-century world of high-rise buildings, hypnotists who can help remove unpleasant ideas and undesirable impulses, and a multimedia culture, the story concerns a man named Mwres (Morris) who is worried about his daughter. She wants to marry for love, which consigns her and her lover to a menial existence in the lower levels, figuratively and literally, of their society. Thus, "A Story of the Days to Come" could be seen as a possible intermediate step on the way to the division between the Eloi and the Morlocks in *The Time Machine*. Also, the Fritz Lang film *Metropolis* (1926) bears a strong resemblance to this story.

The final story in *Tales of Space and Time* is "The Man Who Could Work Miracles." Here the plot is fantastic and the tone is comic. A man who denies the possibility of miracles finds himself with the ability to perform them. After a handful of such miracles, including helping a minister deal with drunkards in his neighborhood, the minister asks him to stop the rotation of the earth so that time can stop and more problems can be solved sooner. Fortunately, the man and the minister are made to realize the disastrous geological consequence of stopping the earth's rotation–among other things, gravity would cease and people would go flying off into space–and so the man undoes all his miracles and everything returns to normal. The story, with some changes, was made into a movie in 1936, with Wells contributing to its production.

Another lesser-known but nonetheless enjoyable scientific romance by Wells, *The First Men in the Moon* (1901), is the last of his early science-fiction novels. In it a scientist named Cavor, along with his friend Bedford (who narrates the story), travel to the moon in a vehicle made of a gravity-defying metal called Cavorite. Again, the concept was not new. In fact, stories of voyages to the moon had been a staple of fantastic literature in Europe for centuries, written even by such diverse figures as Johannes Kepler and Cyrano de Bergerac. More recently, the French writer Jules Verne had written of human explorations of the moon. While Wells's antigravity metal is hardly more convincing than Verne's description of human beings shot to the moon using a gigantic gun, both authors nevertheless attempted to provide at least a somewhat plausible account of how their explorers manage to make it to the lunar surface.

The First Men in the Moon also requires a suspension of disbelief from readers because of its descriptions of the lunar flora and fauna that exist in the moon's slight atmosphere. However, Wells's depictions of the weightlessness of space travel and movement in the reduced gravity of the moon are remarkably accurate, an impressive feat of the scientific imagination considering that no one would experience these sensations in Wells's lifetime.

Readers can also enjoy the variegated lunar society that Wells creates in this book. The hierarchically arranged Selenites, the inhabitants of the moon's interior (the temperature extremes of the surface are too great to sustain all but the most tenuous of life), fulfill different roles much like ants in a colony. They are ruled by the Grand Lunar, like the Martians in *The War of the Worlds* another alien being with a highly developed brain and an underdeveloped body. The Grand Lunar is interested in the two human explorers at first, but his motives are clearly revealed as malevolent as the novel progresses, and Bedford manages to escape.

Wells's literary efforts after this remarkable sequence of scientific romances show a change in direction. Under James's influence, Wells began to devote more of his fictional efforts to contemporary novels of manners as opposed to science fiction or fantasy. Also, as Wells's social and political ideas continued to develop, he gradually moved away from the rather pessimistic portrayal of humanity's place in the universe to a more progressive stance. Thus he began to devote more of his time to nonfiction books on social and political issues and to novels dealing with contemporary issues directly rather than through the lens of fantasy or science fiction. Consequently, Wells tended to dismiss the early scientific romances as products of his youth.

By the beginning of the twentieth century Wells was firmly established as one of the most successful authors of the day, both critically and financially. He built his next home, Spade House in Sandgate, Kent, to his specifications in 1900, and the following year he and Jane had their first child, George Philip, called G. P. or "Gip." (A second son, Frank Richard, was born in 1903.) In 1901 he also published *Anticipations of the Reaction of Mechanical and Scientific Progress upon Human Life and Thought* (1901), his first major nonfiction work of a type he became known for, and began a brief association with the socialist Fabian Society. His circle of friends grew to include writers and thinkers such as Bernard Shaw (whom Wells had met when both were drama critics in the 1890s); Violet Paget, who wrote under the pseudonym Vernon Lee; and socialists Sidney and Beatrice Webb.

Another fantasy novel, *The Sea Lady: A Tissue of Moonshine,* appeared in 1902. As in *The Wonderful Visit,* this work uses a marvelous outsider–in this case a mermaid–to satirize Victorian conventionality. Also, the tone in both novels is largely comic, as opposed to the seriousness that governs most of the scientific romances of the same period, including most of those that compose *Twelve Stories and a Dream* (1903), his last original collection of short fiction.

As in his previous collections, the stories here fall into different categories. Some are stories of monsters, such as the beasts encountered by a group of medieval knights in "The Valley of the Spiders"; others are stories of invention, such as "Filmer," about flight; and "The New Accelerator," about a stimulant for the nervous system that, instead of helping people adapt to the pace of modern life as its inventor had hoped, makes people capable of performing at more than double their normal speed. Another characteristically Wellsian work in *Twelve Stories and a Dream* is "A Dream of Armageddon," about a man's apocalyptic vision of the future.

Two stories deal with the effects of interactions between "civilized" and "uncivilized" worlds. "Jimmy Goggles the God," first published in the *Graphic* (December 1898), concerns a man who lands on an island in the South Pacific in his diving gear. His gear helps him convince the natives that he is a god, but then he has to deal with the discomfort of wearing his equipment in a tropical environment. Also comic in tone, "The Truth about Pyecroft," from the April 1903 issue of *The Strand Magazine,* is about an overweight Englishman who tries an Indian elixir in hopes of losing weight. Lose weight he does; however, he does not decrease any in size. Moreover, he actually becomes buoyant and must wear lead undergarments to keep from floating up to the ceiling. As with "Pollock and the Porroh Man" in *The Plattner Story, and Others,* some critics have viewed such stories as metaphors for the consequences of colonialism on the colonizers.

Like much of *The First Men in the Moon,* the mood of Wells's scientific romance *The Food of the Gods, and How It Came to Earth* (1904) is comic as well. Here a marvelous food, the result of experimentation with improving growth, turns a group of people into giants, who are then attacked by "normal" human beings. The message is obvious, and the book is in some ways closer to Wells's more didactic fictions than his earlier scientific romances.

Despite the proliferation of such futurological works as *Anticipations, The Discovery of the Future* (1902), and *Mankind in the Making* (1903), occasionally Wells used fantastic or science-fiction methods to explore his scientific, social, and political ideas. One prominent example is his *A Modern Utopia* (1905). Influenced by utopian literature ranging from Plato's *Republic* through Sir Thomas More's *Utopia* (1516) to recent examples such as Samuel Butler's *Erewhon* (1872) and William Morris's *News from Nowhere* (1890), Wells still tries to distinguish his work from earlier efforts as a distinctly modern utopia. Interestingly, the novel is more fantastic than science fiction in nature–the narrator and his companion discover the utopian world they then explore as if it were in an alternate dimension–but this modern utopia is definitely organized on scientific principles.

Most important for its status as a modern utopia, this world is predicated not on an ideal order that, once achieved, is always to be maintained, but rather on the realization, fostered by evolutionary theory, that change is not only inevitable but desirable. Nor is the utopia the narrator and his friend investigate perfect in every respect. Indeed, its supporters admit, they are constantly working toward improvement basically through trial and error–in other words, through the scientific method of experimentation.

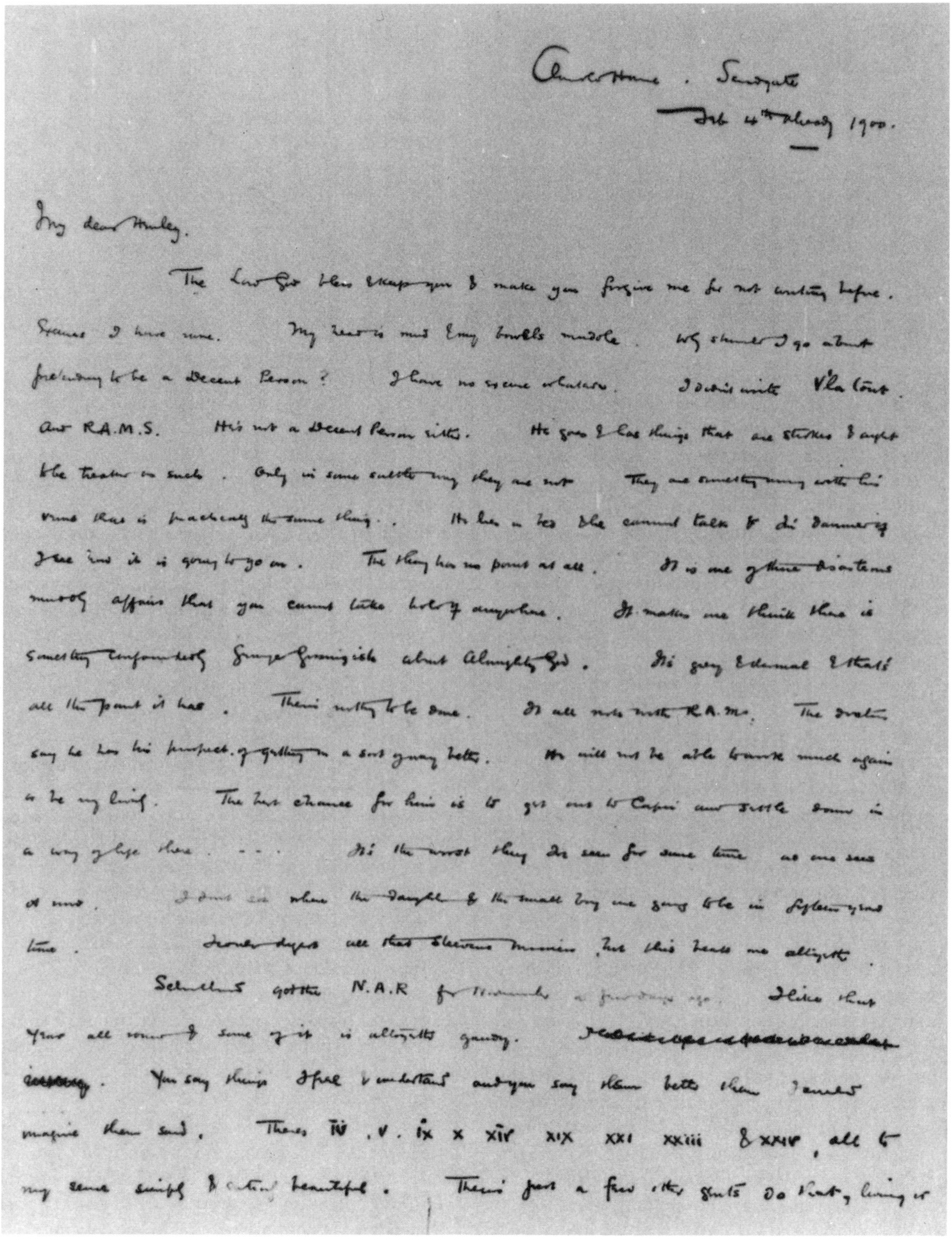

[illegible] . Sandgate
Feb 4th already 1900.

My dear Henley.

The Lord God bless & keep you & make you forgive me for not writing before. Excuses I have none. My head is mud & my bowels muddle. Why should I go about pretending to be a Decent Person? I have no excuse whatever. I didn't write. V'la tout. And R.A.M.S. He's not a Decent Person either. He goes & has things that are strokes & ought to be treated as such. Only in some subtle way they are not. They are something wrong with his veins that is practically the same thing.. He has a leg[illegible] he cannot talk & it's damning to see how it is going to go on. The thing has no point at all. It is one of these [illegible] muddy affairs that you cannot take hold of anywhere. It makes me think there is something confoundedly [illegible] about Almighty God. [illegible] all the point it has. There's nothing to be done. It all rests with R.A.M.S. The doctors say he has his prospect of getting in a sort of way better. He will not be able to write much again in his life. The best chance for him is to get out to Capri and settle down in a way of life there. . . . It's the worst thing I've seen for some time as one sees it now. I don't see where the daughter & the small boy are going to be in fifteen years time. [illegible] all that Stevenson business, but this beats me altogether.

Schwellers got the N.A.R for November a few days ago. I like that Year all round & some of it is altogether gaudy. [illegible] You say things I feel & understand and you say them better than James [illegible] them said. There's IV. V. IX X XIV XIX XXI XXIII & XXIV, all to my sense simply & entirely beautiful. There's just a few other spots [illegible], living or

A letter from Wells to W. E. Henley, 4 February 1900 (MA 1617, Pierpont Morgan Library)

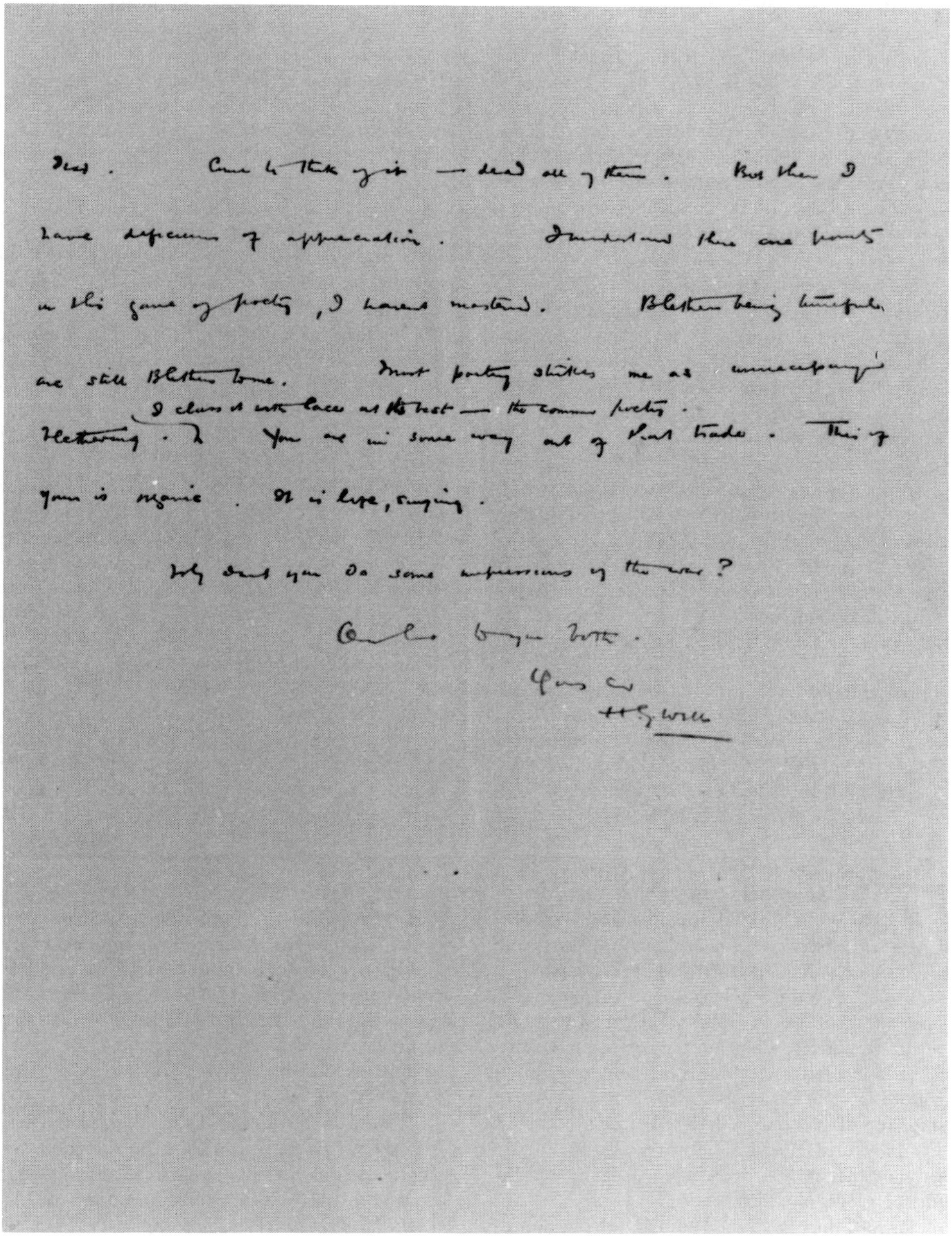

dead. Come to think of it — dead all of them. But then I have defects of appreciation. I understand there are points in this game of poetry, I haven't mastered. Blether being beautiful are still Blether to me. Most poetry strikes me as unnecessary blethering. I class it with lace at the best — the common poetry. You are in some way out of that trade. This of yours is organic. It is life, singing.

Why don't you do some impressions of the war?

Our love to you both.

Yours ever
H.G.Wells

The society described in *A Modern Utopia* has other intriguing characteristics as well. Just as Plato's Republic is a meritocracy rather than a democracy, so Wells's modern utopia is governed by an intellectual elite known as the Samurai, even while it supports a considerable degree of democracy and personal freedom. Part of this freedom extends to sexuality. Here and in many of his realistic novels, Wells speaks out against puritanical limitations on sexual freedom, just as he rejected conventional sexual mores in his own life–as his many extramarital affairs would demonstrate.

His next scientific romance, *In the Days of the Comet* (1906), is like *The Food of the Gods* in serving as a vehicle not only for speculative ideas about the future but also about contemporary issues such as socialism and free love. The story concerns the changes wrought on earth after it passes through a cloud of gases created by a passing comet. Another Wells novel that resembles his realistic works while including scientific speculation is *Tono-Bungay* (1908), which is properly identified as a novel of science rather than a science-fiction novel.

By the middle of the 1900s Wells was extremely well known not only as a writer of fiction but as an important thinker on social and political matters. He broke with the Fabians, finding their brand of socialism at once elitist and too provincial, too limited for the broad vision of the future that Wells had once applied solely to his fiction. In 1906 he visited the United States for the first time on a successful speaking engagement, during which he met President Theodore Roosevelt, and published social critiques and prophetic books such as *The Future in America: A Search after Realities* (1906), *Socialism and the Family* (1906), *New Worlds for Old* (1908), and *First and Last Things: A Confession of Faith and Rule of Life* (1908).

In 1908 Wells also published his novel *The War in the Air, and Particularly How Mr. Bert Smallways Fared While It Lasted.* While in setting and characterization it has more in common with his novels of manners of this period, the book also bears some similarity to his scientific romances, albeit on a more modest scale. Here the speculative element is the common future-war motif, and Wells realistically depicts the use of airships in war less than a decade before it became a reality in World War I. The novel also includes a worldwide plague that over time has decimated humanity to a level of barbarism.

Toward the end of the 1900s and the beginning of the 1910s the Wellses moved twice, first to Church Row in Hampstead, then to Easton Glebe in Essex. At this time Wells shocked readers with his novel *Ann Veronica: A Modern Love Story* (1909), which depicted a young woman's unconventional romantic career. At the same time those familiar with Wells's personal life would have been shocked by his own romantic career. With Jane's tacit if unhappy consent, Wells had been having affairs as early as the mid 1900s, and in the late 1900s and early 1910s he became involved with other women–most notably with Amber Reeves, with whom he had a daughter, Anne Jane, in 1909, and with the young writer Rebecca West, who bore their son, Anthony, in 1914.

Most of the twenty-nine stories in *The Country of the Blind, and Other Stories* (1911) had been published in earlier collections, but a few here appeared in book form for the first time. Two of these, "The Country of the Blind" and "The Door in the Wall," are among Wells's best works of short fiction. The title story of the collection is an ironic twist on the saying, "In the country of the blind the one-eyed man is king." Here an explorer comes upon an isolated valley in the Andes whose inhabitants had been rendered blind generations before by illness. Rather than being treated like a king, the protagonist is considered mad for his talk of a world of sight. Having fallen in love with a native of the valley, he considers having his eyes removed, but after seeing a glorious sunrise he instead decides to leave the valley. The images and themes of the story have offered critics ample material for interpretation.

Another thematically rich story, "The Door in the Wall," is about a man who, after entering into a wonderful magical world through a door he sees in a wall when he is five, periodically sees the door again throughout his life. However, despite his intentions to return, he gets so caught up in his life that he never does. The piece was the title story of an illustrated collection published in 1911 in the United States and in 1915 in Britain.

In 1914 Wells visited Russia for the first of many visits. Like many intellectuals sympathetic to socialist ideas, Wells watched developments in Russia and the new Soviet Union with interest, and later he would meet with leaders such as Lenin and Joseph Stalin. However, unlike many socialists, Wells was quick to observe how the Soviet experiment with communism fell short of his evolving vision of an efficient and humane world state.

Wells's optimism was challenged by the outbreak of war in Europe in 1914. In a book on the subject he tried to view the Great War, during which he worked for the Ministry of Information, as *The War That Will End War* (1914), hoping that the war would precipitate a series of political and social upheavals that would lead to the ideal world he envisioned. In this case his prophesies were wrong, and World War II, which thus turned the Great War into World War I, shook his view of human progress even further at the end of his life. Moreover, World War II saw the realization of a threat he describes in his novel *The World Set Free: A Story of Mankind* (1914): the development of atomic weapons. Written before the start of World War I, the novel presents a world war ended by the use of atomic weapons, after which a more enlightened world order emerges.

Another significant development in Wells's career took place in the mid 1910s. Wells had been a friend of Henry James for several years, but it had always been an unusual relationship. As he often did with younger writers, James assumed the role of the master, encouraging writers in their efforts but also pointing out what he thought were weaknesses in their work. For James, fiction was an art, not to be compromised by either cheapening it for entertainment purposes or by using it for nonartistic purposes. While James largely admired Wells's scientific romances for their craftsmanship, he found fault with many of his novels of manners, openly criticizing them in letters to Wells for their poor characterization and loose structure. He also criticized Wells for using the novel as a vehicle for ideas rather than, as James advocated, as an art sufficient unto itself.

Naturally, Wells began to take offense at what he increasingly believed was James's condescending attitude. (The difference between the two men in age and social class may also have contributed to Wells's growing discomfort.) Finally, Wells decisively ended their relationship with his mean-spirited attack on James in his novel *Boon* (1915), in which he depicts James the writer as a hippopotamus trying to pick up a peanut in a corner. Many critics found the attack to be in poor taste, which damaged Wells's career somewhat in certain literary circles.

Also, Wells was inadvertently caught in the middle of a shift in modern literature. Essentially an unreconstructed Victorian in style apart from his scientific romances, Wells was thus viewed by emerging writers such as James Joyce and Virginia Woolf as old-fashioned. An indication of this split can be seen in the fact that Wells supported Joyce's *A Portrait of the Artist as a Young Man* (1915) but refused to do the same for Joyce's much more experimental *Ulysses* (1922). By the time the modernist revolution had been accepted in literary circles and university English departments, with James hailed as a forerunner, Wells found his fiction granted secondary status within the Western canon.

Nor was his reputation aided by his association in America with science fiction, a genre which editor Hugo Gernsback named in 1929. Science fiction was then published almost exclusively in cheaply produced pulp magazines such as *Amazing Stories,* which Gernsback founded in 1926 with reprints of works by writers such as Edgar Allan Poe, Verne, and Wells, most of whose early scientific romances were reprinted in the magazine. The pulp magazines were widely viewed as popular if not vulgar literature, and Wells's reputation in part suffered by association.

However, Wells still maintained an extremely wide readership in the 1910s and 1920s. His war novel *Mr. Britling Sees It Through* (1916) was one of the most popular books during World War I, and after the war Wells was one of the most vocal proponents for the creation of the League of Nations. He also produced an unlikely best-seller with *The Outline of History,* a massive work spanning the rise of humanity from prehistoric times to the present.

At the end of the 1910s Wells entered into what many scholars have labeled his "religious" phase, a period in which he was preoccupied with his own unorthodox personal theology. Books such as *God the Invisible King* (1917), *The Soul of a Bishop* (1917), and the revised second edition of *First and Last Things* (1917) are all manifestations of this preoccupation, as is his novel *The Undying Fire* (1919), a contemporary retelling of the biblical book of Job. The God in which the beleaguered headmaster finds his consolation, however, is not the traditional deity of Christianity but a God of Progress.

During the 1920s Wells was active not only as a writer–both with new books and with the publication of the twenty-six-volume Atlantic Edition of his collected works to date–but also as a spokesman for a certain brand of progressivism. He visited Russia again and met with Lenin and Maksim Gorky; he attended the World Disarmament Conference in Washington, D.C.; he spoke frequently on the need of an "Open Conspiracy" of leaders to work toward an enlightened world state; he became associated with the writers' organization PEN, whose president he became in 1933; he began to give radio addresses on the BBC on political matters in 1928; and he ran unsuccessfully for Parliament.

The 1920s were also a decade of great personal changes for Wells. He had affairs with women such as birth-control advocate Margaret Sanger and Odette Kuhn; he also ended his relationship with Rebecca West. He began to spend his winters in the Mediterranean, and he moved once again, to Chiltern Court in London. In October 1927 his wife died, and the following year Wells published a collection of her writings, *The Book of Catherine Wells.*

Men Like Gods (1923) and *The Dream* (1924) are two of Wells's few speculative works from the 1920s. *Men Like Gods* is another utopian novel, but it is far different from his earlier *A Modern Utopia.* In the far future humanity has been superseded by a race of superior human beings, who live peacefully in a world lacking an established political structure. The book is

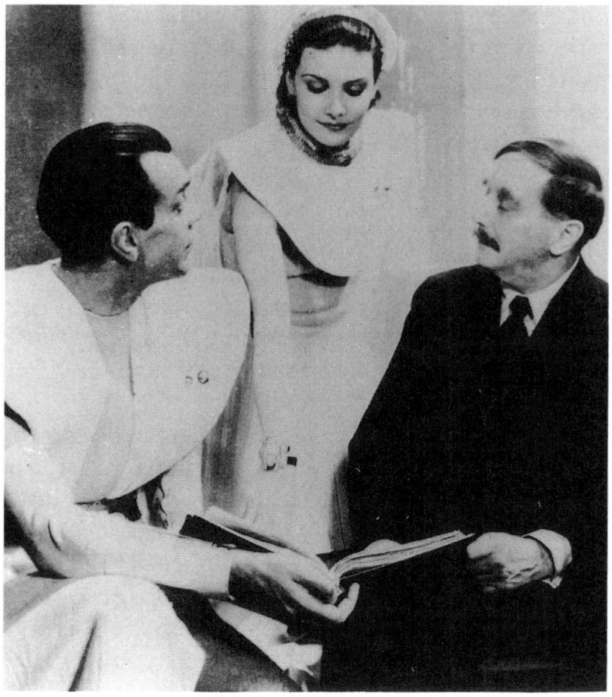

Raymond Massey, Margaretta Scott, and Wells on the set of Things to Come

both a warning–humanity declined because it was unable to use its science wisely–and a visionary image of what the human race may become. *The Dream* is mostly realistic and very much in the mode of the Wellsian novel of ideas, although it uses fantastic elements in a minor way.

The Short Stories of H. G. Wells (1927), massive at 1,150 pages though not complete, collected most of Wells's shorter fiction, including *The Time Machine; The Stolen Bacillus; The Plattner Story, and Others; Tales of Space and Time; Twelve Stories and a Dream;* five stories from *The Country of the Blind;* one of two stories included in *Boon;* and four uncollected stories. One of these uncollected stories, "The Land Ironclads," is noteworthy for predicting the use of tanks in warfare. Another, "The Grisly Folk," returns to prehistoric times with a story of the ascendancy of Homo sapiens over Neanderthals. The title of "The Inexperienced Ghost" provides a good indication of its story, and "The Story of the Last Trump" is about how the trumpet that will signal Judgment Day is misplaced.

Like *The Dream,* both *Mr. Blettsworthy on Rampole Island* (1928) and *The Autocracy of Mr. Parham* (1930) are essentially realistic works that nonetheless contain some fantastic features. In *Mr. Blettsworthy on Rampole Island* a shipwrecked man tries in vain to convince the natives he encounters to forsake their superstitious beliefs in favor of his scientific worldview; at the end he discovers that he has hallucinated the entire experience. *The Autocracy of Mr. Parham* is a political novel in which the protagonist is possessed by a sort of spirit that drives him to seek power.

The Shape of Things to Come (1933), like *A Modern Utopia* and *Men Like Gods,* is strongly utopian in nature. Unlike *A Modern Utopia,* it presents its idealistic picture of how human society could be not as a dreamlike possibility but as likely for humanity in the future. Just as Wells in his many nonfiction works proposed that a radical change in human institutions was necessary for true progress, so here he presents the desired results from such a change. (However, the advanced society he depicts in this future history was also shaped by warfare and disease.) The novel was the basis for the important science-fiction film *Things to Come* (1935), produced by Alexander Korda and directed by William Cameron Menzies with considerable input from Wells.

In 1934 Wells published *Experiment in Autobiography,* which he modestly and inaccurately subtitled "Discoveries and Conclusions of a Very Ordinary Brain (Since 1866)." Extremely readable, it gives his perspective on his various concerns as well as on his relationships with and reactions to contemporary writers. That same year he met with Stalin in the Soviet Union and President Franklin D. Roosevelt in the United States.

America was also where Orson Welles's radio broadcast of *The War of the Worlds* caused widespread panic in October 1938. Ingeniously adapting the novel as a series of radio news bulletins interrupting a musical program, Welles and Howard Koch transposed the Martian invasion from the suburbs of London to the East Coast. Apparently many listeners had both missed the opening announcement that the program was an adaptation of Wells's *The War of the Worlds* and had not read the book, which was not filmed for the first time until 1953.

Wells published little science fiction or fantasy toward the end of his career, devoting his energies instead to political and social matters in his fiction and nonfiction. However, *The Croquet Player* (1936), *Star Begotten: A Biological Fantasia* (1937), *The Camford Visitation* (1937), *The Holy Terror* (1939), and *All Aboard for Ararat* (1940) are all more or less works of speculative fiction. *The Croquet Player* includes an evolutionary allegory within its realistic story. *Star Begotten* is more fantastic, involving the irradiation of the earth, perhaps by Martians who hope to foster an evolutionary leap. In *The Camford Visitation* a bodiless voice haunts a university. *The Holy Terror,* inspired no doubt by contemporary dictatorships, is a political science-fiction novel set in the near future. Like *The Undying Fire,* in *All Aboard for Ararat* Wells again offers a contemporary retelling of a biblical narrative, this time the story of

Noah and the flood. Apparently forgetting his promise to the original Noah never again to flood the earth, God tells Noah Lammock that he is again disappointed with the human race and that he should prepare an ark. Here, though, Noah argues with God about how their plans should be carried out, insisting that the human race should decide its own destiny.

Toward the end of his life Wells also was disappointed with the human race and grew somewhat bitter, his long-standing optimism especially challenged by World War II. Also, his health was beginning to fail. Though he continued to speak on behalf of world peace and the vision of a humanistic world state, his outlook became darker. He lived to see the end of the war but also the beginning of the atomic age he had predicted as early as 1914 in *The World Set Free*. After a lengthy illness, he died on 13 August 1946, a month before his eightieth birthday.

During his lifetime Wells was variously known as a novelist, a social critic, and a political reformer. His legacy, however, has narrowed to his role as a writer of fiction closer to that of Charles Dickens than to that of Henry James or Joseph Conrad, and especially to his unavoidable impact on the development of modern science fiction. His influence is evident in virtually every aspect of the genre, from film adaptations of his work to borrowed concepts in science fiction since his time to his broader themes of evolution and devolution, progress and decline. Interest in Wells's work has grown increasingly since his death, both among scholars of fantasy and science fiction and among literary critics and historians in general, and his reputation will likely continue to rise in years to come.

Letters:

Henry James and H. G. Wells: A Record of Their Friendship, Their Debate on the Art of Fiction and Their Quarrel, edited by Leon Edel and Gordon N. Ray (Urbana: University of Illinois Press, 1958; London: Hart-Davis, 1958);

Arnold Bennett and H. G. Wells: A Record of a Personal and Literary Friendship, edited by Harris Wilson (Urbana: University of Illinois Press, 1960; London: Hart-Davis, 1960);

George Gissing and H. G. Wells: Their Friendship and Correspondence, edited by Royal A. Gettmann (Urbana: University of Illinois Press, 1961; London: Hart-Davis, 1961);

The Correspondence of H. G. Wells, 4 volumes, edited by David C. Smith (London: Pickering & Chatto, 1997).

Bibliographies:

Fred A. Chappell, *Bibliography of H. G. Wells* (Chicago: Covici-McGee, 1924);

Geoffrey H. Wells, *The Works of H. G. Wells, 1887–1925: A Bibliography, Dictionary and Subject-Index* (London: Routledge, 1926);

Gordon N. Ray, ed., *The H. G. Wells Papers at the University of Illinois* (Urbana: University of Illinois Press, 1958);

J. R. Hammond, *Herbert George Wells: An Annotated Bibliography of His Works* (New York & London: Garland, 1977);

H. G. Wells Society, *H. G. Wells: A Comprehensive Bibliography*, fourth edition (London: H. G. Wells Society, 1986);

William J. Scheick and J. Randolph Cox, *H. G. Wells: A Reference Guide* (Boston: G. K. Hall, 1988).

Biographies:

Geoffrey West, *H. G. Wells: A Sketch for a Portrait* (London: Howe, 1930);

Vincent Brome, *H. G. Wells: A Biography* (London & New York: Longmans, Green, 1951);

Julius Kagarlitsky, *The Life and Thought of H. G. Wells* (London: Sedgwick & Jackson, 1966);

Lovat Dickson, *H. G. Wells: His Turbulent Life and Times* (London: Macmillan, 1969; New York: Atheneum, 1969);

Norman Mackenzie and Jeanne Mackenzie, *The Time Traveller: The Life of H. G. Wells* (London: Weidenfeld & Nicolson, 1973); republished as *H. G. Wells: A Biography* (New York: Simon & Schuster, 1973);

Gordon N. Ray, *H. G. Wells and Rebecca West* (New Haven: Yale University Press, 1974; London: Macmillan, 1974);

Frank Wells, *H. G. Wells: A Pictorial Biography* (London: Jupiter, 1977);

J. R. Hammond, ed., *H. G. Wells: Interviews and Recollections* (London: Macmillan, 1980);

Anthony West, *H. G. Wells: Aspects of a Life* (London: Hutchinson, 1984; New York: Random House, 1984);

David C. Smith, *H. G. Wells: Desperately Mortal–A Biography* (New Haven & London: Yale University Press, 1986);

Michael Coren, *The Invisible Man: The Life and Liberties of H. G. Wells* (New York: Random House, 1993).

References:

Brian Ash, *Who's Who in H. G. Wells* (London: Hamilton, 1979);

John Batchelor, *H. G. Wells* (Cambridge: Cambridge University Press, 1985);

Bernard Bergonzi, *The Early H. G. Wells: A Study of the Scientific Romances* (Manchester: Manchester University Press, 1961; Toronto: University of Toronto Press, 1961);

Bergonzi, ed., *H. G. Wells: A Collection of Critical Essays* (Englewood Cliffs, N.J.: Prentice-Hall, 1976);

Robert Bloom, *Anatomies of Egotism: A Reading of the Last Novels of H. G. Wells* (Lincoln: University of Nebraska Press, 1977);

Van Wyck Brooks, *The World of H. G. Wells* (London: Unwin, 1915);

Richard Hauer Costa, *H. G. Wells,* revised edition (Boston: Twayne, 1985);

Robert Crossley, *H. G. Wells* (Mercer Island, Wash.: Starmont, 1984);

Michael Draper, *H. G. Wells* (London: Macmillan, 1987);

Stephen M. Gill, *Scientific Romances of H. G. Wells: A Critical Study* (Cornwall, Ont.: Vesta, 1975);

J. R. Hammond, *H. G. Wells and the Modern Novel* (New York: St. Martin's Press, 1988);

Hammond, *H. G. Wells and the Short Story* (New York: St. Martin's Press, 1992);

Hammond, *An H. G. Wells Companion: A Guide to the Novels, Romances and Short Stories* (London & Basingstoke: Macmillan, 1979; New York: Barnes & Noble, 1979);

Roslynn D. Haynes, *H. G. Wells: Discoverer of the Future–The Influence of Science on His Thought* (New York & London: New York University Press, 1980);

Mark R. Hillegas, *The Future as Nightmare: H. G. Wells and the Anti-Utopians* (New York: Oxford University Press, 1967);

John Huntington, *The Logic of Fantasy: H. G. Wells and Science Fiction* (New York: Columbia University Press, 1982);

Huntington, ed., *Critical Essays on H. G. Wells* (Boston: G. K. Hall, 1991);

Bernard Loing, *H. G. Wells à l'oeuvre: Les débuts d'un écrivain (1894–1900)* (Paris: Didier, 1984);

Frank McConnell, *The Science Fiction of H. G. Wells* (New York & Oxford: Oxford University Press, 1981);

Patrick Parrinder, "From Mary Shelley to *The War of the Worlds*: The Thames Valley Catastrophe," in *Anticipations: Essays on Early Science Fiction and Its Precursors*, edited by David Seed (Liverpool: Liverpool University Press, 1995), pp. 58–74;

Parrinder, *H. G. Wells* (Edinburgh: Oliver & Boyd, 1970; New York: Putnam, 1977);

Parrinder, *Shadows of the Future: H. G. Wells, Science Fiction, and Prophecy* (Syracuse, N.Y.: Syracuse University Press, 1995);

Parrinder, ed., *H. G. Wells: The Critical Heritage* (London & Boston: Routledge & Kegan Paul, 1972);

Parrinder and Christopher Rolfe, eds., *H. G. Wells Under Revision: Proceedings of the International H. G. Wells Symposium*, London, July 1986 (Selinsgrove, Pa.: Susquehanna University Press / London & Toronto: Associated University Presses, 1990);

Ingvald Raknem, *H. G. Wells and His Critics* (London: Allen & Unwin, 1962);

John R. Reed, *The Natural History of H. G. Wells* (Athens: Ohio University Press, 1982);

William J. Scheick, *The Splintering Frame: The Later Fiction of H. G. Wells* (Victoria, B.C.: University of Victoria English Literary Series, 1985);

Scheick, ed., *The Critical Response to H. G. Wells* (Westport, Conn. & London: Greenwood Press, 1995);

David Seed, "Doctor Moreau and His Beast People," *Udolpho,* 17 (June 1994): 8–12;

Leon Stover, *The Prophetic Soul: A Reading of H. G. Wells's* "Things To Come" (Jefferson, N.C. & London: McFarland, 1987);

Darko Suvin and Robert M. Philmus, eds., *H. G. Wells and Modern Science Fiction* (Lewisburg, Pa.: Bucknell University Press / London: Associated University Presses, 1977);

Jack Williamson, *H. G. Wells: Critic of Progress* (Baltimore: Mirage, 1973).

Papers:

The Wells Archive is located in the Rare Book Room, University of Illinois Library, Urbana-Champaign.

Appendix

Documents on British Fantasy and Science Fiction

Horace Walpole
Prefaces to *The Castle of Otranto*

Preface to the First Edition (1765)

The following work was found in the library of an ancient Catholic family in the north of England. It was printed at Naples, in the black letter, in the year 1529. How much sooner it was written does not appear. The principal incidents are such as were believed in the darkest ages of christianity; but the language and conduct have nothing that savours of barbarism. The style is the purest Italian. If the story was written near the time when it is supposed to have happened, it must have been between 1095, the era of the first crusade, and 1243, the date of the last, or not long afterwards. There is no other circumstance in the work, that can lead us to guess at the period in which the scene is laid. The names of the actors are evidently fictitious, and probably disguised on purpose: yet the Spanish names of the domestics seem to indicate that this work was not composed until the establishment of the Arragonian kings in Naples had made Spanish appellations familiar in that country. The beauty of the diction, and the zeal of the author (moderated, however, by singular judgment), concur to make me think, that the date of the composition was little antecedent to that of the impression. Letters were then in their most flourishing state in Italy, and contributed to dispel the empire of superstition, at that time so forcibly attacked by the reformers. It is not unlikely, that an artful priest might endeavour to turn their own arms on the innovators; and might avail himself of his abilities as an author to confirm the populace in their ancient errors and superstitions. If this was his view, he has certainly acted with signal address. Such a work as the following would enslave a hundred vulgar minds, beyond half the books of controversy that have been written from the days of LUTHER to the present hour.

This solution of the author's motives is, however, offered as a mere conjecture. Whatever his views were, or whatever effects the execution of them might have, his work can only be laid before the public at present as a matter of entertainment. Even as such, some apology for it is necessary. Miracles, visions, necromancy, dreams, and other preternatural events, are exploded now even from romances. That was not the case when our author wrote; much less when the story itself is supposed to have happened. Belief in every kind of prodigy was so established in those dark ages, that an author would not be faithful to the manners of the times, who should omit all mention of them. He is not bound to believe them himself, but he must represent his actors as believing them.

If this air of the miraculous is excused, the reader will find nothing else unworthy of his perusal. Allow the possibility of the facts, and all the actors comport themselves as persons would do in their situation. There is no bombast, no similes, flowers, digressions, or unnecessary descriptions. Every thing tends directly to the catastrophe. Never is the reader's attention relaxed. The rules of the drama are almost observed throughout the conduct of the piece. The characters are well drawn, and still better maintained. Terror, the author's principal engine, prevents the story from ever languishing; and it is so often contrasted by pity, that the mind is kept up in a constant vicissitude of interesting passions.

Some persons may, perhaps, think the characters of the domestics too little serious for the general cast of the story; but, besides their opposition to the principal personages, the art of the author is very observable in his conduct of the subalterns. They discover many passages essential to the story, which could not be well brought to light but by their naïveté and simplicity: in particular, the womanish terror and foibles of Bianca, in the last chapter, conduce essentially towards advancing the catastrophe.

It is natural for a translator to be prejudiced in favour of his adopted work. More impartial readers may not be so much struck with the beauties of this piece as I was. Yet I am not blind to my author's defects. I could wish he had grounded his plan on a more useful moral than this: that "the sins of fathers are visited on their children to the third and fourth generation." I doubt whether, in his time, any more than at present, ambition curbed its appetite of dominion from the dread of so remote a punishment.

And yet this moral is weakened by that less direct insinuation, that even such anathema may be diverted, by devotion to St. Nicholas. Here, the interest of the monk plainly gets the better of the judgment of the author. However, with all its faults, I have no doubt but the English reader will be pleased with a sight of this performance. The piety that reigns throughout, the lessons of virtue that are inculcated, and the rigid purity of the sentiments, exempt this work from the censure to which romances are but too liable. Should it meet with the success I hope for, I may be encouraged to re-print the original Italian, though it will tend to depreciate my own labour. Our language falls far short of the charms of the Italian, but for variety and harmony. The latter is peculiarly excellent for simple narrative. It is difficult, in English, to relate without falling too low, or rising too high; a fault obviously occasioned by the little care taken to speak pure language in common conversation. Every Italian or Frenchman, of any rank, piques himself on speaking his own tongue correctly and with choice. I cannot flatter myself with having done justice to my author in this respect: his style is as elegant, as his conduct of the passions is masterly. It is a pity that he did not apply his talents to what they were evidently proper for, the theatre.

I will detain the reader no longer, but to make one short remark. Though the machinery is invention, and the names of the actors imaginary, I cannot but believe, that the ground work of the story is founded on truth. The scene is undoubtedly laid in some real castle. The author seems frequently, without design, to describe particular parts. "The chamber," says he, "on the right hand; the door on the left hand; the distance from the chapel to Conrad's apartment." These, and other passages, are strong presumptions that the author had some certain building in his eye. Curious persons, who have leisure to employ in such researches, may possibly discover in the Italian writers the foundation on which our author has built. If a catastrophe, at all resembling that which he describes, is believed to have given rise to this work, it will contribute to interest the reader, and will make *The Castle of Otranto* a still more moving story.

Preface to the Second Edition (1765)

The favourable manner in which this little piece has been received by the public, calls upon the author to explain the grounds on which he composed it. But, before he opens those motives, it is fit that he should ask pardon of his readers for having offered his work to them under the borrowed personage of a translator. As diffidence of his own abilities, and the novelty of the attempt, were the sole inducements to assume that disguise, he flatters himself he shall appear excusable. He resigned his performance to the impartial judgment of the public; determined to let it perish in obscurity, if disapproved; nor meaning to avow such a trifle, unless better judges should pronounce that he might own it without a blush.

It was an attempt to blend the two kinds of romance, the ancient and the modern. In the former, all was imagination and improbability: in the latter, nature is always intended to be, and sometimes has been, copied with success. Invention has not been wanting; but the great resources of fancy have been dammed up, by a strict adherence to common life. But if, in the latter species, Nature has cramped imagination, she did but take her revenge, having been totally excluded from old romances. The actions, sentiments, and conversations, of the heroes and heroines of ancient days, were as unnatural as the machines employed to put them in motion.

The author of the following pages thought it possible to reconcile the two kinds. Desirous of leaving the powers of fancy at liberty to expatiate through the boundless realms of invention, and thence of creating more interesting situations, he wished to conduct the mortal agents in his drama according to the rules of probability; in short, to make them think, speak, and act, as it might be supposed mere men and women would do in extraordinary positions. He had observed, that, in all inspired writings, the personages under the dispensation of miracles, and witnesses to the most stupendous phenomena, never lost sight of their human character: whereas, in the productions of romantic story, an improbable event never fails to be attended by an absurd dialogue. The actors seem to lose their senses, the moment the laws of nature have lost their tone. As the public have applauded the attempt, the author must not say he was entirely unequal to the task he had undertaken: yet, if the new route he has struck out shall have paved a road for men of brighter talents, he shall own, with pleasure and modesty, that he was sensible the plan was capable of receiving greater embellishments than his imagination, or confuct of the passions, could bestow on it.

With regard to the deportment of the domestics, on which I have touched in the former preface, I will beg leave to add a few words.–The simplicity of their behaviour, almost tending to excite smiles, which, at first, seems not consonant to the serious cast of the work, appeared to me not only not improper, but was marked designedly in that manner. My rule was nature. However grave, important, or even melancholy, the sensations of princes and heroes may be, they do not stamp the same affections on their domestics: at least the latter do not, or should not be made to, express their passions in the same dignified tone. In my humble opinion, the contrast between the sublime of the one and the naïveté of the other, sets the pathetic of the former in a stronger light. The very impatience which a reader feels, while delayed, by the coarse pleasantries of vulgar actors, from arriving at the knowledge of the important catastrophe he expects, perhaps heightens, certainly proves that he has been artfully interested in, the depending event. But I had higher authority than my own opinion for this conduct. The great master of nature, SHAKESPEARE, was the model I copied. Let me ask, if his tragedies of *Hamlet* and *Julius Caesar* would not lose a considerable share of their spirit and wonderful beauties, if the humour of the gravediggers, the fooleries of Polonius, and the clumsy jests of the Roman citizens, were omitted, or vested in heroics? Is not the eloquence of Antony, the nobler and affectedly-unaffected oration of Brutus, artificially exalted by the rude bursts of nature from the mouths of their auditors? These touches remind one of the Grecian sculptor, who, to convey the idea of a Colossus, within the dimensions of a seal, inserted a little boy measuring his thumb.

"No," says Voltaire, in his edition of Corneille, "this mixture of buffoonery and solemnity is intolerable."–Voltaire is a genius*–but not of Shakespeare's magnitude. Without recurring to disputable authority, I appeal from Voltaire to himself. I shall not avail myself of his former encomiums on our mighty poet; though the French critic has twice translated the same speech in *Hamlet,* some years ago in admiration, latterly in derision; and I am sorry to find that his judgment grows weaker when it ought to be farther matured. But I shall make use of his own words, delivered on the general topic of the theatre, when he was neither thinking to recommend or decry Shakespeare's practice; consequently, at a moment when Voltaire was impartial. In the preface to his *Enfant Prodigue,* that exquisite piece, of which i declare my admiration, and which, should I live twenty years longer, I trust I shall never attempt to ridicule, he has these words, speaking of comedy (but equally applicable to tragedy, if tragedy is, as surely it ought to be, a picture of human life; nor can I conceive why occasional pleasantry ought more to be banished from the tragic

Horace Walpole in his library, by J. H. Müntz

scene, than pathetic seriousness from the comic), *"On y voit un mélange de sérieux et de plaisanterie, de comique et de touchant;* souvent même une seule aventure *produit tous ces contrastes. Rien n'est si commun qu'une maison dans laquelle* un père gronde, une fille occupée de sa passion pleure; *le fils se moque des deux, et quelques parents prennent différemment part à la scène, &c. Nous n'inférons pas de là que toute comédie doive avoir des scènes de bouffonnerie et des scènes attendrissantes: il y a beaucoup de très bonnes pièces où il ne règne que de la gaieté; d'autres toutes sérieuses; d'autres mélangées: d'autres où l'attendrissement va jusques aux larmes:* il ne faut donner l'exclusion à aucun genre; *et si on me demandoit, quel genre est le meilleur, je répondrois, celui qui est le mieux traité."* Surely if a comedy may be *toute sérieuse,* tragedy may now and then, soberly, be indulged in a smile. Who shall proscribe it? Shall the critic, who, in self-defence, declares that *no kind* ought to be excluded from comedy, give laws to Shakespeare?

I am aware that the preface from whence I have quoted these passages does not stand in Monsieur de Voltaire's name, but in that of his editor; yet who doubts that the editor and author were the same person? or where is the editor, who has so happily possessed himself of his author's style, and brilliant ease of argument? These passages were indubitably the geniune sentiments of that great writer. In his epistle to Maffei, prefixed to his *Mérope,* he delivers almost the same opinion, though, I doubt, with a little irony. I will repeat his words, and then give my reason for quoting them. After translating a passage in Maffei's *Merope,* Monsieur de Voltaire adds, *"Tous ces traits sont naïfs; tout y est convenable à ceux que vous introduisez sur la scène,* et aux moeurs que vous leur donnez. *Ces familiarités naturelles eussent été, à ce que je crois, bien reçues dans Athènes; mais Paris et notre parterre veulent une autre espèce de simplicité."* I doubt, I say, whether there is not a grain of sneer in this and other passages of that epistle; yet the force of truth is not damaged by being tinged with ridicule. Maffei was to represent a Grecian story: surely the Athenians were as competent judges of Grecian manners, and of the propriety of introducing them, as the parterre of Paris. "On the contrary," says Voltaire (and I cannot but admire his reasoning), "there were but ten thousand citizens at Athens, and Paris has near eight hundred thousand inhabitants, among whom one may reckon thirty thousand judges of dramatic works."–Indeed!–but allowing so numerous a tribunal, I believe this is the only instance in which it was ever pretended that thirty thousand persons, living near two thousand years after the era in question, were, upon the mere face of the poll, declared better judges than the Grecians themselves, of what ought to be the manners of a tragedy written on a Grecian story.

I will not enter into a discussion of the *espèce de simplicité,* which the parterre of Paris demands, nor of the shackles with which *the thirty thousand judges* have cramped their poetry, the chief merit of which, as I gather from repeated passages in the *New Commentary* on Corneille, consists in vaulting in spite of those fetters; a merit which, if true, would reduce poetry from the lofty effort of imagination, to a puerile and most contemptible labour–*difficiles nugae* with a witness! I cannot, however, help mentioning a couplet, which, to my English ears, always sounded as the flattest and most trifling instance of circumstantial propriety, but which Voltaire, who has dealt so severely with nine parts in ten of Corneille's works, has singled out to defend in Racine;

De son appartement cette porte est prochaine,
Et cette autre conduit dans celui de la Reine.

IN ENGLISH.

To Caesar's closet through this door you come,
And t'other leads to the Queen's drawing-room.

Unhappy Shakespeare! hadst thou made Rosencrantz inform his compeer, Guildenstern, of the ichnography of the palace of Copenhagen, instead of presenting us with a moral dialogue between the Prince of Denmark and the grave-digger, the illuminated pit of Paris would have been instructed *a second time* to adore thy talents.

The result of all I have said, is, to shelter my own daring under the canon of the brightest genius this country, at least, has produced. I might have pleaded that, having created a new species of romance, I was at liberty to lay down what rules I thought fit for the conduct of it: but I should be more proud of having imitated, however faintly, weakly, and at a distance, so masterly a pattern, than to enjoy the entire merit of invention, unless I could have marked my work with genius, as well as with originality. Such as it is, the public have honoured it sufficiently, whatever rank their suffrages allot to it.

*The following remark is foreign to the present question, yet excusable in an Englishman, who is willing to think that the severe criticisms of so masterly a writer as Voltaire on our immortal countryman, may have been the effusions of wit and precipitation, rather than the result of judgment and attention. May not the critic's skill, in the force and powers of our language, have been as incorrect and incompetent as his knowledge of our history? of the latter, his own pen has dropped glaring evidence. In his Preface to Thomas Corneille's *Earl of Essex,* Monsieur de Voltaire allows that the truth of history has been grossly perverted in that piece. In excuse he pleads, that when Corneille wrote, the noblesse of France were much unread in English story; but now, says the commentator, that they study it, such misrepresentations would not be suffered—yet forgetting that the period of ignorance is lapsed, and that it is not very necessary to instruct the knowing, he undertakes, from the overflowing of his own reading, to give the nobility of his own country a detail of Queen Elizabeth's favourites—of whom, says he, Robert Dudley was the first, and the Earl of Leicester the second. Could one have believed that it could be necessary to inform Monsieur de Voltaire himself, that Robert Dudley and the Earl of Leicester were the same person?

Mary Wollstonecraft Shelley
Preface to *Frankenstein; or, The Modern Prometheus* (1818)

The event on which this fiction is founded has been supposed, by Dr. Darwin, and some of the physiological writers of Germany, as not of impossible occurrence. I shall not be supposed as according the remotest degree of serious faith to such an imagination; yet, in assuming it as the basis of a work of fancy, I have not considered myself as merely weaving a series of supernatural terrors. The event on which the interest of the story depends is exempt from the disadvantages of a mere tale of spectres or enchantment. It was recommended by the novelty of the situations which it develops; and, however impossible as a physical fact, affords a point of view to the imagination for the delineating of human passions more comprehensive and commanding than any which the ordinary relations of existing events can yield.

I have thus endeavoured to preserve the truth of the elementary principles of human nature, while I have not scrupled to innovate upon their combinations. The *Iliad,* the tragic poetry of Greece,–Shakespeare, in the *Tempest* and *Midsummer Night's Dream,*–and most especially Milton, in *Paradise Lost,* conform to this rule; and the most humble novelist, who seeks to confer or receive amusement from his labours, may, without presumption, apply to prose fiction a licence, or rather a rule, from the adoption of which so many exquisite combinations of human feeling have resulted in the highest specimens of poetry.

The circumstance on which my story rests was suggested in casual conversation. It was commenced, partly as a source of amusement, and partly as an expedient for exercising any untried resources of mind. Other motives were mingled with these, as the work proceeded. I am by no means indifferent to the manner in which whatever moral tendencies exist in the sentiments or characters it contains shall affect the reader; yet my chief concern in this respect has been limited to the avoiding the enervating effects of the novels of the present day, and to the exhibition of the amiableness of domestic affection, and the excellence of universal virtue. The opinions which naturally spring from the character and situation of the hero are by no means to be conceived as existing always in my own conviction; nor is any inference justly to be drawn from the following pages as prejudicing any philosophical doctrine of whatever kind.

Mary Wollstonecraft Shelley (National Portrait Gallery, London)

It is a subject also of additional interest to the author, that this story was begun in the majestic region where the scene is principally laid, and in society which cannot cease to be regretted. I passed the summer of 1816 in the environs of Geneva. The season was cold and rainy, and in the evenings we crowded around a blazing wood fire, and occasionally amused ourselves with some German stories of ghosts, which happened to fall into our hands. These tales excited in us a playful desire of imitation. Two other friends (a tale from the pen of one of whom would be far more acceptable to the public than any thing I can ever hope on to produce) and myself agreed to write each a story, founded on some supernatural occurrence.

The weather, however, suddenly became serene; and my two friends left me on a journey among the Alps, and lost, in the magnificent scenes which they present, all memory of their ghostly visions. The following tale is the only one which has been completed.

Bram Stoker
On Writing *Dracula,* from the Introduction to *Dracula* (1897)

The harbor at Whitby, on the Yorkshire coast, where Bram Stoker began writing Dracula

Having some time at my disposal when in London, I had visited the British Museum, and made search among the books and maps in the library regarding Transylvania; it had struck me that some foreknowledge of the country could hardly fail to have some importance in dealing with a noble of that country. I find that the district he named is in the extreme east of the country, just on the borders of three states, Transylvania, Moldavia and Bukovina, in the midst of the Carpathian mountains; one of the wildest and least known portions of Europe.

I was not able to light on any map or work giving the exact locality of the Castle Dracula, as there are no maps of this country as yet to compare with our own Ordnance Survey maps; but I found that Bistritz, the post town named by Count Dracula, is a fairly well-known place. I shall enter here some of my notes.

In the population of Transylvania there are four distinct nationalities: Saxons in the south, and mixed with them the Wallachs, who are the descendants of the Dacians; Magyars in the west, and Szekelys in the east and north. I am going among the latter, who claim to be descended from Attila and the Huns. This may be so, for when the Magyars conquered the country in the eleventh century they found the Huns settled in it.

I read that every known superstition in the world is gathered into the horseshoe of the Carpathians, as if it were the centre of some sort of imaginative whirlpool; if so my stay may be very interesting.

G. K. Chesterton
"The Ethics of Elfland" (1908)

When the business man rebukes the idealism of his office-boy, it is commonly in some such speech as this: "Ah, yes, when one is young, one has these ideals in the abstract and these castles in the air; but in middle age they all break up like clouds, and one comes down to a belief in practical politics, to using the machinery one has and getting on with the world as it is." Thus, at least, venerable and philanthropic old men now in their honored graves used to talk to me when I was a boy. But since then I have grown up and have discovered that these philanthropic old men were telling lies. What has really happened is exactly the opposite of what they said would happen. They said that I should lose my ideals and begin to believe in the methods of practical politicians. Now, I have not lost my ideals in the least; my faith in fundamentals is exactly what it always was. What I have lost is my old childlike faith in practical politics. I am still as much concerned as ever about the Battle of Armageddon; but I am not so much concerned about the General Election. As a babe I leapt up on my mother's knee at the mere mention of it. No; the vision is always solid and reliable. The vision is always a fact. It is the reality that is often a fraud. As much as I ever did, more than I ever did, I believe in Liberalism. But there was a rosy time of innocence when I believed in Liberals.

I take this instance of one of the enduring faiths because, having now to trace the roots of my personal speculation, this may be counted, I think, as the only positive bias. I was brought up a Liberal, and have always believed in democracy, in the elementary liberal doctrine of a self-governing humanity. If any one finds the phrase vague or threadbare, I can only pause for a moment to explain that the principle of democracy, as I mean it, can be stated in two propositions. The first is this: that the things common to all men are more important than the things peculiar to any men. Ordinary things are more valuable than extraordinary things; nay, they are more extraordinary. Man is something more awful than men; something more strange. The sense of the miracle of humanity itself should be always more vivid to us than any marvels of power, intellect, art, or civilization. The mere man on two legs, as such, should be felt as something more heartbreaking than any music and more startling than any caricature. Death is more tragic even than death by starvation. Having a nose is more comic even than having a Norman nose.

This is the first principle of democracy: that the essential things in men are the things they hold in common, not the things they hold separately. And the second principle is merely this: that the political instinct or desire is one of these things which they hold in common. Falling in love is more poetical than dropping into poetry. The democratic contention is that government (helping to rule the tribe) is a thing like falling in love, and not a thing like dropping into poetry. It is not something analogous to playing the church organ, painting on vellum, discovering the North Pole (that insidious habit), looping the loop, being Astronomer Royal, and so on. For these things we do not wish a man to do at all unless he does them well. It is, on the contrary, a thing analogous to writing one's own love-letters or blowing one's own nose. These things we want a man to do for himself, even if he does them badly. I am not here arguing the truth of any of these conceptions; I know that some moderns are asking to have their wives chosen by scientists, and they may soon be asking, for all I know, to have their noses blown by nurses. I merely say that mankind does recognize these universal human functions, and that democracy classes government among them. In short, the democratic faith is this: that the most terribly important things must be left to ordinary men themselves—the mating of the sexes, the rearing of the young, the laws of the state. This is democracy; and in this I have always believed.

But there is one thing that I have never from my youth up been able to understand. I have never been able to understand where people got the idea that democracy was in some way opposed to tradition. It is obvious that tradition is only democracy extended through time. It is trusting to a consensus of common human voices rather than to some isolated or arbitrary record. The man who quotes

G. K. Chesterton, circa 1900

some German historian against the tradition of the Catholic Church, for instance, is strictly appealing to aristocracy. He is appealing to the superiority of one expert against the awful authority of a mob. It is quite easy to see why a legend is treated, and ought to be treated, more respectfully than a book of history. The legend is generally made by the majority of people in the village, who are sane. The book is generally written by the one man in the village who is mad. Those who urge against tradition that men in the past were ignorant may go and urge it at the Carlton Club, along with the statement that voters in the slums are ignorant. It will not do for us. If we attach great importance to the opinion of ordinary men in great unanimity when we are dealing with daily matters, there is no reason why we should disregard it when we are dealing with history or fable. Tradition may be defined as an extension of the franchise. Tradition means giving votes to the most obscure of all classes, our ancestors. It is the democracy of the dead. Tradition refuses to submit to the small and arrogant oligarchy of those who merely happen to be walking about. All democrats object to men being disqualified by the accident of birth; tradition objects to their being disqualified by the accident of death. Democracy tells us not to neglect a good man's opinion, even if he is our groom; tradition asks us not to neglect a good man's opinion, even if he is our father. I, at any rate, cannot separate the two ideas of democracy and tradition; it seems evident to me that they are the same idea. We will have the dead at our councils. The ancient Greeks voted by stones; these shall vote by tombstones. It is all quite regular and official, for most tombstones, like most ballot papers, are marked with a cross.

I have first to say, therefore, that if I have had a bias, it was always a bias in favour of democracy, and therefore of tradition. Before we come to any theoretic or logical beginnings I am content to allow for that personal equation; I have always been more inclined to believe the ruck of hard-working people than to believe that special and troublesome literary class to which I belong. I prefer even the fancies and prejudices of the people who see life from the inside to the clearest demonstrations of the people who see life from the outside. I would always trust the old wives' fables against the old maids' facts. As long as wit is mother wit it can be as wild as it pleases.

Now, I have to put together a general position, and I pretend to no training in such things. I propose to do it, therefore, by writing down one after another the three or four fundamental ideas which I have found for myself, pretty much in the way that I found them. Then I shall roughly synthesize them, summing up my personal philosophy or natural religion; then I shall describe my startling discovery that the whole thing had been discovered before. It had been discovered by Christianity. But of these profound persuasions which I have to recount in order, the earliest was concerned with this element of popular tradition. And without the foregoing explanation touching tradition and democracy I could hardly make my mental experience clear. As it is, I do not know whether I can make it clear, but I now propose to try.

My first and last philosophy, that which I believe in with unbroken certainty, I learnt in the nursery. I generally learnt it from a nurse; that is, from the solemn and star-appointed priestess at once of democracy and tradition. The things I believed most then, the things I believe most now, are the things called fairy tales. They seem to me to be the entirely reasonable things. They are not fantasies: compared with them other things are fantastic. Compared with them religion and rationalism are both abnormal, though religion is abnormally right and rationalism abnormally wrong. Fairyland is nothing but the sunny country of common sense. It is not earth that judges heaven, but heaven that judges earth; so for me at least it was not earth that criticised elfland, but elfland that criticised the earth. I knew the magic beanstalk before I had tasted beans; I was sure of the Man in the Moon before I was certain of the moon. This was at one with all popular tradition. Modern minor poets are naturalists, and talk about the bush or the brook; but the singers of the old epics and fables were supernaturalists, and talked about the gods of brook and bush. That is what the moderns mean when they say that the ancients did not "appreciate Nature," because they said that Nature was divine. Old nurses do not tell children about the grass, but about the fairies that dance on the grass; and the old Greeks could not see the trees for the dryads.

But I deal here with what ethic and philosophy come from being fed on fairy tales. If I were describing them in detail I could note many noble and healthy principles that arise from them. There is the chivalrous lesson of "Jack the Giant Killer"; that giants should be killed because they are gigantic. It is a manly mutiny against pride as such. For the rebel is older than all the kingdoms, and the Jacobin has more tradition than the Jacobite. There is the lesson of "Cinderella," which is the same as that of the Magnificat—*exaltavit humiles.* There is the great lesson of "Beauty and the Beast"; that a thing must be loved *before* it is lovable. There is the terrible allegory of the "Sleeping Beauty," which tells how the human creature was blessed with all birthday gifts, yet cursed with death; and how death also may perhaps be softened to a sleep. But I am not concerned with any of the separate statutes of elfland, but with the whole spirit of its law, which I learnt before I could speak, and shall retain when I cannot write. I am concerned with a certain way of looking at life, which was created in me by the fairy tales, but has since been meekly ratified by the mere facts.

It might be stated this way. There are certain sequences or developments (cases of one thing following another), which are, in the true sense of the word, reasonable. They are, in the true sense of the word, necessary. Such are mathematical and merely logical sequences. We in fairyland (who are the most reasonable of all creatures) admit that reason and that necessity. For instance, if the Ugly Sisters are older than Cinderella, it is (in an iron and awful sense) *necessary* that Cinderella is younger than the Ugly Sisters. There is no getting out of it. Haeckel may talk as much fatalism about that fact as he pleases: it really must be. If Jack is the son of a miller, a miller is the father of Jack. Cold reason decrees it from her awful throne: and we in fairyland submit. If the three brothers all ride horses, there are six animals and eighteen legs involved: that is true rationalism, and fairyland is full of it. But as I put my head over the hedge of the elves and began to take notice of the natural world, I observed an extraordinary thing. I observed that learned men in spectacles were talking of the actual things that happened—dawn and death and so on—as if *they* were rational and inevitable. They talked as if the fact that trees bear fruit were just as *necessary* as the fact that two and one trees make three. But it is not. There is an enormous difference by the test of fairyland; which is the test of the imagination. You cannot *imagine* two and one not making three. But you can easily imagine trees not growing fruit; you can imagine them growing golden candlesticks or tigers hanging on by the tail. These men in spectacles spoke much of a man named Newton, who was hit by an apple, and who discovered a law. But they could not be got to see the distinction between a true law, a law of reason, and the mere fact of apples falling. If the apple hit Newton's nose, Newton's nose hit the apple. That is a true necessity: because we cannot conceive the one occurring without the other. But we can quite well conceive the apple not falling on his nose; we can fancy it flying ardently through the air to hit some other nose, of which it had a more definite dislike. We have always in our fairy tales kept this sharp distinction between the science of mental relations, in which there really are laws, and the science of physical facts, in which there are no laws, but only weird repetitions. We believe in bodily miracles, but not in mental impossibilities. We believe that a Bean-stalk climbed up to Heaven; but that does not at all confuse our convictions on the philosophical question of how many beans make five.

Here is the peculiar perfection of tone and truth in the nursery tales. The man of science says, "Cut the stalk, and the apple will fall"; but he says it calmly, as if the one idea really led up to the other. The witch in the fairy tale says, "Blow the horn, and the ogre's castle will fall"; but she does not say it as if it were something in which the effect obviously arose out of the cause. Doubtless she has given the advice to many champions, and has seen many castles fall, but she does not lose either her wonder or her reason. She does not muddle her head until it imagines a necessary mental connection between a horn and a falling tower. But the scientific men do muddle their heads, until they imagine a necessary mental connection between an apple leaving the tree and an apple reaching the ground. They do really talk as if they had found not only a set of marvellous facts, but a truth connecting those facts. They do talk as if the connection of two strange things physically connected them philosophically. They feel that because one incomprehensible thing constantly follows another incomprehensible thing the two together somehow make up a comprehensible thing. Two black riddles make a white answer.

In fairyland we avoid the word "law"; but in the land of science they are singularly fond of it. Thus they will call some interesting conjecture about how forgotten folks pronounced the alphabet, Grimm's Law. But Grimm's Law is far less intellectual than Grimm's Fairy Tales. The tales are, at any rate, certainly tales; while the law is not a law. A law implies that we know the nature of the generalisation and enactment; not merely that we have noticed some of the effects. If there is a law that pickpockets shall go to prison, it implies that there is an imaginable mental connection between the idea of prison and the idea of picking pockets. And we know what the idea is. We can say why we take liberty from a man who takes liberties. But we cannot say why an egg can turn into a chicken any more than we can say why a bear could turn into a fairy prince. As *ideas,* the egg and the chicken are further off from each other than the bear and the prince; for no egg in itself suggests a chicken, whereas some princes do suggest bears. Granted, then, that certain transformations do happen, it is essential that we should regard them in the philosophic manner of fairy tales, not in the unphilosophic manner of science and the "Laws of Nature." When we are asked why eggs turn to birds or fruits fall in autumn, we must answer exactly as the fairy godmother would answer if Cinderella asked her why mice turned to horses or her clothes fell from her at twelve o'clock. We must answer that it is *magic.* It is not a "law," for we do not understand its general formula. It is not a necessity, for though we can count on it happening practically, we have no right to say that it must always happen. It is no argument for unalterable law (as Huxley fancied) that we count on the ordinary course of things. We do not count on it; we bet on it. We risk the remote possibility of a miracle as we do that of a poisoned pancake or a world-destroying comet. We leave it out of account, not because it is a miracle, and therefore an impossibility, but because it is a miracle, and therefore an exception. All the terms used in the science books, "law," "necessity," "order," "tendency," and so on, are really unintellectual, because they assume an inner synthesis, which we do not possess. The only words that ever satisfied me as describing Nature are the terms used in the fairy books, "charm," "spell," "enchantment." They express the arbitrariness of the fact and its mystery. A tree grows fruit because it is a *magic* tree. Water runs downhill because it is bewitched. The sun shines because it is bewitched.

I deny altogether that this is fantastic or even mystical. We may have some mysticism later on; but this fairy-tale language about things is simply rational and agnostic. It is the only way I can express in words my clear and definite perception that one thing is quite distinct from another; that there is no logical connection between flying and laying eggs. It is the man who talks about "a law" that he has never seen who is the mystic. Nay, the ordinary scientific man is strictly a sentimentalist. He is a sentimentalist in this essential sense, that he is soaked and swept away by mere associations. He has so often seen birds fly and lay eggs that he feels as if there must be some dreamy, tender connection between the two ideas, whereas there is none. A forlorn lover might be unable to dissociate the moon from lost love; so the materialist is unable to dissociate the moon from the tide. In both cases there is no connection, except that one has seen them together. A sentimentalist might shed tears at the smell of apple-blossom, because, by a dark association of his own, it reminded him of his boyhood. So the materialist professor (though he conceals his tears) is yet a sentimentalist, because, by a dark association of his own, apple-blossoms remind him of apples. But the cool rationalist from fairyland does not see why, in the abstract, the apple tree should not grow crimson tulips; it sometimes does in his country.

This elementary wonder, however, is not a mere fancy derived from the fairy tales; on the contrary, all the fire of the fairy tales is derived from this. Just as we all like love tales because there is an instinct of sex, we all like astonishing tales because they touch the nerve of the ancient instinct of astonishment. This is proved by the fact that when we are very young children we do not need fairy tales: we only need tales. Mere life is interesting enough. A child of seven is excited by being told that Tommy opened a door and saw a dragon. But a child of three is excited by being told that Tommy opened a door. Boys like romantic tales; but babies like realistic tales–because they find them romantic. In fact, a baby is about the only person, I should think, to whom a modern realistic novel could be read without boring him. This proves that even nursery tales only echo an almost pre-natal leap of interest and amazement. These tales say that apples were golden only to refresh the forgotten moment when we found that they were green. They make rivers run with wine only to make us remember, for one wild moment, that they run with water. I have said that this is wholly reasonable and even agnostic. And, indeed, on this point I am all for the higher agnosticism; its better name is Ignorance. We have all read in scientific books, and, indeed, in all romances, the story of the man who has forgotten his name. This man walks about the streets and can see and appreciate everything; only he cannot remember who he is. Well, every man is that man in the story. Every man has forgotten who he is. One may understand the cosmos, but never the ego; the self is more distant than any star. Thou shalt love the Lord thy God; but thou shalt not know thyself. We are all under the same mental calamity; we have all forgotten our names. We have all forgotten what we really are. All that we call common sense and rationality and practicality and positivism only means that for certain dead levels of our life we forget that we have forgotten. All that we call spirit and art and ecstacy only means that for one awful instant we remember that we forget.

But though (like the man without memory in the novel) we walk the streets with a sort of half-witted admiration, still it is admiration. It is admiration in English and not only admiration in Latin. The wonder has a positive element of praise. This is the next milestone to be definitely marked on our road through fairyland. I shall speak in the next chapter about optimists and pessimists in their intellectual aspect, so far as they have one. Here I am only trying to describe the enormous emotions which cannot be described. And the strongest emotion was that life was as precious as it was puzzling. It was an ecstacy because it was an adventure; it was an adventure because it was an opportunity. The goodness of the fairy tale was not affected by the fact that there might be more dragons than princesses; it was good to be in a fairy tale. The test of all happiness is gratitude; and I felt grateful, though I hardly knew to whom. Children are grateful when Santa Claus puts in their stockings gifts of toys or sweets. Could I not be grateful to Santa Claus when he put in my stockings the gift of two miraculous legs? We thank people for birthday presents of cigars and slippers. Can I thank no one for the birthday present of birth?

There were, then, these two first feelings, indefensible and indisputable. The world was a shock, but it was not merely shocking; existence was a surprise, but it was a pleasant surprise. In fact, all my first views were exactly uttered in a riddle that stuck in my brain from boyhood. The question was, "What did the first frog say?" And the answer was, "Lord, how you made me jump!" That says succinctly all that I am saying. God made the frog jump; but the frog prefers jumping. But when these things are settled there enters the second great principle of the fairy philosophy.

Anyone can see it who will simply read "Grimm's Fairy Tales" or the fine collections of Mr. Andrew Lang. For the pleasure of pedantry I will call it the Doctrine of Conditional Joy. Touchstone talked of much virtue in an "if"; according to elfin ethics all virtue is in an "if." The note of the fairy utterance always is, "You may live in a palace of gold and sapphire, *if* you do not say the word 'cow' "; or "You may live happily with the King's daughter, *if* you do not show her an onion." The vision always hangs upon a veto. All the dizzy and colossal things conceded depend upon one small thing withheld. All the wild and whirling things that are let loose depend upon one thing that is forbidden. Mr. W. B. Yeats, in his exquisite and piercing elfin poetry, describes the elves as lawless; they plunge in innocent anarchy on the unbridled horses of the air–

> "Ride on the crest of the dishevelled tide,
> And dance upon the mountains like a flame."

It is a dreadful thing to say that Mr. W. B. Yeats does not understand fairyland. But I do say it. He is an ironical Irishman, full of intellectual reactions. He is not stupid enough to understand fairyland. Fairies prefer people of the yokel type like myself; people who gape and grin and do as they are told. Mr. Yeats reads into elfland all the righteous insurrection of his own race. But the lawlessness of Ireland is a Christian lawlessness, founded on reason and justice. The Fenian is rebelling against something he understands only too well; but the true citizen of fairyland is obeying something that he does not understand at all. In the fairy tale an incomprehensible happiness rests upon an incomprehensible condition. A box is opened, and all evils fly out. A word is forgotten, and cities perish. A lamp is lit, and love flies away. A flower is plucked, and human lives are forfeited. An apple is eaten, and the hope of God is gone.

This is the tone of fairy tales, and it is certainly not lawlessness or even liberty, though men under a mean modern tyranny may think it liberty by comparison. People out of Portland Gaol might think Fleet Street free; but closer study will prove that both fairies and journalists are the slaves of duty. Fairy godmothers seem at least as strict as other godmothers. Cinderella received a coach out of Wonderland and a coachman out of nowhere, but she received a command–which might have come out of Brixton–that she should be back by twelve. Also, she had a glass slipper; and it cannot be a coincidence that glass is so common a substance in folklore. This princess lives in a glass castle, that princess on a glass hill; this one sees all things in a mirror; they may all live in glass houses if they will not throw stones. For this thin glitter of glass everywhere is the expression of the fact that the happiness is bright but brittle, like the substance most easily smashed by a housemaid or a cat. And this fairy-tale sentiment also sank into me and became my sentiment towards the whole world. I felt and feel that life itself is as bright as the diamond, but as brittle as the window-pane; and when the heavens were compared to the terrible crystal I can remember a shudder. I was afraid that God would drop the cosmos with a crash.

Remember, however, that to be breakable is not the same as to be perishable. Strike a glass, and it will not endure an instant; simply do not strike it, and it will endure a thousand years. Such, it seemed, was the joy of man, either in elfland or on earth; the happiness depended on *not doing something* which you could at any moment do and which, very often, it was not obvious why you should not do. Now, the point here is that to *me* this did not seem unjust. If the miller's third son said to the fairy, "Explain why I must not stand on my head in the fairy palace," the other might fairly reply, "Well, if it comes to that, explain the fairy palace." If Cinderella says, "How is it that I must leave the ball at twelve?" her godmother might answer, "How is it that you are going there till twelve?" If I leave a man in my will ten talking elephants and a hundred winged horses, he cannot complain if the conditions partake of the slight eccentricity of the gift. He must not look a winged horse in the mouth. And it seemed to me that existence was itself so very eccentric a legacy that I could not complain of not understanding the limitations of the vision when I did not understand the vision they limited. The frame was no stranger than the picture. The veto might well be as wild as the vision; it might be as startling as the sun, as elusive as the waters, as fantastic and terrible as the towering trees.

For this reason (we may call it the fairy godmother philosophy) I never could join the young men of my time in feeling what they called the general sentiment of *revolt*. I should have resisted, let us hope, any rules that were evil, and with these and their definition I shall deal in another chapter. But I did not feel disposed to resist any rule merely because it was mysterious. Estates are sometimes held by foolish forms, the breaking of a stick or the payment of a peppercorn: I was willing to hold the huge estate of earth and heaven by any such feudal fantasy. It could not well be wilder than the fact that I was allowed to hold it at all. At this stage I give only one ethical instance to show my meaning. I could never mix in the common murmur of that rising generation against monogamy, because no restriction on sex seemed so odd and unexpected as sex itself. To be allowed, like Endymion, to make love to the moon and then to complain that Jupiter kept his own moons in a harem seemed to me (bred on fairy tales like Endymion's) a vulgar anti-climax. Keeping to one woman is a small price for so much as seeing one woman. To complain that I could only be married once was like complaining that I had only been born once. It was incommensurate with the terrible excitement of which one was talking. It showed, not an exaggerated sensibility to sex, but a curious insensibility to it. A man is a fool who complains that he cannot enter Eden by five gates at once. Polygamy is a lack of the realization of sex; it is like a man plucking five pears in mere absence of mind. The æsthetes touched the last insane limits of language in their eulogy on lovely things. The thistledown made them weep; a burnished beetle brought them to their knees. Yet their emotion never impressed me for an instant, for this reason, that it never occurred to them to pay for their pleasure in any sort of symbolic sacrifice. Men (I felt) might fast forty days for the sake of hearing a blackbird sing. Men might go through fire to find a cowslip. Yet these lovers of beauty could not even keep sober for the blackbird. They would not go through common Christian marriage by way of recompense to the cowslip. Surely one might pay for extraordinary joy in ordinary morals. Oscar Wilde said that sunsets were not valued because we could not pay for sunsets. But Oscar Wilde was wrong; we can pay for sunsets. We can pay for them by not being Oscar Wilde.

Well, I left the fairy tales lying on the floor of the nursery, and I have not found any books so sensible since. I left the nurse guardian of tradition and democracy, and I have not found any modern type so sanely radical or so sanely conservative. But the matter for important comment was here: that when I first went out into the mental atmosphere of the modern world, I found that the modern world was positively opposed on two points to my nurse and to the nursery tales. It has taken me a long time to find out that the modern world is wrong and my nurse was right. The really curious thing was this: that modern thought contradicted this basic creed of my boyhood on its two most essential doctrines. I have explained that the fairy tales founded in me two convictions; first, that this world is a wild and startling place, which might have been quite different, but which is quite delightful; second, that before this wildness and delight one may well be modest and submit to the queerest limitations of so queer a kindness. But I found the whole modern world running like a high tide against both my tendernesses; and the shock of that collision created two sudden and spontaneous sentiments, which I have had ever since and which, crude as they were, have since hardened into convictions.

First, I found the whole modern world talking scientific fatalism; saying that everything is as it must always have been, being unfolded without fault from the beginning. The leaf on the tree is green because it could never have been anything else. Now, the fairy-tale philosopher is glad that the leaf is green precisely because it might have been scarlet. He feels as if it had turned green an instant before he looked at it. He is pleased that snow is white on the strictly reasonable ground that it might have been black. Every colour has in it a bold quality as of choice; the red of garden roses is not only decisive but dramatic, like suddenly spilt blood. He feels that something has been *done*. But the great determinists of the nineteenth century were strongly against this native feeling that something had happened an instant before. In fact, according to them, nothing ever really had happened since the beginning of the world. Nothing ever had happened since existence had happened; and even about the date of that they were not very sure.

The modern world as I found it was solid for modern Calvinism, for the necessity of things being as they are. But when I came to ask them I found they had really no proof of this unavoidable repetition in things except the fact that the things were repeated. Now, the mere repetition made the things to me rather more weird than more rational. It was as if, having seen a curiously shaped nose in the street and dismissed it as an accident, I had then seen six other noses of the same astonishing shape. I should have fancied for a moment that it must be some local secret society. So one elephant having a trunk was odd; but all elephants having trunks looked like a plot. I speak here only of an emotion, and of an emotion at once stubborn and subtle. But the repetition in Nature seemed sometimes to be an excited repetition, like that of an angry schoolmaster saying the same thing over and over again. The grass seemed signalling to me with all its fingers at once; the crowded stars seemed bent upon being understood. The sun would make me see him if he rose a thousand times. The recurrences of the universe rose to the maddening rhythm of an incantation, and I began to see an idea.

All the towering materialism which dominates the modern mind rests ultimately upon one assumption; a false assumption. It is supposed that if a thing goes on repeating itself it is probably dead; a piece of clockwork. People feel that if the universe was personal it would vary; if the sun were alive it would dance. This is a fallacy even in relation to known fact. For the variation in human affairs is generally brought into them, not by life, but by death; by the dying down or breaking off of their strength or desire. A man varies his movements because of some slight element of failure or fatigue. He gets into an omnibus because he is tired of walking; or he walks because he is tired of sitting still. But if his life and joy were so gigantic that he never tired of going to Islington, he might go to Islington as regularly as the Thames goes to Sheerness. The very speed and ecstasy of his life would have the stillness of death. The sun rises every morning. I do not rise every morning; but the variation is due not to my activity, but to my inaction. Now, to put the matter in a popular phrase, it might be true that the sun rises regularly because he never gets tired of rising. His routine might be due, not to a lifelessness, but to a rush of life. The thing I mean can be seen, for instance, in children, when they find some game or joke that they specially enjoy. A child kicks his legs rhythmically through excess, not absence, of life. Because children have abounding vitality, because they are in spirit fierce and free, therefore they want things repeated and unchanged. They always say, "Do it again"; and the grown-up person does it again until he is nearly dead. For grown-up people are not strong enough to exult in monotony. But perhaps God is strong enough to exult in monotony. It is possible that God says every morning, "Do it again" to the sun; and every evening, "Do it again" to the moon. It may not be automatic necessity that makes all daisies alike; it may be that God makes every daisy separately, but has never got tired of making them. It may be that He has the eternal appetite of infancy; for we have sinned and grown old, and our Father is younger than we. The repetition in Nature may not be a mere recurrence; it may be a theatrical *encore*. Heaven may *encore* the bird who laid an egg. If the human being conceives and brings forth a human child instead of bringing forth a fish, or a bat, or a griffin, the reason may not be that we are fixed in an animal fate without life or purpose. It may be that our little tragedy has touched the gods, that they admire it from their starry galleries, and that at the end of every human drama man is called again and again before the curtain. Repetition may go on for millions of years, by mere choice, and at any instant it may stop. Man may stand on the earth generation after generation, and yet each birth be his positively last appearance.

This was my first conviction; made by the shock of my childish emotions meeting the modern creed in mid-career. I had always vaguely felt facts to be miracles in the sense that they are wonderful: now I began to think them miracles in the stricter sense that they were *wilful*. I mean that they were, or might be, repeated exercises of some will. In short, I had always believed that the world involved magic: now I thought that perhaps it involved a magician. And this pointed to a profound emotion always present and sub-conscious; that this world of ours has some purpose; and if there is a purpose, there is a person. I had always felt life first as a story: and if there is a story there is a story-teller.

But modern thought also hit my second human tradition. It went against the fairy feeling about strict limits and conditions. The one thing it loved to talk about was expansion and largeness. Herbert Spencer would have been greatly annoyed if any one had called him an imperialist, and therefore it is highly regrettable that nobody did. But he was an imperialist of the lowest type. He popularized this contemptible notion that the size of the solar system ought to over-awe the spiritual dogma of man. Why should a man surrender his dignity to the solar system any more than to a whale? If mere size proves that man is not the image of God, then a whale may be the image of God; a somewhat formless image; what one might call an impressionist portrait. It is quite futile to argue that man is small compared to the cosmos; for man was always small compared to the nearest tree. But Herbert Spencer, in his headlong imperialism, would insist that we had in some way been conquered and annexed by the astronomical universe. He spoke about men and their ideals exactly as the most insolent Unionist talks about the Irish and their ideals. He turned mankind into a small nationality. And his evil influence can be seen even in the most spirited and honourable of later scientific authors; notably in the early romances of Mr. H. G. Wells. Many moralists have in an exaggerated way represented the earth as wicked. But Mr. Wells and his school made the heavens wicked. We should lift up our eyes to the stars from whence would come our ruin.

But the expansion of which I speak was much more evil than all this. I have remarked that the materialist, like the madman, is in prison; in the prison of one thought. These people seemed to think it singularly inspiring to keep on saying that the prison was very large. The size of this scientific universe gave one no novelty, no relief. The cosmos went on forever, but not in its wildest constellation could there be anything really interesting; anything, for instance, such as forgiveness or free will. The grandeur or infinity of the secret of its cosmos added nothing to it. It was like telling a prisoner in Reading gaol that he would be glad to hear that the gaol now covered half the county. The warder would have nothing to show the man except more and more long corridors of stone lit by ghastly lights and empty of all that is human. So these expanders of the universe had nothing to show us except more and more infinite corridors of space lit by ghastly suns and empty of all that is divine.

In fairyland there had been a real law; a law that could be broken, for the definition of a law is something that can be broken. But the machinery of this cosmic prison was something that could not be broken; for we ourselves were only a part of its machinery. We were either unable to do things or we were destined to do them. The idea of the mystical condition quite disappeared; one can neither have the firmness of keeping laws nor the fun of breaking them. The largeness of this universe had nothing of that freshness and airy outbreak which we have praised in the universe of the poet. This modern universe is literally an empire; that is, it was vast, but it is not free. One went into larger and larger windowless rooms, rooms big with Babylonian perspective; but one never found the smallest window or a whisper of outer air.

Their infernal parallels seemed to expand with distance; but for me all good things come to a point, swords for instance. So finding the boast of the big cosmos so unsatisfactory to my emotions I began to argue about it a little; and I soon found that the whole attitude was even shallower than could have been expected. According to these people the cosmos was one thing since it had one unbroken rule. Only (they would say) while it is one thing it is also the only thing there is. Why, then, should one worry particularly to call it large? There is nothing to compare it with. It would be just as sensible to call it small. A man may say, "I like this vast cosmos, with its throng of stars and its crowd of varied creatures." But if it comes to that why should not a man say, "I like this cosy little cosmos, with its decent number of stars and as neat a provision of live stock as I wish to see?" One is as good as the other; they are both mere sentiments. It is mere sentiment to rejoice that the sun is larger than the earth; it is quite as sane a sentiment to rejoice that the sun is no larger than it is. A man chooses to have an emotion about the largeness of the world; why should he not choose to have an emotion about its smallness?

It happened that I had that emotion. When one is fond of anything one addresses it by diminutives, even if it is an elephant or a lifeguardsman. The reason is, that anything, however huge, that can be conceived of as complete, can be conceived of as small. If military moustaches did not suggest a sword or tusks a tail, then the object would be vast because it would be immeasurable. But the moment you can imagine a guardsman you can imagine a small guardsman. The moment you really see an elephant you can call it "Tiny." If you can make a statue of a thing you can make a statuette of it. These people professed that the universe was one coherent thing; but they were not fond of the universe. But I was frightfully fond of the universe and wanted to address it by a diminutive. I often did so; and it never seemed to mind. Actually and in truth I did feel that these dim dogmas of vitality were better expressed by calling the world small than by calling it large. For about infinity there was a sort of carelessness which was the reverse of the fierce and pious care which I felt touching the pricelessness and the peril of life. They showed only a dreary waste; but I felt a sort of sacred thrift. For economy is far more romantic than extravagance. To them stars were an unending income of halfpence; but I felt about the golden sun and the silver moon as a schoolboy feels if he has one sovereign and one shilling.

These subconscious convictions are best hit off by the colour and tone of certain tales. Thus I have said that stories of magic alone can express my sense that life is not only a pleasure but a kind of eccentric privilege. I may express this other feeling of cosmic cosiness by allusion to another book always read in boyhood, "Robinson Crusoe," which I read about this time, and which owes its eternal vivacity to the fact that it celebrates the poetry of limits, nay, even the wild romance of prudence. Crusoe is a man on a small rock with a few comforts just snatched from the sea: the best thing in the book is simply the list of things saved from the wreck. The greatest of poems is an inventory. Every kitchen tool becomes ideal because Crusoe might have dropped it in the sea. It is a good exercise, in empty or ugly hours of the day, to look at anything, the coal-scuttle or the book-case, and think how happy one could be to have brought it out of the sinking ship on to the solitary island. But it is a better exercise still to remember how all things have had this hair-breadth escape: everything has been saved from a wreck. Every man has had one horrible adventure: as a hidden untimely birth he had not been, as infants that never see the light. Men spoke much in my boyhood of restricted or ruined men of genius: and it was common to say that many a man was a Great Might-Have-Been. To me it is a more solid and startling fact that any man in the street is a Great Might-Not-Have-Been.

But I really felt (the fancy may seem foolish) as if all the order and number of things were the romantic remnant of Crusoe's ship. That there are two sexes and one sun, was like the fact that there were two guns and one axe. It was poignantly urgent that none should be lost; but somehow, it was rather fun that none could be added. The trees and the planets seemed like things saved from the wreck: and when I saw the Matterhorn I was glad that it had not been overlooked in the confusion. I felt economical about the stars as if they were sapphires (they are called so in Milton's Eden): I hoarded the hills. For the universe is a single jewel, and while it is a natural cant to talk of a jewel as peerless and priceless, of this jewel it is literally true. This cosmos is indeed without peer and without price: for there cannot be another one.

Thus ends, in unavoidable inadequacy, the attempt to utter the unutterable things. These are my ultimate attitudes towards life; the soils for the seeds of doctrine. These in some dark way I thought before I could write, and felt before I could think: that we may proceed more easily afterwards, I will roughly recapitulate them now. I felt in my bones; first, that this world does not explain itself. It may be a miracle with a supernatural explanation; it may be a conjuring trick, with a natural explanation. But the explanation of the conjuring trick, if it is to satisfy me, will have to be better than the natural explanations I have heard. The thing is magic, true or false. Second, I came to feel as if magic must have a meaning, and meaning must have some one to mean it. There was something personal in the world, as in a work of art; whatever it meant it meant violently. Third, I thought this purpose beautiful in its old design, in spite of its defects, such as dragons. Fourth, that the proper form of thanks to it is some form of humility and restraint: we should thank God for beer and Burgundy by not drinking too much of them. We owed, also an obedience to whatever made us. And last, and strangest, there had come into my mind a vague and vast impression that in some way all good was a remnant to be stored and held sacred out of some primordial ruin. Man had saved his good as Crusoe saved his goods: he had saved them from a wreck. All this I felt and the age gave me no encouragement to feel it. And all this time I had not even thought of Christian theology.

E. M. Forster
"Fantasy," from *Aspects of the Novel* (1927)

A course of lectures, if it is to be more than a collection of remarks, must have an idea running through it. It must also have a subject, and the idea ought to run through the subject too. This is so obvious as to sound foolish, but anyone who has tried to lecture will realize that here is a genuine difficulty. A course, like any other collection of words, generates an atmosphere. It has its own apparatus—a lecturer, an audience or provision for one, it occurs at regular intervals, it is announced by printed notices, and it has a financial side, though this last is tactfully concealed. Thus it tends in its parasitic way to lead a life of its own, and it and the idea running through it are apt to move in one direction while the subject steals off in the other.

The idea running through these lectures is by now plain enough: that there are in the novel two forces: human beings and a bundle of various things not human beings, and that it is the novelist's business to adjust these two forces and conciliate their claims. That is plain enough, but does it run through the novel too? Perhaps our subject, namely the books we have read, has stolen away from us while we theorize, like a shadow from an ascending bird. The bird is all right—it climbs, it is consistent and eminent. The shadow is all right—it has flickered across roads and gardens. But the two things resemble one another less and less, they do not touch as they did when the bird rested its toes on the ground. Criticism, especially a critical course, is so misleading. However lofty its intentions and sound its method, its subject slides away from beneath it, imperceptibly away, and lecturer and audience may awake with a start to find that they are carrying on in a distinguished and intelligent manner, but in regions which have nothing to do with anything they have read.

It was this that was worrying Gide, or rather one of the things that was worrying him, for he has an anxious mind. When we try to translate truth out of one sphere into another, whether from life into books or from books into lectures, something happens to truth, it goes wrong, not suddenly when it might be detected, but slowly. That long passage from *Les Faux Monnayeurs* already quoted, may recall the bird to its shadow. It is not possible, after it, to apply the old apparatus any more. There is more in the novel than time or people or logic or any of their derivatives, more even than Fate. And by "more" I do not mean something that excludes these aspects nor something that includes them, embraces them. I mean something that cuts across them like a bar of light, that is intimately connected with them at one place and patiently illumines all their problems, and at another place shoots over or through them as if they did not exist. We shall give that bar of light two names, fantasy and prophecy.

The novels we have now to consider all tell a story, contain characters, and have plots or bits of plots, so we could apply to them the apparatus suited for Fielding or Arnold Bennett. But when I say two of their names—*Tristram Shandy* and *Moby Dick*—it is clear that we must stop and think a moment. The bird and the shadow are too far apart. A new formula must be found: the mere fact that one can mention Tristram and Moby in a single sentence shows it. What an impossible pair! As far apart as the poles. Yes. And like the poles they have one thing in common, which the lands round the equator do not share: an axis. What is essential in Sterne and Melville belongs to this new aspect of fiction: the fantastic-prophetical axis. George Meredith touched it: he was somewhat fantastic. So did Charlotte Brontë: she was a prophetess occasionally. But in neither of these was it essential. Deprive them of it, and a book remains which still resembles *Harry Richmond* or *Shirley*. Deprive Sterne or Melville of it, deprive Peacock or Max Beerbohm or Virginia Woolf or Walter de la Mare or William Beckford or James Joyce or D. H. Lawrence or Swift, and nothing is left at all.

Our easiest approach to a definition of any aspect of fiction is always by considering the sort of demand it makes on the reader. Curiosity for the story, human feelings and a sense of value for the characters, intelligence and memory for the plot. What does fantasy ask of us? It asks us to pay something extra. It compels us to an adjustment that is different to an adjustment required by a work of art, to an additional adjustment. The other novelists say "Here is something that might occur in your lives," the fantasist says "Here's something that could not occur. I must ask you first to accept my book as a whole, and secondly to accept certain things in my book." Many readers can grant the first request, but refuse the second. "One knows a book isn't real," they say, "still one does expect it to be natural, and this angel or midget or ghost or silly delay about the child's birth–no, it is too much." They either retract their original concession and stop reading, or if they do go on it is with complete coldness, and they watch the gambols of the author without realizing how much they may mean to him.

No doubt the above approach is not critically sound. We all know that a work of art is an entity, etc., etc.; it has its own laws which are not those of daily life, anything that suits it is true, so why should any question arise about the angel, etc. except whether it is suitable to its book? Why place an angel on a different basis from a stockbroker? Once in the realm of the fictitious, what difference is there between an apparition and a mortgage? I see the soundness of this argument, but my heart refuses to assent. The general tone of novels is so literal that when the fantastic is introduced it produces a special effect: some readers are thrilled, others choked off: it demands an additional adjustment because of the oddness of its method or subject matter–like a sideshow in an exhibition where you have to pay sixpence as well as the original entrance fee. Some readers pay with delight, it is only for the sideshows that they entered the exhibition, and it is only to them I can now speak. Others refuse with indignation, and these have our sincere regards, for to dislike the fantastic in literature is not to dislike literature. It does not even imply poverty of imagination, only a disinclination to meet certain demands that are made on it. Mr. Asquith (if gossip is correct) could not meet the demands made on him by *Lady into Fox*. He should not have objected, he said, if the fox had become a lady again, but as it was he was left with an uncomfortable dissatisfied feeling. This feeling reflects no discredit either upon an eminent politician or a charming book. It merely means that Mr. Asquith, though a genuine lover of literature, could not pay the additional sixpence–or rather he was willing to pay it but hoped to get it back again at the end.

So fantasy asks us to pay something extra.

Let us now distinguish between fantasy and prophecy.

They are alike in having gods, and unlike in the gods they have. There is in both the sense of mythology which differentiates them from other aspects of our subject. An invocation is again possible, therefore on behalf of fantasy let us now invoke all beings who inhabit the lower air, the shallow water, and the smaller hills, all Fauns and Dryads and slips of the memory, all verbal coincidences, Pans and puns, all that is mediæval this side of the grave. When we come to prophecy, we shall utter no invocation, but it will have been to whatever transcends our abilities, even when it is human passion that transcends them, to the deities of India, Greece, Scandinavia and Judæa, to all that is mediæval beyond the grave and to Lucifer son of the morning. By their mythologies we shall distinguish these two sorts of novels.

A number of rather small gods then should haunt us today–I would call them fairies if the word were not consecrated to imbecility. (Do you believe in fairies? No, not under any circumstances.) The stuff of daily life will be tugged and strained in various directions, the earth will be given little tilts mischievous or pensive, spot lights will fall on objects that have no reason to anticipate or welcome them, and tragedy herself, though not excluded, will have a fortuitous air as if a word would disarm her. The power of fantasy penetrates into every corner of the universe, but not into the forces that govern it–the stars that are the brain of heaven, the army of unalterable law, remain untouched–and novels of this type have an improvised air, which is the secret of their force and charm. They may contain solid character-drawing, penetrating and bitter criticism of conduct and civilization; yet our simile of the beam of light must remain, and if one god must be invoked specially, let us call upon Hermes–messenger, thief, and conductor of souls to a not too terrible hereafter.

You will expect me now to say that a fantastic book asks us to accept the supernatural. I will say it, but reluctantly, because any statement as to their subject matter brings these novels into the claws of critical apparatus, from which it is important that they should be saved. It is truer of them than of most books that we can only know what is in them by reading them, and their appeal is specially personal–they are sideshows inside the main show. So I would rather hedge as much as possible, and say that they ask us to accept either the supernatural or its absence.

A reference to the greatest of them–*Tristram Shandy*–will make this point clear. The supernatural is absent from the Shandy ménage, yet a thousand incidents suggest that it is not far off. It would not be really odd, would it, if the furniture in Mr. Shandy's bedroom, where he retired in despair after hearing the omitted details of his son's birth, should come alive like Belinda's toilette in *The Rape of the Lock,* or that Uncle Toby's drawbridge should lead into Lilliput? There is a charmed stagnation about the whole epic–the more the characters do the less gets done,

E. M. Foster, circa 1901

the less they have to say the more they talk, the harder they think the softer they get, facts have an unholy tendency to unwind and trip up the past instead of begetting the future, as in well-conducted books, and the obstinacy of inanimate objects, like Dr. Slop's bag, is most suspicious. Obviously a god is hidden in *Tristram Shandy,* his name is Muddle, and some readers cannot accept him. Muddle is almost incarnate–quite to reveal his awful features was not Sterne's intention; that is the deity that lurks behind his masterpiece–the army of unutterable muddle, the universe as a hot chestnut. Small wonder that another divine muddler, Dr. Johnson, writing in 1776, should remark, "Nothing odd will do long: *Tristram Shandy* did not last!" Doctor Johnson was not always happy in his literary judgments, but the appropriateness of this one passes belief.

Well, that must serve as our definition of fantasy. It implies the supernatural, but need not express it. Often it does express it, and were that type of classification helpful, we could make a list of the devices which writers of a fantastic turn have used–such as the introduction of a god, ghost, angel, monkey, monster, midget, witch into ordinary life; or the introduction of ordinary men into no man's land, the future, the past, the interior of the earth, the fourth dimension; or divings into and dividings of personality; or finally the device of parody or adaptation. These devices need never grow stale; they will occur naturally to writers of a certain temperament, and be put to fresh use; but the fact that their number is strictly limited is of interest; and suggests that the beam of light can only be manipulated in certain ways.

I will select, as a typical example, a recent book about a witch: *Flecker's Magic,* by Norman Matson. It seemed to me good and I recommended it to a friend whose judgment I respect. He thought it poor. That is what is so tiresome about new books; they never give us that restful feeling which we have when perusing the classics. *Flecker's Magic* contains scarcely anything that is new–fantasies cannot: only the old old story of the wishing ring which brings either misery or nothing at all. Flecker, an American boy who is learning to paint in Paris, is given the ring by a girl in a café; she is a witch, she tells him; he has only to be sure what he wants and he will get it. To prove her power, a motor-bus rises slowly from the street and turns upside down in the air. The passengers, who do not fall out, try to look as if nothing was happening. The driver, who is standing on the pavement at the moment, cannot conceal his surprise, but when his bus returns safe to earth again he thinks it wiser to get into his seat and drive off as usual. Motor-buses do not revolve slowly through the air–so they do not. Flecker now accepts the ring. His character, though slightly sketched, is individual, and this definiteness causes the book to grip.

It proceeds with a growing tension, a series of little shocks. The method is Socratic. The boy starts by thinking of something obvious, like a Rolls-Royce. But where shall he put the beastly thing? Or a beautiful lady. But what about her carte d'identité? Or money? Ah, that's more like it–he is almost a beggar. Say a million dollars. He prepares to turn the ring for this wish–except while one's about it two millions seem safer–or ten–or–and money blares out into madness, and the same thing happens when he thinks of long life: to die in forty years–no, in fifty–in one hundred–horrible, horrible. Then a solution occurs. He has always wanted to be a great painter. Well, he'll be it at once. But what kind of greatness? Giotto's? Cézanne's? Certainly not; his own kind, and he does not know what that is, so this wish likewise is impossible.

And now a horrible old woman begins to haunt his days and dreams. She reminds him vaguely of the girl who gave him the ring. She knows his thoughts and she is always sidling up to him in the streets and saying, "Dear boy–darling boy–wish for happiness." We learn in time that she is the real witch–the girl was a human acquaintance whom she used to get into touch with Flecker. The last of the witches–very lonely. The rest have committed suicide during the eighteenth century–they could not endure to survive into the world of Newton where two and two make four, and even the world of Einstein is not sufficiently decentralised to revive them. She has hung on in the hope of smashing this world, and she wants the boy to ask for happiness because such a wish has never been made in all the history of the ring.

> Perhaps Flecker was the first modern man to find himself in this predicament? The people of the old world had so little they knew surely what they wanted. They knew about Almighty God, who wore a beard and sat in an armchair about a mile above the fields, and life was very short and very long too, for the days were so full of unthinking effort.
>
> The people of the recorded olden times wished for a beautiful castle on a high hill and lived therein until death. But the hill was not so high one might see from the windows back along thirty centuries–as one may from a bungalow. In the castle there were no great volumes filled with words and pictures of things dug up by man's relentless curiosity from sand and soil in all corners of the world; there was a sentimental half-belief in dragons, but no knowledge that once upon a time only dragons had lived on the earth–that man's grandfather and grandmother were dragons; there were no movies flickering like thoughts against a white wall, no phonograph, no machinery with which to achieve the sensation of speed; no diagrams of the fourth dimension, no contrasts in life like that of Waterville, Minn., and Paris, France. In the castle the light was weak and flickering, hallways were dark, rooms deeply shadowed. The little outside world was full of shadow, and on the very top of the mind of him who lived in the castle played a dim light–underneath were shadows, fear, ignorance, will-to-ignorance. Most of all, there was not in the castle on the hill the breathless sense of imminent revelation–that today or surely tomorrow Man would at a stroke double his power and change the world again.
>
> The ancient tales of magic were the mumbling thoughts of a distant shabby little world–so, at least, thought Flecker, offended. The tales gave him no guidance. There was too much difference between his world and theirs.
>
> He wondered if he hadn't dismissed the wish for happiness rather heedlessly? He seemed to get nowhere thinking about it. He was not wise enough. In the old tales a wish for happiness was never made! He wondered why.
>
> He might chance it–just to see what would happen. The thought made him tremble. He leaped from his bed and paced the red-tiled floor, rubbing his hands together.
>
> "I want to be happy forever," he whispered, to hear the words, careful not to touch the ring. "*Happy . . . for ever*"–the two syllables of the first word, like hard little pebbles, struck musically against the bell of his imagination, but the second was a sigh. *For ever*–his spirit sank under the soft heavy impact of it. Held in his thought the word made a dreary music, fading. "*Happy for ever*"–NO!!

Thus again and again–the mark of the true fantasist–does Norman Matson merge the kingdoms of magic and common sense by using words that apply to both, and the mixture he has created comes alive. I will not tell the end of the story. You will have guessed its essentials, but there are always surprises in the working of a fresh mind, and to the end of time good literature will be made round this notion of a wish.

To turn from this simple example of the supernatural to a more complicated one—to a highly accomplished and superbly written book whose spirit is farcical: *Zuleika Dobson* by Max Beerbohm. You all know Miss Dobson—not personally, or you would not be here now. She is that damsel for love of whom all the undergraduates of Oxford except one drowned themselves during Eights week, and he threw himself out of a window.

A superb theme for a fantasy, but all will depend on the handling. It is treated with a mixture of realism, wittiness, charm and mythology, and the mythology is most important. Max has borrowed or created a number of supernatural machines—to have entrusted Zuleika to one of them would be inept; the fantasy would become heavy or thin. But we pass from the sweating emperors to the black and pink pearls, the hooting owls, the interference of the Muse Clio, the ghosts of Chopin and George Sand, of Nellie O'Mora; just as one fails another starts, to uphold this gayest and most exquisite of funeral palls.

> Through the square, across the High, down Grove Street they passed. The Duke looked up at the tower of Merton, *os oupor authis alla nun panustaton.* Strange that to-night it would still be standing here, in all its sober and solid beauty—still be gazing, over the roofs and chimneys, at the tower of Magdalen, its rightful bride. Through untold centuries of the future it would stand thus, gaze thus. He winced. Oxford walls have a way of belittling us; and the Duke was loth to regard his doom as trivial.
>
> Aye, by all minerals we are mocked. Vegetables, yearly deciduous, are far more sympathetic. The lilac and laburnum, making lovely now the railed pathway to Christ Church meadow, were all a-swaying and nodding to the Duke as he passed by. "Adieu, adieu, your Grace," they were whispering. "We are very sorry for you, very sorry indeed. We never dared suppose you would predecease us. We think your death a very great tragedy. Adieu! Perhaps we shall meet in another world—that is, if the members of the animal kingdom have immortal souls, as we have."
>
> The Duke was little versed in their language; yet, as he passed between these gently garrulous blooms, he caught at the least the drift of their salutation, and smiled a vague but courteous acknowledgment, to the right and the left alternately, creating a very favourable impression.

Has not a passage like this—with its freedom of invocation—a beauty unattainable by serious literature? It is so funny and charming, so iridescent yet so profound. Criticisms of human nature fly through the book, not like arrows but upon the wings of sylphs. Towards the end—that dreadful end often so fatal to fiction—the book rather flags: the suicide of all the undergraduates of Oxford is not as delightful as it ought to be when viewed at close quarters, and the defenestration of Noaks almost nasty. Still it is a great work—the most consistent achievement of fantasy in our time, and the closing scene in Zuleika's bedroom with its menace of further disasters is impeccable.

> And now with pent breath and fast-beating heart, she stared at the lady of the mirror, without seeing her; and now she wheeled round and swiftly glided to that little table on which stood her two books. She snatched Bradshaw.
>
> We always intervene between Bradshaw and anyone whom we see consulting him. "Mademoiselle will permit me to find that which she seeks?" asked Melisande.
>
> "Be quiet," said Zuleika. We always repulse, at first, anyone who intervenes between us and Bradshaw.
>
> We always end by accepting the intervention. "See if it is possible to go direct from here to Cambridge," said Zuleika, handing the book on. "If it isn't, then—well, see how one *does* get there."
>
> We never have any confidence in the intervener. Nor is the intervener, when it comes to the point, sanguine. With mistrust mounting to exasperation Zuleika sat watching the faint and frantic researches of her maid.
>
> "Stop!" she said suddenly. "I have a much better idea. Go down very early to the station. See the stationmaster. Order me a special train. For ten o'clock, say."
>
> Rising, she stretched her arms above her head. Her lips parted in a yawn, met in a smile. With both hands she pushed back her hair from her shoulders, and twisted it into a loose knot. Very lightly she slipped up into bed, and very soon she was asleep.

So Zuleika ought to have come on to this place. She does not seem ever to have arrived and we can only suppose that through the intervention of the gods her special train failed to start, or, more likely, is still in a siding at Bletchley.

Among the devices in my list I mentioned "parody" or "adaptation" and would now examine this further. The fantasist here adopts for his mythology some earlier work and uses it as a framework or quarry for his own purposes. There is an aborted example of this in *Joseph Andrews*. Fielding set out to use *Pamela* as a comic mythology. He thought it would be fun to invent a brother to Pamela, a pure-minded footman, who should repulse Lady Booby's attentions just as Pamela had repulsed Mr. B.'s, and he made Lady Booby Mr. B.'s aunt. Thus he would be able to laugh at Richardson, and incidentally express his own views of life. Fielding's view of life however was of the sort that only rests content with the creation of solid round characters, and with the growth of Parson Adams and Mrs. Slipslop the fantasy ceases, and we get an independent work. *Joseph Andrews* (which is also important historically) is interesting to us as an example of a false start. Its author begins by playing the fool in a Richardsonian world, and ends by being serious in a world of his own–the world of Tom Jones and Amelia.

Parody or adaptation have enormous advantages to certain novelists, particularly to those who may have a great deal to say and plenty of literary genius, but who do not see the world in terms of individual men and women–who do not, in other words, take easily to creating characters. How are such men to start writing? An already existing book or literary tradition may inspire them–they may find high up in its cornices a pattern that will serve as a beginning, they may swing about in its rafters and gain strength. That fantasy of Lowes Dickinson, *The Magic Flute,* seems to be created thus: it has taken as its mythology the world of Mozart. Tamino, Sarastro, and the Queen of the Night stand in their enchanted kingdom ready for the author's thoughts, and when these are poured in they become alive and a new and exquisite work is born. And the same is true of another fantasy, anything but exquisite–James Joyce's *Ulysses.** That remarkable affair–perhaps the most interesting literary experiment of our day–could not have been achieved unless Joyce had had, as his guide and butt, the world of the *Odyssey.*

I am only touching on one aspect of *Ulysses:* it is of course more than a fantasy–it is a dogged attempt to cover the universe with mud, it is an inverted Victorianism, an attempt to make crossness and dirt succeed where sweetness and light failed, a simplification of the human character in the interests of Hell. All simplifications are fascinating, all lead us away from the truth (which lies far nearer the muddle of *Tristram Shandy*), and *Ulysses* must not detain us on the ground that it contains a morality–otherwise we shall also have to discuss Mrs. Humphrey Ward. We are concerned with it because, through a mythology, Joyce has been able to create the peculiar stage and characters he required.

The action of those 400,000 words occupies a single day, the scene is Dublin, the theme is a journey–the modern man's journey from morn to midnight, from bed to the squalid tasks of mediocrity, to a funeral, newspaper office, library, pub, lavatory, lying-in hospital, a saunter by the beach, brothel, coffee stall, and so back to bed. And it coheres because it depends from the journey of a hero through the seas of Greece, like a bat hanging to a cornice.

Ulysses himself is Mr. Leopold Bloom–a converted Jew–greedy, lascivious, timid, undignified, desultory, superficial, kindly and always at his lowest when he pretends to aspire. He tries to explore life through the body. Penelope is Mrs. Marion Bloom, an overblown soprano, by no means harsh to her suitors. The third character is young Stephen Dedalus, whom Bloom recognizes as his spiritual son much as Ulysses recognizes Telemachus as his actual son. Stephen tries to explore life through the intellect–we have met him before in *The Portrait of the Artist as a Young Man,* and now he is worked into this epic of grubbiness and disillusion. He and Bloom meet half way through in Night Town (which corresponds partly to Homer's Palace of Circe, partly to his Descent into Hell) and in its supernatural and filthy alleys they strike up their slight but genuine friendship. This is the crisis of the book, and here–and indeed throughout–smaller mythologies swarm and pullulate, like vermin between the scales of a poisonous snake. Heaven and earth fill with infernal life, personalities melt, sexes interchange, until the whole universe, including poor, pleasure-loving Mr. Bloom, is involved in one joyless orgy.

Does it come off? No, not quite. Indignation in literature never quite comes off either in Juvenal or Swift or Joyce; there is something in words that is alien to its simplicity. The Night Town scene does not come off except as a superfetation of fantasies, a monstrous coupling of reminiscences. Such satisfaction as can be attained in this direction is attained, and all through the book we have similar experiments–the aim of which is to degrade all things and more particularly civilization and art, by turning them inside out and upside down. Some enthusiasts may think that *Ulysses* ought to be mentioned not here but later on, under the heading of prophecy, and I understand this criticism. But I prefer to mention it today with *Tristram Shandy, Flecker's Magic, Zuleika Dobson,* and *The Magic Flute,* because the raging of Joyce, like the happier or calmer moods of the other writers, seems essentially fantastic, and lacks the note for which we shall be listening soon.

We must pursue this notion of mythology further, and more circumspectly.

*Ulysses (Shakespeare & Co., Paris) is not at present obtainable in England. America, more enlightened, has produced a mutilated version without the author's permission and without paying him a cent.

H. G. Wells
Preface to *The Scientific Romances of H. G. Wells* (1933)

Mr. Gollancz has asked me to write a preface to this collection of my fantastic stories. They are put in chronological order, but let me say here right at the beginning of the book, that for anyone who does not as yet know anything of my work it will probably be more agreeable to begin with *The Invisible Man* or *The War of the Worlds. The Time Machine* is a little bit stiff about the fourth dimension and *The Island of Dr. Moreau* rather painful.

These tales have been compared with the work of Jules Verne and there was a disposition on the part of literary journalists at one time to call me the English Jules Verne. As a matter of fact there is no literary resemblance whatever between the anticipatory inventions of the great Frenchman and these fantasies. His work dealt almost always with actual possibilities of invention and discovery, and he made some remarkable forecasts. The interest he invoked was a practical one; he wrote and believed and told that this or that thing could be done, which was not at that time done. He helped his reader to imagine it done and to realize what fun, excitement or mischief would ensue. Many of his inventions have 'come true'. But these stories of mine collected here do not pretend to deal with possible things; they are exercises of the imagination in a quite different field. They belong to a class of writing which includes the *Golden Ass of Apuleius,* the *True Histories of Lucian, Peter Schlemil* and the story of *Frankenstein.* It includes too some admirable inventions by Mr. David Garnett, *Lady into Fox* for instance. They are all fantasies; they do not aim to project a serious possibility; they aim indeed only at the same amount of conviction as one gets in a good gripping dream. They have to hold the reader to the end by art and illusion and not by proof and argument, and the moment he closes the cover and reflects he wakes up to their impossibility.

In all this type of story the living interest lies in their non-fantastic elements and not in the invention itself. They are appeals for human sympathy quite as much as any 'sympathetic' novel, and the fantastic element, the strange property or the strange world, is used only to throw up and intensify our natural reactions of wonder, fear or perplexity. The invention is nothing in itself and when this kind of thing is attempted by clumsy writers who do not understand this elementary principle nothing could be conceived more silly and extravagant. Anyone can invent human beings inside out or worlds like dumb-bells or a gravitation that repels. The thing that makes such imaginations interesting is their translation into commonplace terms and a rigid exclusion of other marvels from the story. Then it becomes human. 'How would you feel and what might not happen to you', is the typical question, if for instance pigs could fly and one came rocketing over a hedge at you? How would you feel and what might not happen to you if suddenly you were changed into an ass and couldn't tell anyone about it? Or if you became invisible? But no one would think twice about the answer if hedges and houses also began to fly, or if people changed into lions, tigers, cats and dogs left and right, or if everyone could vanish anyhow. Nothing remains interesting where anything may happen.

For the writer of fantastic stories to help the reader to play the game properly, he must help him in every possible unobstrusive way to *domesticate* the impossible hypothesis. He must trick him into an unwary concession to some plausible assumption and get on with his story while the illusion holds. And that is where there was a certain slight novelty in my stories when first they appeared. Hitherto, except in exploration fantasies, the fantastic element was brought in by magic. Frankenstein even, used some jiggery-pokery magic to animate his artificial monster. There was trouble about the thing's soul. But by the end of last century it had become difficult to squeeze even a momentary belief out of magic any longer. It occurred to me that instead of the usual interview with the devil or a magician, an ingenious use of scientific patter might with advantage be substituted. That was no great discovery. I simply brought the fetish stuff up to date, and made it as near actual theory as possible.

As soon as the magic trick has been done the whole business of the fantasy writer is to keep eve-

H. G. Wells in 1915

rything else human and real. Touches of prosaic detail are imperative and rigorous adherence to the hypothesis. Any *extra* fantasy outside the cardinal assumption immediately gives a touch of irresponsible silliness to the invention. So soon as the hypothesis is launched the whole interest becomes the interest of looking at human feelings and human ways, from the new angle that has been acquired. One can keep the story within the bounds of a few individual experiences as Chamisso does in *Peter Schlemil,* or one can expand it to a broad criticism of human institutions and limitations as in *Gulliver's Travels.* My early, profound and lifelong admiration for Swift, appears again and again in this collection, and it is particularly evident in a predisposition to make the stories reflect upon contemporary political and social discussions. It is an incurable habit with literary critics to lament some lost artistry and innocence in my early work and to accuse me of having become polemical in my later years. That habit is of such old standing that the late Mr. Zangwill in a review in 1895 complained that my first book, *The Time Machine,* concerned itself with 'our present discontents.' *The Time Machine* is indeed quite as philosophical and polemical and critical of life and so forth, as *Men like Gods* written twenty-eight years later. No more and no less. I have never been able to get away from life in the mass and life in general as distinguished from life in the individual experience, in any book I have ever written. I differ from contemporary criticism in finding them inseparable.

For some years I produced one or more of these 'scientific fantasies', as they were called, every year. In my student days we were much exercised by talk about a possible fourth dimension of space; the fairly obvious idea that events could be presented in a rigid four dimensional space time framework had occurred to me, and this is used as the magic trick for a glimpse of the future that ran counter to the placid assumption of that time that Evolution was a pro-human force making things better and better for mankind. *The Island of Dr. Moreau* is an exercise in youthful blasphemy. Now and then, though I rarely admit it, the universe projects itself towards me in a hideous grimace. It grimaced that time, and I did my best to express my vision of the aimless torture in creation. *The War of the Worlds* like *The Time Machine* was another assault on human self-satisfaction.

All these three books are consciously grim, under the influence of Swift's tradition. But I am neither a pessimist nor an optimist at bottom. This is an entirely indifferent world in which wilful wisdom seems to have a perfectly fair chance. It is after all rather cheap to get force of presentation by loading the scales on the sinister side. Horror stories are easier to write than gay and exalting stories. In *The First Men in the Moon* I tried an improvement on Jules Verne's shot, in order to look at mankind from a distance and burlesque the effects of specialization. Verne never landed on the moon because he never knew of radio and of the possibility of sending back a message. So it was his shot that came back. But equipped with radio, which had just come out then, I was able to land and even see something of the planet.

The three later books are distinctly on the optimistic side. *The Food of the Gods* is a fantasia on the change of scale in human affairs. Everybody nowadays realizes that change of scale; we see the whole world in disorder through it; but in 1904 it was not a very prevalent idea. I had hit upon it while working out the possibilities of the near future in a book of speculations called *Anticipations.*

The last two stories are Utopian. The world is gassed and cleaned up morally by the benevolent tail of a comet in one, and the reader is taken through a dimensional trap door with a weekend party of politicians, into a world of naked truth and deliberate beauty in the other. *Men like Gods* is almost the last of my scientific fantasies. It did not horrify or frighten, was not much of a success, and by that time I had tired of talking in playful parables to a world engaged in destroying itself. I was becoming too convinced of the strong probability of very strenuous and painful human experiences in the near future to play about with them much more. But I did two other sarcastic fantasies, not included here, *Mr. Blettsworthy on Rampole Island* and *The Autocracy of Mr. Parham,* in which there is I think a certain gay bitterness, before I desisted altogether.

The Autocracy of Mr. Parham is all about dictators, and dictators are all about us, but it has never struggled through to a really cheap edition. Work of this sort gets so stupidly reviewed nowadays that it has little chance of being properly read. People are simply warned that there are ideas in my books and advised not to read them, and so a fatal suspicion has wrapped about the later ones. 'Ware stimulants!' It is no good my saying that they are quite as easy to read as the earlier ones and much more timely.

It becomes a bore doing imaginative books that do not touch imaginations, and at length one stops even planning them. I think I am better employed now nearer reality, trying to make a working analysis of our deepening social perplexities in such labours as *The Work, Wealth and Happiness of Mankind* and *After Democracy.* The work in the presence of cataclysmal realities has no need for fresh cataclysmal fantasies. That game is over. Who wants the invented humours of Mr. Parham in Whitehall, when day by day we can watch Mr. Hitler in Germany? What human invention can pit itself against the fantastic fun of the Fates? I am wrong in grumbling at reviewers. Reality has taken a leaf from my book and set itself to supersede me.

Books for Further Reading

Aldiss, Brian W., with David Wingrove. *Trillion Year Spree: The History of Science Fiction.* London: Gollancz, 1986; New York: Atheneum, 1986.

Alkon, Paul K. *Origins of Futuristic Fiction.* Athens & London: University of Georgia Press, 1988.

Ashley, Mike. *Who's Who in Horror and Fantasy Fiction.* London: Elm Tree, 1977.

Bailey, J. O. *Pilgrims through Space and Time: Trends and Patterns in Scientific and Utopian Fiction.* New York: Argus, 1947.

Barron, Neil, ed. *Anatomy of Wonder: A Critical Guide to Science Fiction,* fourth edition. New Providence, N.J.: R. R. Bowker, 1995.

Barron, ed. *Fantasy Literature: A Reader's Guide.* New York & London: Garland, 1990.

Barron, ed. *Horror Literature: A Reader's Guide.* New York & London: Garland, 1990.

Bayer-Berenbaum, Linda. *The Gothic Imagination: Expansion in Gothic Literature and Art.* Rutherford, N.J.: Fairleigh Dickinson University Press, 1982.

Bleiler, E. F. *The Guide to Supernatural Fiction.* Kent, Ohio: Kent State University Press, 1983.

Bleiler and Richard J. Bleiler. *Science-Fiction: The Early Years.* Kent, Ohio & London: Kent State University Press, 1990.

Bleiler, E. F., ed. *Science Fiction Writers: Critical Studies of the Major Authors from the Early Nineteenth Century to the Present Day.* New York: Scribners, 1982.

Bleiler, E. F., ed. *Supernatural Fiction Writers: Fantasy and Horror,* 2 volumes. New York: Scribners, 1985.

Briggs, Julia. *Night Visitors: The Rise and Fall of the English Ghost Story.* London: Faber & Faber, 1977.

Carter, Lin. *Imaginary Worlds: The Art of Fantasy.* New York: Ballantine, 1973.

Carter, Margaret L. *Specter or Delusion? The Supernatural in Gothic Fiction.* Ann Arbor, Mich.: UMI Research Press, 1987.

Cavaliero, Glen. *The Supernatural and English Fiction.* Oxford & New York: Oxford University Press, 1995.

Cawthorn, James, and Michael Moorcock. *Fantasy: The 100 Best Books.* London: Xanadu, 1988; New York: Carroll & Graf, 1988.

Clarke, I. F. *The Pattern of Expectation 1644–2001.* London: Cape, 1979.

Clute, John, and Peter Nicholls, eds. *The Encyclopedia of Science Fiction.* New York: St. Martin's Press, 1993.

Daniels, Les. *Living in Fear: A History of Horror in the Mass Media.* New York: Scribners, 1975.

Day, William Patrick. *In the Circles of Fear and Desire: A Study of Gothic Fantasy.* Chicago: University of Chicago Press, 1985.

de Camp, L. Sprague. *Literary Swordsmen and Sorcerers: The Makers of Heroic Fantasy.* Sauk City, Wis.: Arkham House, 1976.

Green, Roger Lancelyn. *Into Other Worlds: Space-Flight in Fiction, from Lucian to Lewis.* New York: Arno, 1975.

Gunn, James. *The New Encyclopedia of Science Fiction.* New York: Viking, 1988.

Heller, Terry. *The Delights of Terror: An Aesthetics of the Tale of Terror.* Urbana: University of Illinois Press, 1987.

Hillegas, Mark R. *The Future as Nightmare: H. G. Wells and the Anti-Utopians.* New York: Oxford University Press, 1967.

Hume, Kathryn. *Fantasy and Mimesis: Responses to Reality in Western Literature.* London: Methuen, 1984.

Irwin, W. R. *The Game of the Impossible: A Rhetoric of Fantasy.* Urbana: University of Illinois Press, 1976.

Jackson, Rosemary. *Fantasy: The Literature of Subversion.* London: Methuen, 1981.

Jones, Stephen, and Kim Newman, eds. *Horror: 100 Best Books.* London: Xanadu, 1988; New York: Carroll & Graf, 1988.

Joshi, S. T. *The Weird Tale: Arthur Machen, Algernon Blackwood, M. R. James, Ambrose Bierce, H. P. Lovecraft.* Austin: University of Texas Press, 1990.

Kiely, Robert. *The Romantic Novel in England.* Cambridge, Mass.: Harvard University Press, 1972.

Kroeber, Karl. *Romantic Fantasy and Science Fiction.* New Haven: Yale University Press, 1988.

Little, Edmund. *The Fantasts: Studies in J. R. R. Tolkien, Lewis Carroll, Mervyn Peake, Nicolai Gogol, and Kenneth Grahame.* London: Avebury, 1984.

Lochhead, Marion. *The Renaissance of Wonder in Children's Literature.* Edinburgh: Canongate, 1977. Republished as *Renaissance of Wonder: The Fantasy Worlds of J. R. R. Tolkien, C. S. Lewis, George MacDonald, E. Nesbit and Others.* New York: Harper, 1980.

Lovecraft, H. P. *Supernatural Horror in Literature.* New York: Abramson, 1945.

Magill, Frank N., ed. *Survey of Modern Fantasy Literature,* 5 volumes. Englewood Cliffs, N.J.: Salem Press, 1983.

Magill, ed. *Survey of Science Fiction Literature,* 5 volumes. Englewood Cliffs, N.J.: Salem Press, 1979.

Manlove, C. N. *Christian Fantasy: From 1200 to the Present.* London: Macmillan, 1992.

Manlove. *The Impulse of Fantasy Literature.* Kent, Ohio: Kent State University Press, 1983.

Manlove. *Modern Fantasy: Five Studies.* Cambridge: Cambridge University Press, 1975.

Milbank, Alison. *Daughters of the House: Modes of the Gothic in Victorian Fiction.* New York: St. Martin's Press, 1992.

Parrinder, Patrick. *Science Fiction: Its Criticism and Teaching.* London & New York: Methuen, 1980.

Parrinder, ed. *Science Fiction: A Critical Guide.* London & New York: Longman, 1979.

Penzoldt, Peter. *The Supernatural in Fiction.* London: Nevill, 1952.

Philmus, Robert M. *Into the Unknown: The Evolution of Science Fiction from Francis Godwin to H. G. Wells.* Berkeley & Los Angeles: University of California Press, 1970.

Prickett, Stephen. *Victorian Fantasy.* Bloomington: Indiana University Press, 1979.

Punter, David. *The Literature of Terror: A History of Gothic Fiction from 1765 to the Present Day.* New York: Longman, 1980.

Rabkin, Eric S. *The Fantastic in Literature.* Princeton: Princeton University Press, 1976.

Ruddick, Nicholas. *British Science Fiction: A Chronology, 1478–1990.* New York, Westport, Conn. & London: Greenwood Press, 1992.

Ruddick. *Ultimate Island: On the Nature of British Science Fiction.* Westport, Conn. & London: Greenwood Press, 1993.

Sammons, Martha C. *A Better Country: The Worlds of Religious Fantasy and Science Fiction.* Westport, Conn.: Greenwood Press, 1988.

Sargent, Lyman Tower. *British and American Utopian Literature 1516–1975.* Boston: G. K. Hall, 1979.

Scarborough, Dorothy. *The Supernatural in Modern English Fiction.* New York: Putnam, 1917.

Scholes, Robert, and Eric S. Rabkin. *Science Fiction: History, Science, Vision.* New York: Oxford University Press, 1977.

Searles, Baird, Martin Last, Beth Meacham, and Michael Franklin. *A Reader's Guide to Science Fiction.* New York: Avon, 1979.

Searles, Meacham, and Franklin. *A Reader's Guide to Fantasy.* New York: Avon, 1982.

Seed, David, ed. *Anticipations: Essays on Early Science Fiction and Its Precursors.* Liverpool: Liverpool University Press, 1995.

Senf, Carol A. *The Vampire in Nineteenth-Century English Literature.* Bowling Green, Ohio: Bowling Green University Popular Press, 1988.

Smith, Curtis C., ed. *Twentieth-Century Science-Fiction Writers,* second edition. Chicago: St. James Press, 1986.

Stableford, Brian. *Algebraic Fantasies and Realistic Romances: More Masters of Science Fiction.* San Bernardino, Cal.: Borgo Press, 1995.

Stableford. *Scientific Romance in Britain 1890–1950.* London: Fourth Estate, 1985; New York: St. Martin's Press, 1985.

Sullivan, Jack. *Elegant Nightmares: The English Ghost Story from Le Fanu to Blackwood.* Athens: Ohio University Press, 1978.

Sullivan, ed. *The Penguin Encyclopedia of Horror and the Supernatural.* New York: Viking, 1986.

Suvin, Darko. *Metamorphoses of Science Fiction: On the Poetics and History of a Literary Genre.* New Haven & London: Yale University Press, 1979.

Suvin. *Victorian Science Fiction in the U.K.: The Discourses of Knowledge and of Power.* Boston: G. K. Hall, 1983.

Thompson, Raymond H. *The Return from Avalon: A Study of the Arthurian Legend in Modern Fiction.* Westport, Conn.: Greenwood Press, 1985.

Tuck, Donald H., ed. *The Encyclopedia of Science Fiction and Fantasy: Through 1968,* 3 volumes. Chicago: Advent, 1974–1982.

Tymn, Marshall B., and Mike Ashley, eds. *Science Fiction, Fantasy, and Weird Fiction Magazines.* Westport, Conn.: Greenwood Press, 1985.

Varma, Devendra P. *The Gothic Flame: Being a History of the Gothic Novel in England–Its Origins, Efflorescence, Disintegration, and Residuary Influences.* London: Barker, 1957.

Wagar, W. Warren. *Terminal Visions: The Literature of Last Things.* Bloomington: Indiana University Press, 1982.

Wagenknecht, Edward. *Seven Masters of Supernatural Fiction.* New York: Greenwood Press, 1991.

Waggoner, Diana. *The Hills of Faraway: A Guide to Fantasy.* New York: Atheneum, 1978.

Wilson, Colin. *The Strength to Dream: Literature and the Imagination.* London: Gollancz, 1961.

Wolfe, Gary K. *Critical Terms for Science Fiction and Fantasy: A Glossary and Guide to Scholarship.* Westport, Conn.: Greenwood Press, 1986.

Yoke, Carl B., and Donald M. Hassler, eds. *Death and the Serpent: Immortality in Science Fiction and Fantasy.* Westport, Conn.: Greenwood Press, 1985.

Contributors

Walter Albert .... *University of Pittsburgh*
Richard Bleiler .... *University of Connecticut*
Samuel W. Bruce .... *Columbia, South Carolina*
Corbin Scott Carnell .... *University of Florida*
Gary William Crawford .... *Baton Rouge, Louisiana*
Elizabeth S. Davidson .... *University of South Carolina–Spartanburg*
Kyle W. Friedow .... *Kent State University*
Larry R. Hamm .... *Iowa State University*
Donald M. Hassler .... *Kent State University*
Darren Harris-Fain .... *Shawnee State University*
Johan Heje .... *Greve Gymna, Denmark*
Terry Heller .... *Coe College*
Stephanie Moss .... *University of South Florida*
Deborah D. Rogers .... *University of Maine*
Nicholas Ruddick .... *University of Regina*
Amelia A. Rutledge .... *George Mason University*

Cumulative Index

Dictionary of Literary Biography, Volumes 1-178
Dictionary of Literary Biography Yearbook, 1980-1996
Dictionary of Literary Biography Documentary Series, Volumes 1-14

Cumulative Index

DLB before number: *Dictionary of Literary Biography,* Volumes 1-178
Y before number: *Dictionary of Literary Biography Yearbook,* 1980-1996
DS before number: *Dictionary of Literary Biography Documentary Series,* Volumes 1-14

A

B

C

D

E

F

G

I

J

K

L

N

P

Q

R

S

X

Y

ISBN 0-8103-9941-5
90000
9 780810 399419